Computational Fluid Dynamics

Applications in Environmental Hydraulics

Editors

PAUL D. BATES
University of Bristol

STUART N. LANE
University of Durham

ROBERT I. FERGUSON
University of Durham

John Wiley & Sons, Ltd

Copyright © 2005   John Wiley & Sons Ltd, The Atrium, Southern Gate, Chichester,
West Sussex PO19 8SQ, England

Telephone   (+44) 1243 779777

Email (for orders and customer service enquiries): cs-books@wiley.co.uk
Visit our Home Page on www.wiley.com

Reprinted March 2006

All Rights Reserved. No part of this publication may be reproduced, stored in a retrieval system or transmitted in any form or by any means, electronic, mechanical, photocopying, recording, scanning or otherwise, except under the terms of the Copyright, Designs and Patents Act 1988 or under the terms of a licence issued by the Copyright Licensing Agency Ltd, 90 Tottenham Court Road, London W1T 4LP, UK, without the permission in writing of the Publisher. Requests to the Publisher should be addressed to the Permissions Department, John Wiley & Sons Ltd, The Atrium, Southern Gate, Chichester, West Sussex PO19 8SQ, England, or emailed to permreq@wiley.co.uk, or faxed to (+44) 1243 770620.

Designations used by companies to distinguish their products are often claimed as trademarks. All brand names and product names used in this book are trade names, service marks, trademarks or registered trademarks of their respective owners. The Publisher is not associated with any product or vendor mentioned in this book.

This publication is designed to provide accurate and authoritative information in regard to the subject matter covered. It is sold on the understanding that the Publisher is not engaged in rendering professional services. If professional advice or other expert assistance is required, the services of a competent professional should be sought.

Other Wiley Editorial Offices

John Wiley & Sons Inc., 111 River Street, Hoboken, NJ 07030, USA

Jossey-Bass, 989 Market Street, San Francisco, CA 94103-1741, USA

Wiley-VCH Verlag GmbH, Boschstr. 12, D-69469 Weinheim, Germany

John Wiley & Sons Australia Ltd, 33 Park Road, Milton, Queensland 4064, Australia

John Wiley & Sons (Asia) Pte Ltd, 2 Clementi Loop #02-01, Jin Xing Distripark, Singapore 129809

John Wiley & Sons Canada Ltd, 22 Worcester Road, Etobicoke, Ontario, Canada M9W 1L1

Wiley also publishes its books in a variety of electronic formats. Some content that appears in print may not be available in electronic books.

Library of Congress Cataloging in Publication Data

Computational fluid dynamics : applications in environmental hydraulics / editors,
  Paul D. Bates, Stuart N. Lane, Robert I. Ferguson.
     p. cm.
  Includes bibliographical references and index.
  ISBN-13 978-0-470-84359-8 (HB)
  ISBN-10 0-470-84359-4 (HB)
  1. Environmental hydraulics—Mathematical models. 2. Fluid dynamics—Mathematical models.   I. Bates, Paul D.   II. Lane, Stuart N. III. Ferguson, Robert I.
  TC163.5.C66 2005
  627′.042—dc22

2004028499

British Library Cataloguing in Publication Data

A catalogue record for this book is available from the British Library

ISBN-13 978-0-470-84359-8 (HB)
ISBN-10 0-470-84359-4 (HB)

Typeset in 10/12pt Times by Integra Software Services Pvt. Ltd, Pondicherry, India
Printed and bound in Great Britain by Antony Rowe Ltd, Chippenham, Wiltshire
This book is printed on acid-free paper responsibly manufactured from sustainable forestry in which at least two trees are planted for each one used for paper production.

Contents

List of Contributors   vii

1   **Computational Fluid Dynamics modelling for environmental hydraulics**   1
    P.D. Bates, S.N. Lane and R.I. Ferguson

PART ONE   AN OVERVIEW OF COMPUTATIONAL FLUID DYNAMICS SCHEMES   17

2   **Fundamental equations for CFD in river flow simulations**   19
    D.B. Ingham and L. Ma

3   **Modelling solute transport processes in free surface flow CFD schemes**   51
    I. Guymer, C.A.M.E. Wilson and J.B. Boxall

4   **Basic equations for sediment transport in CFD for fluvial morphodynamics**   71
    E. Mosselman

5   **Introduction to statistical turbulence modelling for hydraulic engineering flows**   91
    F. Sotiropoulos

6   **Modelling wetting and drying processes in hydraulic models**   121
    P.D. Bates and M.S. Horritt

7   **Introduction to numerical methods for fluid flow**   147
    N.G. Wright

8   **A framework for model verification and validation of CFD schemes in natural open channel flows**   169
    S.N. Lane, R.J. Hardy, R.I. Ferguson and D.R. Parsons

9   Parameterisation, validation and uncertainty analysis of CFD
    models of fluvial and flood hydraulics in the natural environment   193
    M.S. Horritt

PART TWO   APPLICATION POTENTIAL FOR FLUVIAL STUDIES   215

10 Modelling reach-scale fluvial flows   217
    S.N. Lane and R.I. Ferguson

11 Numerical modelling of floodplain flow   271
    P.D. Bates, M.S. Horritt, N.M. Hunter, D. Mason and D. Cobby

12 Modelling water quality processes in estuaries   305
    R.A. Falconer, B. Lin and S.M. Kashefipour

13 Roughness parameterization in CFD modelling of
    gravel-bed rivers   329
    A.P. Nicholas

14 Modelling of sand deposition in archaeologically significant
    reaches of the Colorado River in Grand Canyon, USA   357
    S. Wiele and M. Torizzo

15 Modelling of open channel flow through vegetation   395
    C.A.M.E. Wilson, T. Stoesser and P.D. Bates

16 Ecohydraulics: A new interdisciplinary frontier for CFD   429
    M. Leclerc

17 Towards risk-based prediction in real-world applications
    of complex hydraulic models   461
    B.G. Hankin and K.J. Beven

18 CFD for environmental design and management   487
    G. Pender, H.P. Morvan, N.G. Wright and D.A. Ervine

Author index   511

Subject index   519

List of Contributors

Paul D. Bates   School of Geographical Sciences, University of Bristol, UK

K.J. Beven   IENS, Lancaster University, UK

Joby Boxall   Department of Civil and Structural Engineering, University of Sheffield, UK

David Cobby   Environmental Systems Science Centre, University of Reading, UK

D.A. Ervine   Department of Civil Engineering, University of Glasgow, UK

R.A. Falconer   Environmental Water Management Research Centre, Cardiff School of Engineering, Cardiff University, UK

Robert I. Ferguson   Department of Geography, University of Durham, UK

Ian Guymer   Department of Civil and Structural Engineering, University of Sheffield, UK

B.G. Hankin   Hydro-Logic Ltd, UK

R.J. Hardy   Department of Geography, University of Durham, UK

M.S. Horritt   School of Geographical Sciences, University of Bristol, UK

Neil M. Hunter   School of Geographical Sciences, University of Bristol, UK

D.B. Ingham   Department of Applied Mathematics, University of Leeds, UK

S.M. Kashefipour   Cardiff School of Engineering, Cardiff University, UK

Stuart N. Lane   Department of Geography, University of Durham, UK

Michel Leclerc   Institut National de la Recherché Scientifique – Eau, Terre et Environnement, Canada

B. Lin Environmental Water Management Research Centre, Cardiff School of Engineering, Cardiff University, UK

L. Ma Energy and Resource Research Institute, University of Leeds, UK

David Mason Environmental Systems Science Centre, University of Reading, UK

H.P. Morvan School of Civil Engineering, The University of Nottingham, UK

Erik Mosselman WL/Delft Hydraulics, The Netherlands

A.P. Nicholas Department of Geography, University of Exeter, UK

D.R. Parsons School of Earth Sciences, University of Leeds, UK

G. Pender Department of Civil & Offshore Engineering, Heriot-Watt University, UK

F. Sotiropoulos School of Civil and Environmental Engineering, Georgia Institute of Technology, USA

T. Stoesser Institute for Hydromechanics, University of Karlsruhe, Germany

Margaret Torizzo Water Resources Division, US Geological Survey, USA

Stephen Wiele Water Resources Division, US Geological Survey, USA

Catherine Wilson Division of Civil Engineering, Cardiff School of Engineering, Cardiff University, UK

Nigel G. Wright School of Engineering, The University of Nottingham, UK

1
Computational Fluid Dynamics modelling for environmental hydraulics

P.D. Bates, S.N. Lane and R.I. Ferguson

1.1 Introduction

Computational Fluid Dynamics (CFD) was developed over 40 years ago by engineers and mathematicians to solve heat and mass transfer problems in aeronautics, vehicle aerodynamics, chemical engineering, nuclear design and safety, ventilation and industrial design. Whilst the fundamental equations of fluid motion that formed the basis of such codes had been well known since the 19th century, their solution for problems with complex geometry and boundary conditions required the development of efficient numerical solution techniques and the ability to implement these on digital computers. The development of this technology in the 1950s and 1960s made such research possible, and CFD was one of the first areas to take advantage of the newly emergent field of scientific computing. In the process, it was soon realized that CFD could be an alternative to physical modelling in many areas of fluid dynamics, with its advantages of lower cost and greater flexibility.

Computational fluid dynamics is therefore an area of science made possible by, and intrinsically linked to, computing. Its development has paralleled that of computer power and availability, and as we move into an age of cheap, powerful desktop computing it is now possible, with a little knowledge, to run large and complex 3D simulations on an average personal computer. However, most research advances in CFD continue to originate in the aeronautics and industrial design communities as a

Computational Fluid Dynamics: Applications in Environmental Hydraulics
Edited by P.D. Bates, S.N. Lane and R.I. Ferguson   © 2005 John Wiley & Sons, Ltd

result of the significant investment levels available in these areas. In such applications the boundary conditions, problem geometry and material properties of any solid surfaces (e.g. drag coefficients) are typically known very precisely and the code is applied to a closed system. In such cases it may be possible to characterize the complete set of process mechanisms that exist and also obtain good experimental data for model validation. Major research questions, therefore, concern improvements to the quality of the numerical solution, the scales of flow resolved by the model for fixed computational costs and the representation of sub-grid-scale processes such as turbulence. Considerable effort is expended on topics such as numerical analysis, turbulence modelling, grid generation and adaptive meshing. Tolerance of solution errors is also low, and such codes are predominantly used in a deterministic fashion as alternatives to laboratory experimentation. This reductionist epistemology serves industrial engineering applications well, and the techniques thus developed have considerable spin-off benefit in other disciplines such as environmental hydraulics.

Early on in the development of CFD it was realized that the technique could also be applied to environmental problems to simulate heat and mass transfer in rivers, lakes, oceans, atmospheres and porous media such as soil and rock (e.g. Freeze and Harlan, 1969; King and Norton, 1978; Fischer *et al.*, 1979). The potential for using computer models to simulate environmental flows was obvious; however, these early applications adopted the same deterministic methodology used in industrial applications which often proved inappropriate, given the data then available. In reality, the application of CFD to environmental flows leads to a series of problems not encountered in industrial applications: geometry and boundary conditions are rarely known with any precision; drag coefficients vary in time and space as result of complex interactions between the material properties of the surface and the flow itself; the driving forces are highly variable, often at scales smaller than the model grid; and the geometry of the problem rarely approximates to a simple, easily meshable surface. Moreover, environmental systems are open and should be conceived as complex assemblages of many different processes and inputs, not all of which will be well characterized in any given application. Model validation data may not be available which tests all relevant aspects of model performance to a sufficient level of detail. In fact, given that CFD models adopt finite representations of time and space that may be very different to the time and space scales over which observations are obtained, it may actually be very difficult to measure those quantities predicted by a given code. In contrast to industrial applications of CFD, environmental applications are characterized by considerable uncertainty over almost every aspect of the modelling process and it may therefore become very difficult to diagnose why a model is going wrong. For example, a mismatch between a model and available validation data may be the result of a poor choice of conceptual model given the problem in hand, lack of data to characterize the problem geometry and boundary conditions, an incorrect parameterization or just insufficient or inappropriate validation data. Most likely all these factors will apply! Whilst this does not mean that CFD models cannot be used to perform numerical experiments in environmental hydraulics, it does suggest that care is required in interpreting model studies which purport to mimic real flow events and which include comparisons with real data.

Environmental applications of CFD thus have some fundamentally different characteristics from other applications of this technology, and as a consequence such applications may have very different research priorities. This is not to say that environmental CFD modellers should be unconcerned about the numerical techniques they use or about the quality of the numerical solutions they produce. Neither does it imply that more highly resolved model grids and greater levels of process inclusion will not lead to more physically realistic models (even if the utility of this reductionist approach may be difficult to prove in our case). Rather, it suggests that the greatest uncertainties in environmental CFD modelling lie elsewhere, and that the key research challenges relate to the identification, quantification and reduction of these. This new research agenda focuses on such questions as: coupling CFD with complex natural terrain; extending process representation to consideration of coupled sediment-flow, water quality-flow and biotic–abiotic problems; scale and resolution effects, including upscaling; issues over what makes sufficient process representation in terms of model simplification; model validation; complex sensitivity and uncertainty analysis; and possible model equifinality. Uncertainty is a particular challenge as uncertainties may be compensating, interacting and non-linear. Further, the data sets available to understand them may be sparse and contain significant but poorly known errors that vary strongly in time and space. The result is that there may be many combinations of models and parameters that fit the available data equally well.

In proposing solutions to these problems, environmental CFD modellers have a distinctive contribution to make to the overall discipline and there is the potential to contribute significant innovative science that may find application in many fields. Whilst much research in 'mainstream' CFD requires very high level mathematical ability that is typically the preserve of a select group of specialists, solution to the problems outlined above requires a different skill set for which environmental scientists may be well suited. The ability to deal with problems characterized by sparse and uncertain data where there may even be debate over the fundamental process mechanisms at work is a key part of scientific training in environmental engineering and the geosciences. Hence, in environmental applications of CFD there are scientific problems of model development and analysis that are not well anticipated or solved by standard CFD research and to which civil engineers, environmental scientists and geographers can contribute significant insight.

Application of CFD techniques to real-world environmental problems has increased sharply in the last decade due to an improving ability to deal with the uncertainties noted above. In part this has been due to improvements in computer power and storage, which have allowed flows over complex natural topographies to be simulated for the first time, and to wider availability of user-friendly code. However, this alone does not explain the rise of environmental applications of CFD. A further major factor is the increased availability of the necessary digital data sets to set up and to test such models. Instrumentation development in a variety of fields has yielded new technologies for topographic surveying (including airborne laser altimetry, stereo-photogrammetry and the Global Positioning System), bathymetric survey (including sidescan sonar and wide swath sonar) and velocity measurement (including acoustic Doppler current profilers and large-scale particle image velocimetry). Such instruments yield data that are critical for environmental applications of CFD and allow users to at least begin the process of uncertainty characterization and reduction. As a consequence, the extent and scope of environmental research now carried out with CFD models is considerable (Hodskinson and Ferguson, 1998; Lane and Richards, 1998; Meselhe and Odgaard, 1998; Sofialidis and Prinos, 1998; Sinha et al., 1998; Bijvelds et al., 1999; Lane et al., 1999, 2000; Nicholas and Smith, 1999; Bradbrook et al., 2000a,b, 2001; Nicholas, 2001), and likely to rise as a result of the continuance of the trends identified above.

In effect, we are beginning to see the development of a coherent body of research that attempts to address the research challenges identified above, but which also acknowledges the more fundamental aspects of CFD modelling that are long-established within the engineering community (e.g. control of numerical accuracy). The purpose of this book is to document this newly emergent science and to provide an accessible 'primer' to CFD modelling for environmental engineers and geoscientists. Accordingly, the book is split into two parts. In Part One, basic topics in CFD modelling are addressed in a thematic manner to provide the necessary theoretical background for students and researchers in the environmental sciences with material specifically tailored to CFD applications to complex natural systems. In Part Two, reviews of leading-edge research applications are presented that exemplify and add understanding to the themes raised in Part One. Central to these reviews is the demonstration of how good practice in CFD modelling can be achieved, with reference to both established and new applications. The remainder of this introduction provides a brief overview of the structure of the book and tries to identify directions for future research in this area.

1.2 Part One: An overview of computational fluid dynamics schemes

At the heart of any CFD model is a set of governing rules, usually written in the form of simultaneous partial differential equations derived from principles of mass and momentum conservation, which define in mathematical form the physical processes to be represented by the model. This representation may not, and likely will not, encapsulate all the physical mechanisms known to occur in a given application, but rather is a statement of the modeller's assumptions about those processes that are critical to the problem in hand. In Keith Beven's terminology (Beven, 2002), the first stage in model building is therefore to move from a *perceptual* model, representing everything we perceive or know about a given flow problem (and which may still be incomplete), to a *conceptual* model that represents our best estimate of the processes, parameters and forcing functions that control the development of a flow field at a particular scale. This conceptual model may be as simple as a series of logical statements, but at some point this needs to be translated into mathematical notation if it is to be turned into computer code and solved for the problem of interest. This volume begins, therefore, with five chapters that outline the fundamental equations used to build computational fluid dynamics models for flow (Ingham and Ma), solute transport (Guymer et al.), sediment transport (Mosselman), turbulence (Sotiropoulos) and moving boundaries (Bates and Horritt). Readers should note that these distinctions are somewhat artificial, but are a convenient way to organize the book. In reality, flow calculations need to consider turbulence closure, solute and sediment transport are driven by advective (flow) and dispersive (turbulent) processes and so on.

Derek Ingham and Lin Ma (Chapter 2) consider the variety of equations available for the simulation of flow processes in fluvial applications. These are all derived from the well known 3D Navier–Stokes equations (Batchelor, 1967), which are typically simplified for environmental applications by averaging in time, to yield the Reynolds-averaged Navier–Stokes equations, and/or in space, to yield 1D or 2D models. None of the resulting equation sets have general analytic solutions for non-trivial problems, and can only be solved using numerical approximation. As Ingham and Ma point out, neither is the process of simplifying the Navier–Stokes equations straightforward. In fact, averaging in time and space invariably introduces new terms into the controlling equations to represent dispersion processes at scales below the model grid or time step over which the averaging occurs. In the case of time averaging these dispersion terms are dominated by the effect of turbulence and require the introduction of some additional model to represent these effects. In a similar fashion, averaging in space also leads to additional dispersion terms because of velocity gradient variations in the flow field. Ingham and Ma discuss briefly the treatment of these effects using turbulence models, with a fuller treatment being given by Sotiropoulos (Chapter 5). The chapter concludes with a discussion of methods to treat the model boundaries in CFD codes, and in particular covers free surface, open flow and wall boundaries. Unlike in standard CFD application areas, free surface flows are common in the environment and require careful treatment, particularly for dynamic problems. Similar consideration needs to be given to inlet and outlet boundaries of the model and to the treatment of flows in the boundary layer below the scale of the model grid. The latter is particularly important because of the significance of bed roughness to the development of fluvial flows.

Ian Guymer *et al.* (Chapter 3) discuss the modelling of solute transport processes in CFD schemes using numerical solutions of the advection–dispersion equation. Combined environmental flow and transport problems are increasingly being treated with a CFD approach as a result of legislation to regulate point and non-point source pollutant discharges and maintain river ecology. Guymer *et al.* demonstrate the importance of turbulence, dispersion and dead zones to the mixing process in real rivers and outline ways in which these can be treated in CFD codes. The transport theme is continued by Erik Mosselman (Chapter 4) who considers ways to model the transport of bed material by rivers to yield simulations of fluvial morphodynamics over various space and timescales. Again, this is based on consideration of the principles of mass and momentum conservation, but is complicated by the number of possible transport modes (suspended load, bedload, etc.) and by the need to employ empirical closure relationships for many of the key terms, particularly in the momentum equations for sediment transport, to facilitate simulations at scales of practical interest. As a result, Mosselman notes that the sediment transport momentum equation becomes, to a large extent, empirical. To counter this, Mosselman argues that sediment transport formulae should not be derived from sediment transport measurements alone, but also from bed level measurements, and that further research into the effect of bed slope and vegetation on sediment transport should be conducted.

In Chapter 5, Fotis Sotiropoulos provides an in-depth review of the treatment of turbulence in CFD schemes. In principle, with a model grid and time step sufficiently fine to resolve turbulent eddies down to the Kolmogorov scale, all turbulent motions can be simulated directly. However, for problems of practical interest, and particularly for environmental problems with complex topography and roughness, methods for parameterizing the impact of turbulent eddies on the large-scale flow development will continue to be required for the foreseeable future. Sotiropoulos discusses Direct Numerical Simulation (DNS), Large Eddy Simulation (LES) and Reynolds Averaging as solutions to these problems, and reviews the entire hierarchy of turbulence models with emphasis on modelling issues that are relevant to environmental engineering applications.

As the final contribution to this introduction to governing equations for CFD schemes, Paul Bates and Matthew Horritt (Chapter 6) consider methods to treat the dynamic extension and retreat of the flow. Moving boundary problems are common in environmental hydraulics, and include overbank flooding, coastal, estuarine and dam-break flows. Methods to treat such problems can be broadly classified into those that adapt the model grid to track the moving shoreline, and simpler methods which retain a fixed numerical grid, but implement additional algorithms to deal with potential discrepancies in mass and momentum conservation in partially wet cells. In each method the topography can be defined as a continuous function (albeit as discretized on the model grid) or a discontinuous staircase. Different treatments apply in each case and are reviewed in this chapter. The chapter deals predominately with treatments for the flow equations, but concludes with a brief discussion of methods to treat combined flow and transport in the presence of a moving boundary.

Having defined the equation set to be solved to represent a given application, the next task is to formulate a way to solve the resulting system in time and space that can be implemented on a digital computer. This problem is considered by Nigel Wright in Chapter 7. As the resulting equation sets used in CFD codes have no general analytical solution, recourse must be made to approximate numerical methods. The first stage in any of these methods of numerical analysis is to convert the differential equations, which are continuous functions, into a set of algebraic equations that connect values at discrete points. There are various techniques for achieving this, and the main ones encountered in CFD modelling are the Finite Difference (FD), Finite Element (FE) and Finite Volume (FV) methods. Wright provides an introduction to each of these methods, using simple examples, before going on to discuss related issues of grid generation and numerical error. Once a discrete form of the governing equations has been defined over a grid, the resulting non-linear algebraic system of equations can be solved with a technique appropriate to the discretization method. Wright emphasizes that the numerical solution should not be treated as a black box by modellers, and that correct interpretation of CFD code output requires a thorough understanding of the numerical method being used.

This theme of good practice is continued in the final two chapters of Part One by Stuart Lane *et al.* (Chapter 8) and Matthew Horritt (Chapter 9). Chapter 8 by Lane *et al.* proposes a framework for the verification and validation of CFD schemes. This is derived from the guidelines for publication of CFD research proposed by the American Society of Mechanical Engineers (ASME), but extended here for the specific case of open channel flows. These guidelines cover such areas as reporting standards, solution accuracy in space and time, mesh independence testing, convergence testing and comparison with experimental results. Lane *et al.* demonstrate how these guidelines may not be sufficient for many practical fluvial applications, but can provide a minimum framework. In particular, ASME criterion 10 requires that 'reasonable agreement' between the model and experimental results be shown. However, Lane *et al.* argue that it may be difficult, both philosophically and practically, to establish that such evidence provides conclusive validation of a model. The chapter then proceeds to develop this notion of validation and discuss surrounding issues of calibration, sensitivity analysis and benchmarking. As a result of this discussion, a series of guidelines are proposed to replace ASME criterion 10 for environmental applications.

Despite our best efforts, almost all aspects of CFD modelling (choice of conceptual model, boundary conditions, topography, model parameters and validation data) will contain uncertainty. Techniques are therefore required to estimate the impact of these uncertainties on model predictions. This area is addressed by Horritt in his chapter on model parameterization, validation and uncertainty analysis. Horritt discusses emergent techniques for uncertainty analysis with a focus on the use of new technologies, such as remote sensing, to provide essential data to constrain model calibration. He concludes that the spatial heterogeneity of natural environments is a significant challenge for CFD modellers that can only be partially overcome with current technologies. Thus, in addition to criteria proposed by Lane *et al.*, full evaluation of a CFD study requires an evaluation of uncertainty propagation through the model system.

1.3 Part Two: Application potential for fluvial studies

Part Two of the book provides an overview of particular applications of CFD to fluvial and estuarine environments. We specifically chose not to consider marine, aeolian or atmospheric issues due to space limitations, although there is much to be learnt from comparing and contrasting aeolian, atmospheric, marine, fluvial and estuarine studies. For example, all four environments have to address the very severe difficulty of how to represent boundary shear stress effects in near surface cells. Many traditional modelling approaches share methods based upon roughness height specification and associated modification of the turbulent velocity profile in the near wall region. Part Two therefore begins with general reviews of the two most common current applications of CFD to the study of fluvial and estuarine flows: (a) reach-scale modelling of fluvial processes (Lane and Ferguson); and (b) floodplain inundation modelling (Bates *et al.*). Lane and Ferguson (Chapter 10) show how CFD has resulted in a much deeper understanding of flow, sediment transfer and ecological processes at the reach scale. This has now included application of CFD to understanding each of the major river channel facets (riffle–pool sequences, meanders, confluences or tributary junctions) and has been extended to include consideration of instream habitat. They argue that these kinds of applications have required innovation in terms of process representation, notably in terms of the strongly related issues of discretization and boundary roughness. However, they also argue that a major change is still needed to extend research from a consideration of flow to also include sediment transport and channel change, where the presence of a deforming boundary can lead to severe numerical instabilities. It is in relation to deformable boundaries that some of the most innovative applications of reach-scale CFD are being explored.

In Chapter 11, Bates *et al.* provide a similar overview of treatments of floodplain inundation. Floodplains represent an environment of particular importance to the riparian manager, and one that is currently moving through a paradigm change in terms of the practical application of flood inundation modelling methods with adoption of approaches that have a strong grounding in CFD. Bates *et al.* review the complexity of the inundation process, which involves the complex lateral transfer of mass and momentum from the river channel to the floodplain, involving flows that may be strongly three dimensional. Away from the channel, shallow water flows become dominant that are more strongly two dimensional. These flows are a challenge as the inundation front can move rapidly across the floodplain (during wetting and drying) interacting strongly with the floodplain topography and vegetation as it does so. This can result in computational difficulties due to the large number of elements that play no part in the solution for much of the time, until they are inundated. It is also associated with numerical stability issues as during wetting and drying flow depths are commonly very small, which can lead to severe numerical diffusion and sometimes instability. Bates *et al.* develop two important themes in CFD applications to floodplain flows. First, they emphasize the need to nest models of different complexity according to the part of the environment being simulated. For instance, they describe the potential utility of nesting a 1D solution of the St Venant equations in a 2D diffusion wave treatment of floodplain flow, as the former is sensitively dependent upon getting the correct flux from channel to floodplain, which in turn requires a good estimate of in-channel water levels. This kind of approach has much potential and it implies that CFD solutions will need to be developed that recognize our own understanding and observations of the system that is to be modelled in developing innovative modelling strategies. Second, they demonstrate the crucial progress that has been made in CFD modelling of floodplain flows as a result of progress in our ability to measure floodplain topography, notably using laser altimetry. As the flows are generally shallow, and provided the river floodplain system does not convey as a two-stage channel system, the topographic forcing of flow is likely to dominate the inundation process. This is where there has been a fundamental improvement in our ability to model flood inundation, one that has led to a situation where progress in CFD is now being driven by the high-quality topographic data available to apply to models rather than an explicit concern over existing modelling methods: with the barrier to topographic representation removed, 2D modelling approaches have become both cost-effective and preferred to their 1D counterparts.

Falconer *et al.* (Chapter 12) address the issue of estuarine flows. As with floodplains, these are zones of particular importance. For instance, they are commonly depositional zones. As rivers invariably pass through them, they can become the repository of any kind of waste material in transport in the river system. Estuarine flows are distinct for a number of reasons. First, and although this may also apply to very wide floodplains, the spatial scale of estuaries can be larger, and this reintroduces important source terms into the momentum balance equations: the effects of the earth's rotation and surface wind stresses. Second, they may have important buoyancy effects, largely associated with gradients of density in the horizontal and vertical, which can lead to important mixing processes. Indeed, important field observations in estuaries have shown that the freshwater–saltwater interface can be a crucial component of the sediment entrainment-deposition process (e.g. Uncles *et al.*, 1998). Falconer *et al.* describe the key processes that must also be incorporated to represent water quality in estuaries. This involves a series of general transport equations (Chapter 3) but with additional processes and parameters introduced to describe non-conservative aspects of water quality dynamics. For instance, faecal coliform modelling requires treatment of the coliform decay rate. Falconer *et al.* present a series of case studies of estuarine modelling. The choice of the Ribble example is interesting as it contrasts nicely with Hankin and Beven's consideration in Chapter 17 of the same system from an uncertainty perspective. This illustrates two complementary approaches to modelling, one emphasizing process complexity and the other emphasizing process uncertainty. It is important to recognize that complexity and uncertainty are neither positively nor negatively correlated, but that (Chapter 9) the relationship between the two is one that merits much further consideration. Falconer *et al.* show that their model could reproduce daily variation in point faecal coliform counts. The Mersey estuary provided a second case study, focusing on salinity and sediment transport. As is commonly the case, predictions of hydrodynamics were excellent, but those of suspended particulate matter only good, and this emphasizes the very great difficulties associated with modelling the sediment transfer process noted previously in Chapter 4.

In Chapters 13 and 14, consideration turns to two very different classes of river problem: gravel-bed (Nicholas) and sand-bed (Wiele and Torizzo) rivers. Whilst this distinction is not necessarily a useful one in substantive terms, it is crucial in CFD terms as the two environments are associated with very different modelling challenges. Gravel-bed rivers are a particular problem because of the complex geometry of the bed surface, which not only raises data acquisition challenges but also requires innovative representation of surface complexity in CFD models. Consideration of a gravel-bed river immediately turns one's attention to the strong spatial and vertical gradients in flow associated with individual clasts or groups of clasts. However, as Nicholas shows, there is a larger scale of consideration where this process detail has to be sacrificed if reach-scale sediment transport and flow problems are to be resolved. Nicholas emphasizes that a key theme in gravel-bed river modelling, especially when trying to scale up, is the specification of appropriate wall functions. This is a sub-grid-scale problem (Chapter 10) and Nicholas presents a range of alternative roughness models which do not require the assumptions associated with conventional wall treatments. These include random perturbation of bed surface topography as opposed to implicit representation of topography in wall treatments and discrete element models, similar to the porosity treatment described by Lane and Ferguson (Chapter 10) but parameterized using geostatistical methods. The modelling approach was particularly successful as it was able to represent known deviations from predicted wall treatment cases. Nicholas extends consideration to 3D modelling of braided rivers which is of particular importance. He presents a particularly novel method for determining distributed patterns of drag, which merits much further exploration. He also confirms the work of Lane and Ferguson (Chapter 10) which shows that roughness is a much less effective calibration parameter in 3D models as compared with 2D and 1D models.

The challenge for sand-bed river modelling is different. Wiele and Torizzo explore the issue of how sand transport adjusts to a dam release and changes in sediment delivery. Modelling sand-bed rivers is a challenge as sediment transport can occur at almost any flow discharge, resulting in the continual interaction between the bed of the river and its flow. Wiele and Torizzo use a depth-averaged model for flow but with a quasi-3D sand transport treatment, the latter driven by depth-averaged flow predictions combined with a near-bed boundary condition for shear stress. The modelling was supported by high-quality data collection as part of the work of the Grand Canyon Monitoring and Research Centre and the model was applied in steady state form to two flows: the highest post-dam closure flow; and the test release flow. Model results are interesting as they show that deposition volumes vary in relation to not only discharge and sand supply but also channel shape.

Many rivers are influenced by instream and bank-side vegetation. Given the volume of research into bed roughness issues in rivers, it is surprising that vegetation has been given so little attention. Wilson *et al.* (Chapter 15) address this issue. The chapter begins with a review of different means of vegetation representation in hydraulic models using conventional experimental–conceptual approaches. Many of these approaches have been based upon manipulation of conventional roughness parameters to allow vegetative resistance effects to be incorporated. Wilson *et al.* show the importance of carefully designed experiments in providing the conceptual underpinning for development of more effective numerical modelling approaches. They show that this leads to better models for practical application but also the development of CFD as a research tool, which in turn leads to new ways of developing models for practical application. Thus, they apply a drag-force approach to two different 3D FV formulations of the Navier–Stokes equations, and then consider how their approach can be upscaled for practical application. The work emphasizes the importance of vegetation in relation to turbulence, something that is often overlooked when CFD models are applied to vegetated systems and only vegetation-roughness issues are considered.

The importance of considering vegetation in rivers partly stems from a changing legislative context in which protection of the biota is now given as much, if not more, importance as management of the abiota. Thus, Leclerc (Chapter 16) continues the ecological theme introduced by Wilson *et al.* but extends consideration to ecohydraulics and the associated use of CFD in habitat management. Numerical habitat modelling, as with studies of floodplain flows, is going through somewhat of a paradigm shift as the conventional 1D approaches to hydraulics implicit in models like IFIM and PHABSIM are being displaced by 2D analyses. This shift was well illustrated by the work of Leclerc in the 1990s and reflects what we now know about the biology of habitat preference curves where what matters is the range of habitat available in a given spatial unit, which requires some form of 2D characterization. Leclerc sets the problem up by doing something that we do not do sufficiently in CFD applications: he considers the problem from the perspective of the goals of the project (in this case – fish) and then asks what must an associated modelling approach deliver as a result. He demonstrates that the amount of modelling complexity that is required depends sensitively upon the goals of the study. Thus, Leclerc provides an important overview of the state of the art in terms of how habitat preference curves, and derivatives, can be developed. This leads into a critical review of how CFD can be used to deliver the necessary environmental variables for populating these methods. This begins with a convincing review of the inadequacies of approaches based upon 1D treatments. He then explores 2D approaches, emphasizing that they will give more useful results but this is at the expense of additional data requirements. As other chapters have shown (Chapters 10 and 11) remote data acquisition will help here. But so will innovative ways of topographic parameterization (Chapter 13) which may in turn make 2D approaches standard in this kind of application. Leclerc emphasizes that 2D and 3D approaches are crucially underpinned by digital elevation models, and shows that these in themselves represent important methodological challenges but also substantial ecologically relevant process knowledge.

The last two chapters in the book both take up the theme of real-world application of CFD codes. Barry Hankin and Keith Beven (Chapter 17) continue the theme developed by Horritt in Part One of the book, namely how to begin to quantify some of the uncertainties that are inevitably present in practical applications. To date this has been an under-researched area, given the computational cost of simulations and the need to undertake multiple realizations of a model, usually in a Monte Carlo framework, to begin the uncertainty analysis process. Hankin and Beven suggest a risk-based approach to environmental CFD modelling should now be adopted and that both the methods and computational capacity now exist to allow this change to take place. Hankin and Beven illustrate this point by comparing a study of a complex system with uncertain inputs where limited computational time was available to an academic study where uncertainty analysis via the Generalized Likelihood Uncertainty Analysis (GLUE) method was used. In the first study it was only possible given the timescales available to carry out a limited sensitivity analysis, and the chapter makes the limitations of this approach all too apparent. Nevertheless, as Hankin and Beven note, complex CFD models are being used to inform billion pound expenditure programmes in the water industry, yet interpretative tools for assessing the degree of belief in these model predictions are not yet seen as integral to the process. This will clearly be a developing theme over the next decade for environmental applications of CFD codes, and studies of the type outlined by Hankin and Beven are likely, through the impact of legislation such as the EU Water Framework Directive, to become increasingly common.

The book concludes with a chapter by Gareth Pender *et al.* (Chapter 18) on the use of CFD methods for environmental design and management. Pender *et al.* take a different approach from that of Hankin and Beven, and instead of showing what may be possible in the future they seek to provide an insight into the quality of the simulations that can now be achieved when CFD tools are placed in the hands of competent design engineers with access to typical field data. Pender *et al.* summarize the issues involved in setting up a CFD applications (including free surface representation, boundary conditions, roughness, geometry and turbulence) and seek to demonstrate how decisions over how these are represented may impact on simulation quality and show the constraints on practical application.

1.4 Where next for environmental CFD?

The chapters in this book clearly demonstrate the rapid recent development of environmental applications of CFD techniques. Integration of data from newly emergent sources, such as remote sensing, with CFD models and the availability of cheap yet powerful desktop computing has yielded an ability to construct multi-dimensional models of flow and transport at a resolution and level of detail that would, even 10-years ago, have seemed unattainable. With hindsight, however, the fact that the controlling equations of these models have been well known for over a century and that numerical techniques to solve them have been available since the 1960s meant that this was, in large part, a development waiting to happen and was only prevented by logistical not theoretical constraints. An unintended consequence of this development has been the democratization of hydraulics research, and its dissemination beyond the limited group of practitioners who were able to secure access to the large-scale flume facilities that were previously necessary to conduct significant science in this area. Such facilities are expensive to maintain, time consuming to use and can only support a limited number of experiments at a time, and hence can only be found within a limited number of institutions. The ability to do significant hydraulic science on a desktop PC has fundamentally revolutionized just who is able to undertake hydraulics research and led to new interest in the subject from scientists such as geographers, earth and environmental scientists, ecologists and meteorologists who lie well beyond the traditional engineering/mathematical focus of the discipline. This development has the potential to be incredibly important, not just because it may increase the volume of hydraulics research conducted but also because scientists from these different disciplines may bring new insights and skills, such as uncertainty analysis, which may be highly relevant to environmental applications of CFD.

Despite this progress, key research questions remain. A particularly central issue is how we validate these newly emergent models. A newly arrived extraterrestrial might view this as a somewhat strange preoccupation, given that all of our other research approaches (fieldwork, experimental data collection in the laboratory, analytical solution of equations) require us to make simplifications that result in the delimitation of our spatial and temporal horizons and the exclusion of possibly important processes (Lane, 2001). However, these philosophical debates aside, we do need to be able to choose between different model formulations, and to advise when the benefits of a more sophisticated process representation outweigh the associated increase in computational and data collection costs. Data, with all their problems, still help us to do this, although as a number of chapters in this book show, other approaches are also of value. It is perhaps here, where CFD meets the real world, that the new challenges are emergent, as we deal with real rivers and floodplains that have complex structure and hence strong horizontal and vertical process gradients, and where traditional simplification of model geometry removes the critical aspects of the processes that are driving the system.

This leads on to a related theme that is again reflected in many of the above chapters. What is the sufficient physics (and biology and chemistry) when we wish to use CFD to understand and to manage the water environment? The word 'sufficient' will clearly depends on what CFD is to be used for and there remains considerable debate over this issue. This is well illustrated in debates over braided river modelling (e.g. Paola, 2001; Lane, 2005) where a model that is based upon a rudimentary flow routing treatment (that cannot be derived from simplification of the 2D shallow water equations) and mass-conservative, but physically realistic sediment transport rules, appears to reproduce the generic properties of river braiding. Does this mean that CFD is not necessary as a research tool? Some would argue that this is the case for certain classes of problems. When this debate is taken into the management arena, where practical considerations provide additional complexity, debates over CFD move beyond strictly technical considerations. It is in this sense that model validation and verification become crucial, as decisions over sufficiency in process representation become bound with social and economic considerations, including commercial development of code.

One way forward in relation to sufficient process representation is to develop nested modelling strategies. This reflects the idea that process complexity is defined by the components of the system that are being studied and that these vary in space, and potentially in time. Thus, what is sufficient process representation depends where you are in the model's space and time, and process complexity can be allowed to vary. This kind of coupling can be a numerical challenge, but it is probably going to be the main way forward in reconciling our growing process sophistication with the practical constraints imposed by data availability and computational demands.

Despite the progress made to date, many research challenges remain and should prove fruitful grounds for research in the coming years. Different researchers would likely come up with different priorities; however, certain key themes are obvious. First, applications to date have tended to consider only single river reaches and a likely avenue for future development will be the consideration of flow and sediment transport along multiple reaches for catchment-wide risk analysis and management. Legislation such as the EU Water Framework Directive will provide further impetus to the adoption of a holistic approach to river basin management that can only be accomplished by the application of modelling tools at a commensurate scale. Data to drive these and other applications will become an increasingly important focus, and studies that examine the use of remotely sensed data for automatic or near-automatic model discretization and parameterization will become increasingly common. This will lead in turn to a more comprehensive consideration of scale effects in CFD modelling and the impact of scale on dominant processes and parameter sensitivity. Remotely sensed data may also improve our ability to calibrate and validate the distributed predictions made by CFD models and a move to multi-criteria validation will also require an increasing level of sophistication in the techniques applied to evaluate uncertainty in CFD codes. Uncertainty will be a further dominant theme, and despite a reductionist tendency in CFD modelling, whereby increases in computer power are most often used to increase process specification or reduce model resolution, the next decade is likely to see a significant move away from a reliance on single deterministic solutions and a move to the analysis of ensembles of simulations. Further technical developments are likely to come in a number of areas including the use of porosity treatments and other boundary representation methods and in particular their extension to erosion and deposition problems. The generation of turbulence and friction by vegetation also requires much further study if we are truly to represent flows over natural surfaces. Methods are required that will enable turbulence and friction generation by plants to be estimated and these effects aggregated to the model grid scale. Data on plant geometry and biomechanics, possibly acquired from remote sensing instruments, will also be required to parameterize these process models in a physically plausible way (e.g. Mason *et al.*, 2003).

Ways of using CFD models may also change, and in particular as we gain further understanding into how to construct models for particular flow problems to yield results that are adequately realistic we may be able to begin to use our models as 'numerical laboratories'. In this way we will be able to conduct computer-based experiments that analyse, for example, how flow in generic situations depends on parameters of channel geometry, what would happen to particular channel and substrate types in the 100-year flood and so on. Lastly, as advances are made in the component areas of flow, sediment transport and ecology modelling, the barriers to model coupling will reduce and codes which can simulate complex assemblages of environmental processes will become more common. Particular examples of such multi-process approaches may include the linkage of flow and sediment transport models in a more closely coupled way than has hitherto been possible and the linking of flow simulations with ecologic processes.

In summary, much progress has been made in environmental applications of CFD which has posed new research questions for hydraulic modellers and resulted in significant new science. Such work is, however, only the beginning of what we may be able to accomplish with CFD techniques in the coming years and the interplay between computational approaches, data collection and focused experimentation is capable of yielding new insights into environmental hydraulics. The chapters in this book testify to the vitality of environmental CFD research and demonstrate the considerable potential for use of these techniques in the future.

References

Batchelor, G.K. 1967. *An Introduction to Fluid Dynamics*. Cambridge University Press, Cambridge, UK, 615pp.

Beven, K. 2002. Towards a coherent philosophy for modelling the environment. *Proceedings of the Royal Society of London Series A, Mathematical, Physical and Engineering Sciences*, **458**(2026), 2465–2484.

Bijvelds, M.D.J.P., Kranenburg, C. and Stelling, G.S. 1999. Three-dimensional numerical simulation of turbulent shallow-water flow in a rectangular harbour. *ASCE Journal of Hydraulic Engineering*, **125**, 26–31.

Bradbrook, K.F., Lane, S.N. and Richards, K.S. 2000a. Numerical simulation of time-averaged flow structure at river channel confluences. *Water Resources Research*, **36**, 2731–2746.

Bradbrook, K.F., Lane, S.N., Richards, K.S., Biron, P.M. and Roy, A.G. 2000b. Large Eddy Simulation of periodic flow characteristics at river channel confluences. *Journal of Hydraulic Research*, **38**, 207–216.

Bradbrook, K.F., Lane, S.N., Richards, K.S., Biron, P.M. and Roy, A.G. 2001. Role of bed discordance at asymmetrical river confluences. *ASCE Journal of Hydraulic Engineering*, **127**, 351–368.

Fischer, H.B., List, E.J., Koh, R.C.Y., Imberger, J. and Brooks, N.H. 1979. *Mixing in Inland and Coastal Waters*, Academic Press, New York, 483pp.

Freeze, R.A. and Harlan, R.L. 1969. Blueprint for physically-based digitally simulated hydrologic response model. *Journal of Hydrology*, **9**, 237–258.

Hodskinson, A. and Ferguson, R.I. 1998. Numerical modelling of separated flow in river bends: Model testing and experimental investigation of geometric controls on the extent of flow separation at the concave bank. *Hydrological Processes*, **12**, 1323–1338.

King, I.P. and Norton, W.R. 1978. Recent application of RMA's finite element models for two dimensional hydrodynamics and water quality. In C.A. Brebbia, W.G. Gray and G.F. Pinder (eds), *Finite Elements in Water Resources: Proceedings of the Second International Conference on Finite Elements in Water Resources, London*, Pentech Press, London, pp. 2.81–2.99.

Lane, S.N. 2001. Constructive comments on D. Massey Space–time, 'science' and the relationship between physical geography and human geography. *Transactions of the Institute of British Geographers*, **26**, 243–256.

Lane, S.N. 2005. Approaching the system-scale understanding of braided river behaviour. Paper forthcoming in G. Sambrook-Smith and J.L. Best (eds), *Braided Rivers*, Blackwell.

Lane, S.N. and Richards, K.S. 1998. Two-dimensional modelling of flow processes in a multi-thread channel. *Hydrological Processes*, **12**, 1279–1298.

Lane, S.N., Bradbrook, K.F., Richards, K.S., Biron, P.M. and Roy, A.G. 1999. The application of computational fluid dynamics to natural river channels: Three-dimensional versus two-dimensional approaches. *Geomorphology*, **29**, 1–20.

Lane, S.N., Bradbrook, K.F., Richards, K.S., Biron, P.M. and Roy, A.G. 2000. Secondary circulation in river channel confluences: Measurement myth or coherent flow structure? *Hydrological Processes*, **14**, 2047–2071.

Mason, D., Cobby, D.M., Horritt, M.S. and Bates, P.D. 2003. Floodplain friction parameterization in two-dimensional river flood models using vegetation heights derived from airborne scanning laser altimetry. *Hydrological Processes*, **17**, 1711–1732.

Meselhe, E.A. and Odgaard, A.J. 1998. Three-dimensional numerical flow model for fish diversion studies at Wanapum Dam. *ASCE Journal of Hydraulic Engineering*, **124**, 1203–1214.

Nicholas, A.P. 2001. Computational fluid dynamics modelling of boundary roughness in gravel-bed rivers: an investigation of random variability in bed elevations. *Earth Surface Processes and Landforms*, **26**, 345–362.

Nicholas, A.P. and Smith, G.H.S. 1999. Numerical simulation of three-dimensional flow hydraulics in a braided channel. *Hydrological Processes*, **13**, 913–929.

Paola, C. 2001. Modelling stream braiding over a range of scales. In M.P. Mosley (ed.), *Gravel Bed Rivers V*, New Zealand Hydrological Society, New Zealand, pp. 11–46.

Sinha, S.K., Sotiropolous, F. and Odgaard, A.J. 1998. Three-dimensional numerical model for flow through natural rivers. *ASCE Journal of Hydraulic Engineering*, **124**, 2–32.

Sofialidis, D. and Prinos, P. 1998. Compound, open-channel flow modelling with non-linear low-Reynolds, k-ε models. *ASCE Journal of Hydraulic Engineering*, **124**, 253–262.

Uncles, R.J., Easton, A.E., Griffiths, M.L., Harris, C., Howland, R.J.M., King, R.S., Morris, A.W. and Plummer, D.H. 1998. Seasonality of the turbidity maximum in the Humber-Ouse Estuary, UK. *Marine Pollution Bulletin*, **37**, 206–215.

Part One

An Overview of Computational Fluid Dynamics Schemes

2

Fundamental equations for CFD in river flow simulations

D.B. Ingham and L. Ma

2.1 Introduction

Modern CFD techniques emerged between the late 1960s and early 1970s when fluid flow investigations were largely experiment based and only very simple fluid flow problems could be accurately numerically solved. With the rapid development of modern computational techniques and numerical solution methodologies over the last few decades, CFD has now been widely used in various industrial applications for investigating a vast range of industrial and environmental problems. However, compared to industrial applications, using CFD to model morphology and hydrology problems is relatively new, research on river flow CFD modelling has been very active, particularly in recent years (e.g. Nezu and Nakagawa, 1993; Lane, 1998; Ma *et al.*, 2002; Cao *et al.*, 2003, etc.). This can be largely attributed to the availability of adequate computational resources at a reasonably low cost. CFD has increasingly acted as an alternative and/or supplementary tool to the more traditional methods in the systematic investigations of various controls in river morphology, flow structure and sediment transport, and it has increasingly played an important role in river management and flood prediction. However, it should be noted that the flow structures in river flows are inherently very complex due to the effects of irregular bank and bed topographies in natural river flows. Further, considering the large scales that are involved, the modelling of natural rivers has itself become computationally

Computational Fluid Dynamics: Applications in Environmental Hydraulics
Edited by P.D. Bates, S.N. Lane and R.I. Ferguson © 2005 John Wiley & Sons, Ltd

very demanding. However, major issues of river flow modelling are in the appropriate representations of the complex river flow conditions in the CFD model. Issues such as grid resolution, grid dependence, representation of wall roughness, appropriate turbulence models, etc., are all currently under intensive discussion (e.g. Hardy *et al.*, 1999). Nevertheless, with a numerical model and the boundary conditions which provide adequate representations of the key proceses of the river flow investigated, CFD simulations may provide considerable insight into, and clearer explanations of, the structure of the flow and the interactions of the key components of the processes than do the traditional field and/or laboratory measurements.

As a basis for the CFD modelling, this chapter gives an outline of the fundamental governing equations of fluid flow and fine sediment/solute transport which are widely used in CFD simulations of river flows. Various turbulence models adopted in the modelling of turbulent flows, and the boundary conditions which are required to define a specific situation under investigation, are discussed. Then two examples of 3D numerical CFD simulations of river flow and pollutant transport in a river channel are presented.

2.2 Basic equations for river flows

The constituent equations for fluid flows are well established and they are basically in the form of a coupled set of partial differential equations, known as the Navier–Stokes equations (e.g. Batchelor, 1967), which are appropriate to river flow modelling. CFD techniques simulate physical fluid flow by numerically solving these coupled partial differential equations. Different ways of numerically solving these equations give rise to different CFD techniques in which various forms of these equations may be employed. In the framework of the finite difference/volume technique, the most fundamental solution method is referred to as the Direct Numerical Simulation (DNS). In the DNS method, the transient form of the Navier–Stokes equations is solved numerically by means of spectral and pseudospectral techniques. However, because of the complexity of general industrial, as well as environmental, problems and the limitation in the capabilities of present computer systems, DNS is nowadays still primarily limited in its use to the study of some of the very simple but fundamental flow problems, such as simple turbulent channel and pipe flows, flow in plane mixing layers, etc. When DNS is employed for simulations of high Reynolds number turbulence flows, such as river flows, a prohibitively large number of computational cells must be employed in order to resolve the smallest turbulence vortices, and usually this is not practical within the present levels of computer techniques.

A more common approach to model turbulent river flows is to use the Reynolds-averaged Navier–Stokes equations incorporating an appropriate turbulence model. This has the advantage that a relatively coarse computational grid may be employed. However, it is evident that even when using the Reynolds-averaged Navier–Stokes equations, it is sometimes very difficult to solve large scale, complex unsteady river flows in a fully 3D model due to the limitations in computer power. This is particularly true when the problem investigated is part of a real river where the flow is turbulent with irregularly shaped banks and beds. Therefore, various simplifications to the governing equations have to be made in order to reduce the dimensions of the problem.

This section presents the Reynolds-averaged Navier–Stokes equations and the 2D depth-averaged equations which have been commonly employed in CFD models of river flow. In addition, fine sediment/pollutant transport equations are discussed for the cases where their mass transfer is important (for further details see Chapter 3 in this book; the transport of bed material is covered in Chapter 4).

It is noted that there have been some attempts to use the Large Eddy Simulation (LES) techniques to investigate steady and unsteady flows in river channels (e.g. Thomas and Williams, 1995; Bradbrook et al., 2000). Although LES itself is a relatively well-established numerical technique, and has been used in numerous engineering and environmental applications, the use of LES in river flow modelling is still at an early stage of development. Therefore, only a brief introduction to the LES approach is presented in section 2.3.4.

2.2.1 The Reynolds-averaged Navier–Stokes equations

The fundamental parameters required to describe a river flow are the pressure and the velocity of the fluid flow. If the flow is assumed to be incompressible and Newtonian, then these parameters are solely governed by the constitutional Navier–Stokes equations which are based on the basic physical principles of conservation of mass and momentum. For an incompressible and turbulent fluid flow, the Reynolds-averaged form of the Navier–Stokes equations may be written in a Cartesian coordinate system as follows (Hinze, 1975):

Navier–Stokes equation

$$\rho \frac{\partial u^i}{\partial t} + \rho u^j \frac{\partial u^i}{\partial x^j} = -\frac{\partial p}{\partial x^i} + \frac{\partial}{\partial x^j}\left[\mu\left(\frac{\partial u^i}{\partial x^j} + \frac{\partial u^j}{\partial x^i}\right)\right] + \frac{\partial \tau^{ij}}{\partial x^j} + \rho g^i \quad (2.1)$$

Continuity equation

$$\frac{\partial u^i}{\partial x^i} = 0 \quad (2.2)$$

where x^i represents the components of the Cartesian coordinate system ($i = 1, 2, 3$); t represents the time; u^i, p, ρ and μ are the mean fluid velocity components, pressure, density and molecular viscosity, respectively; and g^i represents the gravitational acceleration. The first term on the left side of equation (2.1) represents the rate of change of momentum per unit volume with respect to time, the second term represents the change of momentum resulting from advective motion. On the right side of the equation, the first term is the force resulting from the pressure differences in the fluid flow, the second term combines the viscous shear forces resulting from the motion of the fluid and the third term is the force produced by the turbulent fluctuations of the fluid particles. This third term is a result of the Reynolds-averaging of the original Navier–Stokes equations, known as the Reynolds stress tensor and defined as

$$\tau^{ij} = -\rho \overline{u'^i u'^j} \quad (2.3)$$

and, in practice, it represents the effects of the turbulence on the fluid flow. Here the overbar denotes a Reynolds-average and the dash represents the fluctuating part of the turbulent velocity. Since the Reynolds stress is not known *a priori*, the Reynolds-averaged Navier–Stokes equations (2.1)–(2.2) are not closed, unless a model is provided that relates the Reynolds stress tensor τ^{ij} to the global mean property of the fluid flow in a physically consistent fashion. This has prompted the development of various turbulence models and some of these will be discussed in section 2.3. In principle, with an appropriate turbulence model and a set of properly defined boundary conditions for the fluid flow, the river flow can be modelled numerically by solving the Reynolds-averaged Navier–Stokes equations (2.1)–(2.2) in all three directions of the coordinate system. Therefore, it is the turbulence model and the boundary conditions that define the output of the CFD model and they are the focus of most investigations in river flow modelling.

2.2.2 Depth-averaged equations

Due to the complexity of solving the full 3D Navier–Stokes equations, which often place impractical demands on computer resources, there is a practical requirement to reduce the dimension of the governing fluid flow equations and this results in the depth-averaged fluid flow equations.

For a shallow river flow, and in situations where the vertical variations of the fluid flow are less important than the variations of the longitudinal and transverse flows in the river, or the flow is approximately unidirectional, the dimensions of the governing fluid flow equations can be reasonably accurately reduced by integrating the full 3D equations over the water depth, h. This results in the introduction of the depth-averaged velocity components of the fluid velocity, $\bar{u}^i$ in the x^i-direction of the horizontal Cartesian coordinate system ($i = 1, 2$), as follows:

$$\bar{u}^i = \frac{1}{h} \int_0^h u^i \, dz \qquad (i = 1, 2) \tag{2.4}$$

where z is the vertical axis of the 3D Cartesian coordinate system. Thus in the depth-averaged approach, the fluid velocities are assumed to be constant throughout the depth and equal to the depth-averaged velocity, $\bar{u}^i$. In addition, from the assumption of the hydrostatic pressure distributions in the vertical direction, the pressure forces can be evaluated in terms of the depth. However, it should be noted that the assumption of the hydrostatic pressure distribution limits the accuracy of the model in regions of steep slopes and rapid changes in the bed topography.

The 2D depth-averaged Navier–Stokes equations can be expressed in a horizontal Cartesian coordinate system as follows:

Continuity equation

$$\frac{\partial h}{\partial t} + \frac{\partial}{\partial x^i} \left(h \bar{u}^i \right) = 0 \tag{2.5}$$

Momentum equation

$$\frac{\partial \bar{u}^i}{\partial t} + \frac{\partial(\bar{u}^i\bar{u}^j)}{\partial x^j} = -g\frac{\partial h}{\partial x^i} + \frac{1}{\rho}\frac{\partial \tau^{ij}}{\partial x^j} + \frac{\tau_s^i - \tau_b^i}{\rho h} + \frac{1}{\rho h}\frac{\partial}{\partial x^j}\left(\int_0^h \rho(u^i - \bar{u}^i)(u^j - \bar{u}^j)\mathrm{d}z\right) \quad (2.6)$$

The turbulence shear stress, τ^{ij}, can be determined using an appropriate turbulence model, for example the Bousinessq type eddy-viscosity model that is presented in section 2.3 (e.g. Wilson *et al.*, 2003). The depth-averaging also results in the introduction of two other groups of stresses in the equation, i.e. the water surface and bed stresses (τ_s^i and τ_b^i), and the so-called dispersion terms, namely the last term in equation (2.6), and these terms usually require empirical formulae or models.

2.2.2.1 The bed shear stresses

Water surface shear stresses τ_s^i are usually ignored unless strong winds exist. However, the bed shear stresses, τ_b^i, are very important and usually have to be obtained experimentally. It is assumed that bed shear stresses can be expressed as a quadratic function of the depth-averaged velocity as follows (e.g. Rastogi and Rodi, 1978; Ye and McCorquodale, 1997):

$$\tau_b^i = C_f \rho \bar{u}^i \sqrt{\bar{u}^i \bar{u}^i} \quad (2.7)$$

where C_f is the bed friction coefficient. As the bed shear stresses result primarily from the turbulent flow interactions, there is considerable uncertainty in their evaluation. Typically, they may be determined using various empirical functions (e.g. Nezu and Nakagawa, 1993; Lane, 1998). In many applications, the bed friction coefficient is calculated using the Manning's equation as follows:

$$C_f = \frac{gn^2}{h^{1/3}} \quad (2.8)$$

where the parameter n is not a constant but depends on the fluid flow situation under investigation.

In an equivalent formulation, the bed shear stresses are evaluated using the non-dimensional Chezy coefficient, C_s, as follows:

$$\tau_b^i = \frac{1}{C_s^2}\rho \bar{u}^i \sqrt{\bar{u}^i \bar{u}^i} \quad (2.9)$$

and the Chezy coefficient can be related to the effective bed roughness height, K_s, as follows:

$$C_s = 5.75 \log\left(12\frac{h}{K_s}\right) \quad (2.10)$$

It should be noted that there is considerable uncertainty in how to choose the effective roughness, height, K_s, as well as Manning's n, as it requires information on the grain size (Chapter 10). Therefore, both K_s and n are often used as calibration parameters against, for example, measured water surface elevations and fluid velocity. There are also other empirical formulae that may be used to evaluate the bed shear stresses/frictions (e.g. Rastogi and Rodi, 1978; van Rijn, 1987; Nezu and Nakagawa, 1993), and this is clearly an area in which further research is required.

2.2.2.2 The dispersion terms

The dispersion terms, namely the last term in equation (2.6), are produced because of the non-uniformities of the fluid flow velocity in the vertical direction during the depth-averaging. The determination of these terms requires knowledge of the information in the secondary flows across the depth which is, however, usually not known *a priori*. Therefore, we usually have to produce a mathematical model for the dispersion terms (e.g. Lane, 1998). However, these terms are frequently found to be negligible in cases such as straight channel flows (e.g. Nezu and Nakagawa, 1993). However, the dispersion terms can be very important, particularly when there are strong secondary flows, such as fluid flows through a river bend or over channel junctions (e.g. Bernard and Schneider, 1992; Chapter 10 in this book).

Since there are no additional equations for the dispersion terms arising from the depth-averaging, the treatment of the dispersion terms has been mainly through semi-empirical or empirical models to represent their effects on the transport of momentum. The key element of these models is the representation of the secondary flows resulting from the effects of the lateral curvature of the river and the friction at the bottom of the river. One of the key models for the dispersion terms is that described by Johannesson and Parker (1989), where 2D forms of the mass and momentum equations are combined with an empirical shape function for the distribution of the primary velocity across the depth. Another approach is that of Bernard (1992) in which the dispersion terms are expressed as a function of the shear stresses associated with the secondary flow, and empirically derived terms are used. However, regardless of the relative importance of the dispersion terms, the present available treatment of these terms is limited in its representation of the vast complexity and variety of secondary flows occurring in natural river flows. This is largely attributed to the lack of understanding of the mechanisms of secondary flow generation in 3D natural river flows in various bed topographies and channel irregularities. Clearly, more research on the mathematical representation of the dispersion terms is required when modelling complex river flows using the depth-averaging approach. However, fluid flow problems with strong secondary flows can be more completely addressed through 3D simulations.

2.2.3 Pollutant transport equations

The transport of dissolved contaminants in a fluid flow typically occurs through (i) advection, which is produced by the movement of the mean fluid flow, and (ii) diffusion, which is caused by the effect of turbulent fluctuations of the fluid flow.

According to the law of mass conservation, for a control volume in a domain of the fluid flow we have the following mass balance equation for the contaminants:

$$\frac{\partial c}{\partial t} + \frac{\partial (cu^i)}{\partial x^i} - \frac{\partial}{\partial x^i}\left(\frac{v_t}{Sc}\frac{\partial c}{\partial x^j}\right) = 0 \qquad (2.11)$$

where c is the concentration of the pollutant and Sc is the turbulent Schmidt number. In practice, Sc should be determined experimentally, but in CFD calculation it is usually assumed to take a constant value and, typically, in rivers we have $0.3 < Sc < 1.0$ (e.g. Rutherford, 1994; Chapter 3 in this book).

Equation (2.11) describes the transport processes of dissolved/suspended pollutant transport in river flows where no material is deposited on, or lifted from, the river bed. When modelling fine sediment transport, terms such as those representing the advection due to particles falling and the exchange of materials between the bed and the suspended sediments, i.e. the deposition and the erosion flux of the sediments, may be introduced as a source term into equation (2.11). Since the physical processes of the deposition and erosion are very complex, mathematical models for these processes are, at present, largely based on empirical formulae. For more information on the modelling of sediment transport, sediment deposition and river bed erosion, readers are referred to Krone (1962), Partheniades (1965), van Rijn (1987), Moulin and Slama (1998), Olsen and Kjellesvig (1999), Olsen (2003) and several other chapters in this book.

In its depth-averaged form, equation (2.11) may be written in a horizontal Cartesian coordinate system as follows (e.g. Holly and Usseglio-Polatera, 1984):

$$\frac{\partial (hc)}{\partial t} + \frac{\partial (hc\bar{u}^i)}{\partial x^i} - \frac{\partial}{\partial x^i}\left(hD^{ij}\frac{\partial c}{\partial x^j}\right) = 0 \qquad (2.12)$$

In the depth-averaged formulation, equation (2.12), the pollutant dispersion coefficient D^{ij} should incorporate not only the purely turbulent diffusion but also the effect due to the iteration between the vertical shear of the horizontal component of the velocity and the vertical turbulent diffusion, and it is usually evaluated using empirical formulations.

In CFD calculations, the dispersion term can be thought of as being the combination of longitudinal and transversal components which are parallel and perpendicular, respectively, to the local advective velocity vector. Further, the longitudinal and transversal dispersion coefficients, D_L and D_T, respectively, are assumed to be linearly proportional to the local shear velocity, u^*, and the water depth (e.g. Elder, 1959), and can be expressed as follows:

$$D_L = 6hu^* \quad \text{and} \quad D_T = 1.5u^* \qquad (2.13)$$

in which

$$u^* = \sqrt{g}\frac{\sqrt{\overline{u^i u^i}}}{C_s} \qquad (2.14)$$

and C_s is the Chezy coefficient.

If $\theta = \arctan(\bar{u}^1/\bar{u}^2)$ is the angle between the local flow direction and the x^1-axis, then D^{ij} may be calculated as follows:

$$\begin{cases} D^{11} = D_L \cos^2\theta + D_T \sin^2\theta \\ D^{22} = D_L \sin^2\theta + D_T \cos^2\theta \\ D^{12} = (D_L - D_T)\cos\theta \sin\theta \end{cases} \quad (2.15)$$

It should be noted that other ways of evaluating the dispersion coefficient D^{ij} are in use (e.g. Preston, 1985), and they are primarily based on empirical formulae or experimental calibrations.

2.3 Turbulence modelling

The complexity of solving the Navier–Stokes equations, in the context of river flow simulations, lies mainly in the solution for the turbulence of the flow. A number of different turbulence models are available, as discussed at greater length by Sotiropoulos (Chapter 5). Each involves different levels of complexity, with the mixing-length model being the simplest and the Reynolds-Stress Model (RSM) the most complex. Whilst the most commonly used model is the k-ε turbulence model, the k-ω models are increasingly popular in engineering applications.

It should be noted that the use of an appropriate turbulence model in a particular application for the fluid flow simulation is very important, and the model used may vary from one application to another. Further, the use of a specific model in river flow simulations can only be justified when the fluid flow problem has been well defined so as to reflect the real situation in the field. Therefore, often the accurate representation of the boundary conditions may be more important than the choice of the turbulence model itself.

2.3.1 Mixing-length model

By analogy with the molecular viscosity, Prandtl (1925) linked the turbulent viscosity to the overall velocity gradient and proposed the first model known as the mixing-length model, in which the turbulence viscosity, μ_t, is defined as follows:

$$\mu_t = \rho l^2 \left|\frac{\partial u_L}{\partial z}\right| \quad (2.16)$$

where u_L is the longitudinal fluid velocity component, z is the vertical axis of the Cartesian coordinate system, l is known as the Prandtl mixing length which is related to the scale of the turbulence in the flow and can be expressed as a function of the water depth (e.g. Nezu and Rodi, 1986). Then the Reynolds stress tensor τ^{ij} may be evaluated based on the Boussinesq eddy-viscosity hypothesis (Hinze, 1975), which relates the Reynolds stress to the mean fluid velocity through the equation

$$\tau^{ij} = -\frac{2}{3}\rho k \delta_j^i + \mu_t \left(\frac{\partial u^i}{\partial x^j} + \frac{\partial u^j}{\partial x^i}\right) \qquad (2.17)$$

Here k is the turbulent kinetic energy that may be incorporated into the pressure term in the numerical scheme, and δ_j^i is the well-known Kronecker delta function.

The mixing-length model has the advantage of being simple and it is well suited for simple thin shear-layer flows, e.g. open channel flows where the mixing-length scale has been well defined through years of experience. However, the simplicity of the model also implies limitations in its use, and in particular it does not include any effects of the history of the flow and the turbulence transport on the mixing length. Thus the method lacks accuracy when predicting the turbulent viscosity in complex river flows.

2.3.2 Two-equation models

Two-equation models, such as the k-ε model, the k-ω model and their variations, are the most popularly used models in various engineering and environmental applications. Even though more sophisticated models have been proposed (e.g. the RSMs of Launder et al., 1975), the two-equation models with their various modifications are still the most popular turbulence models in use and this is largely due to their ease in implementation, economy in computation and, most importantly, being able to obtain reasonably accurate solutions with the available computer power.

2.3.2.1 The standard k-ε turbulence model

The k-ε model was first proposed by Launder and Spalding (1972) in the early 1970s and it has enjoyed reasonable success in many engineering and scientific applications. The standard k-ε model is also an eddy-viscosity-type turbulence model and it represents the eddy viscosity by the turbulent kinetic energy k and the dissipation rate ε as follows:

$$\mu_t = c_\mu \rho \frac{k^2}{\varepsilon} \qquad (2.18)$$

where c_μ is an empirically derived constant and it usually takes the value of 0.09 in the standard k-ε model (e.g. Launder and Spalding, 1974).

The turbulent kinetic energy k and the dissipation rate ε in the k-ε model have to be obtained in order to estimate the turbulent viscosity and this is achieved by solving two transport equations for k and ε. At high values of the Reynolds numbers, which is nearly always the case for natural river flows, they are given by:

Kinetic energy equation

$$\rho \frac{\partial k}{\partial t} + \rho u^j \frac{\partial k}{\partial x^j} = \frac{\partial}{\partial x^i}\left[\left(\frac{\mu + \mu_t}{\sigma_k}\right)\frac{\partial k}{\partial x^i}\right] + P - \rho\varepsilon \qquad (2.19)$$

where

$$P = (\mu + \mu_t)\left(\frac{\partial u^i}{\partial x^j} + \frac{\partial u^j}{\partial x^i}\right)\frac{\partial u^i}{\partial x^j} \qquad (2.20)$$

is the term which represents the production of turbulence.

Dissipation rate equation

$$\rho\frac{\partial \varepsilon}{\partial t} + \rho u^j \frac{\partial \varepsilon}{\partial x^j} = \frac{\partial}{\partial x^i}\left[\left(\frac{\mu + \mu_t}{\sigma_\varepsilon}\right)\frac{\partial \varepsilon}{\partial x^i}\right] + c_{\varepsilon 1}\frac{\varepsilon}{k}P - c_{\varepsilon 2}\frac{\varepsilon}{k}\rho\varepsilon \qquad (2.21)$$

where $c_{\varepsilon 1}$, $c_{\varepsilon 2}$, σ_k and σ_ε are constants that are typically assumed to take the values of 1.44, 1.92, 1.0 and 1.3, respectively (e.g. Launder and Spalding, 1974).

However, the main shortcoming of the standard k-ε model is its assumption of isotropy in the turbulence fluctuations when modelling the Reynolds stresses and this can result in a large turbulent viscosity. Therefore this model frequently produces inaccurate predictions in some turbulent fluid flows (e.g. Mohammadi and Pironneau, 1994; Ingham et al., 1997). Several other modifications to the standard k-ε model have been developed by aiming at improving the expressions for the Reynolds stress tensor. For example, in the model proposed by Speziale (1987), the Reynolds stress tensor is represented by the classic Boussinesq hypothesis (equation 2.17) plus a nonlinear correction term. There are other similar nonlinear k-ε models that have been used by many researchers in numerous engineering applications, but occasionally they have been reported to have convergence problems.

2.3.2.2 The renormalization group (RNG) k-ε turbulence model

The RNG model is basically obtained theoretically through a process of statistical eliminations and it is a significant improvement over the standard k-ε turbulence model in simulating 3D fluid flows with a large degree of strain in the fluid and for boundaries with a large curvature. For high Reynolds number fluid flows, the RNG k-ε model employs the same expressions for the turbulence viscosity as in the standard k-ε model, namely equation (2.18), and also a similar set of transport equations to the standard k-ε model for k and ε, namely

Kinetic energy equation

$$\rho\frac{\partial k}{\partial t} + \rho u^j \frac{\partial k}{\partial x^j} = \alpha\frac{\partial}{\partial x^i}\left((\mu + \mu_t)\frac{\partial k}{\partial x^i}\right) + P - \rho\varepsilon \qquad (2.22)$$

Dissipation rate equation

$$\rho\frac{\partial \varepsilon}{\partial t} + \rho u^j \frac{\partial \varepsilon}{\partial x^j} = \alpha\frac{\partial}{\partial x^i}\left((\mu + \mu_t)\frac{\partial \varepsilon}{\partial x^i}\right) + c_{\varepsilon 1}\frac{\varepsilon}{k}P - c_{\varepsilon 2}\frac{\varepsilon}{k}\rho\varepsilon - R \qquad (2.23)$$

where R is a parameter which includes the rate-of-strain and plays an important role in anisotropic large-scale eddies, and it can be expressed as follows:

$$R = \rho \frac{c_\mu \eta^3 \left(\frac{1-\eta}{\eta_0}\right)}{1+\beta\eta^3} \frac{\varepsilon^2}{k}, \quad \eta = \frac{k}{\varepsilon}\sqrt{2S_{ij}S_{ij}}, \quad S_{ij} = \frac{1}{2}\left(\frac{\partial u^i}{\partial x^j} + \frac{\partial u^j}{\partial x^i}\right) \quad (2.24)$$

The values of the constants which appear in the RNG theory usually take the values $c_\mu = 0.0845$, $c_{\varepsilon 1} = 1.42$, $c_{\varepsilon 2} = 1.68$, $\alpha = 1.3$, $\beta = 0.011$–0.015 and $\eta_0 = 4.38$ (e.g. Yakhot and Orszag, 1986; Boysan, 1995). These values should be compared with the values as found in the standard k-ε model (e.g. Launder and Spalding, 1974), where $c_\mu = 0.09$, $c_{\varepsilon 1} = 1.44$, $c_{\varepsilon 2} = 1.92$. It should be noted that although the values of the two sets of constants are very similar, the constants employed in the RNG model are basically obtained theoretically whereas those employed in the standard k-ε model have been found from experimental data and thus determined experimentally. Further, it should be noted that the standard empirical constants in all the k-ε models should not be expected to be universal constants and modifications may be required to suit specific physical situations.

The expression (2.18) for the eddy viscosity μ_t holds for high Reynolds number fluid flow in the RNG k-ε model, but with $c_\mu = 0.0845$. In addition to this expression, the RNG theory also gives a more general form for the expression for the eddy viscosity (e.g. Boysan, 1995),

$$\mu_t = \mu\left[1 + \sqrt{\frac{c_\mu \rho}{\mu}}\frac{k}{\sqrt{\varepsilon}}\right]^2 \quad (2.25)$$

which is valid across the full range of turbulent fluid flow conditions from low to high Reynolds numbers. For more details of the model, the reader should consult, for example, Yakhot and Orszag (1986), Yakhot and Smith (1992) and Boysan (1995).

2.3.2.3 The k-ω models

As mentioned above, the standard k-ε turbulence model often produces a high turbulent viscosity and this frequently results in the delayed onset of separation and the under-prediction of the size of the separation region. The k-ω turbulence model tends to be more accurate for boundary layer flows with adverse pressure gradients, and also in transitional flows, when compared to the results obtained when using the k-ε models. In addition, the k-ω model includes a low Reynolds number extension for the near-wall turbulence so that it does not require wall functions, although wall functions can be incorporated when necessary.

Analogous to the k-ε model, the k-ω models are based on the model transport equations for the turbulence kinetic energy k and the specific dissipation rate ω, which can also be thought of as the ratio of ε to k, or the turbulence frequency. Then the turbulent viscosity may be computed from the following:

$$\mu_t = \alpha\rho\frac{k}{\omega} \quad (2.26)$$

Different versions of the k-ω models have different ways of evaluating the factor α and this may have a significant impact on the representations of the transport processes of the shear stresses.

The first k-ω model was that of Kolmogorov (1942) and several other different and improved versions of the k-ω model have been proposed. However, the k-ω models of Wilcox (1988, 1998) and Menter (1994) have been the most extensively tested and widely used.

The Wilcox (1988) model is known to suffer from sensitivity to the free stream boundary conditions (e.g. Menter, 1992), which leads to significant variations in the model predictions. Thus this model is best limited for use in fully turbulent internal flow problems. The Wilcox (1998) model includes a function which limits the dissipation of the turbulence kinetic energy k. However, it still suffers from the problem of the build-up of turbulent viscosity in the vicinity of stagnation points.

The Menter (1994) model, known as the Shear Stress Transport (SST) k-ω model, is a blend of the Wilcox (1988) model and the standard k-ε model. It attempts to apply the Wilcox (1988) model to the regions near to the wall surface and to apply a transformed k-ε model to the high Reynolds number outer boundary layer regions and the free shear layers so that the sensitivity to the free stream conditions can be removed. This has been achieved by the introduction of a blend function, F, which switches the model between the Wilcox (1988) model and the k-ε model according to the distance from the computational cell to the nearest wall and the state of the local fluid flow. In addition, the model employs a limiter in the formulation of the eddy viscosity so that the transport of the shear stresses can be properly represented.

In the Menter SST model, the transport equations for the turbulent kinetic energy, k, and the specific dissipation rate, ω, may be expressed as follows:

$$\rho \frac{\partial k}{\partial t} + \rho u^i \frac{\partial k}{\partial x^i} = \frac{\partial}{\partial x^j}\left(\left(\mu + \frac{\mu_t}{\sigma_k}\right)\frac{\partial k}{\partial x^j}\right) + G_k - Y_k \qquad (2.27)$$

and

$$\rho \frac{\partial \omega}{\partial t} + \rho u^i \frac{\partial \omega}{\partial x^i} = \frac{\partial}{\partial x^j}\left(\left(\mu + \frac{\mu_t}{\sigma_\omega}\right)\frac{\partial \omega}{\partial x^j}\right) + G_\omega - Y_\omega + D_\omega \qquad (2.28)$$

In these equations, G_k and G_ω represent the generation of k and ω, respectively, due to the mean velocity gradients; Y_k and Y_ω represent the dissipation of k and ω respectively, due to turbulence and D_ω is the cross-diffusion term. For more information on the formulation for these quantities, and for the k-ω model in general, readers are referred to Wilcox (1988, 1998), Menter (1994) or Fluent (2003).

The Menter SST model is nowadays one of the most popular turbulence models used in aeronautics as well as in other areas of aerodynamics. It is available in a number of commercially available CFD software codes, such as Fluent and CFX. The Menter SST model keeps all the advantages of the Wilcox models, such as being capable of fairly accurately predicting adverse pressure-gradient boundary layers and flow separation. It does not require any near-wall modifications and it also eliminates problems of sensitivity to the free shear flow boundary conditions. The potential of this model for river flow simulation is an area worthy of further investigation.

2.3.3 Reynolds-stress models (RSMs)

The RSMs have recently emerged as strong candidates for use in industrial turbulent flow calculations as well as in environmental flow modelling and they are now incorporated in some commercial codes. In order to model the different evolutions of the components of the Reynolds stress, the transport equations for each component of the stress τ^{ij} have been employed in the RSMs, in addition to the turbulent dissipation equation (e.g. Rodi, 1993). As the RSMs are second-order closure models, they are potentially superior to the simpler eddy-viscosity models, such as the k-ε models. However, one of the drawbacks of the RSMs is their very complex formulations and this makes it difficult to both implement and ensure numerical convergence. In addition to the high computational costs when using the RSMs, it is reported that not all the fluid flow simulations can benefit from this model, particularly when the fluid flows are straining (e.g. Rodi, 1993).

To reduce the computational cost when using the RSMs, Rodi (1976) proposed an implicit algebraic expression for the Reynolds stress tensor, known as the implicit Algebraic Stress Model (ASM), which supposes that the turbulent kinetic energy and dissipation rate are known, e.g. from the solution of the k-and ε-equations. The ASM is well suited to account for the directional effects due to the body forces and the walls (e.g. Rodi, 1993). However, computations that use this implicit model suggest that the convergence of the algebraic equations is very slow and sometimes stable numerical solutions can be difficult to obtain (e.g. Mohammadi and Pironneau, 1994), and the results may not be very accurate (e.g. Boysan, 1995). Therefore over the last decade the use of the ASM has significantly decreased.

Based on the implicit ASM of Rodi (1976), Gatski and Speziale (1993) have proposed the Explicit Algebraic Stress Models (EASMs) that are formally obtained from full second-order closures in the homogeneous equilibrium limit by assuming that the advection and diffusion terms in the RSMs may be replaced by an algebraic expression. Rokni (1997) has successfully applied this model in the simulation of the turbulent convective heat transfer in a duct. The EASMs maintain the relative simplicity of the k-ε model, while incorporating more of the physics of the turbulence, such as the anisotropies of the turbulent dissipation. It is reported for turbulent fluid flows, close to equilibrium, that the EASMs yield results that are almost indistinguishable from those obtained when using full second-order closure models.

The various RSMs comprise a group of complicated formulations that require a potentially excessive amount of computational effort for the simulation of 3D fluid flows. However, it should be noted that the use of the RSMs may be the way forward when the anisotropic Reynolds stress has a large effect on the fluid flow structures, such as those that occur when simulating the secondary fluid flow in an open channel which is free of bed topography. Clearly, more research work on RSMs is required.

2.3.4 Large eddy simulation

The turbulent models presented in the previous sections are those for the Reynolds stresses when solving the Reynolds-averaged equations where all spectra of the turbulent eddies, from large to small scales, are to be modelled. With modern computational techniques it has become practical to use much finer grids so that the evolution of some of the large-scale turbulent motions, which are prone to be affected by the boundary conditions, can be resolved and directly simulated. Only the unresolved smaller scale turbulent motions, which are relatively independent of the boundary conditions, need to be modelled through the use of a sub-grid model. This CFD technique is called large eddy simulation. Therefore, LES is potentially more accurate than the Reynolds-averaged equations but it is much more computationally expensive since fine computational grids are still required.

In LES, the governing Navier–Stokes equations are filtered in space by a filter function so that the resolved scales may be separated from the unresolved scales, and the quantities to be solved in the CFD calculations are filtered quantities. A filtered quantity, $\bar{f}$, is defined as follows:

$$\bar{f}(\mathbf{x}) = \int_D f(\mathbf{x}')G(\mathbf{x},\mathbf{x}')\mathrm{d}\mathbf{x}' \tag{2.29}$$

where D is the fluid domain, and G is the filter function that determines the scale of the resolved eddies. In the CFD calculations, the discretization itself can be treated as an implicit filtering operation to the governing equations. The filtered time-dependent Navier–Stokes equations may be expressed as follows, eliminating the over bar for the filtered quantities:

$$\frac{\partial u^i}{\partial x^i} = 0 \tag{2.30}$$

$$\rho\frac{\partial u^i}{\partial t} + \rho u^j\frac{\partial u^i}{\partial x^j} = -\frac{\partial p}{\partial x^i} + \frac{\partial}{\partial x^j}\left[\mu\left(\frac{\partial u^i}{\partial x^j}+\frac{\partial u^j}{\partial x^i}\right)\right] + \frac{\partial \tau^{ij}}{\partial x^j} + \rho g^i \tag{2.31}$$

This equation is similar to the Reynolds-averaged equations (2.1). However, the velocity components, u^i, are the filtered velocity components corresponding to the grid spacing used in the computations, and therefore now they are not the mean fluid velocity components, and the turbulence stresses, τ^{ij}, are the sub-grid stresses and the models for them are different from what they are for the Reynolds stresses.

The filtering process effectively filters out the eddies whose scales are smaller than the filter width or the grid spacing used in the computations. The resulting equations thus govern the dynamics of the large eddies and the smaller sub-grid eddies are to be modelled. The most commonly used model for sub-grid turbulence is the eddy-viscosity-based model given by

$$\tau^{ij} - \frac{1}{3}\tau^{kk}\delta^i_j = 2\mu_t S_{ij} \tag{2.32}$$

where μ_t is the sub-grid turbulent viscosity which can be expressed (e.g. Lilly, 1966; Bradbrook et al., 2000; Fluent, 2003), as follows:

$$\mu_t = \rho L_s^2 \left| \sqrt{2 S_{ij} S_{ij}} \right| \quad (2.33)$$

where L_s is the sub-grid mixing length and it can be expressed as follows:

$$L_s = \min(\kappa d, C_s V^{1/3}) \quad (2.34)$$

where κ is the von Karman constant and takes the value 0.42, d is the distance from the computational cell to the closest wall, $C_s = 0.1$–0.23 and V is the volume of the computational cell.

However, it is noted that the eddy-viscosity-based sub-grid model gives a poor representation of the sub-grid-scale transport (e.g. Temmerman et al., 2003). Whilst this is not important when the sub-grid eddies are small in a fine grid, it can be seriously detrimental to the accuracy when a coarse grid is employed when the transport of the sub-grid-scale eddy is important. There are a number of other sub-grid models in existence and the interested reader is referred to, for example, Abba et al. (2003) and Temmerman et al. (2003).

2.4 Free surface flows

Most of the current CFD investigations on river flows employ a so-called fixed lid surface approach where, for steady fluid flows, an approximate fixed lid may be specified to represent the water surface. Unfortunately, for unsteady flow situations, the instantaneous water surface is not known *a priori* and it may have to be numerically predicted at each time interval as any incorrect specification of the water surface will affect the distribution of the mass and momentum of the fluid flow in the computational domain. This may result in the fluid velocity being over-estimated or underestimated and thus this leads to errors in the prediction of the bed shear stress and consequently bedload transport.

2.4.1 Fixed lid schemes

For steady state river flows, for which the water surface is known, the shape of the free surface may be specified as a fixed lid boundary of the computational domain. On this fixed lid, frictionless conditions are enforced which allow the water to slip, but not to pass through the free surface. In such a situation the water flow may be solved in the usual manner, subject to the other appropriate boundary conditions being specified. It is clear that when a prescribed water surface is used, any incorrect specification of the water surface position may result in over-under-predictions of the fluid velocity and this leads to inaccuracies in the bed shear stress calculations. In order to improve the accuracy of the model, the water surface may be modified by the pressure at the water surface. Two key water surface correction models exist, namely the porosity model of Spalding (1985) and the surface mesh deformation model described in, for example, Olsen and Kjellesvig (1998).

Spalding (1985) proposed a model in which the porosity concept is used to represent the water surface. In this model, a layer of cells with appropriate porosity values may be defined on the top of a given free surface to model the effect of the water surface elevation. The porosity of the cells at the water surface may be defined as follows (e.g. Spalding, 1985):

$$P_{\text{por}} = 1 + \frac{p}{\rho g h_c} \quad (2.35)$$

where P_{por} is the porosity and h_c is the height of the cell in the top layer of the fluid. The dummy porosity cells are present primarily in order to make approximate corrections to the mass continuity equation by effectively changing the cell size at the surface as the porosity varies from 0 to 1.

In the surface mesh deformation model, the water surface may be corrected by deforming the mesh based on the predicted pressure at the water surface (e.g. Olsen and Kjellesvig, 1998; Booker, 2003). The water surface is fixed at the downstream boundary where the pressure is taken as a reference pressure, p_{ref}. A pressure deficit at each cell is then calculated by subtracting this reference pressure from the extrapolated pressure for each cell and used to move the water surface by a vertical height

$$\Delta h = \frac{l}{\rho g}(p - p_{\text{ref}}) \quad (2.36)$$

where l is the difference in the height of the water surface at p and p_{ref}.

The water surface correction models described above are simple and effective in representing the free surface in CFD simulations and they can be used for moderate water surface elevations. However, in situations where large water level changes occur, for example due to a large change in bed elevation, or during a flood event, it may be difficult to obtain accurate predictions of the water surface and the fluid flow when using these water surface correction approaches.

2.4.2 The Volume of Fluid (VOF) model

In flood seasons, the water discharge, and thus the water level, changes rapidly, and the water surface must be free to change instantaneously in the computational domain and also as part of the numerical solution in order to simulate the flood flow. In this case, the unsteady solution procedures of the Volume of Fluid (VOF) technique may be employed. It should be noted that although the VOF technique was designed for solving unsteady fluid flow problems, it may also be used to predict a steady flow when the water surface is not known *a priori*.

The VOF model (e.g. Hirt and Nichols, 1981), can be employed to track the position of the water surface in the computational domain when the control volume method is employed. In each control volume in the computational domain, a water volume fraction, F, can be defined as follows:

$$F = \frac{\delta \Omega_{\text{water}}}{\delta \Omega_{\text{cell}}} \quad (2.37)$$

where $\delta\Omega_{cell}$ is the volume of the computational cell and $\delta\Omega_{water}$ is the fraction of the volume of the cell filled with water. Therefore, we have

The volume fraction equation

$$\begin{cases} F = 1, & \text{cell is full of water} \\ F = 0, & \text{cell is full of air} \\ 0 < F < 1, & \text{cell contains a free surface} \end{cases} \quad (2.38)$$

Thus, according to the law of mass conservation of air and water, the volume fraction of the water satisfies the partial differential equation

$$\frac{\partial F}{\partial t} + u^i \frac{\partial F}{\partial x^i} = 0 \quad (2.39)$$

There are different ways of solving equation (2.39) for the water volume fraction F in a cell. The most original way is to use the so-called donor–acceptor scheme (e.g. Hirt and Nichols, 1981), by solving the equation explicitly. Once the location of the water surface has been obtained, the appropriate boundary conditions may be enforced on the free surface.

However, it should be stressed that implementing boundary conditions on a free surface is not always an easy task, in particular when the free surface is constantly changing its position. In order to avoid the complexity of implementing boundary conditions on the free surface, a multi-flow approach can be employed where some airflow can be introduced into the computation domain above the water cells. Then the water surface is transformed into an interface between the air and the water, and the 'donor–acceptor' scheme may be used to solve equation (2.39) and predict the location of the interface. In this case, the effect of the possible airflow on the water surface may also easily be taken into account if so desired. It should be noted that because the viscosity of air is approximately 50 times smaller than that of water, the airflow above the water has little effect on the water flow unless a large airflow rate is imposed, such as when modelling the effects of a strong wind on the water surface.

In addition to the use of the 'donor–acceptor' scheme, equation (2.39) can also be solved implicitly in the same manner as that used to solve other transport equations. However, due to the interpolation being employed in the solution process of the volume fraction equation, the numerical diffusion occurs near to the interface between the air and the water and this may result in difficulties in accurately defining the location of the water surface. To reduce this numerical diffusivity which is caused by the discretization of the volume fraction equation, a second-order discretization scheme, such as the QUICK or second-order upwind scheme, may be used.

It should be stressed that the use of VOF is computationally expensive. Therefore, the fixed lid approach with the use of pressure corrections for the water surface may be used with some confidence for most river flow simulations where no large water surface elevations occur. Lane and Ferguson (Chapter 10) also discuss this issue.

2.5 Boundary conditions

Since the governing fluid flow equations are elliptic in nature, boundary conditions are required on all the boundaries of the solution domain in order to define a specific fluid flow of interest. For transient flow modelling, initial conditions representing the state of the fluid flow at the time when the computation starts must also be specified. It should be stressed that the boundary conditions imposed must reflect the real situation of the river flow that is being modelled, although this is not always an easy task. However, any deviations from this principle will affect the accuracy, if not the correctness, of the numerical solution, and it may well undermine the effort of choosing a sophisticated turbulence model.

2.5.1 Upstream conditions

On the upstream boundary of the computational domain, the fluid velocity can be specified in a way that usually complies with the rate of water discharge. A uniform velocity profile may be employed when the inlet boundary is located sufficiently far upstream of the region under investigation to allow a fully developed flow to occur in the region of interest. However, this development length may be extremely large and great care should be taken when using this condition. When LES is employed, stochastic components of the fluid velocity may be required to be specified.

2.5.2 Downstream conditions

The downstream boundary of the computational domain should normally be taken in a straight part of the river where the fluid velocity does not change significantly further downstream of the location of the boundary. In this situation, the fluid flow may be treated as being fully developed and the most commonly used zero gradient in the direction of the downstream flow may be applied to all the variables of the fluid flow (except the pressure). When the flow cannot be treated as being fully developed at the downstream boundary, in some situations a pressure condition may be used. Otherwise a fluid velocity profile will have to be given at the outlet boundary and care must be taken so that this is done in such a manner that mass is conserved. The velocity profile can come only from measured data. In either situation, the downstream boundary should not be located where reverse flow may exist, as this will probably cause problems with numerical convergence.

When the water level is to be solved as part of the numerical solution, care must be taken when specifying the condition at the downstream boundary as this will affect the predicted water levels in the whole of the solution domain. When a fixed lid is used for the water surface, i.e. the water surface level is prescribed and it is fixed during the numerical computation, the modelled river flow is implicitly transformed into a duct flow and a pressure condition may be applied at the downstream boundary if the flow is not fully developed.

2.5.3 Wall functions and bed roughness

Fluid flow near to the bed and banks of a river is often very complex in terms of both its mean and turbulent structure. Although research is still required to improve our understanding of these regions, the general principles of fluid mechanics show that the fluid flow near to these boundaries experiences a rapid decrease in velocity over a boundary layer in order to comply with the no-slip condition. Therefore, an extremely fine grid near to these boundaries is usually required in order to simulate accurately their effect on the fluid flow. In the situation where a large number of grid cells are already required to cover the large lengths downstream of the river, use of a fine grid to resolve the near-boundary flow is usually not a realistic option.

In the case where the major interest is in the characteristics of the mean fluid flow region, rather than on the flow near to the solid boundaries, wall functions may be employed and hence the use of fine grids in the vicinity of the wall boundaries will not be necessary. The standard wall function may be expressed as follows:

$$u^+ = \frac{1}{\kappa}\ln(Ez^+) - \Delta B(K_s^+) \tag{2.40}$$

$$u^+ = \frac{u_\parallel}{u^*}, \quad z^+ = \frac{z}{z^*}, \quad K_s^+ = \frac{K_s}{z^*}, \quad z^* = \frac{\nu}{u^*} \quad \text{and} \quad u^* = \sqrt{\frac{\tau_w}{\rho}} \tag{2.41}$$

where $u_\parallel$ represents the fluid velocity component parallel to the solid wall, z is the distance from the wall, u^* is the wall friction velocity, K_s is the wall roughness height, E is a constant, usually taking a value of 9.8, ΔB is an expression related to the hydraulically rough bed, κ is Von Karman's constant, usually taking a value of 0.4187, and ν and τ_w are the dynamic viscosity of the fluid and the shear stress at the solid wall boundaries, respectively. Most river flows are in the fully rough regime, i.e. $K_s^+ > 60$, and equation (2.40) may be employed to calculate the fluid velocity near to the wall inside the boundary layer.

An appropriate representation of the bed roughness in the CFD model is very important as it can markedly affect the overall numerical solutions and, in particular, the predicted near-bed shear stress, as discussed by Lane and Ferguson (Chapter 10). However, it should be noted that there is considerable uncertainty in the selection of an appropriate value for K_s (e.g. Hey, 1979; Bray, 1980; Ferguson et al., 1989). This is an area in which further research is required. For a hydraulically smooth wall, the contribution of grain roughness to K_s is zero, and if form roughness is negligible then $\Delta B = 0$.

2.6 Examples of the CFD modelling of the fluid flow in river channels

As examples of the application of CFD in river flow modelling, this section presents briefly some of the CFD work that the authors have performed in the investigations of fluid flow and pollutant transport in river channels. The purpose of the examples is to give the reader some ideas of how CFD can be used in the investigation of river

38 Computational Fluid Dynamics

flow problems. For more detailed descriptions of the models presented, readers are referred to Hancu *et al.* (2002) and Ma *et al.* (2002).

2.6.1 Three-dimensional model for an upland river channel flow

This section presents some results from a combined numerical- and physical-scale modelling project that investigates the flow and sediment dynamics of the River Calder as it passes through a straight reach with non-erodible banks and minimal spatial variation in bed topography. The location of the field site is on the River Calder, Todmorden, at the base of the Pennine Hills in the UK. The study reach is 175 m long and 10 m wide with near-vertical, stone-sided walls of about 3.5 m high and an average bed slope of 0.0035. Although the fixed sidewalls prevent the flow spreading laterally, the channel is free to rework the coarse alluvial bed material (mean surface particle size $D_{50} = 0.067$ m) and this creates a subtle, but clear pool–riffle sequence (Figure 2.1). The walls of the channel are almost vertical and therefore in the computational model investigated here we assumed that both walls are vertical and smooth.

A flume has been set up to physically model the flow and it provides the boundary conditions, for both the upstream and downstream fluid flows, and validation for the numerical model. The 3D numerical simulation of the fluid flow was performed with a steady water discharge, held constant at 12.4×10^{-3} m^3 s^{-1} (scale discharge of 89.9 m^3 s^{-1}) which is approximately the scaled equivalent of the 100-year flood conditions. The control volume method, incorporating the RNG k-ε turbulence closure model, has been employed in order to solve the Reynolds-averaged form of the Navier–Stokes equations and the VOF method is used to predict the water surface elevation. The model calculation was performed using the commercial software package FLUENT 4.4.4.

Figure 2.2 shows the comparison between the numerically predicted and the experimentally measured water surface positions viewing downstream along the centreline of the channel and across the channel at a location half way down

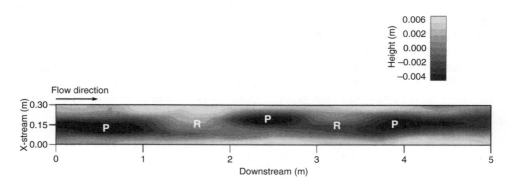

Figure 2.1 Flume bed topography produced from interpolation of 717 EDM (electronic distance measurement) survey points of the Todmorden study reach and scaled by 1:35. Note the limited range in bed topography but also the clear demarcation of shallow pool (labelled P) and riffle (labelled R) sites.

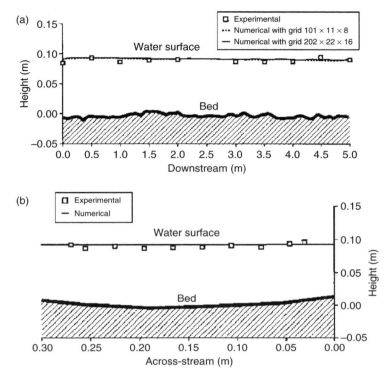

Figure 2.2 Position of the water surface measured in the 1:35 flume model and produced by the numerical model: (a) view downstream along the centreline of the channel and using two different grid sizes in the numerical model; and (b) view across the channel (downstream is into the paper) at a location half way down the 5 m long model.

the model channel, respectively. The water surface across the channel is, in general, quite flat. This is a consequence of the large relative roughness being modelled (in this case 'form' roughness only). It is noted that the numerically predicted water surface using the VOF model is very close to the flume measurements and this is essential for the accurate prediction of the overall magnitude and variations of the downstream fluid velocity component as well as the secondary flow. In the computation using VOF for a steady flow, the initial water surface position and shape can be set arbitrarily and when numerical convergence has been achieved, the correct water surface may be obtained. When simulating an unsteady flow using VOF, the instantaneous water surface profile will be numerically predicted based on a prescribed initial position of the water surface.

Figure 2.3 shows comparisons between the numerically predicted contours of the streamwise fluid velocity and those of the flume data at two different downstream locations. The two cross sections are located at 2 and 3 m from the location of the upstream boundary conditions in a 5 m long model channel. These two positions correspond to the poolhead and the downstream riffle, respectively (Figure 2.1). Reasonably good agreement has been obtained between the numerical and experimental data for the streamwise flow in terms of the general patterns of the fluid flow and the magnitudes of the fluid velocity. This implies that, with the RNG turbulence model and the VOF method, the CFD model can reproduce the main structure of the fluid flow over a real undulating bed topography. The shift of the core of high-speed fluid flow towards the left-hand side of the channel and to below the water surface in both the measured and the predicted results in Figure 2.3 is suggested to be a consequence of the asymmetric profile of the input fluid velocity which is generated by the asymmetry of the bed topography in the study reach model and *prior* to the upstream end of the scale model, and it appears to dominate the fluid flow structure throughout the flume. Undoubtedly, this is partly because the pool–riffle topography is so subdued (although it has a maximum amplitude of 0.35 m in the field, see Figure 2.1), but also because the 1:100 recurrence interval flood drowns out the bed form roughness.

Figure 2.4 shows the numerically obtained secondary flow vectors for the sections 2 and 3 m, respectively. The magnitude of the secondary flows is very small, reaching a maximum of about 10% of the downstream flow, with many points being characterized by values of less than 5% of the primary velocity. The secondary flow is strongest near the bed and in regions near the bed where there is the greatest change in topography and this reflects the impact of the bed shape on the secondary motion of the fluid flow.

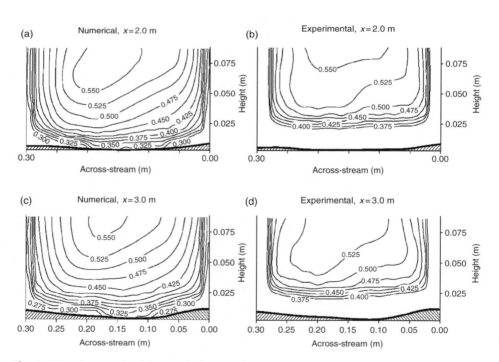

Figure 2.3 Streamwise (u) isovels for two locations in the 5 m long flume and numerical model. Isovels are shown for 0.025 m s^{-1} intervals. The view shown is looking downstream at the sections. The downstream positions of 2 m (a and b) and 3 m (c and d) correspond to a poolhead and riffle, respectively (Figure 2.1).

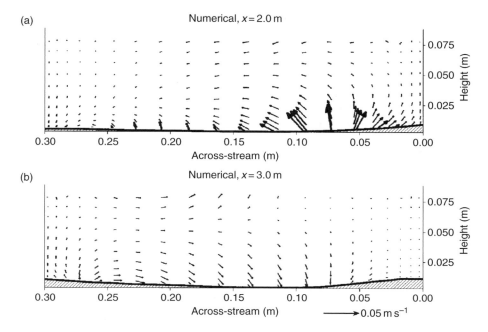

Figure 2.4 Secondary flow vectors for sections (a) 2 m and (b) 3 m from the numerical model. Primary flow direction is into the page.

2.6.2 Pollutant transport in a partially blocked channel flow

This section presents results of a simulation of the fluid and mass transport processes with flow control obstructions being placed in a straight channel. The channel studied is 2 m wide and 26 m long with plane blocking blades being placed in the channel perpendicular to the sidewalls and the bottom surface. This forms a cell between the two blades and the sides of the channel. Water is then guided into the cell upstream from one side and then forced to leave the cell from the slot, which is one-eighth of the channel width, adjacent to the opposite side of the channel. The rate of water discharge is $0.009 \text{ m}^3 \text{ s}^{-1}$ and the water depth in all the cases considered is 0.2 m.

In order to investigate the behaviour of the dispersion of the contaminant in the neighbourhood of the blocking blades, a given amount of salt is released into the steady water flow at an upstream position of the channel. Then the contaminant concentrations inside the cell are measured and modelled numerically as the contaminant passes through the blocking blades.

In the fluid flow simulations, the governing 3D Reynolds-averaged form of the Navier–Stokes equations, incorporating the RNG k-ε turbulence model, has been solved using the control volume numerical technique. The modelling of the transport of the contaminant in the channel has been accomplished by numerically solving the unsteady contaminant transport equation that includes the effect of the advection and turbulent diffusion of the contaminant in the water flow.

42   Computational Fluid Dynamics

Experimental results showed that the water surface elevation in the channel is extremely small. Therefore, in this investigation the water surface in the numerical model has been prescribed according to the experimental data and it acts as a frictionless fixed lid at the surface of the water flow in the numerical simulations.

Figure 2.5 shows the numerically predicted fluid flow patterns in the cell, formed by the two blocking blades, at the vertical position half way below the water surface. As can be seen in Figure 2.5, a strong recirculating flow, which is produced by the redirection of the fluid flow by the blocking blade, has been

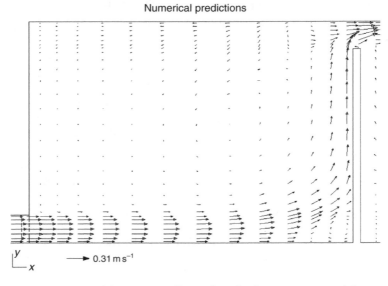

Figure 2.5 *Comparisons of the numerically predicted velocity vectors and the experimental observations for the fluid flow halfway below the water surface. The blade length is seven-eighths the channel width and the rate of water discharge is $0.009\,m^3\,s^{-1}$.*

predicted and it dominates the fluid flow in the space between the blades. In addition to this large recirculating region, a small vortex is also predicted and it is located in the vicinity of the bottom right-hand corner of the cell. These predictions are consistent with the experimental observations using floats, which is shown in the top of Figure 2.5.

Further, the blocking blades and the channel bed impose a strong resistance to the fluid flow and this generates strong 3D flows. This is clearly seen through the differences in the directions of the fluid flows at different depths below the water surface. At its extreme, Figure 2.6 shows the numerically predicted fluid flow

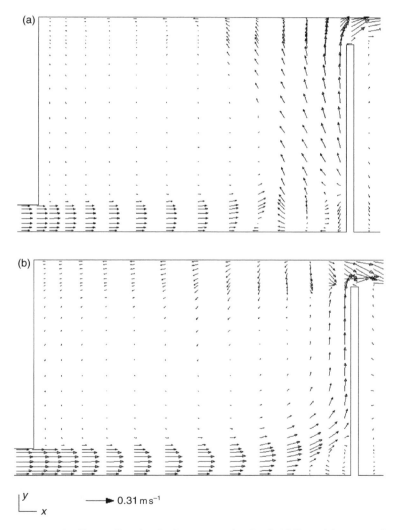

Figure 2.6 *Numerically predicted velocity vectors for the fluid flow; (a) near to the channel bed; and (b) near to the water surface. The blade length is seven-eighths the channel width and the rate of water discharge is $0.009\,m^3\,s^{-1}$.*

patterns in the planes which are near to the channel bed and the water surface, respectively. The main fluid flow near to the bed is generally slower than it is near to the surface. However, near to the corner between the bed and the second blade, reverse currents develop where the blade forces the main fluid flow to change its direction. In addition, the vertical, z-component of the fluid flow velocity is also observed in the vicinity of the corners between the second blade and the walls of the channel and this leads to the fluid flowing away from the walls of the channel near to these corners and these secondary flows can be observed in Figure 2.6.

Figure 2.7 shows the profile of the salt concentration as a function of time at the upstream position of the channel where this has been set as a time-dependent inlet condition to the computational domain. This profile has been obtained experimentally. The initial condition has been specified such that the salt concentration before time $t = 0$ is zero everywhere in the channel. Then, the CFD model simulations have been performed and Figure 2.8 shows the numerically predicted contours for the salt concentration distribution in the channel at $t = 100$, 165, 300 and 500 s after the salt concentration has been specified according to Figure 2.7. It is evident from Figure 2.8 that the movement of the salt in the channel is the combined effect of both advection and diffusion processes but it is mainly through the advection of the main recirculating fluid flows. Therefore the distribution of the salt concentration also partially reflects the flow patterns in the channel where recirculating flows are clearly seen in Figure 2.8.

Figure 2.9 shows the variation of the salt concentration in the space between the blocking blades at times between 100 and 700 s after the release of the salt at the upstream location. Also shown on the figure are the measured salt concentrations. It can be seen that the numerical simulations are in good agreement with the experimental data. In this simulation, it is found that the accuracy of the boundary

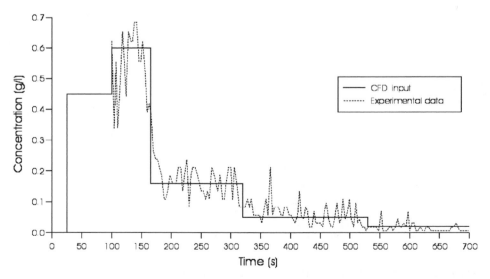

Figure 2.7 Variations in the salt concentration that has been specified at the upstream location of the channel based on the experimental data.

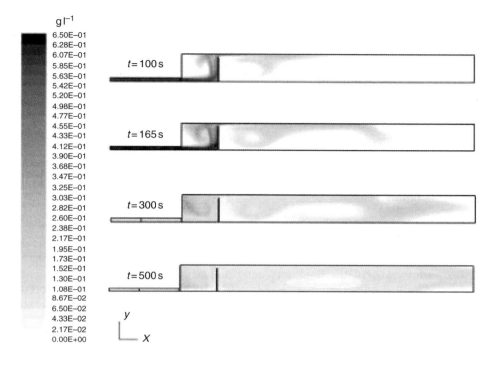

Figure 2.8 *Numerically predicted salt concentrations (kg m^{-3}) at t = 100, 165, 300 and 500 s after the release of the salt solution at the upstream location of the channel.*

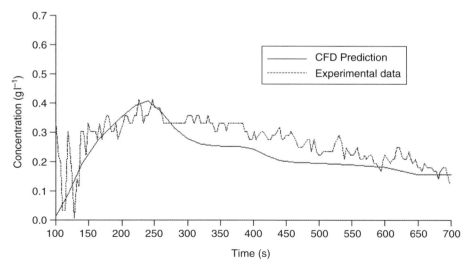

Figure 2.9 *Numerically predicted salt concentrations inside the cell as a function of time after the release of the salt solution at the upstream location of the channel.*

conditions for the salt concentration at the upstream boundary is vital for the correct predictions of the concentration of the salt as a function of time in the space between the blocking blades.

Channel flows containing particles and contaminant materials commonly exist in natural rivers and man-made channels and many of them have some kind of obstructions. The model presented here is an example that shows that CFD can be used to model the pollutant/contaminant transportation in river channels and it assists in gaining a better understanding of how various obstructions in the channel, or river, may influence the fluid and mass transport.

2.7 Concluding remarks

It is noted from the CFD simulations of the river flows presented in section 2.6 that with the modern CFD techniques that are now available it is possible to numerically model very complex river flows and pollutant transport in rivers with reasonable accuracy provided that the boundary conditions can be properly represented and specified. Through numerical experiments, CFD can also provide information on the interaction between the flow and the boundary conditions in an efficient and cost-effective manner. CFD simulation of river flows is not only financially cost-effective and very efficient, but it can also provide us with a much deeper insight into the structure of river flows than those that may be provided using experimental/field measurements. However, it should be noted that there is still a great deal of uncertainty in the accuracy of the representation of various flow boundary conditions that exist in natural rivers, particularly in the models of bed form, bed roughness and turbulence generation. These are not only a numerical issue but also an issue of the understanding of the mechanisms through which these boundary conditions affect the fluid flow. It should also be stressed that CFD can never replace experimentation or field measurements. Rather, the results obtained using CFD simulations need to be validated by performing experimental/field measurements. In some situations, CFD simulations may not even be able to provide accurate predictions of some characteristics of the river flow. This may be due to the limitations of the models employed, the simplification of the specified boundary conditions, or even the capacity of the computer employed. Therefore, the CFD simulations can only act as a supplement to experimental field measurements, and clearly CFD simulations, experimental measurements and field surveys should be carried out simultaneously in order to obtain a good understanding of some of the complex river flows.

References

Abba, A., Cercignani, C. and Valdettaro, L. (2003). Analysis of subgrid scale models. *Computers & Mathematics with Applications*, **46**(4), 521–535.

Batchelor, G.K. (1967). *An Introduction to Fluid Dynamics*. Cambridge University Press, Cambridge, England.

Bernard, R.S. (1992). STREMR: Numerical model for depth-averaged incompressible flow. *Technical Report HY-105*. US Army Corps Engineers Waterways Experiment Research Station, Vicksburg, Mississippi.

Bernard, R.S. and Schneider, M.L. (1992). Depth-averaged numerical modelling for curved channels. *Technical Report HL-92-9*. US Army Corps Engineers Waterways Experiment Research Station, Vicksburg, Mississippi.

Booker, D.J. (2003). Hydraulic modelling of fish habitat in urban rivers during high flows. *Hydrological Processes*, **17**, 577–599.

Boysan, F. (1995). *Advanced Turbulence Modelling*. Short Course Notes at the University of Leeds, Fluent Europe Ltd, pp. 1279–1298.

Bradbrook, K.F., Lane, S.N., Richards, K.S., Biron, P.M. and Roy, A.G. (2000). Large eddy simulation of periodic flow characteristics at river channel confluences. *Journal of Hydraulic Research*, **38**, 207–216.

Bray, D.I. (1980). Evaluation of effective boundary roughness for gravel-bed rivers. *Canadian Journal of Civil Engineering*, **7**, 392–397.

Cao, Z.X., Carling, P. and Oakey, R. (2003). Flow reversal over a natural pool–riffle sequence: A computational study. *Earth Surface Processes and Landforms*, **28**(7), 689–705.

Elder, J.W. (1959). The dispersion of marked fluid in turbulent shear flow. *Journal of Fluid Mechanics*, **5**, 544–560.

Ferguson, R.I., Prestegaard, K.L. and Ashworth, P.J. (1989). Influence of sand on hydraulics and gravel transport in a braided gravel bed river. *Water Resources Research*, **25**, 635–643.

Fluent (2003). *Fluent 6.1 User's Guide*. Fluent Incorporated, Lebanon, NH 03766.

Gatski, T.B. and Speziale, C.G. (1993). On explicit algebraic stress models for complex turbulent flows. *Journal of Fluid Mechanics*, **254**, 59–78.

Hancu, S., Ghinda, T., Ma, L., Lesnic, D. and Ingham, D.B. (2002). Numerical and physical modelling of the fluid and mass transfer in a channel. *International Journal of Heat and Mass Transfer*, **45**, 2707–2718.

Hardy, R.J., Bates, P.D. and Anderson, M.G. (1999). The importance of spatial resolution in hydraulic models for floodplain environments. *Journal of Hydrology*, **216**, 124–136.

Hey, R.D. (1979). Flow resistance in gravel-bed rivers. *Journal of Hydraulic Division*, ASCE, **105**, 365–379.

Hinze, J.O. (1975). *Turbulence*. Second edition, McGraw-Hill, New York.

Hirt, C.W. and Nichols, B.D. (1981). Volume of fluid (VOF) method for the dynamics of free boundaries. *Journal of Computational Physics*, **39**, 201–225.

Holly, F.M. and Usseglio-Polatera, J.M. (1984). Dispersion simulation in two-dimensional tidal flow. *Journal of Hydraulic Engineering*, ASCE, **110**, 905–926.

Ingham, D.B., Bloor, M.I.G., Wen, X. and Ma, L. (1997). *Report for the British Gas Research and Technology Centre on Low and High Pressure Modelling of the Uniflow Hydrocyclone Separation Behaviour and Flow Patterns*. Internal Report, University of Leeds, Leeds, UK.

Johannesson, H. and Parker, G. (1989). Linear theory of river meanders, in G. Parker (ed.), *River Meandering*, American Geophysical Union Monograph, **12**, 181–213.

Kolmogorov, A.N. (1942). Equations of turbulent motion of an incompressible fluid. *Izv Akad Nauk SSR Ser Phys*, 6, Vol. 1/2, 56.

Krone, R.B. (1962). *Flume Studies of the Transport of Sediment in Esturarial Shoaling Processes: Final Report*, HESER Laboratory, University of Berkeley, California.

Lane, S.N. (1998). Hydraulic modelling in hydrology and geomorphology: A review of high resolution approaches. *Hydrological Processes*, **12**, 1131–1150.

Launder, B.E. and Spalding, D.B. (1972). *Lectures in Mathematical Models of Turbulence*. Academic Press, London.

Launder, B.E. and Spalding, D.B. (1974). The numerical computation of turbulent flow. *Computer Methods in Applied Mechanics and Engineering*, **3**, 269–289.

Launder, B.E., Reece, G.J. and Rodi, W. (1975). Progress in the development of a Reynolds-stress turbulence closure. *Journal of Fluid Mechanics*, **68**, 537–566.

Lilly, D.K. (1966). On the application of the eddy viscosity concept in inertial subrange of turbulence. *NCAR* Manuscript 123.

Ma, L., Ashworth, P.J., Best, J.L., Ingham, D.B., Elliott, L. and Whitcombe, L. (2002). Computational fluid dynamics (CFD) and physical modelling of an upland urban river. *Geomorphology*, **44**, 375–391.

Menter, F.R. (1992). Improved two-equation k-ω turbulence model for aerodynamic flows, NASA TM-103975.

Menter, F.R. (1994). Two-equation eddy-viscosity turbulence models for engineering applications. *AIAA Journal*, **32**(8), 1598–1605.

Mohammadi, B. and Pironneau, O. (1994). *Analysis of the k-Epsilon Turbulence Model*. Wiley, Chichester.

Moulin, C. and Ben Slama, E. (1998). The two-dimensional transport module SUBIEF. Applications to sediment transport and water quality processes. *Hydrological Process*, **12**, 1183–1195.

Nezu, I. and Nakagawa, H. (1993). *Turbulence in Open Channel Flows*. Balkema, Rotterdam, The Netherlands, p. 279.

Nezu, I. and Rodi, W. (1986). Experimental study on secondary currents in open channel flows. *Proceedings of the 21st IAHR Congress*, **2**, 115–119.

Olsen, N.R.B. (2003). Three-dimensional CFD modeling of self-forming meandering channel. *Journal of Hydraulic Engineering*, ASCE, **129**, 366–372.

Olsen, N.R.B. and Kjellesvig, H.M. (1998). Three-dimensional numerical flow modelling for estimation of maximum local scour depth. *Journal of Hydraulic Research*, **36**, 579–590.

Olsen, N.R.B. and Kjellesvig, H.M. (1999). Three-dimensional numerical modelling of bed changes in a sand trap. *Journal of Hydraulic Research*, **37**, 189–198.

Partheniades, E. (1965). Erosion and deposition of cohesive soils. *Journal of Hydraulic Division*, ASCE, **91**, 105–138.

Prandtl, L. (1925). Über die ausgebildete Turbulenz. *Zeitschrift fur Angewandte Mathematik und Mechanik*, **5**, 136.

Preston, R.W. (1985). The representation of dispersion in two-dimensional water flow. *Report No. TPRD/L.2783/N84*, Central Electricity Research Laboratories, Leatherhead, England, May 1985, 1–13.

Rastogi, A.K. and Rodi, W. (1978). Predictions of heat and mass transfer in open channels. *Journal of Hydraulic Engineering*, **104**, 397–420.

Rodi, W. (1976). A new algebraic relation for calculating the Reynolds stresses. *Zeitschrift faddotur Angewandte Mathematik und Mechanik*, **56**, T219–T221.

Rodi, W. (1993). *Turbulence Models and Their Application in Hydraulics*. A.A. Balkema, Netherlands.

Rokni, M. (1997). Application of an explicit algebraic stress model for turbulent convective heat transfer in ducts, *Numerical Methods for Thermal Problems*. Pineridge Press, Swansea, UK, Vol. X, 102–114.

Rutherford, J.C. (1994). *River Mixing*. John Wiley & Sons, Chichester.

Spalding, D.B. (1985). The computation of flow around ships with allowance for free-surface and density-gradient effects, *Proceedings of the First International Maritime Simulation Symposium*, Munich, pp. 101–113.

Speziale, C.G. (1987). On nonlinear k-1 and k-epsilon models of turbulence. *Journal of Fluid Mechanics*, **178**, 459–475.

Temmerman, L., Leschziner, M.A., Mellen, C.P. and Frohlich, J. (2003). Investigation of wall-function approximations and subgrid-scale models in large eddy simulation of separated flow in a channel with streamwise periodic constrictions. *International Journal of Heat and Fluid Flow*, **24**(2), 157–180.

Thomas, T.G. and Williams, J.J.R. (1995). Large eddy simulation of turbulent flow in an asymmetric compound channel. *Journal of Hydraulic Research*, **33**, 27–41.

van Rijn, L.C. (1987). *Mathematical Modelling of Morphological Processes in the Case of Suspended Sediment Transport*. PhD Thesis, Delft University of Technology, Holland.

Wilcox, D.C. (1988). Reassessment of the scale determining equation for advanced turbulence models. *American Institute of Aeronautics and Astronautics Journal*, **26**(11), 1299–1310.

Wilcox, D.C. (1998). *Turbulence Modeling for CFD*. Second edition, DCW Industries Inc., La Cañada, California.

Wilson, C.A.M.E., Stoesser, T., Olsen, N.R.B. and Bates, P.D. (2003). Application and validation of numerical codes in the prediction of compound channel flows. *Proceedings of the Institution of Civil Engineers-Water and Maritime Engineering*, **156**, 117–128.

Yakhot, V. and Orszag, S.A. (1986). Renormalization group analysis of turbulence: I. Basic theory. *Journal of Scientific Computing*, **1**, 1–51.

Yakhot, V. and Smith, L.M. (1992). The renormalization group, the ε-expansion and derivation of turbulence models. *Journal of Scientific Computing*, **3**, 35–61.

Ye, J. and McCorquodale, J.A. (1997). Depth-averaged hydrodynamic model in curvilinear collocated grid. *Journal of Hydraulic Engineering*, **127**, 380–388.

3

Modelling solute transport processes in free surface flow CFD schemes

I. Guymer, C.A.M.E. Wilson and J.B. Boxall

3.1 Introduction

There is a growing need to understand better the fate of soluble pollutants once they have entered natural watercourses. An engineer must be able to quantify effluent mixing to be able to design and operate discharges to comply with current water quality regulations. The regulatory authorities have a need to understand river mixing to monitor outfalls and ensure that they comply with their discharge permits and to design appropriate chemical and biological monitoring systems. Within the natural sciences, an understanding of the mixing of effluents and tributary inflows, Figure 3.1, is required to describe their effects on fish stocks and to quantify the rate of mass transfer across the hyporheic zone and to predict concentrations in pore water and overlying river water. Even short-term exposure to relatively low pollution concentrations could be damaging to macroinvertebrates. Hence, intermittent discharges, such as Combined Sewer Overflows (CSO) or highway runoff events containing soluble pollutants and contaminated fine sediments, can have an impact on river ecology. The understanding of the dominant mixing processes and the ability to describe them mathematically can aid understanding and interpretation of effects.

Any comprehensive description of mixing processes in natural watercourses must include a number of complex factors such as turbulence, velocity gradients,

Figure 3.1 Channel mixing at river confluence. (Photograph with permission of I. Guymer.)

secondary flow and other non-uniformities that are associated with natural open channel flows. The following section aims to describe the physical processes that contribute to solute transport in natural environmental flows and provide a background to the source and scale of the processes. This chapter will describe and develop the basic equation used for simulating solute transport, introduce the effect of turbulence and comment on the importance of spatial averaging. It concludes with an example application to simulate transient solute concentration fields created from a short-duration pulse in a meandering channel. The predicted spatial and temporal concentration distributions are compared with temporal concentration measurements performed at points within cross sections of a laboratory channel (Boxall, 2000).

3.2 Advection, diffusion and dispersion

When soluble material is introduced into a flow, advection carries it away from the source. Advection is the bodily movement of a parcel of fluid resulting from an imposed current. The tracer is spread by the combined actions of diffusion and differential advection. Differential advection is when parcels of material in a flow experience different velocities, whether at different spatial positions or at the same location at different times. This differential advection produces a spreading that is commonly termed 'shear dispersion'. Diffusion at the molecular scale is due to small-scale localised fluctuations, the effects of Brownian motion. Often in environmental hydraulics problems, the effects of the fluctuations in flow velocity caused by turbulence are considered to produce a similar effect. However, the turbulent fluctuations are much greater than those of molecular Brownian motion and typically produce one million times greater effects (Smith, 1992).

3.3 Derivation of advection diffusion equation

Consider a body of fluid (the receiving fluid) into which a small volume of a soluble neutrally buoyant pollutant is introduced. If the receiving fluid is at rest, the pollutant will gradually spread due to the effects of molecular diffusion. This process is slow, for example a source of dye a few millimetres in diameter placed into a still body of water could take about 24 hours to spread through a 1 m diameter. Molecular diffusion coefficients are a property of the fluid and of the order 10^{-10}–10^{-9} m^2/s (Smith, 1992). The coefficients are constant for a set of parameters: solvent, solute and temperature.

In 1855, the German scientist Fick drew an analogy between the diffusion of salt in water and the diffusion of heat along a metal rod. He hypothesised that the rate of transfer of tracer should be proportional to the concentration gradient between two regions. In a 1D form this is:

$$J_x = -e_m \frac{\partial c}{\partial x} \tag{3.1}$$

where J_x is the mass flux (kg/m^2 s) in the x-direction, c is concentration (kg/m^3), and e_m (m^2/s) is the molecular diffusion coefficient. This is a mathematical statement of Fick's first law. It can be expanded in all co-ordinate directions. The negative sign denotes diffusion from areas of high concentration to areas of lower concentration. If the ideas of conservation of mass are incorporated into the diffusion process, it is possible to derive a second descriptive relationship. For simplicity consider only a 1D situation and a volume of fluid within a stationary body of water (Figure 3.2). As the tracer passes through the surfaces, the concentration within the volume changes. Over a small time period this change can be written as:

$$\frac{(c_{t+\Delta t} - c_t)}{\Delta t} \Delta x \equiv \frac{\partial c}{\partial t} \Delta x \tag{3.2}$$

where c_t and $c_{t+\Delta t}$ are the tracer concentrations within the volume at times t and Δt, respectively. By considering conservation of mass along the x-axis, it must hold true

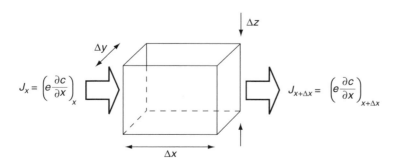

Figure 3.2 Diffusive flux into and out of small fluid element.

that the rate of change of mass given by equation (3.2) must equal the net diffusive flux into and out of the volume. This change in flux can be written as:

$$J_x - J_{x+\Delta x} \equiv -\frac{\partial J}{\partial x}\Delta x \qquad (3.3)$$

Equating equations (3.2) and (3.3) yields:

$$\frac{\partial c}{\partial t} + \frac{\partial J}{\partial x} = 0 \qquad (3.4)$$

This combined with Fick's first law gives:

$$\frac{\partial c}{\partial t} - \frac{\partial}{\partial x}\left(e_m \frac{\partial c}{\partial x}\right) = 0 \qquad (3.5)$$

If the diffusion coefficient is assumed constant with distance, then equation (3.5) becomes:

$$\frac{\partial c}{\partial t} = e_m \frac{\partial^2 c}{\partial x^2} \qquad (3.6)$$

This relationship defines the transport of mass by a Fickian diffusion process and is known as the 'diffusion' equation. However, equation (3.6) is not sufficient to describe solute transport in free surface environmental flows. Considering the same volume within a steady, laminar, open channel flow, in all the three co-ordinate directions, including an advective transport term, the expression for total flux in each co-ordinate direction can be written as:

$$J_x = uc - e_m \frac{\partial c}{\partial x} \qquad (3.7a)$$

$$J_y = vc - e_m \frac{\partial c}{\partial y} \qquad (3.7b)$$

$$J_z = wc - e_m \frac{\partial c}{\partial z} \qquad (3.7c)$$

where u, v and w are velocities (m/s) in the x-, y- and z-co-ordinate directions, respectively, and uc, vc and wc are the advective fluxes in each direction (kg/m^2 s). The change in mass within the parcel can now be expressed as:

$$\frac{\partial c}{\partial t}\Delta x \Delta y \Delta z \qquad (3.8)$$

Next, consider the net flux in each co-ordinate direction:

$$(\text{netflux})_x = -\left(\frac{\partial J_x}{\partial x}\Delta x\right)\Delta y\Delta z \tag{3.9a}$$

$$(\text{netflux})_y = -\left(\frac{\partial J_y}{\partial y}\Delta y\right)\Delta x\Delta z \tag{3.9b}$$

$$(\text{netflux})_z = -\left(\frac{\partial J_z}{\partial z}\Delta z\right)\Delta x\Delta y \tag{3.9c}$$

The total flux is equal to the summation of the terms in equation 3.9. As previously stated this change in net flux must be equal to the change in concentration within the parcel. Therefore, combining equations (3.8) and (3.9) yields:

$$\frac{\partial c}{\partial t} + \frac{\partial J_x}{\partial x} + \frac{\partial J_y}{\partial y} + \frac{\partial J_z}{\partial z} = 0 \tag{3.10}$$

Then substituting the expressions for flux from equation (3.7) and dividing through by the elemental volume produces:

$$\frac{\partial c}{\partial t} + u\frac{\partial c}{\partial x} + v\frac{\partial c}{\partial y} + w\frac{\partial c}{\partial z} = e_\text{m}\left(\frac{\partial^2 c}{\partial x^2} + \frac{\partial^2 c}{\partial y^2} + \frac{\partial^2 c}{\partial z^2}\right) \tag{3.11}$$

This is known as the 'advection–diffusion' equation, and is the form of the equation presented by Fischer et al. (1979), and Rutherford (1994) in their authoritative texts on environmental and river mixing. There are a number of analytical solutions for this equation (Carslaw and Jaeger, 1959; Crank, 1975). It should be noted that in this form, no account has yet been made for any turbulence effects, and that the molecular diffusion coefficient, e_m, is assumed homogeneous in each co-ordinate direction. As the majority of open channel flows are turbulent in nature, except in the viscous sub-layer near the bed, the role of turbulence must be addressed.

3.4 Turbulence

Turbulent flow is the most common flow type in environmental hydraulics. It is important as it is the velocity fluctuations characteristic of the turbulent nature of the flow that govern the energy dissipation rates, the transport of matter and the processes of dispersion or mixing.

Batchelor, in his book *The Life and Legacy of G.I. Taylor* (1996), presents a chapter on 'Turbulence: A challenge'. He describes turbulence as a 'flow in which the velocity of the fluid is a random function of position and time and (where) inertia forces are significant'. He quotes the opening remark from Lamb (1959) on turbulence: 'It remains to call attention to the chief outstanding difficulty of our subject.'

Batchelor suggests that even today there is some justification for regarding the remark as still appropriate.

If the receiving fluid is turbulent, then considering the turbulent fluctuations without any associated flow field increases the magnitude of the diffusion coefficients by several orders of magnitude, to 10^{-3}–10^{-1} m^2/s (Smith, 1992). The turbulent fluctuations may be considered as a small-scale, short-term, differential advection, whereby if any numerical scheme could describe these fluctuations at the necessary temporal and spatial resolution, only a diffusion coefficient would be required.

Reynolds first examined turbulence and its relevance to mixing problems. He noted that a steady point source of dye released into a laminar pipe flow resulted in a single ribbon of dye along the pipe. The same source in a turbulent pipe flow resulted in rapid transverse and vertical mixing within the pipe. This is due to turbulent flow possessing random unsteady components of velocity. Turbulent flow may be defined by a Reynolds number, Re, equal to uL/ν where L is a turbulence length scale (m), often taken as the hydraulic radius or flow depth for open channel flow, and ν is the kinematic viscosity (m^2/s).

Turbulent flow is characterised by random short-term fluctuations. Turbulence is non-self-sustaining, in the absence of energy input the flow will revert to laminar flow conditions. In open channel flow, energy is primarily introduced through gravity. Turbulence is generated from velocity shear and hence turbulence is large in regions of high shear, such as around bed irregularities and obstructions. Energy enters at the largest length scales and dissipates down to the smallest eddies, finally being dissipated as heat energy. The largest eddies have the greatest effect on momentum and mass transfer within the flow. Depending on the length and time-scales being considered, it can be argued that turbulence alone does not produce reductions in concentrations. The presence of turbulence causes the fluctuations in flow direction and magnitude resulting in differential advection. This creates magnified concentration gradients, accelerating the effect of molecular diffusion.

It is theoretically possible to model turbulent flows numerically with equation (3.11) since the mechanism causing a reduction in concentration is still driven by molecular diffusion. This requires that all instantaneous velocity fluctuations are known throughout the body of the fluid. If this is the case then only a diffusion coefficient is required. However, a very detailed evaluation of velocities both temporally and spatially would be essential. Each particle would need to be modelled individually, and then the mean concentration distribution calculated by means of superposition. This is due to the random nature of the flow, which causes each particle to follow a different path. Advances in computational power means that this is becoming possible, but requires very detailed and accurate parameter definition, and simulations are time consuming to run. In most applications this approach is still considered impractical.

One way to overcome this was suggested by Reynolds, who decomposed the instantaneous turbulent flow velocities, u, as:

$$u = u_m + u' \qquad (3.12)$$

where u_m is a temporal mean velocity, and u' is a deviation from the mean velocity. The same decomposition can also be made for the instantaneous concentrations.

With these decompositions it is necessary to define the mean as the ensemble mean, that is, the mean of a number of observations such that the results are independent of each other, and of the previous result. Good approximations of the ensemble mean can be obtained from temporal measurements, provided that the time period for averaging is much greater than the largest eddy scale. This definition can be applied to both velocity and concentration data. It is important to note that concentrations predicted by most turbulence models are ensemble mean values, and that the peak instantaneous concentration may be significantly higher due to the fluctuating component. This may be vital in assessing ecological impacts: Fischer et al. (1979) commented that 'if a few high concentrations kill an organism, then an even larger number of low concentrations will not bring it back to life'. It should also be noted that the fluctuations from the temporal mean value of a concentration are a function of the position within the plume.

The expression given in equation (3.12) can be substituted into equation (3.11) to yield:

$$\frac{\partial(c_m + c')}{\partial t} + (u_m + u')\frac{\partial(c_m + c')}{\partial x} + (v_m + v')\frac{\partial(c_m + c')}{\partial y} + (w_m + w')\frac{\partial(c_m + c')}{\partial z}$$
$$= e_m\left(\frac{\partial^2(c_m + c')}{\partial x^2} + \frac{\partial^2(c_m + c')}{\partial y^2} + \frac{\partial^2(c_m + c')}{\partial z^2}\right) \qquad (3.13)$$

If equation (3.13) is then simplified by taking ensemble means and considering continuity it can be shown that:

$$\overset{(1)}{\frac{\partial c_m}{\partial t}} + \overset{(2)}{\left(u_m\frac{\partial c_m}{\partial x} + v_m\frac{\partial c_m}{\partial y} + w_m\frac{\partial c_m}{\partial z}\right)}$$
$$= e_m\overset{(3)}{\left(\frac{\partial^2 c_m}{\partial x^2} + \frac{\partial^2 c_m}{\partial y^2} + \frac{\partial^2 c_m}{\partial z^2}\right)} - \overset{(4)}{\left(\frac{\partial(u'c')}{\partial x} + \frac{\partial(v'c')}{\partial y} + \frac{\partial(w'c')}{\partial z}\right)} \qquad (3.14)$$

where expression (1) represents rate of change of mean concentration with time, (2) the advection of ensemble mean concentration by mean velocity, (3) the molecular diffusion, and (4) the turbulent diffusion. In rivers turbulent or eddy diffusion causes a tracer to spread far more rapidly than could happen if spreading occurred due to molecular diffusion alone. The cross product terms in expression (4), that is $u'c'$, $v'c'$ and $w'c'$, are equivalent to the flux of tracer produced by the turbulent velocity fluctuations. It is necessary to obtain detailed records of velocity and concentration data to evaluate these terms.

Taylor (1921) presented a paper on the theoretical analysis of the spreading of a cloud of tracer particles released into stationary homogeneous turbulence. This has long remained the basis for the analysis of all turbulent diffusion. Taylor utilised a Lagrangian co-ordinate system, in which the system moves at the mean flow rate. Thus, the effect of turbulence is isolated. The result of this approach is that, after some time, the rate at which the variance of the tracer cloud or distribution increases is linear. Taylor (1954) went on to confirm his hypothesis by the investigation of longitudinal dispersion in turbulent pipe flow. His work was extended by Elder (1959) to open channels of infinite width.

Making the analogy between turbulent diffusion and Fick's first law, equation (3.7) can be re-written as:

$$J_x = \overline{u'c'} = -e_t \frac{\partial c_m}{\partial x} \quad (3.15a)$$

$$J_y = \overline{v'c'} = -e_t \frac{\partial c_m}{\partial y} \quad (3.15b)$$

$$J_z = \overline{w'c'} = -e_t \frac{\partial c_m}{\partial z} \quad (3.15c)$$

where e_t is a turbulent diffusion coefficient, or eddy diffusivity (m^2/s). Following the same derivation used for equation (3.11), an equation for turbulent advection–diffusion can now be written:

$$\frac{\partial c_m}{\partial t} + u_m \frac{\partial c_m}{\partial x} + v_m \frac{\partial c_m}{\partial y} + w_m \frac{\partial c_m}{\partial z} = (e_m + e_t)_x \frac{\partial^2 c_m}{\partial x^2} + (e_m + e_t)_y \frac{\partial^2 c_m}{\partial y^2} + (e_m + e_t)_z \frac{\partial^2 c_m}{\partial z^2} \quad (3.16)$$

Although this analogy is supported by Taylor's analysis it should be noted that this is not a conclusive proof of its validity. However, Rutherford (1994) reports a large body of empirical evidence to support the use of equation (3.16) to describe turbulent mixing situations. Due to the nature of the derivation of this equation it should be applied with care in practical situations, particularly in the near field when it is imperative that sufficient time must elapse such that the effects of buoyancy and discharge momentum have been negated.

In equation (3.16) the two separate coefficients for molecular and turbulent diffusion are included. However, it is common practice to include only one term. This is because molecular diffusion is normally several orders of magnitude smaller than turbulent diffusion and hence can be neglected, or assumed incorporated into the turbulent diffusion term, leaving:

$$\frac{\partial c}{\partial t} + u \frac{\partial c}{\partial x} + v \frac{\partial c}{\partial y} + w \frac{\partial c}{\partial z} = e_x \frac{\partial^2 c}{\partial x^2} + e_y \frac{\partial^2 c}{\partial y^2} + e_z \frac{\partial^2 c}{\partial z^2} \quad (3.17)$$

It should be noted that all concentrations and velocities in equation (3.17) are ensemble mean values. Subscript m has been omitted for the sake of simplicity. The diffusion coefficient is stipulated separately in each co-ordinate direction. This reflects the importance of different governing factors in each direction. As noted by Rutherford (1994) the key to the successful application of this equation is that the averaging period is large compared with the timescale of the turbulent fluctuations, but is small compared with the timescale of the changes in concentration that are under investigation.

3.5 Dispersion

When considering mixing over a length or area, which could be up to the depth or width of the channel, or over the scale of the numerical grid used for CFD simulations, it is necessary to take account of the effects of velocity gradients over that length scale. Depending on the dimensionality of the numerical code this is performed through the inclusion of a term or coefficient. In 1D and 2D codes this is a dispersion term that accounts for the velocity gradients, the turbulent diffusion and the spatial resolution of the grid. In a 3D code using a spatial grid resolution that is sufficiently fine to capture the spatial variations in velocity, this is a diffusion term. This accounts solely for the molecular and turbulent diffusion that occurs over the time step of the calculations. If the grid is not refined sufficiently then this term will also account for the spatial averaging of the velocity in each cell.

Early ideas on dispersion of solutes within flows arose from Taylor (1953) who was initially asked to aid the understanding of how drugs are dispersed when injected into blood vessels of animals. His paper describes how in steady laminar flow in a tube, the different longitudinal velocities throughout the cross section create shear effects and 'stretch' a cloud of tracer. The paper shows that the length of an initial cloud of solute increases linearly with the square root of the distance travelled. It also shows that the longitudinal dispersion coefficient, describing the spreading along the axis of the flow, can be determined from $a^2 U^2/48D$ where a is the pipe radius, U is the mean velocity and D is the molecular diffusivity of the solute.

The result presented in this early publication illustrates the often-misunderstood concept of the inverse relationship between the longitudinal shear-induced spreading and the transverse mixing, in this case molecular diffusion. It shows that when the molecular diffusivity is reduced, the longitudinal dispersion increases. This can be explained by considering the relative position of two parcels of fluid, one in the region of maximum velocity at the centre of the pipe and one in the low velocity near the boundary. If there is no diffusion, they do not move throughout the cross section, but stay in the same cross-sectional location and move only with the longitudinal velocity. The distance between the parcels during a fixed period is simply a result of their relative velocity, that is the difference in their velocities, differential advection. If on the other hand they are subject to any process causing them to move throughout the cross section, they will experience a variety of velocities. The average velocity of the parcel, previously fixed at the maximum velocity at the centre of the pipe, over the same period of time must decrease. Similarly, the mean velocity of the parcel previously located in the slow flow near the boundary will increase. This results in a reduction in the distance between the two parcels over the time period, reducing the longitudinal dispersion.

If spatial averaging is undertaken, possibly up to the channel dimensions, then shear dispersion coefficients need to be included to represent the effects caused by differential advection over the averaged distance. Depending on the integrated length scales, shear dispersion coefficients in the range $1-10^3 \, m^2/s$ (Smith, 1992) can be obtained.

3.6 Dead zones

Most solute transport concepts are derived from consideration of a stationary fluid (molecular diffusion), the effects of turbulent fluctuations (turbulent diffusion or small-scale differential advection) or the effects of spatial averaging over velocity profiles (differential advection or shear dispersion). In addition to these situations, natural river channels often exhibit regions of recirculating flow or stagnant water. These features are often termed 'dead zones' and effectively trap soluble pollutants. Wallis *et al*. (1989) state 'The term dead zone is often misunderstood: although it implies a form of pocket that is separated from the main flow, it should be considered in a wider context as a bulk parameter that describes not only the effect of segregated regions of flow, but also other dispersive catalysts such as eddies, viscous sub-layers and velocity profile.' Figure 3.3 illustrates a dead zone trapping effect. It shows flow downstream of a waterfall where the main body of water is flowing directly downstream. Adjacent to the near bank there is an area of recirculation, clearly illustrated by the buoyant natural tracers, camellia petals. The effect is highlighted by the long exposure photograph.

Valentine and Wood (1977) looked at the effect of the dead zone on the 1D mixing or longitudinal dispersion coefficient. Where effects are averaged across the cross section and along the reach, the effects of a boundary dead zone may simply be considered as an additional 'differential advective' effect and hence incorporated into the standard average solution. The advent of CFD and its application to natural river channels allows conditions within dead zones and areas of recirculation to be explicitly incorporated within the solution and investigated as an individual process.

Figure 3.3 'Camellias in a Whirl' by Hiromu Hirayama. (Photograph with permission of Hiromu Hirayama.)

3.7 Estimating turbulent diffusion coefficients

Solution of equation (3.17) at each computational node requires values to be specified for each of the turbulent diffusion coefficients. It is often suggested that an estimate of the magnitude of the turbulent diffusion coefficient can be obtained by employing a turbulent Schmidt number. The dimensionless turbulent Schmidt number is used to convert the turbulent viscosity (calculated from a turbulence model) into the turbulent diffusion coefficient.

$$Sc_t = \frac{\nu_t}{e_t} \qquad (3.18)$$

where Sc_t is the turbulent Schmidt number and ν_t is the turbulent viscosity. Reynolds (1975) compares more than 30 ways of predicting the turbulent Schmidt number. Launder (1976) describes the findings of a conference paper that reported values of Sc_t in the range 0.5–1.0.

Grimm (2003) investigated the effect of spatial variations in the value of the turbulent Schmidt number and showed that for a straight pipe it had a significant effect on predicting longitudinal dispersion. Increasing the value of the turbulent Schmidt number produced a proportional increase in the predicted longitudinal dispersion. However, he concluded that in both natural environmental flows and engineered structures, predictions of solute transport were insensitive to changes in the coefficient. This suggests that it is the large-scale cross-sectional shape changes and roughness variations that dominate the processes.

3.8 Application to meandering channel

Similar challenges are faced when describing solute transport as when describing the full 3D flow field. The resolution in prediction is dependent on the scale of the mesh used and the magnitude of the mixing coefficient required is dependent on what processes are being temporally and spatially averaged. For example, predicting the flow field in a natural river requires inclusion of a boundary roughness coefficient. This must not only describe the texture of the material and the energy loss of the fluid, but must also include any shape or form roughness effects at the same scale as the computational mesh. Wilson *et al.* (2003) performed a numerical simulation of the flows in a meandering laboratory channel with a naturally formed bed and compared predicted velocity distributions with those measured. This study included a section on determining the appropriate magnitude of the boundary roughness coefficient to include both material and form roughnesses.

Similarly for describing solute transport in CFD schemes, the magnitude of the time steps for calculations (temporal averaging) and the physical size of the computational mesh (spatial averaging) will govern the magnitude of the mixing coefficient required. Only by defining the scale of the averaging and understanding the magnitude of the physical processes will it be possible to accurately incorporate solute transport processes. The following section presents results from a study undertaken to illustrate the potential of using CFD techniques to describe solute mixing processes and to compare the results to a comprehensive laboratory study (Boxall, 2000).

The CFD output presented are from simulating channel flow within the planform configuration shown in Figure 3.4. These results are compared with laboratory investigations that formed the initial part of an EC-funded project investigating the influence of bends on transverse mixing (Boxall and Guymer, 2003; Boxall et al., 2003). For this study the Flood Channel Facility (FCF), sited at HR Wallingford, UK, comprised two complete meander cycles, with transition channels of approximately a quarter meander cycle at inlet and outlet. Initially, the meandering main channel, constructed within concrete flood plains, was filled with sand ($d_{50} = 0.84$ mm) screeded to a trapezoidal profile. The discharge was fixed at 0.0255 m^3/s and run for one hour to produce the resultant bed features. This shape, which was not in equilibrium as the migration of the meanders caused the channel to impinge on the concrete flood plains (section I–K, Figure 3.4), was then fixed (Khalil, 1972) and a uniform flow established.

Turbulence measurements were recorded using a 2D laser Doppler anemometer (in two orientations, x, z and x, y) at several cross sections (Boxall, 2000). These were used to inform the application of a 3D numerical model (Wilson et al., 2003) with a standard k-ε turbulence closure model (Launder and Spalding, 1974; Chapter 2 in this book). The standard k-ε model was chosen because it is the most widely used turbulence model in CFD codes and requires no empirical input. The 3D finite volume program SSIIM was used (Olsen, 2003). This model solves the Reynolds-averaged Navier–Stokes equations for each cell, and uses the standard k-ε model for turbulence closure. Recommended values for the five constants in the k-ε model as given by Rodi (1980) were used and hard-coded into the software. The grid is structured, three dimensional and non-orthogonal as shown in Figure 3.5. The mesh size is 97:46:12, longitudinal:transverse:vertical gridlines, generating a total of 47 520 cells.

For the velocity simulations presented, the Power-Law Scheme (Patankar, 1980) was used for discretisation of the convection terms. The SIMPLE method (Patankar, 1980) is used for the coupling between the pressure and velocity. SSIIM uses the wall law (Schlichting, 1979) for rough boundaries in the cell bordering the river boundary.

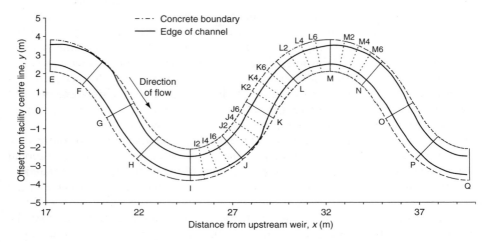

Figure 3.4 Experimental planform configuration (Boxall and Guymer, 2003).

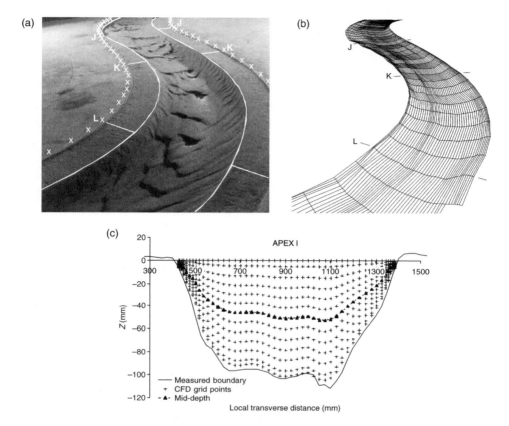

Figure 3.5 CFD grid geometry: (a) photograph of channel bed looking upstream; (b) bed boundary grid geometry; and (c) example of cross-section grid geometry.

The water velocities and turbulence parameters are prescribed as a Dirichlet boundary condition at the upstream inlet. Zero-gradient boundary conditions are used for all the variables at the downstream outlet. At the water surface, zero-gradient boundary conditions are used for all variables except the turbulent kinetic energy, which was set to zero. The convergence criterion employed is that the residuals from the continuity and momentum equations, non-dimensionalised by dividing by the characteristic parameter at the upstream boundary, should be less than 10^{-3}. For further details regarding the velocity simulations, in particular the development of the secondary circulations and the convergence procedure, see Wilson *et al.* (2003).

A comparison between the measured and predicted velocity fields is given in Figure 3.6. This study indicates the accuracy with which the velocity distribution can be modelled using a generalised 3D approach, with a widely available turbulence model. At cross sections I and K (Figure 3.6) the model correctly predicts the existence of observed complex features such as the presence and rotational sense of secondary currents. As expected in the presence of planform curvature, at apex section I, the maximum longitudinal velocity lies towards the outer bank. This is

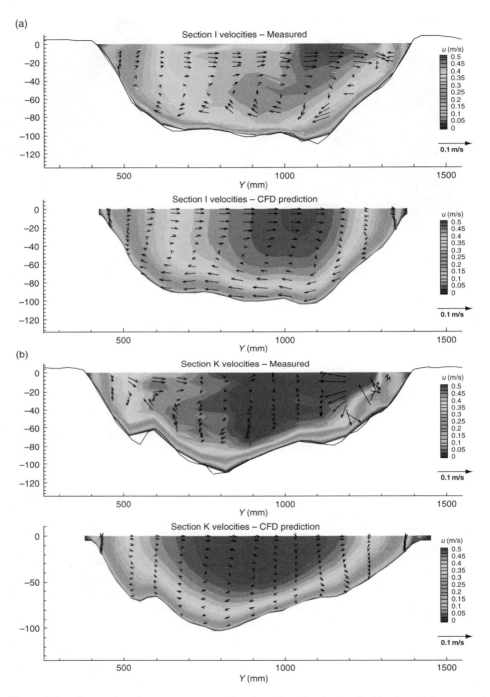

Figure 3.6 Comparison between measured and predicted velocity distributions: (a) apex and (b) cross over.

caused by the convective transport of high momentum fluid from inner to outer regions by the secondary circulation. The predicted secondary cell is weaker in strength than that measured, and the transverse velocities are slightly underpredicted. This leads to reduced momentum transfer towards the outer region of the bend and consequently a discrepancy in the position of the longitudinal velocity maximum. At cross section K, Figure 3.6, the predicted secondary velocity field is more uniform throughout the cross section compared with the measured secondary velocities. This is caused by the limited spatial resolution of the local bathymetric measurements.

Extending the application of CFD modelling to describe solute transport, numerical predictions of the spatial and temporal concentration fields resulting from a short pulse input of soluble tracer have been considered. In the laboratory study, the fundamental procedure followed that undertaken in previous studies of longitudinal dispersion using the FCF (Guymer, 1998). Fluorescent tracer was introduced as a line source across the bed and five fluorimeters were employed to simultaneously measure the temporal variation of solute concentrations. Results from one injection on this channel configuration are presented in Figure 3.7 and further details of the results may be found in Guymer *et al.* (1999).

Most numerical evaluation studies for simulating the transport of solute have used the data from the experimental study conducted by Chang (1971), who performed a continuous injection of a solute tracer. Demuren and Rodi (1986) applied a fully 3D model with a modified version of the k-ε model to account for the effect of streamline curvature on the turbulent transport mechanisms. A value of 0.5 for the turbulent Schmidt number was recommended for a fully 3D model with k-ε turbulence closure (Demuren and Rodi, 1986) to compensate for the anisotropic turbulence structure.

To model the transport of the solute tracer, the velocity field is first computed using steady conditions whereby the momentum and continuity equations (Navier–Stokes equations) are solved and then the resulting velocity field is used in the

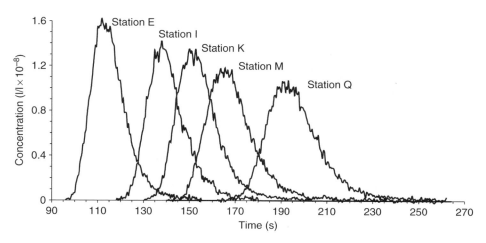

Figure 3.7 Examples of measured temporal tracer concentration distributions (Guymer et al., 1999).

advection–diffusion simulation. This simulation computes the concentration at each time step. A model of the longitudinal transport and mixing of solute in the channel has employed the second-order upwind scheme for the advective term in the solution of the advection–diffusion equation as this provides a more accurate solution compared with a first-order scheme. For the tracer concentration, zero gradients were used at the downstream boundary, the water surface and the walls, and Dirichlet conditions were used at the upstream boundary. An implicit scheme was used for the transient term.

Outputs from the numerical simulation of the solute transport using a 0.5 s time step are shown in Figures 3.8 and 3.9. Whereas for the laboratory study, in which

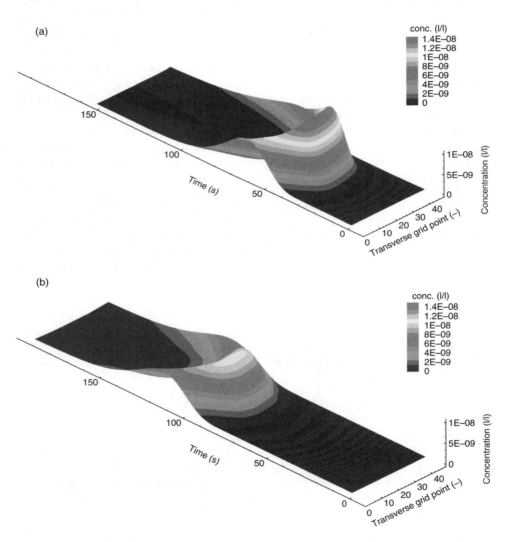

Figure 3.8 Predicted temporal concentrations across channel sections: (a) section I (mid-depth) and (b) section Q (mid-depth).

Modelling solute transport processes 67

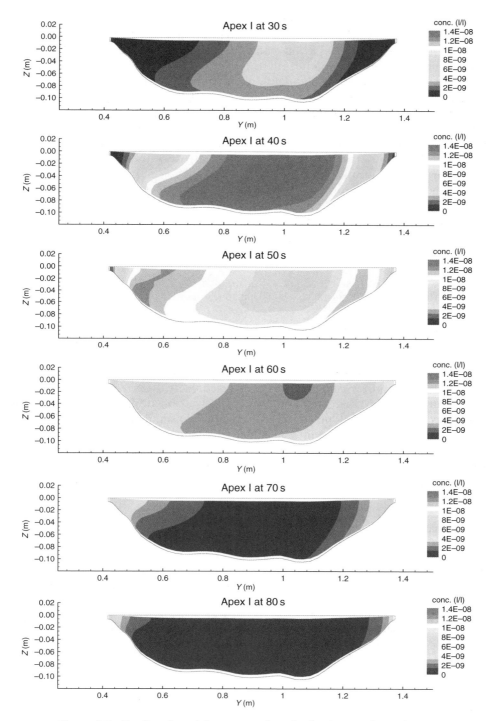

Figure 3.9 *Predicted spatial concentration distributions at channel apex.*

within each cross section only single-point measurements were available, CFD simulations predict values of the solute concentration at each computational node at every time step. Figure 3.8 illustrates the predicted transverse distribution of solute concentration at the mid-depth at two cross sections, I and Q. To aid interpretation, the position of the mid-depth plane is highlighted in Figure 3.5c.

These results clearly illustrate the major mixing process occurring in open channel flow, namely transverse differential advection. The reduction in peak concentration and the transverse variation in solute concentrations are caused by the effects of transverse velocities and turbulent diffusion. The sequence of cross-sectional concentration fields presented in Figure 3.9 show clearly how the solute tracer initially arrives around the location of the peak longitudinal velocity (30 s). At the time of peak concentration, around 40 s, these results illustrate that the tracer has not achieved complete cross-sectional mixing. The plots from 50 s onwards show clearly how the solute is retained adjacent to the channel boundaries due to the low velocities.

3.9 Conclusions

The major physical processes governing the mixing of solutes in free surface environmental flows have been described and the equations to describe the resulting effects within a CFD scheme have been developed. The equations have been employed to predict the spreading of a pulse input of solute tracer within a channel. The solute transport simulation was conducted in two stages: a hydrodynamic simulation was first conducted and then the advection–diffusion simulation was performed using the resulting converged velocity field simulation. This study shows that the transport of solute in a meandering channel with a physically realistic bathymetry can be simulated with a 3D numerical model using a k-ε turbulence closure model. The simulation is plausible and has been shown to capture known physical mechanisms.

References

Batchelor, G. (1996) *The Life and Legacy of G.I. Taylor*, Cambridge University Press, Cambridge.
Boxall, J.B. (2000). 'Dispersion of Solutes in Sinuous Open Channel Flows', PhD Thesis, University of Sheffield, Sheffield, UK.
Boxall, J.B. and Guymer, I. (2003) 'On the analysis and prediction of transverse mixing coefficients in natural channels', *Journal of Hydraulic Engineering*, ASCE, 129, 129–139.
Boxall, J.B., Guymer, I. and Marion, A. (2003) 'Transverse mixing in sinuous natural open channel flows', *Journal of Hydraulic Research*, 41, 153–165.
Carslaw, H.S. and Jaeger, J.C. (1959) *Conduction of Heat in Solids*, 2nd Edition, Clarendon Press, Oxford.
Chang, Y.C. (1971) 'Lateral Mixing in Meandering Channels', PhD Thesis, University of Iowa, USA.
Crank, J. (1975) *The Mathematics of Diffusion*, 2nd Edition, Clarendon Press, Oxford, 347pp.

Demuren, A.O. and Rodi, W. (1986) 'Calculation of flow and pollutant dispersion in meandering channels', *Journal of Fluid Mechanics*, 172, 63–92.

Elder, J.W. (1959) 'The dispersion of marked fluid in turbulent shear flow', *Journal of Fluid Mechanics*, 5, 544–560.

Fischer, H.B., List, E.J., Koh, R.C.Y., Imberger, J. and Brooks, N.H. (1979) *Mixing in Inland and Coastal Waters*, Academic Press, New York.

Grimm, J.P. (2003) 'An Evaluation of Alternative Methodologies for the Numerical Simulation of Solute Transport', PhD Thesis, University of Sheffield, Sheffield, UK.

Guymer, I. (1998) 'Longitudinal dispersion in a sinuous channel with changes in shape', *Journal of Hydraulic Engineering*, ASCE, 124, 33–40.

Guymer, I., Boxall, J., Marion, A., Potter, R., Trevisan, P., Bellinello, M. and Dennis, P. (1999) 'Longitudinal dispersion in a natural formed meandering channel', *IAHR 28th Conference Congress*, Theme D2, Graz, Austria.

Khalil, M.B. (1972) 'On preserving the sand patterns in river models', *Journal of Hydraulic Research*, IAHR, 10, 291–303.

Lamb, H. (1959) *Hydrodynamics*, Cambridge University Press, Cambridge.

Launder, B.E. (1976) 'Heat and mass transport', in Bradshaw, P. (ed.), *Topics in Applied Physics*, Vol. 12, *Turbulence*, Springer, Berlin, Germany.

Launder, B.E. and Spalding, D.B. (1974) 'The numerical computation of turbulent flows', *Computational Methods in Applied Mechanics and Engineering*, 3, 269–289.

Olsen, N.R.B. (2003) *SSIIM User's Manual*, The Norwegian University of Science and Technology (www.bygg.ntnu.no/~nilsol/ssiimwin).

Patankar, S.V. (1980) *Numerical Heat Transfer and Fluid Flow*, McGraw-Hill Book Company, New York.

Reynolds, A.J. (1975) 'The prediction of turbulent Prandtl and Schmidt numbers', *International Journal of Heat and Mass Transfer*, 18, 1055–1069.

Rodi, W. (1980) *Turbulence Models and Their Application in Hydraulics – A State of the Art Review*, IAHR, Delft, The Netherlands.

Rutherford, J.C. (1994) *River Mixing*, John Wiley & Sons, UK.

Schlichting, H. (1979) *Boundary Layer Theory*, 7th Edition, McGraw-Hill Book Company, New York.

Smith, R. (1992) 'Physics of Dispersion' Coastal and Estuarine Pollution – Methods and Solutions' Technical sessions, Scottish hydraulics study group, One day seminar 3rd April, Glasgow.

Taylor, G.I. (1921) 'Diffusion by continuous movements', *Proceedings of the London Mathematical Society*, 20, 196–212.

Taylor, G.I. (1953) 'Dispersion of soluble matter in solvent flowing slowly through a tube', *Proceedings of the Royal Society of London, Series A*, 219, 186–203.

Taylor, G.I. (1954) 'The dispersion of matter in turbulent flow through a pipe', *Proceedings of the Royal Society of London, Series A*, 223, 446–468.

Valentine, E.M. and Wood, I.R. (1977) 'Longitudinal dispersion with dead zones', *Journal of the Hydraulics Division*, ASCE, 103, 975–990.

Wallis, S.G., Young, P.C. and Beven, K.J. (1989) 'Experimental investigation of the aggregated dead zone model for longitudinal solute transport in stream channels', *Proceedings of the Institution of Civil Engineers*, Part 2, 87, 1–22.

Wilson, C.A.M.E., Boxall, J.B., Guymer, I. and Olsen, N.R.B. (2003) 'Validation of a 3D numerical code in the simulation of pseudo-natural meandering flows', *Journal of Hydraulic Engineering*, ASCE, 129, 758–768.

4

Basic equations for sediment transport in CFD for fluvial morphodynamics

E. Mosselman

4.1 Introduction

Alluvial rivers flow through a bed of their own sediments, continually forming and deforming that bed by transporting the sediments from one place to another. Equations for sediment transport are hence a key element in the mathematical modelling of fluvial morphodynamics. This chapter treats those equations from the perspective of modelling fluvial phenomena, such as bars, riffles, pools and longitudinal river bed profiles, which occur on essentially larger scales than ripples and dunes. These fluvial phenomena are relevant morphological types in physical habitat assessment methods (e.g. Maddock, 1999). Their modelling allows a depth-averaged approach.

Analogous to the equations for the motion of water, sediment transport equations can be conceived as mathematical expressions for the principles of conservation of mass and conservation of momentum. The conservation of mass is expressed straightforwardly by a sediment balance equation, which reads in depth-averaged form:

$$(1-\varepsilon)\frac{\partial z_b}{\partial t} + \frac{\partial q_{sx}}{\partial x} + \frac{\partial q_{sy}}{\partial y} = 0 \quad (4.1)$$

in which q_{sx} and q_{sy} are volumetric sediment transports per unit width (excluding pores) in the x- and y-direction, respectively (m²/s), t is time (s), x and y are horizontal space co-ordinates (m), z_b denotes bed level (m + datum) and ε is the porosity of the bed (-). The porosity term translates the net volume of sediment grains into the bulk volume which corresponds to the volumetric changes of bed topography due to erosion and sedimentation. Usually $\varepsilon = 0.4$ is assumed (e.g. Jansen *et al.*, 1979). Similar balance equations can be written for individual size fractions of a mixture of graded sediment, involving the relative volume of each size fraction in the mixture as an additional parameter.

The derivation of sediment momentum equations for the larger-scale fluvial phenomena is not as straightforward as the derivation of the sediment mass balance equations because it requires so many empirical closure relationships that the equations become to a large extent empirical. Accordingly, mathematical models for fluvial morphodynamics commonly use simple empirical predictors that relate the rate of sediment transport to local flow conditions. This chapter presents the basic principles of sediment transport equations in fluvial morphodynamics together with some recent advances and remaining problems regarding transport over sloping beds, graded sediments, transport over floodplains and elementary transport processes on a smaller scale. In particular, it seeks to demonstrate that assessment of the quality of these predictors should be based not only on statistical methods to fit certain formulae to measured sediment transport rates, but also on an analysis of the performance of these formulae in mathematical models. That is why this chapter focuses on generic features of sediment transport equations rather than on the specific equations themselves. Detailed presentations of specific sediment transport equations abound in other publications (e.g. van Rijn, 1993).

4.2 Transport modes

Figure 4.1 shows how sediment transport can be divided on the basis of origin and on the basis of transport mechanism. Washload can have origins far upstream and does not respond to variations in flow conditions while travelling through the river. As its rate depends on the supply from the catchment upstream, it is also called 'supply-limited transport'. Usually a diameter somewhere between 50 and 70 μm is taken as a practical distinction between washload and bed material load, but one should not overlook the fact that smaller particles may still settle in response to flow deceleration in stagnant water bodies or lee sides behind islands. Washload may also settle in estuaries where cohesive particles may flocculate under the influence of salt and an associated density current. Bed material load originates from the alluvial bed

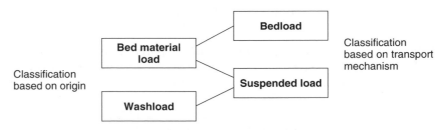

Figure 4.1 *Classification of sediment transport.*

and interacts with it. Its rate of transport depends on the transport capacity of the flow and therefore it is also often called 'capacity-limited transport'. It is this sediment transport which is considered in mathematical models for fluvial morphodynamics.

Bed material load can be either bedload or suspended load. In bedload, the sediment is transported by rolling, sliding and small jumps (saltation). Bedload responds more quickly to variations in flow conditions. Although small lag effects in time and space can be identified (Phillips and Sutherland, 1989, 1990), immediate adaptation is commonly assumed in practical applications. Suspended load consists of sediments that are held in suspension by turbulent upward water motions counteracting gravity. It exhibits clear lag effects in the sense that its adaptation to variations in flow conditions requires a certain time and distance. The adaptation can be modelled using a vertical convection–diffusion equation with a boundary condition at the bed. Elimination of the vertical equation by Galappatti and Vreugdenhil's (1985) asymptotic solution leads to depth-averaged equations for depth-averaged concentrations of suspended sediment. These equations, implemented in the Delft3D modelling system, are relaxation equations with characteristic time and length scales for temporal and spatial adjustments to varying hydraulic conditions. Temporal lag effects are usually neglected in fluvial applications by assuming quasi-steady sediment concentrations, based on the idea that changes in sediment concentrations are much faster than the evolution of the river bed. This is analogous to the commonly applied quasi-steady flow assumption, which holds well for Froude numbers below 0.6–0.8 (de Vries, 1965; Jansen *et al.*, 1979) and still allows the modelling of varying discharges as long as the discharges are kept constant within each bed evolution step. Spatial lag effects are important if the corresponding adaptation lengths are large with respect to the computational cells. This corresponds roughly to $\Delta x \gg u \cdot h/w_s$, where Δx denotes the length of a computational cell (m), u is the flow velocity (m/s), h is the water depth (m) and w_s is the fall velocity of a sediment particle (m/s). A sharper criterion is that spatial lag effects should be accounted for if the adaptation lengths are equal to five times the length of a computational cell or more. Adaptation lengths up to two or three computational cells can be neglected because the grid is too coarse for a good numerical representation of the adaptation and the resulting differences are small. The adaptation length can be obtained from van Rijn's (1987) graph in Figure 4.2, where the ratio of adaptation length to water depth, L/h, can be derived as a function of the degree of overloading or underloading, s/s_e, and the ratio of fall velocity to shear velocity, w_s/u_*. Van Rijn defines his adaptation length as the length over which 95% of the total adaptation has occurred instead of the napierian percentage of 63 ($=1 - 1/e$). This definition implies that a length equal to one-third of van Rijn's adaptation length can be used in evaluations of the necessity to include spatial lag effects.

4.3 Sediment transport formulae

The sediment transport formulae in the remainder of this chapter are local predictors which assume immediate adaptation of the transport to the local transport capacity. They may represent pure bedload, but also total bed material load with substantial transport in suspension if the corresponding adaptation lengths are smaller than two

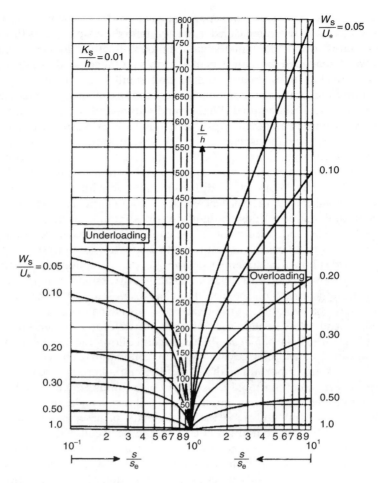

Figure 4.2 Adaptation lengths for suspended sediment transport. (Reproduced by permission of van Rijn, 1987.)

or three computational cell lengths. The sediment transport formulae can generally be expressed as a functional relationship between an Einstein parameter, Φ, for sediment transport and a Shields parameter, θ, for sediment mobility:

$$\Phi = \Phi(\theta) \quad \text{with} \quad \Phi = \frac{q_s}{\sqrt{g\Delta D^3}} \quad \text{and} \quad \theta = \frac{\tau}{\rho g \Delta D} = \frac{u^2}{C^2 \Delta D} \qquad (4.2)$$

in which C is the Chézy coefficient for hydraulic roughness (m$^{1/2}$/s), D is a characteristic diameter of sediment particles (m), g is the acceleration due to gravity (m/s^2), q_s is the volumetric sediment transport per unit width (excluding pores) (m^2/s), Δ is the relative submerged density of sediment particles (-) defined by $\Delta = (\rho_s - \rho)/\rho$, θ is the Shields parameter (-), ρ is the mass density of water (kg/m^3), ρ_s is the mass density of the sediment (kg/m^3), τ is the bed shear stress exerted by the flow (N/m^2) and Φ is the Einstein parameter (-). The Einstein parameter represents the non-dimensional volumetric sediment transport per unit width and the Shields parameter represents the non-dimensional flow strength, relative to both weight and size of sediment particles.

The functional relationships $\Phi(\theta)$ may contain parameters for the gradation of the sediment, but they do not account for the independent behaviour and mutual interaction of the different size fractions within a sediment mixture. These formulae are therefore primarily valid for uniform sediment, i.e. unimodal, narrowly graded sediment.

Most sediment transport formulae can be written as a specific case of the generic formula

$$\Phi = \alpha \theta^{n-\gamma}(\mu\theta - \theta_c)^\gamma \quad (4.3)$$

where α is a non-dimensional coefficient, n and γ are non-dimensional exponents, μ is a ripple factor and θ_c is a critical Shields parameter value for incipient motion. Two major subsets can be distinguished: simpler formulae with a threshold for incipient motion and formulae without threshold. The simpler formulae with a threshold are found by substituting $n = \gamma$:

$$\Phi = \alpha(\mu\theta - \theta_c)^\gamma \quad (4.4)$$

Examples are the sediment transport predictors of Meyer-Peter and Müller (1948) and van Rijn (1984), referred to as MPM-type formulae. The original Meyer-Peter and Müller predictor has the following values: $\alpha = 8$, $\theta_c = 0.047$, $\gamma = 1.5$ and $\mu = (C/C')^{3/2}$, where C is the Chézy coefficient related to the total flow resistance and C' is the Chézy coefficient related to the skin friction of sediment grains only, without the effect of bed-form drag. The value of the critical Shields parameter for incipient motion often results from calibration against data of sediment transport well away from incipient motion and hence it does not necessarily represent a true criterion for this threshold. Buffington and Montgomery (1997) review eight decades of incipient motion studies, whereas recent advances are presented by Cheng and Chiew (1999), Statzner et al. (1999), Wallbridge et al. (1999), Pilotti and Menduni (2001), Shvidchenko et al. (2001), Dancey et al. (2002) and Papanicolaou et al. (2002).

There are two slightly different ways in which the formulae without threshold for incipient motion are obtained from the generic formula, depending on whether the ripple factor is maintained or not. The ripple factor is maintained by substituting $\gamma = n$ and $\theta_c = 0$:

$$\Phi = \alpha \mu^n \theta^n \quad (4.5)$$

The ripple factor is eliminated by substituting $\gamma = 0$:

$$\Phi = \alpha \theta^n \quad (4.6)$$

An example is the sediment transport predictor of Engelund and Hansen (1967) for which $n = 2.5$. These formulae are called EH-type formulae. If the Engelund and Hansen predictor satisfies equation (4.6), α is given by $\alpha = 0.05\, C^2/g$. Sometimes $(C^2/g)^{2/5}$ is interpreted as Engelund and Hansen's ripple factor, which leads to equation (4.5) with $\alpha = 0.05$.

In fluvial hydraulics the sediment transport formulae are utilized to assess the morphological evolution of an alluvial river bed under natural conditions or in response to imposed changes. The coefficient α affects mainly the timescale of the morphological changes, whereas the parameters γ, n and θ_c have a major influence on the resulting morphological forms and patterns. A key parameter in this respect is the degree of non-linearity, b, of the transport formula. It is defined as

$$b = \frac{\theta}{\Phi}\frac{d\Phi}{d\theta} \tag{4.7}$$

(cf. Figure 4.3). The higher this degree of non-linearity, the stronger the response of longitudinal river profiles to river bed constriction, water extraction and sediment mining (Jansen *et al.*, 1979) and the stronger the tendency to form bars and islands (Struiksma *et al.*, 1985; Colombini *et al.*, 1987; Struiksma and Crosato, 1989). For instance, the following relation arises from Jansen *et al.* (1979) for the response of the equilibrium river gradient to changes in width, water discharge and equilibrium sediment transport:

$$i = \frac{B^{\frac{2b-3}{2b}} Q_s^{\frac{3}{2b}}}{\alpha^{\frac{3}{2b}} Q C^2} \tag{4.8}$$

where i denotes the equilibrium river gradient (-), B is the river width (m), Q is the water discharge (m^3/s) and Q_s is the sediment load (m^3/s). Wang *et al.* (1995) demonstrate that the stability of river bifurcations is also sensitive to the degree of non-linearity of the sediment transport formula. In sum, the degree of non-linearity has a strong effect on morphological phenomena. It may be evident from Figure 4.4, however, that a degree of non-linearity derived from field measurements by curve

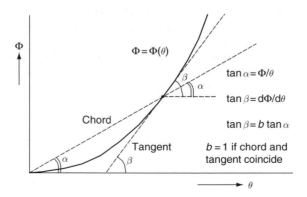

Figure 4.3 *Degree of non-linearity of sediment transport predictor.*

Basic equations for sediment transport in CFD 77

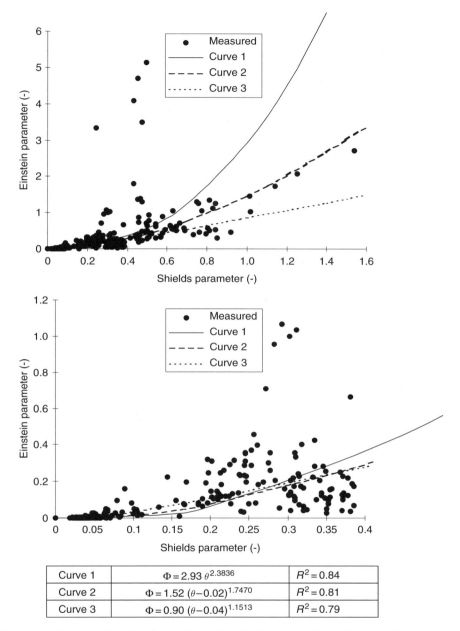

Curve 1	$\Phi = 2.93\, \theta^{2.3836}$	$R^2 = 0.84$
Curve 2	$\Phi = 1.52\, (\theta - 0.02)^{1.7470}$	$R^2 = 0.81$
Curve 3	$\Phi = 0.90\, (\theta - 0.04)^{1.1513}$	$R^2 = 0.79$

Figure 4.4 Measured bedload rates in the Dutch Rhine branches, with best-fit curves based on linear regression between log (Φ) and log ($\theta - \theta_c$) for different imposed values of θ_c: (a) data; and (b) data for Shields parameters below 0.4.

fitting is not very accurate. Several transport formulae with different degrees of non-linearity can be fitted to the measured data, but a decision that one formula fits better than another is hardly justifiable. This is the first example of what this chapter seeks to demonstrate: Assessment of the quality of sediment transport predictors should be based not only on statistical analysis of data, but also on an analysis of the performance of these formulae in mathematical models.

Application of the definition of the degree of non-linearity to EH-type formulae yields simply $b = n$. Application to MPM-type formulae yields

$$b = \frac{\gamma \mu \theta}{\mu \theta - \theta_c} \qquad (4.9)$$

This relation implies that the degree of non-linearity decreases monotonically to γ as θ increases. For large values of the Shields parameter ($\mu\theta \gg \theta_c$), the degree of non-linearity equals γ. For small Shields parameters ($\mu\theta \approx \theta_c$), the degree of non-linearity tends to extremely high values (cf. Paintal, 1971). Yet extremely high degrees of non-linearity do not lead to extreme responses because close to incipient motion the transport rates are so low that the speeds of morphological change are low as well.

A minimum value for b, and hence γ and n, occurs for high sediment transport rates and can be identified on physical grounds. This is demonstrated as follows. The definition of specific discharge ($q = h \cdot u$) and Chézy's equation ($u^2 = C^2 \cdot h \cdot i$) can be combined into

$$q = \frac{u^3}{C^2 i} = \frac{(\Delta D)^{3/2} C \theta^{3/2}}{i} \qquad (4.10)$$

where the water depth, h, has been eliminated, u denotes the depth-averaged flow velocity (m/s) and q is the water discharge per unit width (m²/s). As q_s is proportional to θ^b if $\mu\theta \gg \theta_c$ and q is proportional to $\theta^{3/2}$ according to equation (4.10), the depth-averaged concentration of transported sediment, c, obeys

$$c = \frac{q_s}{q} \sim \theta^{b-3/2} \qquad (4.11)$$

It is very unlikely that the concentration could decrease as the Shields mobility parameter increases. This leads to the condition $b \geq 1.5$. Equation (4.8) confirms independently that $b \geq 1.5$ because a lower value would imply the unrealistic behaviour that a constriction of the river bed leads to steeper slopes. Caution is mandatory when curve fitting yields a sediment transport formula with a degree of non-linearity below 1.5. This is another example that the performance of a formula in mathematical models should be considered as well when evaluating its quality. Measurements in large sand-bed rivers usually result in values for b between 1.5 and 2.5.

4.4 Transverse bed slope effects

Two effects cause deviations between the directions of bedload transport and depth-averaged flow. The first effect is that curved flow induces a secondary, helical water motion that leads to a difference between near-bed flow direction and depth-averaged flow direction. The second effect is that sediment particles experience a downward acceleration along transverse bed slopes due to gravity. In the Delft3D modelling system, these effects are modelled by

$$\tan \alpha_s = \frac{\sin \alpha_\tau - \frac{1}{f}\frac{\partial z_b}{\partial y}}{\cos \alpha_\tau - \frac{1}{f}\frac{\partial z_b}{\partial x}} \quad (4.12)$$

where α_s is the direction of sediment transport (rad), α_τ is the direction of bed shear stress (rad) and f is a dimensionless parameter. Equation (4.12) has two complications. First of all, it assumes that all sediment is transported close to the bed, which holds for bedload but not for suspended load. This is another aspect of suspended sediment transport than the lag effects. Only the latter can be neglected if the adaptation lengths are sufficiently small. A justification to apply a local sediment transport predictor, without lag effects, to predict total loads consisting of bedload and suspended load does not imply that the suspended character can be neglected in the formula for the effect of transverse bed slopes. Talmon (1992) corrected the transverse bed slope effect empirically for suspended sediment transport on the basis of flume experiments.

The second complication of equation (4.12) is that longitudinal and transverse bed slopes have the same effect. From a mathematical modelling point of view, this isotropic formulation has the advantage that the slopes can be expressed in a co-ordinate system which does not depend on flow directions. In reality, however, the effect of bed slopes is anisotropic. The slope effect must be seen as a resultant from integration over subgrid bed slope variations due to ripples and dunes. As these bed forms are often arranged perpendicular to the flow lines, the integrated formulation in cross-stream direction can be expected to yield a relation which differs from the one in streamwise direction.

Mathematical expressions for f have been proposed by van Bendegom (1947), Engelund (1974, 1981), Engelund and Fredsøe (1976), Zimmermann and Kennedy (1978), Ikeda (1982), Ikeda and Nishimura (1985), Kovacs and Parker (1994) and Talmon et al. (1995). Most of these relations have been derived theoretically from a balance of forces under the assumption of a smooth bed, free of bed forms. As real alluvial beds do not satisfy this assumption, Talmon et al. (1995) derived an empirical relationship for f from mobile-bed flume experiments. They found that f is proportional to $\sqrt{\theta}$. They also found a scale effect in the proportionality constant $(f/\sqrt{\theta})$ when calibrating the relation for laboratory experiments as well as field conditions. To account for this scale effect, they added a dependence on the ratio of median grain size to water depth. The resulting relation reads

$$f = 9\left(\frac{D_{50}}{h}\right)^{0.3}\sqrt{\theta} \quad (4.13)$$

where D_{50} denotes the median diameter of the sediment grain-size distribution (m).

The bed slope effect is diffusive, as can be shown by substituting equation (4.12) without helical flow ($\alpha_\tau = 0$) and without bed slopes in x-direction ($\partial z_b/\partial x = 0$) into the term for transverse sediment transport gradients in the sediment balance:

$$\frac{\partial q_{sy}}{\partial y} = \frac{\partial q_{sx} \tan \alpha_s}{\partial y} = \frac{\partial}{\partial y}\left(-\frac{q_{sx}}{f}\frac{\partial z_b}{\partial y}\right) = -\frac{\partial z_b}{\partial y}\frac{\partial}{\partial y}\left(\frac{q_{sx}}{f}\right) - \frac{q_{sx}}{f}\frac{\partial^2 z_b}{\partial y^2} \qquad (4.14)$$

The last term represents diffusion with diffusion coefficient q_{sx}/f. This means that the bed slope effect can be used to make a numerical model more stable or vice versa, numerical stability parameters affect the representation of bed slope effects. It is important to be aware of this when interpreting model results. Again an intimate link is found between a sediment transport relation and its performance in a mathematical model.

4.5 Graded sediment

The simple MPM- and EH-type sediment transport predictors do not account for the independent behaviour and mutual interaction of sediment particles with differing physical characteristics. Segregation of these particles during processes of erosion, transport and deposition gives rise to spatial grain sorting phenomena such as coarse bed surface layers, downstream fining (with corresponding decrease in river gradient), coarser-grained riffles in straight channels and finer-grained point bars in meandering rivers. Downstream fining results also from abrasion, but usually grain sorting is the main cause. Grain sorting also affects the patterns and dimensions of bars, riffles and pools (Lanzoni and Tubino, 1999; Mosselman *et al.*, 1999; Lanzoni, 2000). The sorting may be due to differences in mass density (mineralogical composition), grain size and grain shape. For the sorting of coarse sediments, differences in grain size are the dominant cause and it is for these differences that graded-sediment functionality has been implemented in mathematical models. It has become customary in fluvial morphodynamics to use the term 'graded sediment' not only for widely graded sediment but also for narrowly graded bimodal or trimodal sediment distributions because those can be treated in the same way in mathematical models. Thus unimodal, narrowly graded sediment is usually referred to as 'uniform sediment', whereas widely graded sediment as well as bimodal and trimodal sediments are referred to as 'graded sediment'.

Powell (1998) reviews key concepts of graded sediment such as size-selective entrainment, hiding and exposure and the equal-mobility hypothesis. In principle, the entrainment of larger grains requires a higher shear stress. This can readily be seen from the Shields criterion for incipient motion

$$\tau_c = \theta_c \rho g \Delta D \qquad (4.15)$$

where the subscript 'c' refers to the threshold for incipient motion. This relation implies that the minimum shear stress for entrainment, τ_c, is proportional to the characteristic particle diameter, D: $\tau_c \propto D$. Certain flows may hence entrain only the finer grain-size fractions of a mixture (size-selective entrainment). However, entrainment thresholds for individual size fractions in coarse, heterogeneous sediments appear to differ significantly from those associated with the movement of uniform sediment of the same size. Relatively fine grains in a mixture are less mobile than they would otherwise be in uniform sediment because they protrude less into the flow than surrounding coarser grains, which also act to shelter them from the mobilizing influence of the flow. This phenomenon is called 'hiding' or 'shielding'. Likewise, relatively coarse grains in a mixture are more easily entrained because they protrude more into the flow. This phenomenon is called 'exposure'. Hiding and exposure are accounted for in MPM-type sediment transport formulae for individual fractions by a correction of the critical shear stress. The most commonly used correction is the one of Egiazaroff (1965), modified by Ashida and Michiue (1972, 1973), but for clarity the simplified correction from Powell's (1998) review is used here:

$$\tau_{ci} = \left(\frac{D_i}{D_{50}}\right)^{\beta} \theta_c \rho g \Delta D_i \qquad (4.16)$$

where D_i is the characteristic particle diameter of sediment fraction i (m), β is the exponent of the hiding-and-exposure correction (-) and τ_{ci} is the critical shear stress of sediment fraction i (N/m^2).

Hiding and exposure thus level out the differences in size-dependent mobility. The result is a weaker dependence of the minimum entrainment shear stress on particle diameter: $\tau_{ci} \propto D_i^{1+\beta}$ with $-1 \leq \beta < 0$. Values for β were found to be very close to -1. If β equals -1, τ_{ci} no longer depends on D_i, which implies that entrainment is no longer size-selective. This is the basis of the much-debated equal-mobility hypothesis, which states that all grain sizes have the same critical shear stress for incipient motion.

The above hiding-and-exposure corrections hold for widely graded sediment. Their applicability to bimodal sediment mixtures is questionable. Bimodal sand–gravel mixtures can exhibit partial transport with a sand fraction in motion and a gravel fraction below the threshold of motion, which does not satisfy equal mobility at all. Much insight into the bedload transport processes of a sand–gravel mixture has been gained in recent years from experiments in the 98 m long and 1.5 m wide sand flume of WL/Delft Hydraulics (Kleinhans, 2002; Kleinhans and van Rijn, 2002; Kleinhans et al., 2002; Blom et al., 2003). One of the findings is that, in contrast to the downward fining under an armoured flat bed, upward fining arises because gravel accumulates in the troughs between sand dunes that may propagate eventually over a deeper static armour layer of gravel. Previously Wilcock and Southard (1989) and Klaassen (1990) had observed the same phenomenon.

The present description of graded sediment in numerical models such as SOBEK and Delft3D assumes widely graded sediment. It is based on (i) division of the sediment mixture into separate fractions, (ii) transport formulae and mass conservation equations for each of the separate fractions, (iii) hiding-and-exposure corrections to the critical shear stress of each of the fractions and (iv) an active layer or transport layer affected by erosion and sedimentation (Hirano, 1972). The thickness of this layer is equal to half the height of the bed forms. An exchange layer may be added under the transport layer (Ribberink, 1987). The exchange of sediment between the active layer and the substrate is thought to be associated with the extension of sporadic deeper troughs of the bed forms. Figure 4.5 shows a typical result from a 2D morphological model with graded sediment (from Sloff *et al.*, 2001). For the time being, the effect of transverse bed slopes has been implemented by extending Equation (4.13) in a way similar to the hiding-and-exposure correction of equation (4.16). The resulting relation reads

$$f = A_{sh} \theta_i^{B_{sh}} \left(\frac{D_i}{h}\right)^{C_{sh}} \left(\frac{D_{50}}{D_i}\right)^{D_{sh}} \qquad (4.17)$$

where A_{sh}, B_{sh}, C_{sh} and D_{sh} are calibration coefficients with values $B_{sh}=0.5$ and $C_{sh}=0.3$ as first estimates. The present knowledge of this relation is still unsatisfactory. It is not clear, for instance, whether the proper mobility parameter to be used in the equation is the Shields parameter for each individual grain-size fraction (θ_i) or an overall Shields parameter for an average grain-size (θ). Furthermore, the large number of calibration parameters causes a problem of equifinality. Experimental research is still needed to improve this equation. In the experiments, it is recommended that the morphological evolution of a transversely tilted bed be measured rather than mere bedload directions. This will allow a better evaluation of how

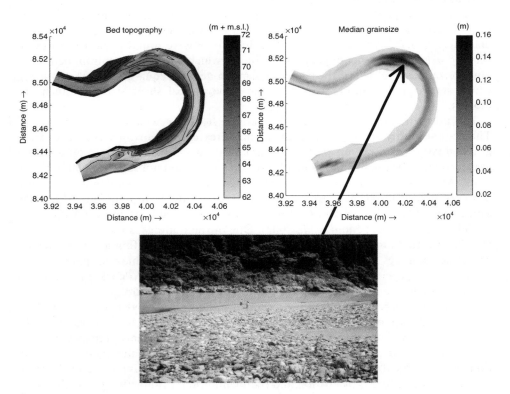

Figure 4.5 Results from Delft3D model of Tenryuu River, Japan: bed topography and spatial grain-size distribution, with reproduction of complex cobble cluster.

various equations perform in the mathematical reproduction of the morphological development.

Ribberink (1987) finds that the present set of equations for graded sediment in numerical models can become elliptic. That is unrealistic for systems with time derivatives because it implies that present conditions depend on the future. The unrealistic elliptic character has been one of the rationales for developing a new model concept for graded sediment, viz. the depth-continuous continuity concept of Parker et al. (2000) and Blom (2003). This new concept also has other advantages over the standard layer approaches. It accounts for vertical sorting processes in the absence of net aggradation or degradation, including upward fining within bed forms.

4.6 Sediment transport on floodplains

Traditionally CFD for fluvial morphodyamics focused on the main channel, dealing with problems of scour and navigability. Recent trends for ecological rehabilitation of rivers and the lowering of floodplains for flood mitigation call for tools to predict morphological processes in floodplains as well. Existing morphological models for main channels are not automatically applicable to floodplains because the conditions in floodplains are different. Cohesive and vegetated floodplain surfaces may be non-erodible, vegetation may alter sediment transport capacities and levees and steep banks may form sheer obstacles to the transport of sediment. SOBEK and Delft3D contain satisfactory functionality for the representation of sediment transport over non-erodible layers, but the effects of vegetation and obstacles are still subject to fundamental process research.

Sediment transport over non-erodible layers has been implemented in SOBEK and Delft3D on the basis of Struiksma's (1999) model concept. Usually bedload is modelled with a simple predictor in which sediment transport is a function of the local transport capacity of the flow. When transported over a non-erodible layer, however, no sediment can be supplied from the bed when the capacity increases. The sediment transport thus becomes supply limited rather than capacity limited. At first glance, this might seem to necessitate an entirely different computational approach in which volumes of sediment supplied at the boundaries are followed as they travel over the non-erodible layer. Struiksma's model concept, however, retains the relation with local transport capacity. It accounts for transport limitations due to non-erodible layers through a reduction of the transport layer thickness on the bed. The volumetric sediment transport per unit width according to a customary transport predictor, q_{sf}, is modified with a correction factor, ψ, to obtain the actual sediment transport per unit width, q_s, over the non-erodible layer:

$$q_s = \psi\left(\frac{\delta}{\delta_a(h)}\right) q_{sf}(u) \qquad (4.18)$$

in which δ is the thickness of alluvium on the non-erodible bed (m) and δ_a is the maximum thickness of alluvium at which the non-erodible layer affects the sediment transport (m). The bed is non-alluvial ($\psi < 1$) if $\delta/\delta_a < 1$ and alluvial ($\psi = 1$), if $\delta/\delta_a \geq 1$ (Figure 4.6). The correction factor, ψ, accounts for the reduction in the transport capacity due to a thinner transport layer and hindered bed-form development. Basically, the information on the reduction is carried by the sediment layer thickness, δ. Local increases in q_{sf} are counteracted by decreases in δ and ψ in such a way that the resulting erosion cannot reach below the surface of the non-erodible layer. This limits the supply of sediment from the bed. The supply of sediment from upstream is limited by the occurrence of similar reductions upstream. The maximum alluvial thickness for influence of the non-erodible layer, δ_a, corresponds to the thickness of the transport layer on an alluvial bed. It can be taken equal to half the bed-form height and is therefore primarily a function of the local water depth.

Flow resistance due to vegetation is a much researched topic, quite in contrast to the effect of vegetation on erosion, transport and deposition of sediment. Some effects of vegetation on sediment transport may be known qualitatively, but there are hardly any quantitative data from field measurements or laboratory experiments. As a consequence, it is difficult to develop or to test sub-models. One of the problems of customary sediment transport equations is that they contain a flow resistance parameter which may be closely related to sediment transport through bed shear stresses and bed forms, but is totally unrelated to bedload when the flow resistance is primarily due to vegetation. This is illustrated conceptually in Figure 4.7. For instance, the equivalent Nikuradse roughness that is currently being applied for floodplain vegetation in hydraulic models of the Dutch rivers can be much larger than the water depth.

As a result, the Chézy coefficient becomes extremely small (even negative when applying White–Colebrook's logarithmic formula). Using this Chézy coefficient in customary sediment transport formulae implies, unrealistically, that vegetated zones become zones of slow currents with very high sediment transport rates (Baptist, 2001). Karanxha (2002) therefore proposes a decomposition of the total shear stress into bed skin friction, bed-form drag and vegetation-form drag, linking bedload to the first two components only. Thannbichler's (2002) experiments on sediment transport with flexible, submerged vegetation show a case in which all sediments are transported in suspension above the vegetation, whereas bedload and the development of bed forms remain absent.

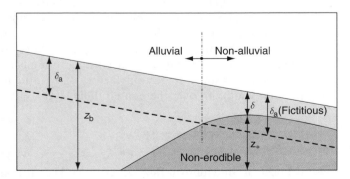

Figure 4.6 *Thickness of alluvium in Struiksma's (1999) model concept for bedload over non-erodible layers. (Published by IAHR.)*

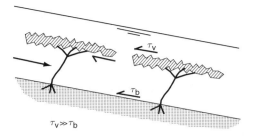

Figure 4.7 Uncoupling of bed roughness and vegetation roughness.

Equation (4.12) for the effect of transverse bed slopes on sediment transport directions expresses that gravity produces an extra downslope sediment transport component. By inference, longitudinal bed slopes affect the magnitude of sediment transport in such a way that downward slopes increase and upward slopes decrease the transport. This picture is valid for mild bed slopes, but Lauchlan's (2001) experiments show that steeper slopes, with more 3D flow, may have the opposite effect. All arriving sediments were transported over the weir in her flume, whereas sedimentation occurred downstream of the weir. In models for fluvial phenomena on the scales of bars, riffles, pools and longitudinal river bed profiles, steeper slopes are subgrid elements. Proper parameterization of their effects still requires further research.

4.7 Concluding remarks

This chapter has sought to demonstrate that the goodness of sediment transport equations should be based not only on sediment transport measurements, but also on the performance of these equations in the mathematical reproduction of morphological phenomena. The more general implication is that sediment transport equations should not be derived from sediment transport measurements alone, but also from bed level measurements. Ideally laboratory experiments and field measurement campaigns are designed in such a way that the time evolution of some morphological phenomenon can be followed. An example is Havinga's (1982) assessment of the sediment transport in the IJssel River by studying the downstream migration and gradual filling of a trench dredged across the river bed. Similar approaches are presented and advocated by McLean and Church (1999), Ham and Church (2000) and Duizendstra (2001).

The emphasis of this chapter on phenomena on a larger scale might yield an impression of anti-reductionism. This is not the intention. Indeed sediment transport equations cannot be derived straightforwardly from basic conservation principles, because upscaling requires so many empirical closure relationships that the resulting equations on a larger scale become to a large extent empirical. Nonetheless, those equations should comply with the descriptions of elementary processes on a smaller scale and progress in elementary processes will have spin-offs for the parameterizations on a larger scale. Important research avenues in this respect are turbulence modelling including direct numerical simulation, probabilistic description of turbulent bed shear stresses (Einstein, 1950; van Rijn, 1987; Zanke, 1990; Bridge and Bennett, 1992; Kleinhans and van Rijn, 2002; Papanicolaou *et al.*, 2002), modelling of the motion of individual sediment particles (Sekine and Kikkawa, 1992; Sekine and Parker, 1992; Nino and Garcia, 1998; Lee *et al.*, 2002; Malmaeus and Hassan, 2002; Nino *et al.*, 2002), modelling of bed-form development and modelling of sediment grain sorting (Parker *et al.*, 2000; Blom, 2003).

The knowledge gaps regarding the influence of bed slopes and vegetation continue to call for experimental research under controlled conditions in laboratory flumes. Field studies remain mandatory as well. After all, results from laboratory flumes exhibit considerable scale effects when they are extrapolated to large rivers where flow depths and Reynolds numbers are up to 100 times larger than those found in the flumes (Molinas and Wu, 2001).

Acknowledgements

Many colleagues at WL/Delft Hydraulics and Delft University of Technology are to be credited for the insights described in this chapter.

References

Ashida, K. and M. Michiue (1972), Study on hydraulic resistance and bedload transport rate in alluvial streams. *Transactions*, JSCE, 206, pp. 59–69.

Ashida, K. and M. Michiue (1973), *Studies on Bed-load Transport in Open Channel Flows*. Proceedings of the International Symposium on River Mechanics, IAHR, Bangkok, Thailand, paper A36, pp. 407–418.

Baptist, M.J. (2001), Numerical modelling of the biogeomorphological developments of secondary channels in the Waal river. *CFR Project Report 11*, Delft University of Technology.

Blom, A. (2003), A continuum vertical sorting model for rivers with non-uniform sediment and dunes. PhD Thesis, Twente University, Enschede, The Netherlands.

Blom, A., J.S. Ribberink and H.J. de Vriend (2003), Vertical sorting in bed forms: Flume experiments with a natural and a tri-modal sediment mixture. *Water Resources Research*, AGU, Vol. 39, No. 2, p. 1025 (13pp.).

Bridge, J.S. and S.J. Bennett (1992), A model for the entrainment and transport of sediment grains of mixed sizes, shapes and densities. *Water Resources Research*, AGU, Vol. 28, pp. 337–363.

Buffington, J.M. and D.R. Montgomery (1997), A systematic analysis of eight decades of incipient motion studies, with special reference to gravel-bedded rivers. *Water Resources Research*, AGU, Vol. 33, No. 8, pp. 1993–2029.

Cheng, N.S. and Y.M. Chiew (1999), Incipient sediment motion with upward seepage. *Journal of Hydraulic Research*, IAHR, Vol. 37, No. 5, pp. 665–681.

Colombini, M., G. Seminara and M. Tubino (1987), Finite-amplitude alternate bars. *Journal of Fluid Mechanics*, Vol. 181, pp. 213–232.

Dancey, C.L., P. Diplas, A. Papanicolaou and M. Bala (2002), Probability of individual grain movement and threshold condition. *Journal of Hydraulic Engineering*, ASCE, Vol. 128, No. 12, pp. 1069–1075.

de Vries, M. (1965), *Consideration about Non-steady Bed-load Transport in Open Channels*. 11th IAHR Congress, Leningrad, Vol. III, Paper No. 3.8.

Duizendstra, H.D. (2001), Determination of the sediment transport in an armoured gravel-bed river. *Earth Surface Processes and Landforms*, BGRG, Vol. 26, No. 13, pp. 1381–1393.

Egiazaroff, I.V. (1965), Calculation of non-uniform sediment concentrations. *Journal of Hydraulic Division*, ASCE, Vol. 91, No. HY4, pp. 225–247.

Einstein, H.A. (1950), The bed-load function for sediment transportation in open channel flow. *Tech. Bull. No. 1026*, US Department of Agriculture, Washington DC.

Engelund, F. (1974), Flow and bed topography in channel bends. *Journal of Hydraulic Division*, ASCE, Vol. 100, No. HY11, pp. 1631–1648.

Engelund, F. (1981), The motion of sediment particles on an inclined bed. *Prog. Rep. 53*, Institute Hydrodynamic and Hydraulic Engineering, Technical University of Denmark.

Engelund, F. and J. Fredsøe (1976), A sediment transport model for straight alluvial channels. *Nordic Hydrology*, Vol. 7, pp. 293–306.

Engelund, F. and E. Hansen (1967), *A Monograph on Sediment Transport in Alluvial Streams*. Teknisk Forlag, Copenhagen.

Galappatti, R. and C.B. Vreugdenhil (1985), A depth-integrated model for suspended sediment transport. *Journal of Hydraulic Research*, IAHR, Vol. 23, No. 4, pp. 359–377.

Ham, D.G. and M. Church (2000), Bed-material transport estimated from channel morphodynamics: Chilliwach River, British Columbia. *Earth Surface Processes and Landforms*, BGRG, Vol. 25, No. 10, pp. 1123–1142.

Havinga, H. (1982), Deventer baggerproef. Meetverslag, Rijkswaterstaat Directie Waterhuishouding en Waterbeweging (District Zuidoost), nota 82.4 (in Dutch; measurement report of dredging field experiment in the river IJssel at Deventer).

Hirano, M. (1972), Studies on variation and equilibrium state of a river bed composed of non-uniform material. *Transactions*, JSCE, Vol. 4.

Ikeda, S. (1982), Lateral bed load transport on side slopes. *Journal of Hydraulic Division*, ASCE, Vol. 108, No. HY11, pp. 1369–1373.

Ikeda, S. and T. Nishimura (1985), Bed topography in bends of sand-silt rivers. *Journal of Hydraulic Engineering*, ASCE, Vol. 111, No. 11, pp. 1397–1411.

Jansen, P.Ph., L. van Bendegom, J. van den Berg, M. de Vries and A. Zanen (1979), *Principles of River Engineering: The Non-tidal Alluvial River*, Pitman, London.

Karanxha, A. (2002), Interaction Between Flow – Sediment Transport – Morphology and Vegetation. MSc Thesis, IHE, Delft.

Klaassen, G.J. (1990), *Experiments on the Effect of Gradation and Vertical Sorting on Sediment Transport Phenomena in the Dune Phase*. Proceedings of International Grain Sorting Seminar, IAHR, Monte Verità, Ascona, Switzerland, pp. 127–145.

Kleinhans, M.G. (2002), Sorting out sand and gravel: Sediment transport and deposition in sand-gravel bed rivers. *Netherlands Geographical Studies 293*, Utrecht, ISSN 0169-4839.

Kleinhans, M.G. and L.C. van Rijn (2002), Stochastic prediction of sediment transport in sand-gravel bed rivers. *Journal of Hydraulic Engineering*, ASCE, Vol. 128, No. 4, pp. 412–425.

Kleinhans, M.G., A.W.E. Wilbers, A. de Swaaf and J.H. van den Berg (2002), Sediment supply-limited bedforms in sand-gravel bed rivers. *Journal of Sedimentary Research*, Vol. 72, No. 5, September.

Kovacs, A. and G. Parker (1994), A new vectorial bedload formulation and its application to the time evolution of straight river channels. *Journal of Fluid Mechanics*, Vol. 267, pp. 153–183.

Lanzoni, S. (2000), Experiments on bar formation in a straight flume. 2. Graded sediment. *Water Resources Research*, AGU, Vol. 36, No. 11, pp. 3351–3363.

Lanzoni, S. and M. Tubino (1999), Grain sorting and bar instability. *Journal of Fluid Mechanics*, Vol. 393, pp. 149–174.

Lauchlan, C. (2001), Sediment transport over steep slopes: An experimental investigation; Results and analysis. *Report for Rijkswaterstaat-RIZA & Delft Cluster*, August, Delft University of Technology.

Lee, H.Y., J.Y. You and Y.T. Lin (2002), Continuous saltating process of multiple sediment particles. *Journal of Hydraulic Engineering*, ASCE, Vol. 128, No. 4, pp. 443–450.

Maddock, I. (1999), The importance of physical habitat assessment for evaluating river health. *Freshwater Biology*, Vol. 41, pp. 373–391.

Malmaeus, J.M. and M.A. Hassan (2002), Simulation of individual particle movement in a gravel streambed. *Earth Surface Processes and Landforms*, Vol. 27, No. 1, pp. 81–97.

McLean, D.G. and M. Church (1999), Sediment transport along lower Fraser River – 2. Estimates based on the long-term gravel budget. *Water Resources Research*, AGU, Vol. 35, No. 8, pp. 2549–2559.

Meyer-Peter, E. and R. Müller (1948), *Formulas for Bed-Load Transport*. Proceedings of 2nd Congress IAHR, Stockholm, Paper No. 2, pp. 39–64.

Molinas, A. and B.S. Wu (2001), Transport of sediment in large sand-bed rivers. *Journal of Hydraulic Research*, IAHR, Vol. 39, No. 2, pp. 135–146.

Mosselman, E., A. Sieben, K. Sloff and A. Wolters (1999), *Effect of Spatial Grain Size Variations on Two-dimensional River Bed Morphology*. Proceedings of IAHR Symposium on River, Coastal and Estuarine Morphodynamics, Genova, 6–10 September, Vol. I, pp. 499–507.

Nino, Y. and M. Garcia (1998), Using Lagrangian particle saltation observations for bedload sediment transport modelling. *Hydrological Processes*, Vol. 12, No. 8, pp. 1197–1218.

Nino, Y., A. Atala, M. Barahona and D. Aracena (2002), Discrete particle model for analyzing bedform development. *Journal of Hydraulic Engineering*, ASCE, Vol. 128, No. 4, pp. 381–389.

Paintal, A.S. (1971), Concept of critical shear stress in loose boundary open channels. *Journal of Hydraulic Research*, IAHR, Vol. 9, No. 1, pp. 91–108.

Papanicolaou, A.N., P. Diplas, N. Evaggelopoulos and S. Fotopoulos (2002), Stochastic incipient motion criterion for spheres under various bed packing conditions. *Journal of Hydraulic Engineering*, ASCE, Vol. 128, No. 4, pp. 369–380.

Parker, G., C. Paola and S. Leclair (2000), Probabilistic Exner sediment continuity equations for mixtures with no active layer. *Journal of Hydraulic Engineering*, ASCE, Vol. 126, No. 11, pp. 818–826.

Phillips, B.C. and A.J. Sutherland (1989), Spatial lag effects in bed load sediment transport. *Journal of Hydraulic Research*, IAHR, Vol. 27, No. 1, pp. 115–133.

Phillips, B.C. and A.J. Sutherland (1990), Temporal lag effect in bed load sediment transport. *Journal of Hydraulic Research*, IAHR, Vol. 28, No. 1, pp. 5–23.

Pilotti, M. and G. Menduni (2001), Beginning of sediment transport of incoherent grains in shallow shear flows. *Journal of Hydraulic Research*, IAHR, Vol. 39, No. 2, pp. 115–124.

Powell, D.M. (1998), Patterns and processes of sediment sorting in gravel-bed rivers. *Progress in Physical Geography*, Vol. 22, No. 1, pp. 1–32.

Ribberink, J.S. (1987), Mathematical modelling of one-dimensional morphological changes in rivers with non-uniform sediment. *Communications on Hydrology and Geotechnological Engineering*, No. 87-2, Delft University of Technology, ISSN 0169-6548.

Sekine, M. and H. Kikkawa (1992), Mechanics of saltating grains. II. *Journal of Hydraulic Engineering*, ASCE, Vol. 118, No. 4, pp. 536–558.

Sekine, M. and G. Parker (1992), Bed-load transport on transverse slope. I. *Journal of Hydraulic Engineering*, ASCE, Vol. 118, No. 4, pp. 513–535.

Shvidchenko, A.B., G. Pender and T.B. Hoey (2001), Critical shear stress for incipient motion of sand/gravel streambeds. *Water Resources Research*, AGU, Vol. 37, No. 8, pp. 2273–2283.

Sloff, C.J., H.R.A. Jagers, Y. Kitamura and P. Kitamura (2001), *2D Morphodynamic Modelling with Graded Sediment*. Proceedings of 2nd IAHR Symposium on River, Coastal and Estuarine Morphodynamics, 10–14 September, Obihiro, Japan, pp. 535–544.

Statzner, B., M.F. Arens, J.Y. Champagne, R. Morel and E. Herouin (1999), Silk-producing stream insects and gravel erosion: Significant biological effects on critical shear stress. *Water Resources Research*, AGU, Vol. 35, No. 11, pp. 3495–3506.

Struiksma, N. (1999), *Mathematical Modelling of Bedload Transport over Non-erodible Layers*. Proceedings of IAHR Symposium on River, Coastal and Estuarine Morphodynamics, Genova, 6–10 September, Vol. I, pp. 89–98.

Struiksma, N. and A. Crosato (1989), Analysis of a 2-D bed topography model for rivers. In: S. Ikeda and G. Parker (eds), *River Meandering*, AGU, Water Resources Monograph 12, pp. 153–180.

Struiksma, N., K.W. Olesen, C. Flokstra and H.J. de Vriend (1985), Bed deformation in curved alluvial channels. *Journal of Hydraulic Research*, IAHR, Vol. 23, No. 1, pp. 57–79.

Talmon, A.M. (1992), Bed topography of river bends with suspended sediment transport. *Communications on Hydrology and Geotechnological Engineering*, No. 92-5, Delft University of Technology, ISSN 0169-6548.

Talmon, A.M., N. Struiksma and M.C.L.M. van Mierlo (1995), Laboratory measurements of the direction of sediment transport on transverse alluvial-bed slopes. *Journal of Hydraulic Research*, IAHR, Vol. 33, No. 4, pp. 495–517.

Thannbichler, C. (2002), A Flume Experiment on Sediment Transport with Flexible, Submerged Vegetation. MSc Thesis, Delft University of Technology, University of Stuttgart, Institut für Wasserbau.

van Bendegom, L. (1947), Some considerations on river morphology and river improvement. *De Ingenieur*, Vol. 59, No. 4 (in Dutch; English trans.: Natl. Res. Council Canada, Tech. Translation 1054, 1963).

van Rijn, L.C. (1984), Sediment transport, Part I: Bed load transport. *Journal of Hydraulic Engineering*, ASCE, Vol. 110, No. 10, pp. 1431–1456.

van Rijn, L.C. (1987), Mathematical modelling of morphological processes in the case of suspended sediment transport. Dissertation, Delft University of Technology.

van Rijn, L.C. (1993), *Principles of Sediment Transport in Rivers, Estuaries and Coastal Seas*. Aqua Publications, Amsterdam, ISBN 90-800356-2-9.

Vries, M. de (1965), *Considerations about Non-steady Bed-load Transport in Open Channels*. Proceedings of 11th Congress IAHR, Leningrad, pp. 3.8.1–3.8.8 (also Delft Hydr. Lab. Publ. No. 36).

Wallbridge, S., G. Voulgaris, B.N. Tomlinson and M.B. Collins (1999), Initial motion and pivoting characteristics of sand particles in uniform and heterogeneous beds: Experiments and modelling. *Sedimentology*, Vol. 46, No. 1, pp. 17–32.

Wang, Z.B., R.J. Fokkink, M. de Vries and A. Langerak (1995), Stability of river bifurcations in 1D morphodynamic models. *Journal of Hydraulic Research*, IAHR, Vol. 33, No. 6, pp. 739–750.

Wilcock, P.R. and J.B. Southard (1989), Bed load transport of mixed size sediment: Fractional transport rates, bed forms, and the development of a coarse bed surface layer. *Water Resources Research*, AGU, Vol. 25, pp. 1629–1641.

Zanke, U. (1990), Der Beginn der Sedimentbewegung als Wahrscheinlichkeitsproblem. Wasser + Boden, Vol. 42, No. 1, pp. 40–43 (in German; The start of sediment motion as a probability problem).

Zimmermann, C. and J.F. Kennedy (1978), Transverse bed slopes in curved alluvial streams. *Journal of Hydraulic Division*, ASCE, Vol. 104, No. HY1, pp. 33–48.

5
Introduction to statistical turbulence modelling for hydraulic engineering flows

F. Sotiropoulos

5.1 Introduction

Environmental hydraulics flows take place in geometrically complex, natural and man-made domains bounded by surfaces that may be rough and/or evolve with time, are turbulent and three dimensional in the mean, and are often complicated due to stratification, multi-phase and mass transfer phenomena. Numerical modelling of such flows poses a grand challenge to even the most advanced CFD models available today. Perhaps the leading modelling challenge stems from the highly turbulent character of such flows and the need to develop reliable and efficient engineering turbulence models that account for these effects.

In principle, Direct Numerical Simulation (DNS) can be used to model turbulent flows free of any modelling uncertainties and errors other than those inherent in the truncation error of the numerical scheme. In DNS, the unsteady, 3D Navier–Stokes equations are solved directly using spatial and temporal resolutions sufficiently fine to resolve the dynamics of the entire spectrum of turbulent eddies in the flow: from the energy-producing largest eddies, whose size is comparable to the flow domain, to the smallest, Kolmogorov scales, at which turbulence energy is dissipated into heat by molecular action. DNS has yielded important insights into the structure of relatively simple, low Reynolds number flows but the computational resources it requires increase dramatically with Reynolds number – the number of grid nodes required for accurate DNS is roughly of order $O(Re^2)$. Consequently, DNS is not a practical modelling alternative for simulating flows at engineering-relevant Reynolds numbers, at least not within the foreseeable future (Spalart, 2000). DNS will instead continue to provide important databases of instantaneous turbulence fields that will guide refinement and development of simpler engineering calculation models.

A more realistic modelling tool, from a computational standpoint, is the so-called Large Eddy Simulation (LES) approach, which has attracted considerable attention during the past decade. In LES, a spatially filtered form of the 3D, unsteady Navier–Stokes equations is solved to resolve directly all scales of motion that are larger than the width of the filter, which is typically equated with the grid spacing. Smaller-scale motions, collectively referred to as sub-grid scales, are typically modelled via a sub-grid turbulence model. The reader is referred to Piomelli (1999) and Chapter 2 in this book for overviews of the LES approach and to Thomas and Williams (1995), Bradbrook *et al.* (2000) and Zedler and Street (2001) for recent applications of LES to hydraulic engineering problems. In spite of encouraging results for complex flows at moderate Reynolds numbers, meaningful LES at Reynolds numbers of engineering interest would still require grid resolutions that are beyond the capabilities of existing and projected computer capacity (Spalart, 2000).

This state of affairs leaves statistical turbulence models that rely on the so-called Reynolds decomposition as the most viable alternative for practical engineering computations. According to the Reynolds decomposition, which was introduced by Osborne Reynolds some 100-years ago (Reynolds, 1895), instantaneous flow quantities are decomposed in terms of a statistically stationary mean value and a zero-mean, random turbulent fluctuation. Governing equations for the mean flow quantities are derived by introducing the Reynolds decomposition into the instantaneous Navier–Stokes equations and applying time averaging. Due to the averaging operation, the resulting system of equations, known as the Reynolds-averaged Navier–Stokes (RANS), is not a closed system as it contains six more unknowns than equations. The unknowns are the so-called Reynolds stresses, which involve time-averaged products of fluctuating velocity components and need to be modelled for closure. In this chapter, common turbulence modelling strategies for the RANS equations will be reviewed with emphasis on issues and challenges that are relevant to hydraulic engineering flows. This chapter is by no means intended to be an exhaustive treatment of the large and rapidly expanding body of literature on statistical turbulence modelling. It is rather intended to introduce the fundamentals of the subject and point readers interested in a more in-depth treatment to relevant recent literature.

The chapter is organized as follows. Section 5.2 presents the governing equations, introduces the concept of Reynolds decomposition, discusses the closure problem of turbulence, and explains the need for turbulence closure models. Section 5.3 reviews the entire hierarchy of statistical turbulence closure models. Isotropic, eddy-viscosity models, including algebraic mixing-length models and two-equation models, are discussed first (section 5.3.1), followed by more advanced, non-isotropic models that resolve the anisotropy of the turbulence stresses (section 5.3.2). Issues related to the implementation of boundary conditions near solid walls and free surfaces are also discussed. Finally, section 5.4 briefly reviews common modelling practices for stratified flows.

5.2 Governing equations, the Reynolds decomposition and the closure problem

Turbulent flow is inherently unsteady, three dimensional, rotational and chaotic. Instantaneous motions in a turbulent flow of an incompressible, Newtonian fluid are governed by the 3D, time-dependent, incompressible Navier–Stokes (N-S) equations. Using compact tensor notation (repeated indices imply summation) the governing equations can be formulated as follows:

Continuity

$$\frac{\partial \tilde{U}_i}{\partial x_i} = 0 \tag{5.1}$$

Momentum

$$\frac{\partial \tilde{U}_i}{\partial t} + \frac{\partial}{\partial x_j}\left(\tilde{U}_i \tilde{U}_j\right) = -\frac{1}{\rho}\frac{\partial \tilde{P}}{\partial x_i} + \frac{1}{\rho}\frac{\partial \tilde{\tau}_{ij}}{\partial x_i} \tag{5.2}$$

where $\tilde{U}_i$ ($i=1, 2, 3$) are the instantaneous velocity components, $\tilde{P}$ is the instantaneous pressure and $\tilde{\tau}_{ij}$ ($i,j = 1, 2, 3$) are the components of the viscous stress tensor. For a Newtonian fluid the stress tensor is related to the rate of strain tensor as follows:

$$\tilde{\tau}_{ij} = \mu \left(\frac{\partial \tilde{U}_i}{\partial x_j} + \frac{\partial \tilde{U}_j}{\partial x_i}\right) \tag{5.3}$$

Statistical turbulence models rely on the Reynolds decomposition to express the instantaneous velocity components and pressure in equations (5.1)–(5.3) into a mean value plus a random, fluctuating part. For statistically stationary turbulent flows the Reynolds decomposition reads as follows:

$$\tilde{U}_i(\vec{x}, t) = U_i(\vec{x}) + u_i(\vec{x}, t) \tag{5.4}$$

$$U_i(\vec{x}) = \lim_{T \to \infty} \frac{1}{T} \int_t^{t+T} \tilde{U}_i(\vec{x}, t) dt \tag{5.5}$$

where U_i is the mean velocity, u_i is the random (zero-mean) turbulent contribution and the averaging interval T is taken to be much greater that the maximum period of the turbulent fluctuations.

By substituting the Reynolds decomposition into the instantaneous N-S equations and applying time averaging (equation 5.5), the RANS equations are obtained as follows:

$$\frac{\partial U_i}{\partial x_i} = 0$$

$$\frac{\partial U_i}{\partial t} + \frac{\partial}{\partial x_j}(U_i U_j) = -\frac{1}{\rho}\frac{\partial P}{\partial x_i} + \frac{1}{\rho}\frac{\partial}{\partial x_i}(\tau_{ij} - \rho\overline{u_i u_j}) \quad (5.6)$$

The nine quantities $-\rho\overline{u_i u_j}$ express the mean rate of transport of momentum by turbulent velocity fluctuations and comprise the so-called Reynolds stress tensor. Since $u_i u_j$ equals $u_j u_i$, the Reynolds stress tensor is symmetric and, thus, has only six independent components. Hence, the Reynolds-averaging of the N-S equations introduced six new unknowns into the governing equations.

Transport equations for these six new unknowns can be derived by taking moments of the instantaneous N-S equations and applying time-averaging:

$$\underbrace{\frac{\partial \overline{u_i u_j}}{\partial t} + U_k \frac{\partial \overline{u_i u_j}}{\partial x_k}}_{(a)} = \underbrace{P_{ij}}_{(b)} + \underbrace{\varepsilon_{ij}}_{(c)} - \underbrace{\Phi_{ij}}_{(d)} + \frac{\partial}{\partial x_k}\left(\underbrace{\nu \frac{\partial \overline{u_i u_j}}{\partial x_k}}_{(e)} + \underbrace{C_{ijk}}_{(f)}\right) \quad (5.7)$$

The various terms in the above equations can be interpreted as follows:

(a) *Convection* of Reynolds stresses by the mean flow.
(b) *Production* of turbulence by mean shear. This term sustains turbulence by extracting energy from the mean flow and is given as follows:

$$P_{ij} = -\overline{u_i u_k}\frac{\partial U_j}{\partial x_k} - \overline{u_j u_k}\frac{\partial U_i}{\partial x_k} \quad (5.8)$$

(c) *Viscous dissipation* of Reynolds stresses (six new unknowns)

$$\varepsilon_{ij} = 2\mu \overline{\frac{\partial u_i}{\partial x_k}\frac{\partial u_j}{\partial x_k}} \quad (5.9)$$

(d) *Pressure-redistribution* term (six new unknowns). This term acts to drive the turbulence toward an isotropic state by redistributing the Reynolds stresses.

$$\Phi_{ij} = \overline{p\left(\frac{\partial u_i}{\partial x_j} + \frac{\partial u_j}{\partial x_i}\right)} \quad (5.10)$$

(e) *Diffusion* of Reynolds stresses via molecular mixing.
(f) *Turbulent diffusion* (ten new unknowns) due to velocity and pressure fluctuations

$$C_{ijk} = \overline{\rho u_i u_j u_k} + \overline{p u_i}\delta_{jk} + \overline{p u_j}\delta_{ik} \quad (5.11)$$

Terms (a), (b) and (e) contain only mean velocity components and the Reynolds stresses and, thus, do not require modelling when equation (5.7) is used to close the mean flow equation (5.6). Terms (c), (d) and (f), on the other hand, introduce 22 new unknowns into the governing equations. Because of the nonlinearity of the N-S equations, the higher moments we take to derive transport equations for unknowns such as those given by equations (5.9)–(5.11), the more new unknowns we generate. At no point of this process will we have enough equations for all the unknowns. This is the so-called *closure problem of turbulence*, whose solution necessitates the development of turbulence closure models.

5.3 Turbulence models for the RANS equations

Two general closure strategies are typically adopted to develop practical turbulence models for engineering calculations. In the first strategy, the RANS equations (5.6) are closed by employing the so-called eddy-viscosity concept to express the components of the Reynolds stress tensor in terms of the gradients of the mean velocity field via a "turbulence" constitutive law. Models in this category are typically denoted as eddy-viscosity models. Depending on the form of the "turbulence" constitutive law such models adopt, they can be classified either as isotropic (linear) or non-isotropic (non linear) eddy-viscosity models. Depending on the approach adopted to determine the eddy viscosity, on the other hand, such models can be classified as either algebraic or differential models. The second modelling strategy relies on Second-Moment Closure (SMC) models or Reynolds-Stress Transport (RST) models. In such models the components of the Reynolds stresses needed to close the RANS equations (5.6) are obtained by solving modelled versions of the RST equations (5.7). In principle, RST models comprise a more physics-based turbulence modelling framework than their eddy viscosity counterparts at the expense of being more cumbersome to implement numerically and computationally more expensive.

5.3.1 Isotropic eddy-viscosity models

Following Boussinesq's assumption, the components of the Reynolds stress tensor are assumed to vary linearly with the mean rate of strain tensor as follows:

$$-\overline{u_i u_j} = \nu_t \left(\frac{\partial U_i}{\partial x_j} + \frac{\partial U_j}{\partial x_i} \right) - \frac{2}{3} k \delta_{ij} \qquad (5.12)$$

where k is the turbulence kinetic energy ($\equiv \overline{u_i u_i}/2$), δ_{ij} is the unit tensor (Kronecker's delta), and ν_t is the so-called eddy viscosity. The underlying assumption behind equation (5.12) is that the transport of momentum by turbulent fluctuations is similar to that via random molecular motion in laminar flows. It is important to emphasize that the eddy-viscosity concept is purely phenomenological and has no mathematical basis. Unlike its molecular counterpart (equation 5.3), which is a property of the fluid, the eddy viscosity is a property of the flow and, in general, it should depend on flow quantities and vary in space and time.

Eddy-viscosity models express ν_t as the product of a turbulence length scale, ℓ_t, and a turbulence velocity scale, u_t:

$$\nu_t = \ell_t u_t \tag{5.13}$$

Depending on the approach adopted to calculate these scales, we can classify isotropic eddy-viscosity models in three general categories: (1) *Zero-equation* or algebraic eddy-viscosity models, which specify both scales in terms of an explicit algebraic relation; (2) *One-equation* models, which employ one additional Partial Differential Equation (PDE) for the velocity scale and specify the length scale algebraically; and (3) *Two-equation* models, which employ one PDE for the velocity scale and one PDE for the length scale. The following sections will review briefly the zero-equation models and focus on two representative two-equation models, which are among the most widely used turbulence models in engineering practice today.

5.3.1.1 Algebraic (zero-equation) models

Algebraic models rely on Prandtl's famous *mixing-length* hypothesis. By drawing an analogy with the molecular momentum transport process and replacing the molecular thermal velocity and mean free path with characteristic turbulent velocity and length scales, Prandtl proposed the following model for the shear Reynolds stress:

$$-\overline{u_1 u_2} = \underbrace{\ell_t^2 \left| \frac{dU_1}{dx_2} \right|}_{\nu_t} \frac{dU_1}{dx_2} \tag{5.14}$$

Equation (5.14) is valid only for a simple 2D boundary layer flow where $-\overline{u_1 u_2}$ is the primary Reynolds stress. For 3D flows, the mixing-length model, used in conjunction with equation (5.12), can be generalized as follows:

$$\nu_t = 2\ell_t^2 \sqrt{S_{ij} S_{ij}}, \quad S_{ij} = \frac{1}{2}\left(\frac{\partial U_j}{\partial x_i} + \frac{\partial U_i}{\partial x_j}\right) \tag{5.15}$$

The mixing-length, ℓ_t, is an empirical quantity that is typically assumed to be proportional to some characteristic length scale of the flow and needs to be specified using input from experiments.

The mixing-length model is attractive because of its simplicity and computational expedience. The mixing-length assumption, however, is valid only when the turbulence is in local equilibrium and, thus, the model is suitable for very simple, 2D in-the-mean flows.

5.3.1.2 Two-equation models

Two-equation models are the simplest complete turbulence models in the sense that they can be used to predict a given turbulent flow without requiring prior empirical input about the structure of turbulence – as do, for example, the algebraic or one-equation models, which require the specification of the mixing length or the turbulence length scale empirically. As implied by their name, two-equation models employ two additional PDEs to calculate the turbulence velocity and/or length scales. Most such models employ the same transport equation for the Turbulence Kinetic Energy (TKE) to calculate a local turbulence velocity scale ($u_t \approx k^{1/2}$).

The exact TKE transport equation can be derived by contracting the RST equations (5.7). Three terms arise in the resulting equation that require modelling: (1) the production of TKE by mean shear; (2) the diffusion of TKE by pressure fluctuations; and (3) the diffusion of TKE by velocity fluctuations. The first term can be closed by adopting the Boussinesq hypothesis, equation (5.12), while the last two terms are typically lumped together and modelled via a gradient-diffusion model (Jones and Launder, 1972; Durbin and Pettersson Reif, 2001). The final modelled equation for the TKE reads as follows:

$$\frac{\partial k}{\partial t} + U_m \frac{\partial k}{\partial x_m} = P_k - \varepsilon + \frac{\partial}{\partial x_m}\left[\left(\nu + \frac{\nu_t}{\sigma_k}\right)\frac{\partial k}{\partial x_m}\right] \quad (5.16)$$

where P_k is the production of TKE by mean shear:

$$P_k = \nu_t \left(\frac{\partial U_i}{\partial x_\ell} + \frac{\partial U_\ell}{\partial x_i}\right)\frac{\partial U_i}{\partial x_\ell} \quad (5.17)$$

ε is the rate of viscous dissipation of the TKE and σ_k is a closure constant. What differentiates the various two-equation closures is the variable that is used to determine the dissipation, ε, or equivalently the turbulence length scale. The most commonly used such variables are: (a) the actual rate of dissipation ε; and (b) the specific dissipation rate, $\omega (=\varepsilon/k$, dissipation per unit kinetic energy).

The k-ε model

The most popular and widely used turbulence model is the so-called standard k-ε model. The model, which was originally proposed by Jones and Launder (1972) and later retuned by Launder and Sharma (1974), employs the modelled TKE equation (5.16) in conjunction with a transport equation for the rate of viscous dissipation ε. The exact ε-equation, however, turns out to be far more complicated than the exact TKE equation as it involves several new unknown double and triple correlations of fluctuating velocity components that are very difficult, if not impossible, to measure. Consequently, *the modelled ε-equation is based entirely on empirical physical reasoning and dimensional arguments*. The equations of the standard k-ε model read as follows:

$$\begin{aligned}\frac{\partial k}{\partial t} + U_m \frac{\partial k}{\partial x_m} &= P_k - \varepsilon + \frac{\partial}{\partial x_m}\left[\frac{\nu_t}{\sigma_k}\frac{\partial k}{\partial x_m}\right] \\ \frac{\partial \varepsilon}{\partial t} + U_m \frac{\partial \varepsilon}{\partial x_m} &= \frac{\varepsilon}{k}(C_{\varepsilon 1} P_k - C_{\varepsilon 2}\varepsilon) + \frac{\partial}{\partial x_m}\left[\left(\nu + \frac{\nu_t}{\sigma_\varepsilon}\right)\frac{\partial \varepsilon}{\partial x_m}\right]\end{aligned} \quad (5.18)$$

$$\nu_t = C_\mu \left(k^{1/2}\right)\left(\frac{k^{3/2}}{\varepsilon}\right) = C_\mu \frac{k^2}{\varepsilon} \quad (5.19)$$

The "standard" values of the constants in the above equations are given as follows (Launder and Sharma, 1974):

$$C_{\varepsilon 1} = 1.44, \quad C_{\varepsilon 2} = 1.92, \quad C_\mu = 0.09, \quad \sigma_k = 1.0, \quad \sigma_\varepsilon = 1.3 \qquad (5.20)$$

The above k-ε model equations are valid only sufficiently far from solid boundaries. Due to the highly empirical nature of the ε-equation, equation (5.18), the model equations become ill-conditioned ("blow-up") as the wall is approached and need to be modified to account for the effects of the wall on the local structure of turbulence as well as to allow stable and robust implementation of the model in numerical computations (Durbin and Pettersson Reif, 2001). There are three primary effects imparted by the wall on the structure of turbulence: (1) turbulent eddies are distorted and constrained in size, being compressed in the wall-normal direction and elongated in the streamwise direction; (2) production of turbulence increases due to the no-slip condition and intensification due to stretching of fluctuating vorticity; and (3) turbulence energy is damped and dissipated into heat via viscous action.

One approach to incorporate near-wall effects in the k-ε model is to modify the model equations so that they reproduce wall effects and are numerically well behaved in the vicinity of the wall. A k-ε model that is valid all the way to the wall will be denoted as a near-wall model or a low Reynolds number model – the term Reynolds number in this context refers to a local Reynolds number based on turbulence length and velocity scales (section "Reynolds-stress modelling near walls"), which approaches zero inside the laminar sublayer. Adopting a low Reynolds number modelling approach necessitates the generation of a very fine mesh in the vicinity of the wall so that adequate resolution of the laminar sublayer is achieved. The governing equations can then be integrated all the way to the wall where no-slip conditions for the mean velocity components and the Reynolds stresses can be applied. Several low Reynolds number versions of the k-ε model have been proposed in the literature and the reader is referred to Patel *et al.* (1984) for a comprehensive review. Such models, however, have not found wide use in hydraulic engineering simulations, presumably because they increase numerical stiffness, due to both the very fine mesh resolution near the wall and the presence of rapidly varying source terms in the model equations, and reduce robustness. Another limitation of such models, which is critical in hydraulic engineering problems, is that no straightforward approach has yet been determined for incorporating wall roughness effects.

An alternative to low Reynolds number modelling is the so-called two-layer approach. The flow domain is divided into two zones: the inner layer – which includes the sublayer, the buffer layer and a portion of the fully turbulent region; and the outer layer – which encompasses the remainder of the flow domain. The standard k-ε model, equations (5.16)–(5.19), is used in the outer layer. In the inner layer only the k-equation is solved while the ill-posed ε-equation is replaced by an algebraic equation that prescribes the turbulence length scale in terms of k, the distance from the wall, and the viscosity of the fluid (Chen and Patel, 1988). In general, two-layer models are numerically more robust than low Reynolds number k-ε models. Their application, however, in geometrically complex domains is hindered by the need to determine the location of the interface between the inner and outer layers, which depends on the local wall shear stress and is not known *a priori* (see Sotiropoulos and Patel, 1995b, for application of a two-layer model to complex flows). Moreover, attempts to implement roughness effects in the two-layer model, by making the length scale equation used in the inner layer dependent upon the roughness length scale, have not been very successful (Yoon and Patel, 1993).

The most common approach for avoiding the previously discussed modelling difficulties near the wall and accounting for wall effects is to adopt *wall functions*. Since the structure of turbulence near the wall is dominated by the wall itself, it is assumed that the near-wall flow depends on local (inner) variables. Dimensional analysis arguments supported by experiments lead to the so-called law of the wall:

$$U^+ \equiv \frac{U}{u_\tau} = F\left(\frac{u_\tau y}{\nu}\right) \quad \text{or} \quad U^+ = F(y^+) \tag{5.21}$$

where y is the distance from the wall, u_τ is the friction or wall shear velocity, $u_\tau = \sqrt{\tau_w/\rho}$, and τ_w is the wall shear stress. Note that the non-dimensional variable y^+ can be interpreted as the ratio of the distance from the wall, y, to the viscous length scale, ν/u_τ. Experiments in pipes, flat plate boundary layers and wide open channels indicate that F has a three-layer structure:

$$U^+ = \begin{cases} y^+ & \text{in the sublayer, } y^+ < 5 \\ y^+ - e^{-\kappa B}\left\{e^{-\kappa U^+} - 1 - \kappa U^+ - \frac{(\kappa U^+)^2}{2} - \frac{(\kappa U^+)^3}{2}\right\} & \text{in the buffer layer, } 5 < y^+ < 60 \\ \frac{1}{\kappa}\ln(y^+) + B & \text{in the fully turbulent layer, } y^+ > 60 \end{cases}$$
(5.22)

where $\kappa = 0.418$ and $B = 5.45$. The wall functions approach, which was proposed by Launder and Spalding (1974), assumes that for all flows the near-wall velocity profile obeys the above law. Rather than integrating the governing equations through the laminar sublayer and applying boundary conditions at the wall, the computational grid is constructed such that the first grid point off the wall is located within the fully turbulent region, typically $y^+ > 60$. Boundary conditions are applied at this first layer of nodes off the wall by assuming that the mean velocity field is logarithmic. Boundary conditions for k and ε are obtained by assuming equilibrium between turbulence production and energy dissipation. These assumptions finally result in the following set of boundary conditions:

$$U^+ = \frac{1}{\kappa}\ln(y^+) + B, \quad k^+ = \frac{k}{u_\tau^2} = \frac{1}{\sqrt{C_\mu}} = 3.33 \quad \text{and} \quad \varepsilon^+ \equiv \frac{\nu\varepsilon}{u_\tau^4} = \frac{1}{\kappa y^+} \tag{5.23}$$

Equations (5.23) are valid for a hydraulically smooth surface and need to be modified for a rough wall. In the presence of roughness, the turbulence structure near the wall depends on the viscous length scale ν/u_τ, and the roughness length scale k_s. For a rough bed composed of uniform sand grains densely packed together, the grain height can be used to define k_s. For characterizing roughness other than sand grains, however, k_s is taken to be an *equivalent sand grain* height. These two parameters can be combined into a single non-dimensional quantity:

$$k_s^+ = \frac{u_\tau k_s}{\nu} \tag{5.24}$$

which expresses the ratio of the roughness to the viscous length scales and is sometimes referred to as the roughness Reynolds number. This non-dimensional parameter can be used to classify the near-wall flow in three regimes: (1) hydraulically smooth ($k_s^+ < 5$); (2) incompletely rough ($5 < k_s^+ \leq 70$); and (3) fully rough ($k_s^+ < 70$).

Experiments (Nikuradse, 1933; Tani, 1987) have shown that the presence of roughness modifies the law of the wall as follows:

$$U^+ = \frac{1}{\kappa}\ln(y^+) + B - \Delta B \qquad (5.25)$$

where ΔB is the so-called velocity shift, which, following Cebeci and Bradshaw (1977), depends on k_s^+ as follows:

$$\Delta B = \begin{cases} 0 & \text{for } k_s^+ < 2.25 \\ \left[B - 8.5 + \frac{1}{\kappa}\ln(k_s^+)\right]\sin\{0.4258[\ln(k_s^+) - 0.811]\} & \text{for } 2.25 \leq k_s^+ \leq 90 \\ B - 8.5 + \frac{1}{\kappa}\ln(k_s^+) & \text{for } k_s^+ \geq 90 \end{cases}$$

(5.26)

A unique difficulty encountered in the numerical modelling of natural river flows is that the size distribution of roughness elements, i.e. the value of k_s^+ at every point on the river bed, is heterogeneous and not known *a priori*. The same difficulty is also encountered in laboratory-scale models, which require careful calibration with field measurements to ensure that the model roughness distribution corresponds to that in the field. Sinha *et al.* (1998) developed a novel and simple-to-implement algorithm for calibrating the roughness distribution in the numerical model. Rather than assuming a homogeneous equivalent roughness height throughout the river reach, they divided the reach into three zones of varying roughnesses based on experience gained during the calibration of the laboratory model. The equivalent roughness height in each zone was determined iteratively via the following numerical calibration procedure. The calculation starts assuming hydraulically smooth boundary conditions everywhere. The so-computed near-bed velocity magnitude is compared with laboratory measurements throughout the reach and a local velocity deficit ΔB is calculated. With ΔB known, equations (5.26) are solved via the Newton–Raphson scheme to determine the local k_s^+ value. With the equivalent roughness height now given, the governing equations can be solved using the wall functions approach described earlier to determine a new velocity field. The computed velocity field near the bed is once again compared to the available measurements to calculate a new ΔB distribution, which in turn is used to further adjust the k_s^+ values and so on (see Sinha *et al.*, 1998, for the numerical implementation details).

The wall functions approach is widely used in engineering calculations because: (1) it eliminates the need for very fine computational meshes that are required for direct integration of the governing equations all the way to the wall; (2) it alleviates numerical stiffness due to stiff source terms in the turbulence closure equations and large aspect ratio meshes; and (3) it is the only feasible approach for simulations of complex engineering flows at Reynolds numbers of practical interest. In spite of its computational expedience, however, this approach has disadvantages, which should be kept in mind when applying wall functions to calculate complex, 3D flows. First, strictly speaking, wall functions are valid only for "simple" equilibrium flows such as flows in a straight pipe and past flat plates. The standard law-of-the-wall breaks down in flows with strong pressure gradients, flow reversal, streamline curvature, streamline convergence and divergence, swirl and rotation, and highly 3D flows with complex secondary motions. Second, extension to three dimensions is ambiguous and highly empirical. Third, the location and extent of the logarithmic region is not known *a priori* in complex flows. And, fourth, the numerical solutions could be sensitive to the location of the first grid node off the wall where wall functions are applied.

In an open channel simulation, boundary conditions for k and ε are also required at the free surface (fs). Like a solid wall, the free surface causes the flattening of the turbulent eddies, thus acting to diminish velocity fluctuations in the vertical direction. There is, however, a significant difference between the structure of turbulence near a wall and that near a free surface: in the latter case, the mean velocity gradients are very small and, thus, the rate of production of k near the free surface is negligible. The simplest and most common set of boundary conditions can be derived by treating the free surface as a rigid plane of symmetry (the so-called rigid lid assumption) and k and ε can be obtained by setting their normal derivatives equal to zero:

$$\left.\frac{\partial k}{\partial z}\right|_{\text{fs}} = \left.\frac{\partial \varepsilon}{\partial z}\right|_{\text{fs}} = 0 \qquad (5.27)$$

A more sophisticated treatment was proposed by Rodi (1993) who suggested to specify the value of ε at the free surface as follows:

$$\varepsilon_{\text{fs}} = 2.32 \frac{k_{\text{fs}}^{3/2}}{h} \qquad (5.28)$$

where k_{fs} is the free surface value of k obtained from the zero-gradient condition. This boundary condition yields a higher value of ε_{fs} than that obtained from the zero-gradient condition. The higher rate of dissipation reduces the turbulence length scale, thus attempting to mimic the damping of turbulent fluctuations by the surface. A different approach to account for the same effect is to multiply the value of k_{fs} determined by the zero-gradient condition by a damping function (Nezu and Nakagawa, 1993). This treatment also reduces the eddy viscosity as the surface is approached.

The predictive capabilities of the standard k-ε model with wall functions in hydraulic engineering applications have been demonstrated in a number of studies (Demuren and Rodi, 1986; Demuren, 1993; Sinha *et al.*, 1998; Meselhe and Sotiropoulos, 2000; Wu *et al.*, 2000). Experience gained from these and other studies suggests that the model performs well for moderately curved open channels of complex bathymetry. For flows with strong secondary motions and vortices, such as flows in strongly curved channels, wall functions tend to underpredict the strength of the secondary motion (Sotiropoulos and Patel, 1995a; Meselhe and Sotiropoulos, 2000). In such cases, models that resolve the details of the near-wall flow and/or account for anisotropic effects may be required for reliable predictions (Sotiropoulos and Patel, 1995a).

The k-ω *model*

The ω-equation was originally proposed by Kolmogorov (1942) who employed similar physical reasoning and dimensional arguments as those involved in the derivation of the ε-equation. The modern version of the k-ω model was proposed by Wilcox (1988).

$$\frac{\partial k}{\partial t} + U_\ell \frac{\partial k}{\partial x_\ell} = \nu_t \left(\frac{\partial U_i}{\partial x_\ell} + \frac{\partial U_\ell}{\partial x_i}\right) \frac{\partial U_i}{\partial x_\ell} + \frac{\partial}{\partial x_\ell}\left[(\nu + \sigma^* \nu_t)\frac{\partial k}{\partial x_\ell}\right] - \beta^* k \omega \qquad (5.29)$$

$$\frac{\partial \omega}{\partial t} + U_\ell \frac{\partial \omega}{\partial x_\ell} = \alpha \frac{\omega}{k} \nu_t \left(\frac{\partial U_i}{\partial x_\ell} + \frac{\partial U_\ell}{\partial x_i}\right) \frac{\partial U_i}{\partial x_\ell} - \beta \omega^2 + \frac{\partial}{\partial x_\ell}\left[(\nu + \sigma \nu_t)\frac{\partial \omega}{\partial x_\ell}\right] \qquad (5.30)$$

$$\nu_t = \frac{k}{\omega} \qquad (5.31)$$

$$\alpha = \frac{5}{9}, \quad \beta = \frac{3}{40}, \quad \beta^* = \frac{9}{100}, \quad \sigma = \frac{1}{2}, \quad \sigma^* = \frac{1}{2}$$

A more recent version of the model that improves the prediction of the turbulence statistics in the buffer layer and the laminar sublayer has been proposed by Wilcox (1993).

Unlike the ε-equation, the ω-equation is well posed inside the laminar sublayer and, thus, the model equations can be integrated all the way to the wall. A difficulty with implementation of the model arises, however, because of the asymptotic behaviour of ω near a solid boundary, which can be shown to be as follows:

$$\omega \approx \frac{2\nu}{\beta^* y^2} \quad \text{as} \quad y \to 0 \qquad (5.32)$$

Therefore, at the wall $\omega \to \infty$. To circumvent this singularity, Wilcox (1994) suggested generating a very fine mesh in the near-wall region, with 7–10 points within the sublayer ($y^+ < 2.5$), and explicitly imposing that these nodes obey equation (5.32) instead of solving the ω-equation. The main advantage of the ω variable is that its wall value can be calibrated to account for wall roughness effects. Wilcox (1988) showed that effects of sand grain roughness of height k_s can be adequately accounted for by using the following boundary conditions for ω at $y = 0$:

$$\omega_{\text{wall}} = \begin{cases} \frac{2500\nu}{k_s^2} & \text{for } k_s^+ < 25 \\ \frac{100 u_\tau}{k_s} & \text{for } k_s^+ \geq 25 \end{cases} \qquad (5.33)$$

Equations (5.33) also lead to a very convenient and efficient approach for specifying wall boundary conditions for ω at smooth walls by assuming the wall to be *slightly rough*. Menter (1993) has shown that the smooth wall flow characteristics are correctly reproduced if the first grid node off the wall at $y^+ \approx 3$ and ω at $y = 0$ is set as follows:

$$\omega = \frac{60\nu}{\beta(\Delta y)^2} \tag{5.34}$$

where Δy is the spacing of the first grid node off the wall. Equation (5.34) is very useful for complex 3D calculations and in essence allows the use of the k-ω model for such simulations. The approach suggested by Wilcox, as discussed above, required 7–10 points within $y^+ < 2.5$, which results in prohibitively fine computational meshes. For applications of the k-ω model to hydraulic engineering flows, the reader is referred to Yoon and Patel (1993, 1996), Yoon et al. (1996), Neary et al. (1999), Constantinescu and Patel (2000), and Chrisohoides et al. (2003).

5.3.2 Non-isotropic turbulence models

For the most part, turbulence models based on the Boussinesq approximation, equation (5.12), work reasonably well for 3D flows with secondary motions driven by pressure gradients in the flow – curved open channels, meandering rivers, etc. Such secondary motions can be quite intense, often becoming as large as 20–30% of the streamwise flow, occur in both laminar and turbulent flows, and were termed by Prandtl as *secondary currents of the first kind* (Bradshaw, 1987). Secondary currents are also present in straight open channels. These are called *secondary currents of the second kind* and exist only in turbulent flows. Their magnitude is usually small, typically 1–5% of the bulk flow, but they can have a profound effect on distributing momentum, heat, and sediment within the channel cross section. Secondary currents of the second kind are driven by the anisotropy of the normal Reynolds stresses and strong spatial gradients of the Reynolds shear stresses (Bradshaw, 1987). In open channel flows these currents are known to give rise to a number of complex flow features (Nezu and Nakagawa, 1993). In narrow open channels, for instance, the well-known *velocity dip* phenomenon, where the maximum mean streamwise velocity in the cross section occurs below the free surface, is the result of transverse transport of streamwise momentum by secondary currents of the second kind driven by the free surface-induced anisotropy of the Reynolds stresses (Nezu and Nakagawa, 1993). Heterogeneous bed roughness (Naot, 1984), rapid lateral changes of the channel cross section, as in compound channels in flood plains (Naot et al., 1993; Sofialidis and Prinos, 1999) and even small spanwise disturbances in wide open channels (Colombini, 1993) tend to enhance turbulence anisotropy and can give rise to multicellular secondary currents and the formation of complex bed forms, such as sand ridges and troughs. Models based on the Boussinesq assumption cannot reproduce the correct anisotropy of the Reynolds stresses and have been shown to fail to resolve even qualitatively such complex flow features (Demuren and Rodi, 1984). Recent work has further shown that Boussinesq models also fail in flows involving intense secondary motions of the first kind, such as flows in strongly curved ducts and channels (Sotiropoulos and Patel, 1995a; Sotiropoulos and Ventikos, 1998. In such flows, the strong three dimensionality of the mean flow tends to intensify the Reynolds stresses, which in turn affect the development of the secondary motion.

This section briefly describes turbulence models that do not adopt the linear Boussinesq model given by equation (5.12). Such models are denoted as non-Boussinesq models and can be classified into two broad categories: (1) Eddy-viscosity models with nonlinear constitutive relations; and (2) Reynolds-stress models.

5.3.2.1 Nonlinear eddy-viscosity models

Nonlinear eddy-viscosity models assume the Boussinesq approximation to be only the leading term in a series expansion of functionals. That is, the expression for the Reynolds stresses consists of the Boussinesq terms given in equation (5.12) plus additional higher-order terms:

$$-\overline{u_i u_j} = -\frac{2}{3} k \delta_{ij} + \nu_t S_{ij} + f\left[(S_{ij})^n;\ (\Omega_{ij})^n;\ k;\ \varepsilon\right] \qquad (5.35)$$

where

$$S_{ij} = \frac{\partial U_i}{\partial x_j} + \frac{\partial U_j}{\partial x_i} \quad \text{and} \quad \Omega_{ij} = \frac{\partial U_i}{\partial x_j} - \frac{\partial U_j}{\partial x_i} \qquad (5.36)$$

and n is an exponent greater than one. Such a constitutive relation is usually employed in conjunction with one of the two-equation models we discussed above to obtain the necessary velocity and length scales. What differentiates the various models is the selection of the functional f. Two of the most popular constitutive relations are presented below.

Speziale (1987) proposed a constitutive relation which is quadratic in the mean rate of strain and incorporates memory effects through a generalized time derivative of the strain rate tensor. The f functional for this model reads as follows:

$$f = C_D C_\mu \frac{k^2}{\varepsilon} \left(S_{ij} S_{kj} - \frac{1}{3} S_{mn} S_{mn} \delta_{ij} \right) + 2 C_E C_\mu^2 \frac{k^3}{\varepsilon^2} \left(\overset{\circ}{S}_{ij} - \frac{1}{3} \overset{\circ}{S}_{mm} \delta_{ij} \right) \qquad (5.37)$$

where $\overset{\circ}{S}_{ij}$ is the frame-indifferent Oldroyd derivative of S_{ij}

$$\overset{\circ}{S}_{ij} = \frac{\partial S_{ij}}{\partial t} + U_k \frac{\partial S_{ij}}{\partial x_k} - \frac{\partial U_i}{\partial x_k} S_{kj} - \frac{\partial U_j}{\partial x_k} S_{ki} \qquad (5.38)$$

and $C_D = C_E = 1.68$. Colombini (1993) employed this model in conjunction with the k-ε model to investigate the role of secondary currents of the second kind in the formation of sand ridges in wide open channels.

A cubic model that depends both on the strain rate and rotation tensors was proposed by Craft *et al.* (1995).

$$f = \nu_t \left\{ \frac{k}{\varepsilon} \left[c_1 \left(S_{ik}S_{kj} - \frac{1}{3} S_{pq}S_{pq} \delta_{ij} \right) + c_2 \left(\Omega_{ik}S_{kj} + \Omega_{jk}S_{ki} \right) \right. \right.$$

$$\left. + c_3 \left(\Omega_{ik}\Omega_{kj} - \frac{1}{3} \Omega_{pq}\Omega_{pq} \delta_{ij} \right) \right] - \left(\frac{k}{\varepsilon} \right)^2 \left[c_4 C_\mu \left(S_{ki}\Omega_{lj} + S_{kj}\Omega_{li} - \frac{2}{3} S_{km}\Omega_{lm} \delta_{ij} \right) S_{kl} \right.$$

$$\left. \left. + c_6 C_\mu S_{ij}S_{kl}S_{kl} + c_7 C_\mu S_{ij}\Omega_{kl}\Omega_{kl} \right] \right\}$$

(5.39)

The coefficients $c_1, c_2, \ldots, c_7$ and C_μ are very complicated functions of invariants of the mean rate of strain and rotation tensors and the Reynolds-stress anisotropy tensor ($a_{ij} = \overline{u_i u_j}/k - 2\delta_{ij}/3$). The reader is referred to Craft *et al.* (1995) and Sofialidis and Prinos (1999) for complete models using this constitutive relation. A comprehensive evaluation of various linear and nonlinear, two-equation models for a flow in a strongly curved bend can be found in Sotiropoulos and Ventikos (1998).

5.3.2.2   *Reynolds-stress models*

Full Reynolds-stress transport (RST) models

In RST models the components of the Reynolds-stress tensor are calculated directly by solving transport equations for the individual Reynolds stresses, equations (5.7). The major difficulty associated with RST models is that in order to achieve closure at the second-moment level, all processes involving interactions among fluctuating quantities (turbulence and pressure diffusion, C_{ijk}, pressure redistribution, Φ_{ij}, and viscous dissipation, ε_{ij}) need to be modelled in terms of the mean rate of strain and the Reynolds stresses. A brief overview of common modelling practices for these terms is given below.

The diffusion tensor C_{ijk}
In modelling the diffusion tensor, the pressure diffusion terms are not explicitly modelled and are either neglected or absorbed in the model for turbulent diffusion. Daly and Harlow (1970) proposed a very simple gradient-diffusion model, which in compact tensor form reads as:

$$-\overline{u_i u_j u_k} = c_s \frac{k}{\varepsilon} \overline{u_k u_\ell} \frac{\partial \overline{u_i u_j}}{\partial x_\ell}$$

(5.40)

A more complicated expression, also based on the gradient-diffusion hypothesis, was proposed by Hanjalic and Launder (1972).

The re-distribution tensor
The re-distribution tensor Φ_{ij} is typically modelled as the sum of a turbulent or "slow" part ($\Phi_{ij,1}$), a mean-strain or "rapid" part ($\Phi_{ij,2}$) and a "wall-reflection" part (Φ_{ij}^w) as follows:

$$\Phi_{ij} = \Phi_{ij,1} + \Phi_{ij,2} + \Phi_{ij}^w$$

(5.41)

The slow part represents the effect of fluctuating quantities and is modelled, following Rotta (1951), as a linear function of the anisotropy tensor a_{ij} so that it always tends to equalize the normal stresses and diminish the shear stresses:

$$\Phi_{ij,1} = -C_1 \varepsilon \left(\frac{\overline{u_i u_j}}{k} - \frac{2}{3} \delta_{ij} \right) \qquad (5.42)$$

The second ("rapid") term arises from the mean rate of strain and its interaction with the turbulence. The simplest model for this term was proposed by Naot *et al.* (1970):

$$\Phi_{ij,2} = C_2 \left(P_{ij} - \frac{2}{3} \delta_{ij} P \right) \qquad (5.43)$$

A more general expression was given by Launder *et al.* (1975). The wall-reflection term Φ_{ij}^w accounts for the effect of pressure fluctuations at the wall or free surface, which tends to impede the transfer of turbulence energy from the streamwise to the wall-normal (free surface) components. Gibson and Launder (1978) proposed to account for these effects by including wall-reflection or wall-echo terms for both the slow and rapid parts of the redistribution tensor:

$$\Phi_{ij}^w = \Phi_{ij,1}^w + \Phi_{ij,2}^w$$

$$\Phi_{ij,1}^w = C_1' \frac{\varepsilon}{k} \left(\overline{u_\ell u_m} n_\ell n_m \delta_{ij} - \frac{3}{2} \overline{u_\ell u_i} n_\ell n_j - \frac{3}{2} \overline{u_\ell u_j} n_\ell n_i \right) f \qquad (5.44)$$

$$\Phi_{ij,2}^w = C_2' \frac{\varepsilon}{k} \left(\Phi_{\ell m,2} n_\ell n_m \delta_{ij} - \frac{2}{3} \Phi_{\ell i,2} n_\ell n_j - \Phi_{\ell j,2} n_\ell n_i \right) f \qquad (5.45)$$

where n_i ($i = 1, 2, 3$) are the components of the unit vector normal to the wall. The function f depends on a generalized distance from the wall (Naot and Rodi, 1982). When two walls intersect to form a corner, there exist a number of alternative treatments to account for their combined damping effects. One can assume, for instance, that the combined damping of the two walls is equal to the damping exerted by each wall alone. This requires the definition of two damping functions f_y and f_z as follows (flow is along the x-direction):

$$f_y = \frac{L_y^2}{\langle y \rangle^2} \qquad f_z = \frac{L_z^2}{\langle z \rangle^2} \qquad (5.46)$$

where $\langle y \rangle$ and $\langle z \rangle$ are the generalized wall-distance functions (Naot and Rodi, 1982) and L_y and L_z are turbulence length scales defined as (Cokljat and Younis, 1995):

$$L_y = \frac{1}{\kappa} \left| \frac{\overline{uv}}{k} \right|_{\text{wall}}^{3/2} \frac{k^{3/2}}{\varepsilon} \qquad L_z = \frac{1}{\kappa} \left| \frac{\overline{uw}}{k} \right|_{\text{wall}}^{3/2} \frac{k^{3/2}}{\varepsilon} \qquad (5.47)$$

The effects of the free surface are accounted for by introducing another damping function f_{fs}, as follows (Naot and Rodi, 1982):

$$f_{\text{fs}} = \frac{L^2}{[\langle y_{\text{fs}}\rangle + 0.16L]^2} \tag{5.48}$$

where

$$L = \frac{C_\mu^{3/4}}{\kappa}\frac{k^{3/2}}{\varepsilon}$$

and y_{fs} is the generalized distance from the free surface. For the general case, including both multiple walls and a free surface, the following damping function is employed:

$$f = f_y + f_z + f_{\text{fs}} \tag{5.49}$$

It should be apparent from the above discussion that due to the dependence of equations (5.44) on wall proximity (distance from the wall) and orientation (normal vectors) parameters, its implementation in complex, 3D, multi-connected geometries is not only cumbersome but also arbitrary. For that reason there has been considerable recent work aimed at developing models which account for wall-reflection effects without involving proximity and orientation parameters. The reader is referred to Craft (1998), Launder and Li (1994), Durbin (1993) and Speziale et al. (1991) for such models.

The dissipation rate tensor
The dissipation rate tensor, ε_{ij}, is typically modelled by invoking the assumption of local isotropy:

$$\varepsilon_{ij} = \frac{2}{3}\varepsilon\delta_{ij} \tag{5.50}$$

The assumption of local isotropy is not valid near walls where turbulence approaches a 2D state. A non-isotropic dissipation model was proposed by Rotta (1951):

$$\varepsilon_{ij} = \frac{\overline{u_i u_j}}{k}\varepsilon \tag{5.51}$$

The rate of dissipation of the TKE, ε, is calculated by solving an equation similar to the ε-equation (equation 5.18) used in standard isotropic models. At the RST level of closure the modelled ε-equation reads as follows:

$$\frac{\partial \varepsilon}{\partial t} + U_\ell \frac{\partial \varepsilon}{\partial x_\ell} = C_{\varepsilon 1}\frac{\varepsilon}{k}P_k - C_{\varepsilon 2}\frac{\varepsilon^2}{k} + C_\varepsilon \frac{\partial}{\partial x_i}\left[\frac{k}{\varepsilon}\overline{u_i u_\ell}\frac{\partial \varepsilon}{\partial x_\ell}\right] \tag{5.52}$$

where P_k is the rate of production of TKE:

$$P_k = -\overline{u_i u_\ell}\frac{\partial U_i}{\partial x_\ell} \tag{5.53}$$

To account for the effects of the free surface in the equation (5.53), Gibson and Rodi (1989) proposed replacing the P_k term in equation (5.52) with the following expression:

$$P_k^* = 0.47 P_k + 3.9\varepsilon \frac{\alpha_{ij}\alpha_{ij}}{1 + 1.5\sqrt{\alpha_{ij}\alpha_{ij}}} \tag{5.54}$$

The first term in the right-hand side of this generalized production term is the standard rate of production of k by mean shear, which is negligible near the free surface. The second term in the above equation acts to produce ε in regions of high anisotropy of the Reynolds stresses (recall that α_{ij} is the anisotropy tensor). That is, this term tends to increase ε near the free surface, thus reducing the length scale.

Algebraic Reynolds-stress (ARS) models

The ARS models are essentially similar to nonlinear two-equation models. They employ a standard two-equation model along with algebraic expressions that relate explicitly the Reynolds stresses with mean velocity gradients. The main difference between ARS and nonlinear models is that in the former the expressions for the Reynolds stresses are derived by simplifying the full RST equations.

For flows where convection and diffusion terms balance each other (local equilibrium), the full RST equations (equation 5.7) can be simplified as follows:

$$0 = P_{ij} - \frac{2}{3}\delta_{ij}\varepsilon - \Phi_{ij} \tag{5.55}$$

By substituting the production tensor (equation 5.8) and a suitable model for Φ_{ij}, we obtain a set of six algebraic equations for the Reynolds stresses. For non-equilibrium flows convection and diffusion effects cannot be neglected. Rodi (1976) suggested that because $\overline{u_i u_j} \propto k$, one may approximate these terms in the RST equations (5.7) as follows:

$$\left(\frac{\partial \overline{u_i u_j}}{\partial t} + U_k \frac{\partial \overline{u_i u_j}}{\partial x_k} \right) - \frac{\partial}{\partial x_k}\left(\nu \frac{\partial \overline{u_i u_j}}{\partial x_k} + C_{ijk} \right) \approx \frac{\overline{u_i u_j}}{k}(P_k - \varepsilon) \tag{5.56}$$

Using this assumption, an ARS model suitable for non-equilibrium flows can be derived as follows:

$$\frac{\overline{u_i u_j}}{k}(P_k - \varepsilon) = P_{ij} - \frac{2}{3}\delta_{ij}\varepsilon - \Phi_{ij} \tag{5.57}$$

The ARS models given by either equation (5.56) or (5.57) are implemented in conjunction with the k-ε model.

Reynolds-stress modelling near walls

The RST and ARS models discussed earlier are valid only within the fully turbulent region and, as was the case with the isotropic k-ε model discussed earlier, need to be modified to account for near-wall effects. This can be accomplished by either adopting wall functions to bridge the gap between the laminar sublayer and the fully turbulent region or modifying the above models so that they are valid all the way to the wall in the low (turbulent) Reynolds number limit.

Wall functions for Reynolds-stress models
Wall functions for Reynolds-stress models are implemented in a similar manner as discussed previously for the standard k-ε model but additional equations are needed to specify the individual Reynolds stresses at the first grid node off the wall. One approach to accomplish this is to use empirical values based on experimental data from simple channel or boundary layer flows. Gibson *et al.* (1981) proposed the following values for the ratios of each Reynolds stress to the TKE (also known as structural parameters):

$$\frac{\overline{u^2}}{k} = 1.1 \quad \frac{\overline{v^2}}{k} = 0.25 \quad \frac{\overline{w^2}}{k} = 0.65 \quad -\frac{\overline{uv}}{k} = 0.26 \tag{5.58}$$

A more general procedure, which does not rely on experimental data but rather determines the Reynolds stresses from an algebraic version of the parent RST model, was proposed by Sotiropoulos and Patel (1995a). It is outlined as follows:

- Generate a computational mesh so that the first three grid lines off the wall (say $\zeta = 1, 2$ and 3, with the wall at $\zeta = 0$) are located within the fully turbulent region.
- Solve the mean flow and RST equations up to $\zeta = 3$ and then use the resulting velocity field, in conjunction with the law of the wall (equation (5.23) or (5.25)), to determine the friction velocity u_τ.
- With u_τ known, the law of the wall along with assumptions about the orientation of the velocity vector (it is assumed to be parallel to the wall and its direction is determined by extrapolating from interior nodes) are used to calculate the three velocity components at $\zeta = 1$ and 2.
- The velocity components at $\zeta = 2$ provide boundary conditions for the mean flow equations at $\zeta = 3$, while those at $\zeta = 1$ are used to discretize, at $\zeta = 2$, terms involving mean velocity gradients in the RST equations.
- The Reynolds stresses at $\zeta = 2$, needed to solve the RST equations at $\zeta = 3$, are obtained in accordance with local equilibrium assumptions by setting all transport terms in the RST equations to zero and solving the resulting algebraic system, equation (5.55).

This procedure is general and applicable to complex geometries and allows the computation of the near-wall stress levels in a manner that is compatible and consistent with the parent RST model.

Near-wall RST modelling
RST models that are valid all the way to the wall should be capable of reproducing the observed flow features in the fully turbulent region (i.e. be compatible with their parent high-Re model away from the wall) and the asymptotic variations of various flow quantities required by the no-slip condition at the wall. Asymptotic values of various turbulence quantities in the vicinity of the wall can be derived by expanding the velocity fluctuations in Taylor series in the direction normal to the wall, imposing the no-slip conditions, and determining the leading order terms in the series expansion

so that the continuity equation is satisfied. The following asymptotic expressions valid as $y \to 0$ are then obtained:

$$u \to A(x, z, t)y + O(y^2)$$
$$v \to B(x, z, t)y^2 + O(y^3)$$
$$w \to C(x, z, t)y + O(y^2)$$
$$k \to \frac{1}{2}\left(\overline{A^2 + C^2}\right)y^2 + O(y^3) \qquad (5.59)$$
$$\varepsilon \to \nu\left(\overline{A^2 + C^2}\right) + O(y)$$
$$-\overline{uv} \to -\overline{AB}Yy^3 + O(y^4)$$

where A, B and C are unknown functions of time and the spatial directions parallel to the wall. Near-wall models should duplicate these limits in order to be asymptotically consistent. Several near-wall models have been proposed in the literature and the reader is referred to So *et al.* (1991) for a fairly recent review of various proposals. Here we summarize the near-wall RST closure by Launder and Shima (1989), as revised by Shima (1993), which was the first near-wall RST model to be applied with a great deal of success to a variety of complex internal and external flows (Sotiropoulos and Patel, 1993, 1994, 1995a,b).

The Launder–Shima model is an extension of the parent high-Re Gibson and Launder model (equations (5.41)–(5.44)) to near-wall flows. The pressure redistribution tensor Φ_{ij} is modelled according to equations (5.41)–(5.44) but the constants in these equations are made functions of the turbulent Reynolds number

$$Re_\mathrm{T} = \frac{k^2}{\varepsilon \nu} \qquad (5.60)$$

and the second and third invariants of Reynolds stress anisotropy tensor α_{ij}:

$$A_2 = \alpha_{ij}\alpha_{ij} \quad A_3 = \alpha_{ik}\alpha_{kj}\alpha_{ji}$$

as follows:

$$C_1 = 1 + 2.58 A A_2^{1/4}\left[1 - \mathrm{e}^{-(0.0067 Re_\mathrm{T})^2}\right]$$
$$C_2 = 0.75 A^{1/2}$$
$$C_1^w = -\frac{2}{3}C_1 + 1.67 \qquad (5.61)$$
$$C_2^w = \max\left[\frac{1}{C_2}\left(\frac{2}{3}C_2 - \frac{1}{6}\right),\ 0\right]$$

where A is the flatness parameter: $A = 1 - 9(A_2 - A_3)/8$.

The following ε-equation is solved:

$$\frac{\partial \varepsilon}{\partial t} + U_\ell \frac{\partial \varepsilon}{\partial x_\ell} = -(C_{\varepsilon 1} + \psi_1 + \psi_2)\frac{\varepsilon}{k}P_k \\ - C_{\varepsilon 2}\frac{\varepsilon \tilde{\varepsilon}}{k} + \frac{\partial}{\partial x_i}\left[\left(\nu \delta_{i\ell} + C_\varepsilon \frac{k}{\varepsilon}\overline{u_i u_\ell}\right)\frac{\partial \varepsilon}{\partial x_\ell}\right] \quad (5.62)$$

where

$$\tilde{\varepsilon} = \varepsilon - 2\nu\left(\frac{\partial \sqrt{k}}{\partial x_j}\right)^2 \\ \psi_1 = 2.5A\left(\frac{P_k}{\varepsilon} - 1\right) \\ \psi_2 = 0.3(1 - 0.3A_2)\mathrm{e}^{(-0.002 Re_T)^2} \quad (5.63)$$

At the wall, the no-slip conditions are applied to set all velocity components and Reynolds stresses equal to zero. The rate of dissipation is obtained by setting

$$\tilde{\varepsilon} = 0 \quad \text{or equivalently} \quad \varepsilon = 2\nu\left(\frac{\partial \sqrt{k}}{\partial x_j}\right)^2 \quad (5.64)$$

Equation (5.64) is designed to ensure that the exact asymptotic behaviour of ε is recovered as the wall is approached. It can be easily shown from the asymptotic expansions given in equation (5.59) that in the limit $y \to 0$:

$$\varepsilon \to \frac{2\nu k}{y^2} \quad (5.65)$$

We should finally note that the specific forms of the various coefficients given in the above equations have been calibrated such that the model reproduces correctly experimental and DNS data for simple channel and boundary layer flows (see Launder and Shima, 1989; Shima, 1993, for details).

Application of RST and ARS models

The RST models have yet to be applied to hydraulic engineering flows, presumably due to their complexity and associated increased computational overhead. Applications of such models to various complex, internal and external flows have been reported by Sotiropoulos and Patel (1994, 1995a,b). ARS models on the other hand have been extensively applied to simulate stress-driven secondary currents in straight open channels. The reader is referred to Cokljat and Younis (1995), Naot et al. (1993), Demuren and Rodi (1984) and Naot and Rodi (1982) for such applications.

5.4 Turbulence modelling for stratified flows

Many flows occurring in natural environments, such as in the atmosphere, in oceans, lakes and reservoirs, are stratified due to temperature or salinity gradients. Under stratified flow conditions, buoyancy forces become important and could greatly impact the structure of turbulence and the transport and mixing of temperature, pollutants, nutrients and sediments. The physical phenomena to be simulated are very complex and are associated with a wide range of length and timescales. An excellent review of the physics of stratified flows can be found in Rodi (1987). Here, the basic modelling strategies for simulating buoyancy-dominated flows are summarized.

5.4.1 The governing equations

For problems in which the density gradients are not large, the governing equations can be greatly simplified by invoking the so-called Boussinesq approximation and accounting for density variations only in the gravity term. The Boussinesq form of the Reynolds-averaged transport equations for mass, momentum and the relevant stratification-inducing scalar (temperature or salinity) read as follows:

$$\frac{\partial U_i}{\partial x_i} = 0 \tag{5.66}$$

$$\frac{\partial U_i}{\partial t} + \frac{\partial}{\partial x_j}(U_i U_j) = -\frac{1}{\rho_r}\frac{\partial P}{\partial x_i} + \frac{\partial}{\partial x_j}\left(\nu\frac{\partial U_i}{\partial x_j} - \overline{u_i u_j}\right) + g_i\frac{\rho - \rho_r}{\rho_r} \tag{5.67}$$

$$\frac{\partial \Theta}{\partial t} + \frac{\partial}{\partial x_j}(\Theta U_j) = \frac{\partial}{\partial x_j}\left(\alpha\frac{\partial \Theta}{\partial x_j} - \overline{\theta u_j}\right) + S_\Theta \tag{5.68}$$

where ρ is the local fluid density, ρ_r is a reference density, α is the molecular (heat or mass) diffusivity coefficient, Θ is stratification-inducing scalar, θ is the corresponding fluctuating quantity and S_Θ is a source or sink of Θ. The local density is related to the temperature or salinity via an equation of state of the following form:

$$\frac{\rho - \rho_r}{\rho_r} = -\beta(\Theta - \Theta_r) \tag{5.69}$$

In addition to the six Reynolds stresses $\overline{u_i u_j}$, equations (5.67)–(5.69) include three new unknowns: the three components of the scalar flux vector $\overline{u_i \theta}$, which express the transport of the fluctuating scalar due to turbulent velocity fluctuations.

Transport equations for the Reynolds stresses can be derived from above equations (5.67)–(5.69) following the same procedure used to derive equation (5.7) for non-stratified flows (see Chen and Jaw, 1998, for details). The resulting equations are identical to equation (5.7) with some additional terms representing production of Reynolds stresses due to buoyancy:

$$\frac{\partial \overline{u_i u_j}}{\partial t} + U_k\frac{\partial \overline{u_i u_j}}{\partial x_k} = \quad \cdots \quad - \underbrace{\beta\left(g_i\overline{u_j\theta} + g_j\overline{u_i\theta}\right)}_{\text{Buoyancy production terms}} \tag{5.70}$$

For the sake of brevity dots in equation (5.70) imply terms that are identical to those in the right-hand side of equation (5.7). Similarly, one can derive transport equations for the turbulent scalar fluxes, which read as follows:

$$\frac{\partial \overline{u_i\theta}}{\partial t} + U_\ell \frac{\partial \overline{u_i\theta}}{\partial x_\ell} = \frac{\partial}{\partial x_\ell}\left(-\overline{u_i u_\ell \theta} - \delta_{i\ell}\frac{\overline{P\theta}}{\rho} + \alpha\overline{u_i\frac{\partial \theta}{\partial x_\ell}} + \nu\overline{\theta\frac{\partial u_i}{\partial x_\ell}}\right) - \left(\overline{u_i u_\ell}\frac{\partial \Theta}{\partial x_\ell} + \overline{u_\ell \theta}\frac{\partial U_i}{\partial x_\ell}\right)$$

$$- (\alpha + \nu)\overline{\frac{\partial \theta}{\partial x_\ell}\frac{\partial u_i}{\partial x_\ell}} + \frac{\overline{P}}{\rho}\frac{\partial \theta}{\partial x_i} + \overline{s_\theta u_i} - \beta g_i \overline{\theta^2} \qquad (5.71)$$

The term $\overline{\theta^2}$ in the above equation is called the scalar variance and an exact transport equation for it can be derived as follows:

$$\frac{\partial \overline{\theta^2}}{\partial t} + U_\ell \frac{\partial \overline{\theta^2}}{\partial x_\ell} = \frac{\partial}{\partial x_\ell}\left(-\overline{u_\ell \theta^2} + \alpha\frac{\partial \overline{\theta^2}}{\partial x_\ell}\right) - 2\overline{u_\ell \theta}\frac{\partial \Theta}{\partial x_\ell} - \varepsilon_\theta + \overline{s_\theta \theta} \qquad (5.72)$$

where ε_θ is the rate of dissipation of $\overline{\theta^2}$, a quantity analogous to the rate of dissipation of the TKE, defined as follows:

$$\varepsilon_\theta = 2\alpha\overline{\frac{\partial \theta}{\partial x_\ell}\frac{\partial \theta}{\partial x_\ell}} \qquad (5.73)$$

It is readily apparent from equations (5.71)–(5.73) that stratification complicates considerably the problem of closing the Reynolds-averaged transport equations. Terms such as the scalar variance $\overline{\theta^2}$ and rate of dissipation ε_θ are new in stratified flows and need to be modelled because $\overline{\theta^2}$ appears in the equations for the turbulent scalar flux equation $\overline{u_i \theta}$, which in turn appears in the $\overline{u_i u_j}$ equations. Thus, for stratified flows the turbulent momentum transfer and the turbulent scalar transfer are strongly coupled.

5.4.2 Turbulence models for stratified flows

5.4.2.1 *Isotropic eddy-viscosity models*

The simplest approach for closing the Reynolds-averaged equations for stratified flows is to adopt the isotropic eddy-viscosity/diffusivity concept. The Reynolds stresses are related to the mean rate of strain via equation (5.12) while a similar model is adopted to express the turbulent scalar fluxes in terms of the mean scalar gradients:

$$-\overline{u_i u_j} = \nu_t\left(\frac{\partial U_i}{\partial x_j} + \frac{\partial U_j}{\partial x_i}\right) - \frac{2}{3}k\delta_{ij} \quad \text{and} \quad -\overline{u_i \theta} = \Gamma_t\frac{\partial \Theta}{\partial x_i} \qquad (5.74)$$

Γ_t is the eddy-diffusivity coefficient, a quantity analogous to the eddy viscosity, which is related to the eddy viscosity by assuming a constant Schmidt number σ_t, as follows:

$$\Gamma_t = \frac{\nu_t}{\sigma_t} \qquad (5.75)$$

With these modelling assumptions the governing equations, equations (5.66)–(5.69), can be closed via a model for the eddy viscosity. Such a model can be based on a simple mixing-length approach or on a k-ε type model, similar to those discussed in previous sections for non-stratified flows.

Algebraic mixing-length model

For stratified flows the generalized form of Prandtl's mixing-length model given by equation (5.15) can also be applied but needs to be modified to account for the effect of stratification on the local structure of turbulence. For example, for stable stratification turbulence is suppressed while for unstable stratification turbulence is enhanced. Stratification effects are typically parameterized in terms of the gradient Richardson number, the ratio of the local production of turbulence due to buoyancy effects to that due to mean shear:

$$Ri = -\frac{g}{\rho}\frac{\partial \rho/\partial z}{(\partial U/\partial z)^2} \tag{5.76}$$

where z is the direction of stratification. For stable and unstable stratification this quantity will be positive (denser fluid at the bottom) and negative (denser fluid at the top), respectively. By setting a critical level of Ri, Ri_c, beyond which turbulence cannot be sustained, a mixing-length model modified for stratification effects can be formulated as follows (Mason and Sykes, 1982):

Stable stratification ($Ri > 0$)

$$\nu_t = \begin{cases} 2\ell_t^2\sqrt{S_{ij}S_{ij}}(1 - Ri/Ri_c)^2 & 0 \le Ri \le Ri_c \\ 0 & Ri > Ri_c \end{cases} \tag{5.77}$$

Unstable stratification

$$\nu_t = 2\ell_t^2\sqrt{S_{ij}S_{ij}}(1 - Ri)^{1/2} \tag{5.78}$$

Choosing Ri_c to be a small value (say one or less) suppresses the eddy viscosity to a very small level in regions of the flow away from shear zones. A more extensive discussion of the various empirical correlations of the form given by equations (5.77) and (5.78) for describing the influence of Ri of the mixing length can be found in Rodi (1993).

The k-ε model

A k-ε model suitable for stratified flow computations reads as follows:

$$\frac{\partial k}{\partial t} + U_\ell \frac{\partial k}{\partial x_\ell} = \underbrace{\nu_t\left(\frac{\partial U_i}{\partial x_\ell} + \frac{\partial U_\ell}{\partial x_i}\right)\frac{\partial U_i}{\partial x_\ell}}_{P_k} + \frac{\partial}{\partial x_\ell}\left[\frac{\nu_t}{\sigma_k}\frac{\partial k}{\partial x_\ell}\right] + \underbrace{\beta g_i \frac{\nu_t}{\sigma_t}\frac{\partial \Theta}{\partial x_i}}_{G=\text{buoyancy production}} - \varepsilon \tag{5.79}$$

$$\frac{\partial \varepsilon}{\partial t} + U_\ell \frac{\partial \varepsilon}{\partial x_\ell} = C_{\varepsilon 1} \frac{\varepsilon}{k}(P_k + C_{\varepsilon 3}G) - C_{\varepsilon 2}\frac{\varepsilon^2}{k} + \frac{\partial}{\partial x_\ell}\left[\frac{\nu_t}{\sigma_\varepsilon}\frac{\partial \varepsilon}{\partial x_\ell}\right] \tag{5.80}$$

$$\nu_t = C_\mu \frac{k^2}{\varepsilon}$$

$$C_{\varepsilon 1} = 1.44, \quad C_{\varepsilon 2} = 1.92, \quad C_\mu = 0.09, \quad \sigma_k = 1.0, \quad \sigma_\varepsilon = 1.3$$

Stratification effects enter in the above equations through the G term (buoyancy production/destruction). The value of the $C_{\varepsilon 3}$ coefficient of this term, however, in the ε-equation has been found to be dependent upon the flow. For flows where G is a source term (unstably stratified flows), $C_{\varepsilon 3}$ should be 1, while in stably stratified flows (G is a sink), $C_{\varepsilon 3}$ should be near zero (see Rodi, 1987, for details).

5.4.2.2 Reynolds-stress models

The RST models for stratified flows require solving transport equations for $\overline{u_i u_j}$, $\overline{u_i \theta}$, $\overline{\theta^2}$, ε and possibly even for ε_θ. That is, in addition to the five transport equations for the mean flow, equations (5.66)–(5.68), up to twelve additional transport equations may need to be solved for turbulence closure at the second-moment level.

Modelling of the $\overline{u_i u_j}$ equations

The buoyancy production tensor (terms in the right-hand side of equation (5.70)) is exact at this closure level and does not require modelling. The turbulent diffusion and dissipation tensors are modelled by adopting the same approaches as for non-stratified flows (section 5.3.2.2). The pressure redistribution term Φ_{ij} requires special care since it can be shown that the instantaneous fluctuations of the pressure depend on the instantaneous fluctuations of the scalar. Gibson and Launder (1978) proposed the following model:

$$\Phi_{ij} = \Phi_{ij,1} + \Phi_{ij,2} + \Phi_{ij,3} \tag{5.81}$$

where the terms $\Phi_{ij,1}$ and $\Phi_{ij,2}$ are modelled as previously discussed for non-stratified flows (equations (5.42) and (5.43)) and $\Phi_{ij,3}$ is modelled in terms of the buoyancy production terms as follows:

$$\Phi_{ij,3} = -C_3\left(G_{ij} - \frac{2}{3}\delta_{ij}G\right) \tag{5.82}$$

where

$$G_{ij} = \beta\left(g_i \overline{u_j \theta} + g_j \overline{u_i \theta}\right) \tag{5.83}$$

$G = G_{11} + G_{22} + G_{33}$ and $C_3 = 0.3 \div 0.5$. The above model needs to be enhanced with wall-reflection terms (as in the non-stratified case, equations (5.44)) if it is to be applied to wall-bounded flows.

Modelling of the equations for $\overline{u_i\theta}$ and $\overline{\theta^2}$

Gibson and Launder (1978) proposed the following modelled equation for $\overline{u_i\theta}$ and $\overline{\theta^2}$:

$$\frac{\partial \overline{u_i\theta}}{\partial t} + U_\ell \frac{\partial \overline{u_i\theta}}{\partial x_\ell} = \underbrace{\frac{\partial}{\partial x_\ell}\left[\left(C_T \frac{k^2}{\varepsilon} + \nu\right)\frac{\partial \overline{u_i\theta}}{\partial x_\ell}\right]}_{\text{Diffusion}} - \underbrace{\left(\overline{u_i u_\ell}\frac{\partial \Theta}{\partial x_\ell} + \overline{u_\ell \theta}\frac{\partial U_i}{\partial x_\ell}\right)}_{\text{Mean production}} \quad (5.84)$$

$$- \underbrace{\beta g_i \overline{\theta^2}}_{\text{Buoyancy production}} \underbrace{- C_{T1}\frac{\varepsilon}{k}\overline{u_i\theta} + C_{T2}\frac{\partial U_i}{\partial x_m}\overline{u_m\theta} - C_{T3}\beta g_i \overline{\theta^2}}_{\text{Pressure redistribution}}$$

with $C_T = 0.07$, $C_{T1} = 3.2$, $C_{T2} = 0.5$, $C_{T3} = 0. \div 0.5$, and

$$\frac{\partial \overline{\theta^2}}{\partial t} + U_\ell \frac{\partial \overline{\theta^2}}{\partial x_\ell} = \frac{\partial}{\partial x_\ell}\left(C_\theta \frac{k^2}{\varepsilon}\frac{\partial \overline{\theta^2}}{\partial x_\ell} + \alpha \frac{\partial \overline{\theta^2}}{\partial x_\ell}\right) - 2\overline{u_\ell \theta}\frac{\partial \Theta}{\partial x_\ell} - \varepsilon_\theta \quad (5.85)$$

with $C_\theta = 0.13$. The ε_θ term is modelled following the proposal of Launder (1980):

$$\varepsilon_\theta = 2\alpha \overline{\frac{\partial \theta}{\partial x_\ell}\frac{\partial \theta}{\partial x_\ell}} \approx \text{constant} \times \left(\frac{\theta^2}{u^2}\right) \times \varepsilon = C_{\theta 1}\frac{\overline{\theta^2}}{k}\varepsilon \quad (5.86)$$

with $C_{\theta 1} = 0.62$.

ARS models

Following Gibson and Launder (1978), a simple algebraic model can be derived by:

1. Relating all transport terms in the RST equations to those in the k-equation by the following isotropic expression:

$$\left(\frac{\partial \overline{u_i u_j}}{\partial t} + U_k \frac{\partial \overline{u_i u_j}}{\partial x_k}\right) - \frac{\partial}{\partial x_k}\left(\nu \frac{\partial \overline{u_i u_j}}{\partial x_k} + C_{ijk}\right) = \frac{2}{3}\delta_{ij}\left\{\frac{\partial k}{\partial t} + U_\ell \frac{\partial k}{\partial x_\ell} - \frac{\partial}{\partial x_\ell}\left[\left(\nu + \frac{\nu_t}{\sigma_k}\right)\frac{\partial k}{\partial x_\ell}\right]\right\}$$
$$= \frac{2}{3}\delta_{ij}\{P_k + G - \varepsilon\}$$

(5.87)

2. Neglecting all transport terms in the $\overline{u_i\theta}$ and $\overline{\theta^2}$ equations.

Then, the following explicit algebraic expressions can be derived:

$$\overline{u_i u_j} = \frac{2}{3}\delta_{ij}k + \frac{k}{C_1 \varepsilon}(1 - C_2)\left[P_{ij} + G_{ij} - \frac{2}{3}\delta_{ij}(P + G)\right] \quad (5.88)$$

$$\overline{u_i\theta} = \frac{1}{C_{T1}}\frac{k}{\varepsilon}\left[\overline{u_i u_j}\frac{\partial \Theta}{\partial x_\ell} + (1 - C_{T2})\left(\overline{u_i\theta}\frac{\partial U_i}{\partial x_\ell} + \beta g_i \overline{\theta^2}\right)\right] \quad (5.89)$$

$$\overline{\theta^2} = -2C_{\theta 1}\frac{k}{\varepsilon}\overline{u_i\theta}\frac{\partial\Theta}{\partial x_i} \tag{5.90}$$

The modelled k- and ε-equations, equations (5.79) and (5.80), need to be solved along with the algebraic system of equations to obtain all turbulent quantities.

5.5 Closing remarks

This chapter has attempted to provide a brief introduction to the fundamentals of statistical turbulence modelling with special emphasis on modelling issues that are relevant to flow problems in environmental and hydraulic engineering applications. For a more in depth treatment the interested reader can resort to the recent textbooks on the subject by Chen and Jaw (1998), Wilcox (1997) and Durbin and Pettersson Reiff (1999). The monographs by Rodi (1994) and Nezu and Nakagawa (1998) are also excellent resources for the overview and application of turbulence models to hydraulic engineering problems.

Acknowledgements

This work was supported by NSF Career grant 9875691 and a grant from Oak Ridge National Laboratory and DOE monitored by Dr Michael Sale.

References

Bradshaw, P. (1987), "Turbulent Secondary Flows", *Ann. Rev. Fluid Mech.*, 19, 53–74.
Bradbrook, K.F., Lane, S.N., Richards, K.S., Biron, P.M. and Roy, A.G. (2000), "Large Eddy Simulation of Periodic Flow Characteristics at River Channel Confluences", *IAHR J. Hydr. Res.*, 38(3), 207–215.
Cebeci, T. and Bradshaw, P. (1977), "Momentum Transfer in Boundary Layers", Hemisphere Publishing Corporation, Washington.
Celik, I. and Rodi, W. (1984), "Simulation of Free-Surface Effects in Turbulent Channel Flow", *Physiochem. Hydrodyn.*, 5, 217–227.
Chen, C.J. and Jaw, S.Y. (1998), "Fundamentals of Turbulence Modelling", Taylor & Francis, Washington, DC.
Chen, H.C. and Patel, V.C. (1988), "Near-Wall Turbulence Models for Complex Flows Including Separation", *AIAA J.*, 26, 641–648.
Choi, Y.D., Iacovides, H. and Launder, B.E. (1989), "Numerical Computation of Turbulent Flow in a Square-Sectioned 180 deg Bend", *ASME J. Fluids Eng.*, 111, 59–68.
Chrisohoides, A., Sotiropoulos, F. and Sturm, T.W. (2003), "Coherent Structures in Flat-Bed Bridge Abutment Flows: Experiments and CFD Simulations", *ASCE J. Hydr. Eng.* 129(3), 171–249.
Cokljat, D. and Younis, B.A. (1995), "Second-Order Closure Study of Open-Channel Flows", *ASCE J. Hydr. Eng.*, 121(2), 94–107.
Colombini, M. (1993), "Turbulence-Driven Secondary Flows and Formation of Sand Ridges", *J. Fluid Mech.*, 254, 701–719.

Constantinescu, G.S. and Patel, V.C. (2000), "Role of Turbulence Model in Prediction of Pump-Bay Vortices", *ASCE J. Hydr. Eng.*, 126(5), 387–391.

Craft, T.J. (1998), "Developments in a Low-Reynolds-Number Second-Moment Closure and Its Application to Separating and Reattaching Flows", *Int. J. Heat Fluid Fl.*, 19(5), 541–548.

Craft, T.J., Launder, B.E. and Suga, K. (1995), "A Non-Linear Eddy-Viscosity Model Including Sensitivity to Stress Anisotropy", Proceedings of 10th Symposium on Turbulent Shear Flows, Pennsylvania State University, State College, PA.

Daly, B.J. and Harlow, F.H. (1970), "Transport Equations in Turbulence", *Phys. Fluids*, 13, 2634–2649.

Demuren, A.O. (1993), "A Numerical Model for Flow in Meandering Channels with Natural Bed Topography", *Water Resour. Res.*, 29(4), 1269–1277.

Demuren, A.O. and Rodi, W. (1984), "Calculations of Turbulence-Driven Secondary Motion in Non-Circular Ducts", *J. Fluid Mech.*, 140, 189–222.

Demuren, A.O. and Rodi, W. (1986), "Calculation of Flow and Pollutant Dispersion in Meandering Channels", *J. Fluid Mech.*, 172, 63–92.

Durbin, P.A. (1993), "A Reynolds-Stress Model for Near-Wall Turbulence", *J. Fluid Mech.*, 249, 465–498.

Durbin, P.A. and Pettersson Reif, B.A. (2001), "Statistical Theory and Modelling for Turbulent Flows", John Wiley & Sons Ltd, West Sussex, England.

Gatski, T.B. and Speziale, C.G. (1993), "On Explicit Algebraic Stress Models for Complex Turbulent Flows", *J. Fluid Mech.*, 254, 59–78.

Gibson, M.M. (1978), "An Algebraic Stress and Heat-Flux Model for Turbulent Shear Flow with Streamline Curvature", *Int. J. Heat Mass Trans.*, 21, 1609.

Gibson, M.M. and Launder, B.E. (1978), "Ground Effects on Pressure Fluctuations in the Atmospheric Boundary Layers", *J. Fluid Mech.*, 86, 491–511.

Gibson, M.M. and Rodi, W. (1989), "Simulation of Free-Surface Effects of Turbulence with a Reynolds Stress Model", *J. Hydr. Res., IAHR*, 27, 233–244.

Gibson, M.M., Jones, W.P. and Younis, B.A. (1981), "Calculation of Turbulent Boundary Layers on Curved Surfaces", *Phys. Fluids*, 24, 386–395.

Hanjalic, K. and Launder, B.E. (1972), "A Reynolds-Stress Model of Turbulence and its Application to Thin Shear Flows", *J. Fluid Mech.*, 52, 609–638.

Hanjalic, K. and Launder, B.E. (1976), "Contribution Towards a Reynolds-Stress Closure for Low-Reynolds-Number Turbulence", *J. Fluid Mech.*, 74, 593–610.

Hebbert, B., Imberger, J., Loh, I. and Patterson, J. (1979), "Collie River Underflow into the Wellington Reservoir", *ASCE J. Hydr. Div.*, 105(HY5), 533–545.

Hossain, M.S. and Rodi, W. (1982), "A Turbulence Model for Buoyant Flows and Its Application to Vertical Buoyant Jets", in Turbulent Buoyant Jets and Plumes, HMT Ser., 6, 121–178.

Jones, W.P. and Launder, B.E. (1972), "The Prediction of Laminarization with a Two-Equation Model of Turbulence", *Int. J. Heat Mass Trans.*, 15, 301–314.

Kolmogorov, A.N. (1942), "Equations of Turbulent Motion of an Incompressible Fluid", Izvestia Academy of Sciences, USSR, *Physics*, 6(1, 2), 56–58.

Krishnappan, B.G. and Lau, Y.L. (1986), "Turbulence Modelling of Flood Plain Flows", *ASCE J. Hydr. Eng.*, 112(4), 251–266.

Launder, B.E. and Li, S.P. (1994), "On the Elimination of Wall-Topography Parameters from Second-Moment Closure", *Phys. Fluids*, 6, 999–1006.

Launder, B.E. and Sharma, B.I. (1974), "Application of the Energy-Dissipation Model of Turbulence to the Calculation of Flow Near a Spinning Disk", *Lett. Heat Mass Trans.*, 1, 131–138.

Launder, B.E. and Shima, N. (1989), "Second-Moment Closure for the Near-Wall Sublayer: Development and Application", *AIAA J.*, 27, 1319–1325.

Launder, B.E. and Spalding, D.B. (1974), "The Numerical Computation of Turbulent Flow", *Comp. Meth. Appl. Mech. Eng.*, 3, 269–289.

Launder, B.E., Reece, G.J. and Rodi, W. (1975), "Progress in the Development of a Reynolds-Stress Turbulence Closure", *J. Fluid Mech.*, 68, 537–566.

Lumley, J.L. (1978), "Computational Modelling of Turbulent Flows", in *Advances in Applied Mechanics* (ed. C.S. Yih), Academic, Orlando, Fla.

Lumley, J.L. (1980), "Second Order Modelling of Turbulent Flows", in *Prediction Methods for Turbulent Flows* (ed. W. Kollmann), Hemisphere.

Lumley, J.L. and Newman, G.R. (1977), "Return to Isotropy of Homogeneous Turbulence", *J. Fluid Mech.*, 82, 161–178.

Mason, P.J. and Sykes, R.I. (1982), "A Two-Dimensional Numerical Study of Horizontal Roll Vortices in an Inversion Capped Planetary Boundary Layer", *Quart. J. R. Met. Soc.*, 108, 801.

Menter, F.R. (1993), "Zonal Two Equation k-ω Turbulence Models for Aerodynamic Flows", *24th Fluid Dynamics Conference*, Orlando, Florida, 1–21, July 6–9.

Meselhe, E.A. and Sotiropoulos, F. (2000), "Three-Dimensional Numerical Model for Open-Channels with Free-Surface Variations", *J. Hydr. Res.*, 38(2), 115–121.

Naot, D. (1984), "Response of Channel Flow to Roughness Heterogeneity", *ASCE J. Hydr. Eng.*, 110, 1568–1587.

Naot, D. and Rodi, W. (1982), "Calculation of Secondary Currents in Channel Flow", *ASCE J. Hydr. Eng.*, 108(8), 948–968.

Naot, D., Shavit, A. and Wolfshtein, M. (1970), "Interactions Between Components of the Turbulent Velocity Correlation Tensor", *Israel J. Tech.*, 8, 259.

Naot, D., Nezu, I. and Nakagawa, H. (1993), "Hydrodynamic Behavior of Compound Rectangular Open Channels", *ASCE J. Hydr. Eng.*, 119(3), 390–408.

Neary, V.S., Sotiropoulos, F. and Odgaard, A.J. (1999) "Three-Dimensional Numerical Model of Lateral-Intake flows", *J. Hydr. Eng.*, 125(2), 126–140.

Nezu, I. and Nakagawa, H. (1993), *Turbulence in Open-Channel Flows*, IAHR/AIRH Monograph, Balkema, Rotterdam, The Netherlands.

Nezu, I. and Rodi, W. (1986), "Open-Channel Flow Measurements with a Laser Doppler Anemometer", *ASCE J. Hydr. Eng.*, 112, 335–155.

Nikuradse, J. (1933), "Gesetzmäßigkeit der turbulenten strömung im glatten rohren", Forschungsarbeiten, Ing.-Wesen, Heft 356.

Olsen, N.R.B. and Stoksteth, S. (1995), "Three-Dimensional Numerical Modelling of Water Flow in a River with Large Bed Roughness", *J. Hydr. Res., IAHR*, 33, 571–581.

Patel, V.C., Rodi, W. and Scheuerer, G. (1984), "Turbulence Models for Near-Wall and Low Reynolds Number Flows: A Review", *AIAA J.*, 23(9), 1308–1319.

Piomelli, U. (1999), "Large-Eddy Simulation: Achievements and Challenges", *Prog. Aerosp. Sci.*, 35(4), 335–362.

Rastogi, A. and Rodi, W. (1978), "Predictions of Heat and Mass Transfer in Open Channels", *ASCE J. Hydr. Eng.*, 104(3), 397–420.

Reynolds, O. (1895), "On the Dynamical Theory of Incompressible Viscous Fluids and the Determination of the Criterion", *Phil. Trans. Roy. Soc. A*, 186, 123.

Rodi, W. (1976), "A New Algebraic Relation for Calculating Reynolds Stresses", *ZAMM*, 56, 219.

Rodi, W. (1987), "Examples of Calculation Methods for Flow and Mixing in Stratified Fluids", *J. Geophys. Res.*, 92(C5), 5305–5328.

Rodi, W. (1993), *Turbulence Models and Their Applications in Hydraulics*, 3rd edn, IAHR Monograph, Delft, The Netherlands.

Rotta, J.C. (1951), "Statistische Theorie Nichthomogener Turbulenz", *Z. Phys.*, 129, 547–572.

Rotta, J.C. (1968), "Uber eine Methode zur Berechnung Turbulenter Scherströmungen", Aerodynamische Versuchanstalt, Göttingen, Rep. 69A14.

Shima, N. (1993), "Prediction of Turbulent Boundary Layers with Second-Moment Closure: Part II – Effects of Streamline Curvature and Spanwise Rotation", *J. Fluids Eng.*, 115, 56–63.

Sinha, S., Sotiropoulos, F. and Odgaard, A.J. (1998), "Three-Dimensional Numerical Model for Turbulent Flows Through Natural Rivers of Complex Bathymetry", *ASCE J. Hydr. Eng.*, 124(1), 13–24.

So, R.M.C., Lai, Y.G., Zhang, H.S. and Hwang, B.C. (1991), "Second Order Near-Wall Turbulence Closure: A Review", *AIAA J.*, 29, 1819.

Sofialidis, D. and Prinos, P. (1999), "Numerical Study of Momentum Exchange in Compound Open Channel Flow", *ASCE J. Hydr. Eng.*, 125(2), 33–39.

Sotiropoulos, F. and Patel, V.C. (1993), "Evaluation of Some Near-wall Models for the Reynolds-stress Transport Equations in a Complex 3-D Shear Flow". Proceedings of International Conference on Near-Wall Turbulent Flows, So, C.G. Speziale and B.E. Launder (eds), Tempe, AZ, 987–998.

Sotiropoulos, F. and Patel, V.C. (1994), "Prediction of Turbulent Flow Through a Transition Duct Using a Second-Moment Closure", *AIAA J.*, 32(11), 2194–2204.

Sotiropoulos, F. and Patel, V.C. (1995a), "On the Role of Turbulence Anisotropy and Near-Wall Modelling in Predicting Complex, 3D, Shear Flows", *AIAA J.*, 33(3), 504–514.

Sotiropoulos, F. and Patel, V.C. (1995b), "Application of Reynolds-Stress Transport Models to Stern and Wake Flows", *J. Ship Res.*, 39(4), 263–283.

Sotiropoulos, F. and Ventikos, Y. (1998), "Flow Through a Curved Duct Using Nonlinear Two-Equation Turbulence Models", *AIAA J.*, 36(7), 1256–1262.

Spalart, P.R. (2000), "Strategies for Turbulence Modelling and Simulations", *Int. J. Heat Fluid Fl.*, 21(3), 252–263.

Speziale, C.G. (1987), "On Nonlinear k-l and k-ε Models of Turbulence", *J. Fluid Mech.*, 178, 459–475.

Speziale, C.G., Sarkar, S. and Gatski, T.B. (1991), "Modelling the Pressure-Strain Correlation of Turbulence – An Invariant Dynamical Systems Approach", *J. Fluid Mech.*, 227, 245–272.

Tani, J. (1987), "Turbulent Boundary Layer Development over Rough Surfaces", *Perspectives in Turbulent Studies*, Springer, Verlag.

Thomas, T.G. and Williams, J.J.R. (1995), "Large-Eddy Simulation of Turbulent Flow in an Asymmetric Compound Channel", *J. Hydr. Res.*, 33(1), 27–41.

Viollet, P.L. (1985), "The Modelling of Turbulent Recirculating Flows for the Purpose of Reactor Thermal-Hydraulic Analysis", Rep. E44/R05, Elec. de France, Chatou, France.

Wilcox, D.C. (1988), "Reassessment of the Scale Determining Equation for Advanced Turbulence Models", *AIAA J.*, 26, 1299–1310.

Wilcox, D.C. (1993), Turbulence Modelling for CFD, DCW Ind., La Cañada, CA.

Wilcox, D.C. (1994), "Simulation of Transition with a Two-Equation Turbulence Model", *AIAA J.*, 32(2), 247–255.

Wu, W., Rodi, W. and Wenka, T. (2000), "3D Numerical Modelling of Flow and Sediment Transport in Open Channels", *ASCE J. Hydr. Eng.*, 126(1), 4–15.

Yoon, J.Y. and Patel, V.C. (1993), "A Numerical Model of Flow in Channels with Sand-Dune Beds and Ice Covers", IIHR Report No. 362, Iowa Institute of Hydraulic Research, University of Iowa, Iowa City, IA.

Yoon, J.Y. and Patel, V.C. (1996), "Numerical Model of Turbulent Flow Over Sand Dune", *ASCE J. Hydr. Eng.*, 122(1), 10–18.

Yoon, J.Y., Patel, V.C. and Ettema, R. (1996), "Numerical Model of Flow in Ice-Covered Channel", *ASCE J. Hydr. Eng.*, 122(1), 19–26.

Zedler, E.A. and Street, R.L. (2001), "Large-Eddy Simulation of Sediment Transport: Currents Over Ripples", *ASCE J. Hydr. Eng.*, 127(6), 444–452.

6
Modelling wetting and drying processes in hydraulic models

P.D. Bates and M.S. Horritt

6.1 Introduction

Dynamic changes in flow field extent are a critical aspect of many environmental flows including floods, estuarine flows, coastal flows and dam breaks. Such problems all involve a moving shoreline for transient problems or one whose location is not known *a priori* in the steady state case. They therefore belong to a general class of moving boundary problems in CFD (see Löhner, 2001, pp. 225–237, for a discussion of approaches in aeronautics), which also includes simulating the displacement of the free surface of water in unsteady 3D Navier–Stokes or 2D vertical computations. Both horizontal and vertical moving boundaries typically require additional algorithms to be incorporated into the code to ensure their correct treatment. A number of such algorithms are available, and the method chosen can have significant consequences for both computational cost and physical realism of the resulting simulations. In particular, previous studies in environmental hydraulics have shown that the manner in which dynamic moving boundary problems are treated may be significant to representing adequately the mass and momentum conservation (King and Roig, 1988; Bates and Hervouet, 1999), frictional losses (Defina, 2000) and sediment and pollutant transport (Moulin and Ben Slama, 1998) for such flows. As well as being prevalent in environmental hydraulics, moving boundary flows are often some of the most significant practical problems tackled by CFD methods. Floods and dam breaks, for example, are significant environmental hazards which result in loss of life and considerable socio-economic damage. Better prediction of dynamic changes in flood extent is therefore a key issue in environmental management, as well as raising fundamental scientific issues and numerical challenges.

6.2 Fluid dynamics of moving boundary problems in environmental hydraulics

Specific evidence from laboratory and field studies detailing the fluid dynamics of horizontal wetting and drying processes for fluvial flows is generally lacking. Some laboratory investigations of 1D and 2D dam-break flows have been conducted for simple topography (e.g. Shige-Eda and Akiyama, 2003); however, no experimental studies of flooding over complex natural topography have yet been conducted. Field evidence of wetting and drying is even more difficult to obtain. This is especially difficult in the case of dam-break flows, and here some of the few quantitative data are the wave propagation speeds for the Malpasset dam break reported by Hervouet (2000). These were obtained because the catastrophic flood wave resulting from the near-instantaneous failure of the Malpasset dam in southern France destroyed three electricity substations in the downstream valley and the timing of their shutdown (and hence the arrival of the flood wave) could be established with a high degree of precision. For field observations of wide-area flooding, the best available data consist of maps of flood extent derived from aerial photography or satellite and airborne synthetic aperture radar (Bates *et al.*, 2004). However, in all cases, only a single inundation image per event is available and such data therefore provide little evidence of the dynamic propagation of the flood wave across the floodplain. To the authors' knowledge, the only detailed field studies of flooding processes have been conducted by Simm (1993) and Nicholas and Mitchell (2003). Both studies were conducted on the same reach of River Culm, Devon, UK, albeit at different locations. The River Culm is a relatively narrow channel approximately 10 m wide meandering across a topographically complex floodplain which varies in width between 200 and 500 m. Overbank flooding occurs on about six occasions per year making the site ideal for this kind of investigation. Simm (1993) produced field sketches of flood extent during three flood events, whilst Nicholas and Mitchell (2003) constructed a Digital Elevation Model (DEM) from detailed ground survey for a channel–floodplain section approximately 1 km long and then mapped inundation sequences and water depths for 12 floods that occurred in the period 1999–2001. Nicholas and Mitchell (2003) showed for this site that the sequence of inundation did not vary between events, although greater discharge did lead to increased inundation. Inundation patterns were complex and strongly dependent on local topography, with some evidence of hysteresis between the wetting and drying phases of the floods studied. However, this study is the exception and our understanding of the fluid dynamics of wetting and drying is very limited. In particular, almost no field measurements of wave propagation speeds over floodplains or nearshore flow velocities have been conducted.

A rather better understanding of wetting and drying processes has been achieved in coastal studies for the case of the run-up of non-breaking waves on beaches. As with dam break-type flows (and unlike flooding over fluvial or tidal floodplains), beach run-up lends itself to laboratory experimentation and many such tests have been conducted in order to provide guidelines for designing coastal structures (e.g. Owen, 1980; De Waal and Van Der Meer, 1992; Briggs *et al.*, 1995; Liu *et al.*, 1995). Several analytical solutions for beach run-up of non-breaking waves also exist including the work of Carrier and Greenspan (1958), Ryrie (1983), Thacker (1981), Synolakis (1987) and Brocchini and Peregrine (1996) based on simplified solution of the 1D or 2D shallow water equations (Chapter 2). However, beach run-up has only limited parallels with wave propagation over fluvial or tidal floodplains. Beach slopes are generally greater, and the duration of a wave running up a beach is several orders of magnitude smaller than that of a wave propagating over a floodplain. Moreover, as with all laboratory experiments and analytical solutions, the domain topography in these studies is typically highly idealised and uniform. This lack of similarity to the conditions found in real-world fluvial systems means that such tests provide only limited insight into the fluid dynamics of wetting and drying for fluvial problems, although they may be a good choice with which to begin validation of different wetting and drying algorithms.

Unlike the horizontal case, vertically moving boundaries resulting from displacement of the free surface of water are much more straightforward as this is just a function of the fluid movement itself, rather than the product of the fluid's interaction with a complex boundary. Nevertheless, many unsteady fluvial flows (floods, tides, tsunamis, dam breaks, etc.) involve substantial surface displacement and thus here too there is a requirement for accurate and robust tracking of the domain deformation.

This lack of empirical evidence means that most modelling approximations for horizontal moving boundary shallow water flows are justified on purely theoretical grounds. For example, floods, estuarine and dam-break flows are all typified by high width-to-depth ratios and the shallow water approximation (Chapter 2) is often therefore justified. Vertical pressure distributions can thus be assumed hydrostatic and the variation in vertical velocity (relative to the horizontal variation) is deemed negligible. Consequently, moving boundary flows in environmental hydraulics are often assumed to be two dimensional and to only vary significantly in the x and y cartesian directions. This is not to imply that 3D flow processes do not exist in particular regions of a shallow water flow, merely that their impact is assumed not to dominate the evolution of the moving shoreline or is subsumed into another term in the governing equations (e.g. bed friction or turbulence). For example, predicting correctly the shoreline location in the far field of a floodplain flow may not be strongly dependent on detailed flow dynamics in the near field region close to the channel where processes are known to be strongly three dimensional (Chapter 11). Partly, this is a consequence of scale, as most moving boundary problems involve the lateral extension of the flow field over significant distances (perhaps 1–10 km) away from the river or estuarine main channel or dam, and regions of strong 3D flow or large-scale coherent turbulence may therefore be relatively limited in spatial extent. For this reason too, Reynolds-averaging is often used to treat turbulence, with the Boussinesq assumption and a relatively simple closure model invoked to deal with the resulting unknown turbulent stresses (for a more detailed discussion see Chapter 5). However, further simplification to a 1D treatment is generally not possible with shallow water flows over complex topography because of the difficulty of predicting flow paths *a priori*.

As well as perhaps being over-specified for the problem in hand, certain types of 3D models may also suffer from numerical stability problems when used to simulate dynamic moving boundary flows. In particular, models which use a deformable mesh (Soulaimani and Saad, 1998) or the σ-transform (Stoesser *et al.*, 2003) to discretise the grid in the vertical may suffer from stability problems during dynamic shallow water flows as with these methods the upper boundary of the computational domain moves with the water free surface. If the number of vertical grid layers remains fixed, then cell height-to-length ratios may become highly distorted as water depth, h, approaches zero. Whilst the Volume of Fluid (VOF) method (Ma *et al.*, 2002) can be used to track the horizontal moving boundary in 3D models whilst retaining fixed vertical grid increments (Chapter 11), avoidance of distorted elements for flows with horizontal extent 100–10 000 m and depths generally much less than 1–10 m still requires numerical grids that are horizontally very highly resolved and which are likely to incur a prohibitive computational cost. Thus, to date most 3D numerical models of compound channel flow have not been applied to problems with significant changes in domain extent over low-gradient floodplains. Neither may 3D approaches be necessary, as for many scales of compound channel flows the shallow water approximation may be adequate. For these reasons, the rest of this chapter is largely concerned with the treatment of moving boundaries in 2D codes, although some of the approximations to wetting and drying processes could also be used in 3D codes and these are therefore mentioned where appropriate.

Modelling studies have shown that for dynamic shallow water flows there will be a complex and scale-dependent relationship between flow hydraulics, frictional losses and topography (Marks and Bates, 2000; Bates *et al.*, 2003). The best field evidence for this actually comes from studies of spatial variability of floodplain sediment deposition (Walling and He, 1996) but the precise hydraulic mechanisms have yet to be clearly established. In particular, Balzano (1998) notes that there is an insufficient knowledge of resistance laws which describe flow of a thin film of water over topography with irregularities approximately the same magnitude as the flow depth. We can also note that as most regions over which shallow water flows extend are vegetated (floodplains, salt marshes, etc.), correct treatment of frictional losses may also be complex (Chapter 15).

Given the above perceptual model of moving boundary flows in environmental hydraulics, the shallow water approximation is typically invoked and numerical modelling of such problems largely makes use of 2D Reynolds-averaged codes. Here the vertical momentum equation is not solved. Instead, the Reynolds-averaged Navier–Stokes equations are integrated over the flow depth to yield the water depth and the two components of the depth-averaged velocity (for further details see Chapter 2). The resulting equations can be given in non-conservative form as:

Continuity equation

$$\frac{\partial h}{\partial t} + h\nabla \mathbf{u}_\mathrm{d} + \mathbf{u}_\mathrm{d} \cdot \nabla h = 0 \tag{6.1}$$

Momentum equation

$$\frac{\partial \mathbf{u}_d}{\partial t} + (\mathbf{u}_d \cdot \nabla)\mathbf{u}_d + \frac{\nu_t}{h}\nabla(h\nabla \cdot \mathbf{u}_d) + g\nabla h = \mathbf{S} - g\nabla Z_f \qquad (6.2)$$

where $\mathbf{u}_d$ is the depth-averaged velocity vector with components u_d and v_d (with dimensions [LT^{-1}]) in the x and y Cartesian directions [L]; Z_f is the bed elevation [L]; ν_t is the kinematic turbulent viscosity [L^{2}T^{-1}]; **S** is a vector of source terms (friction, Coriolis force and wind stress) and g is the gravitational acceleration [LT^{-2}]. Of these, Ip et al. (1998) state that for many shallow water problems the primary force balance is between friction and the pressure or free surface gradient, and hence the source term is dominated by the flow friction, for example given by Manning's equation and a roughness specified by Manning's n:

$$\mathbf{S} = \frac{gn^2|\mathbf{u}_d|\mathbf{u}_d}{h^{4/3}} \qquad (6.3)$$

Shoreline dynamics in 2D models can be considered by defining the shoreline as the contour at which $h = 0$, and therefore ∇h is normal to the shoreline. From (6.1) we see that the second term vanishes at the shoreline. For the steady state case this means that the velocity is normal to ∇h and hence to the shoreline. Given a shoreline location (x, y) at time t, so that $h^t(x, y) = 0$, the location at time $t + dt$ will be given by:

$$h^{t+dt}(x + dx, y + dy) = h^t(x, y) + \frac{\partial h}{\partial t}dt + \frac{\partial h}{\partial x}dx + \frac{\partial h}{\partial y}dy = 0 \qquad (6.4)$$

Substituting for the time derivative using (6.4) and dividing by dt we obtain:

$$\mathbf{u}_d \cdot \nabla h = \frac{\partial h}{\partial x}\frac{\partial x}{\partial t} + \frac{\partial h}{\partial y}\frac{\partial y}{\partial t} \qquad (6.5)$$

Since the shoreline velocity is itself normal to the shoreline (and hence parallel to ∇h), we conclude that the shoreline moves with a velocity given by the component of $\mathbf{u}_d$ normal to the shoreline. This produces a fundamental difference between the hydraulics of the advancing and receding shorelines and provides a theoretical explanation for the field observations of Nicholas and Mitchell (2003); although for complex river reaches topography may be a more important factor in determining differences between wetting and drying behaviour. For a wetting front, this non-zero velocity introduces a singularity into the momentum equation (6.2) through the friction term (6.3), as $h \to 0$ and this will tend to generate large free surface gradients in the shoreline region in order to overcome the large friction forces. For a drying front, the exact nature of the flow near the shoreline becomes unclear. Since the free surface gradient cannot exceed the bed slope, a driving term capable of generating the fluid velocity required to move the shoreline despite the shallow depth is absent. Fluid near the shoreline will therefore tend to form into a thin 'film' rather than a definite drying front and become difficult to treat with standard resistance laws (cf. Balzano, 1998). The shallow depth of this film will also tend to make it susceptible to topographic features smaller than the grid scale of the model, and to other flow processes like infiltration not included in the shallow water equations.

As noted above, correct treatment of such a moving boundary in hydraulic models is not straightforward, and a modeller must typically introduce some additional approximations to deal with the problems generated. Such approaches can be classified according to whether they deform the grid to follow the moving shoreline (a Lagrangian or Lagrangian–Eulerian approach) or employ a fixed grid (a Eulerian approach) but implement some further approximations to the controlling equations for computational cells at the boundary. These methods are discussed in sections 6.3 and 6.4.

6.3 Deforming grid methods

A moving computational grid effectively removes many of the problems associated with the treatment of wetting and drying problems in hydraulic models by ensuring that the boundary of the computational mesh coincides exactly with the shoreline (Lynch and Gray, 1980). Thus the whole model domain can be considered wet, with depth going to zero only at boundary nodes. The mesh is deformed appropriately to maintain this condition: the boundary nodes moving with the velocity component normal to the shoreline. Examples of deforming grid approaches in fluvial hydraulics can be found in Jamet and Bonnerot (1975), Lynch and Gray (1980), Johns et al. (1982) Gopalakrishnan and Tung (1983), Kawahara and Umetsu (1986), Akanbi and Katopodes (1988), Gopalakrishnan (1989), Roig and Evans (1993), Petera and Nassehi (1996), Christian and Palmer (1997), Soulaimani and Saad (1998), Behr (2001), Tezduyar (2001), Beckett et al. (2002), Cha et al. (2002) and Feng and Perić (2003).

The general approach taken by such codes is to rewrite the controlling equations into an Arbitrary Lagrangian–Eulerian (ALE) reference frame (Hirt et al., 1974; Löhner, 2001, p. 225). This is usually chosen over a fully Lagrangian approach (e.g. Feng and Perić, 2003) as this latter method tends to result in serious numerical difficulties in the interior of the flow domain (Bottasso, 1997). In the ALE reference frame one defines a mesh velocity vector $\mathbf{w} = (w^x, w^y)$, which then becomes an additional velocity term in equations (6.1)–(6.3). In the case of no element movement ($\mathbf{w} = 0$) we recover the Eulerian form of the equations of motion and in the case where elements move with the fluid velocity ($\mathbf{w} = \mathbf{u}_d$) we recover the Lagrangian form (Löhner, 2001, p. 225). Calculation of the mesh velocity field then proceeds as normal, but with modified source and flux terms that account for $\mathbf{w}$. This allows node movement to be tracked over time and for node locations in particular regions of the domain, such as the flow interior, to remain fixed at their initial positions (e.g. Lynch and Gray, 1980). According to Lynch and Gray (1980), simulation begins with the specification of an initial grid and initial/boundary conditions. The solution then marches forward in time with nodes moved according to $\mathbf{w}$ while maintaining the initial grid connectivity. At some time the mesh may become unacceptably distorted due to the cumulative movement and remeshing of the domain will be required. Lynch and Gray (1980) note that remeshing will become necessary when one of two problems arises:

1. Computational cells become distorted or the node distribution becomes either too sparse or dense, thereby leading to instability, low accuracy or unnecessary computational cost.
2. The grid becomes incapable of representing accurately key spatial fields such as topography or friction.

Once the domain has been remeshed to an acceptable quality, and the model state variables and parameters interpolated to the new grid, the simulation again proceeds forward in time with incremental node movement. The whole process therefore requires calculation of some spatially distributed measure of mesh quality, both in terms of the numerical solution (Chapter 8) and its ability to represent key spatial fields (e.g. Bates et al., 2003) at each time step where node movements occur. Within this general framework, various deforming mesh schemes can be implemented. For example, the method of Lynch and Gray (1980) is based on finite elements which are allowed to deform continuously, whilst Gopalakrishnan and Tung (1983) use elements which split into two parts when the element length becomes greater than 1.2 times the initial length.

Deforming grid approaches have also been used to track the moving surface boundary in 2D vertical models (e.g. Feng and Perić, 2003), and in 3D models to track all domain boundaries as the fluid evolves (e.g. Soulaimani and Saad, 1998; Feng and Perić, 2003). For example, Feng and Perić (2003) describe a 3D Lagrangian model that uses four-node tetrahedral elements to give a fully unstructured grid. However, even here testing has only been attempted for limited time periods on simple hypothetical domains, with the simulations terminated before remeshing to overcome element distortion was required. Significant development of fully unstructured numerical approaches and remeshing algorithms is therefore required before 3D deforming grid approaches can be applied to practical fluvial problems. Moreover, while the numerical problem of deforming the grid to follow a moving surface boundary is practically identical to, or perhaps even simpler than, the 2D horizontal approaches described earlier, the need to represent very shallow flow depths over large horizontal regions of floodplain or tidal flat without distortion of cell height-to-length ratios may incur a prohibitive computational cost in fully 3D codes.

While providing a rigorous numerical solution to wetting and drying problems, dealing with a deformable mesh represents a considerable computational burden. First, the need to track exactly an irregular shoreline and allow greater element density in particular parts of the domain gives an advantage to methods based on unstructured grids such as Finite Volume (FV) and Finite Element (FE) approaches. These methods necessarily incur a greater computational burden because of the more complex grid topology and more involved mathematics than methods based on structured grids such as the Finite Difference (FD) approach (Chapter 7). Second, testing for mesh quality (Chapter 8), remeshing and interpolation are all non-trivial programming problems (Lynch and Gray, 1980). Remeshing will also affect both the algebraic systems to be solved, as it defines the FE basis and weighting functions or the fluxes between control volumes in the FV method, and the interpolated topography. The latter may be a particular problem for environmental studies where, unlike industrial engineering applications, the geometry may not be known exactly and therefore interpolation errors may vary across the domain in an uncontrolled way during the simulation as the grid changes. It may therefore be very difficult to ensure grid-independent solutions for real-world applications of such models, particularly in the case of unstructured grids. For example, Hardy *et al.* (1999) show that for typical FE solutions of the shallow water equations that mesh independence may be almost impossible to obtain even for a hypothetical smooth floodplain topography. Lastly, a relatively high-resolution mesh may be required in the vicinity of the shoreline in order to predict accurately boundary node velocities even though velocity and water surface gradients in such regions may actually be relatively low. Unstructured grids, remeshing and high levels of grid refinement at the shoreline will all lead to significantly increased computational costs for deforming grid approaches.

Given finite computational resources, the additional computational burden of adaptive gridding can be compensated for only by using a smaller number of computational nodes than would be possible for the same cost in a fixed grid model. The consequences for the modeller are then a restriction to either smaller domain sizes, simpler topography or more coarse numerical grids which allow a lower degree of topographic complexity to be included in the model. This is critical as topography is clearly a major control on wetting front propagation over previously dry areas (Nicholas and Mitchell, 2003). As Tchamen and Kawahita (1998) note, for these reasons previous applications of deforming grid approaches in fluvial hydraulics have been restricted to relatively small hypothetical domains with simple topography. At present we do not know whether shoreline tracking or topographic inclusion is the more critical factor for accurate modelling of dynamic wetting and drying processes. However, benchmark studies of various flood inundation models (Bates and De Roo, 2000; Horritt and Bates, 2001) have shown that model performance in predicting inundation extent is more likely to be a function of grid resolution (and hence the representation of topography) than the physical or numerical sophistication of the model used. Despite the fact that it may be more difficult to treat moving boundaries in fixed grid models, the need to include topographic complexity down to relatively small scales, in order to model shoreline evolution accurately, may continue to give an advantage to fixed grid (Eulerian) codes for the foreseeable future. Accordingly, these are discussed in section 6.4.

6.4 Fixed grid methods

Given the computational disadvantages associated with deforming meshes, most current hydraulic models of moving boundary shallow water flow adopt a fixed grid approach, whereby the discontinuity in hydrodynamic behaviour occurring at the shoreline is dealt with within fixed elements. The consequences of this treatment for a given model varies according to (1) the flow hydraulics and (2) the specific topographic discretisation implemented. First, during gradually varied flow elements usually wet from the bottom up and drain under gravity, whilst during dam-break flows of water may enter an element from the top as the wave flows down the slope, and this will impact on the type of correction required. Second, topography may be

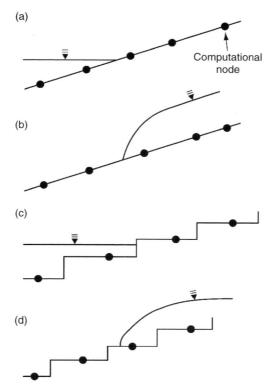

Figure 6.1 *Topographic discretisation in computational fluid dynamics models: continuous slope approximation during gradual flooding (a) and dam-break flows (b); staircase approximation during gradual flooding (c) and dam-break flows (d).*

discretised as either continuous or discontinuous (Figure 6.1). For continuous topography (Figure 6.1a and b), the bed gradient varies continuously, and usually linearly, between computational nodes to represent the topography in one dimension as a series of piecewise segments or in two dimensions as a series of surface patches with non-zero slope. This is typical of many unstructured FE models (e.g. MacArthur *et al.*, 1990; Hervouet and Van Haren, 1996) where the computations are performed at the element vertices. In the second method (Figure 6.1c and d), the bed slope across each computational cell is zero and the topography is represented as a series of planar horizontal elements forming a discontinuous 'staircase'. This representation is typically used in structured grid models, including many FD, FV and storage cell codes (e.g. Bates and De Roo, 2000). For clarity, wetting and drying approaches relevant to continuous and discontinuous topography will be discussed separately in sections 6.4.1 and 6.4.2.

6.4.1 Fixed grid methods with continuous topography

Figure 6.1a makes clear that representation of moving boundary problems in fixed grid models with continuously sloping topography leads unavoidably to the

130   Computational Fluid Dynamics

existence of partially wet elements at the shoreline. This leads to undesirable numerical consequences that need to be corrected with additional model approximations. For continuously sloping topography three broad classes of wetting/drying treatment are available:

1. To initialise the model with a thin layer of water everywhere in the domain;
2. To exclude partially wet elements from the computational domain;
3. To include partially wet elements within the computational domain.

However, none of these methods are ideal. For example, initialising a model with water everywhere and then maintaining this thin film as the solution progresses (e.g. Zhang and Cundy, 1989) may result in an incorrect mass of water in the domain and may also affect wave propagation, as wave behaviour is known to vary markedly depending on whether it occurs over wet or dry beds (Bradford and Sanders, 2002). Similarly, inclusion or exclusion of partially wet elements from the numerical solution both result in discrepancies in mass and momentum conservation which can have consequences for both model stability and predicted hydraulics over the entire domain. By excluding elements we effectively assume that a vertical barrier exists at the first wet node back from the boundary and that, if required, the shoreline location during gradual flooding can be estimated by horizontal inland extrapolation (Figure 6.2a). Lynch and Gray (1980) point out that where boundary motion is significant this approximation is likely to fail in a number of ways. First, continuity is violated as we exclude some mass of water from the computational domain. Whilst the model may thus still be mass conserving, the wrong physical quantity is being conserved. Second, there will also be a corresponding deficit in momentum due to non-representation of shallow water frictional losses, the impact of which may be felt across the whole domain. Lastly, waves incident to the vertical boundary may be

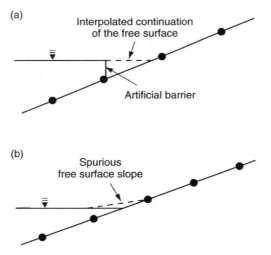

Figure 6.2   Possible boundary treatments in fixed grid models with continuous topography: (a) element exclusion approximation during gradual flooding; and (b) element inclusion approximation during gradual flooding.

reflected perfectly, whereas in reality frictional losses would attenuate and delay these motions. Not taking this factor into account may lead to significant error in transient simulations. Inclusion of partially wet elements within the numerical solution (Figure 6.2b) will also lead to the wrong physical quantities being conserved as we now treat a volume of air as if it were water. Moreover, in such elements water depth is zero at one or more nodes of a computational element but non-zero at the remainder. In this case the numerical model will interpolate a spurious water surface gradient across the element that may lead to the generation of unrealistic velocities in nearshore areas (Hervouet and Janin, 1994).

Both inclusion and exclusion methods also suffer from the fact that small changes in water depth can lead to large step changes in the horizontal position of the predicted wet/dry boundary. Not only are these movements unrealistic, but they are also potentially numerically unstable and can generate spurious disturbances or oscillations which propagate from the boundary. The only solution to these problems may be to refine the mesh at the boundary of the domain, despite the fact that velocity and water surface gradients there may be low and not require such high resolution grids in order to develop realistic simulations. A fixed grid wetting and drying algorithm for continuous topography therefore needs to overcome the above problems in an efficient manner to enable an element density appropriate to the process gradient to be employed.

Tchamen and Kawahita (1998) also note difficulties with numerical solvers for shallow water flows over both continuous and discontinuous topographies. They suggest numerical instabilities may originate in frontal cells in the form of negative values of the flow depth, h, and/or velocity values that are too large. These result because either the manner in which to perform a rigorous discretisation of the basic conservation laws over frontal cells is not readily apparent or because of the use of a non-monotone scheme. Unphysical negative depths may also result from round-off errors in regions with small depth and large velocities as a result of divisions by h when h is very small. Clearly, these may have adverse impacts on the mass conservation properties and dynamics of the numerical solution.

Solutions to the above problems generally consist of a number of key steps. First, one needs to employ a numerical solver that is robust for shallow water flows and minimises the occurrence of unphysical negative depths or oscillations in the solution. Such oscillations occur frequently with Galerkin FE solvers used for advection-dominated flows (Brookes and Hughes, 1982) and would clearly compound problems with impulsive boundary movement during dynamic wetting and drying on fixed grids as elements switch into and out of the solution. With this basic requirement in place, the next step in dealing with moving boundary problems in fixed grid models is to correctly identify the partially wet or frontal cells. At this point a decision needs to be taken to follow either an exclusion or inclusion approach. With exclusion methods, the partially wet cells can then merely be masked from the solution, a strategy that has been used to some effect despite the potential drawbacks noted previously (Horritt, 2002; Quecedo and Pastor, 2002). Exclusion methods have also been used in 3D fixed grid models which use a vertical grid discretisation that tracks the movements of the surface boundary (e.g. Zheng *et al.*, 2003) as this avoids the generation of a singularity in elements where one or more nodes have a depth of zero. However, for shallow water depths over low gradient and complex topography, this may still lead to unacceptable element distortion in near-shore areas. An alternative solution for

3D modelling is presented by Lin and Falconer (1997) who chose the height of the surface layer of the model grid for their 3D estuarine model to be greater than the tidal range, thereby confining wetting and drying processes to that layer and allowing them to use the 2D wetting and drying treatment of Falconer and Chen (1991) to deal with moving boundary effects. For inclusion methods, it is common to try to correct for the spurious effects so generated. Accordingy, elements are then classified according to whether they are of 'flooding' or 'dam break' type (see Figure 6.1) and some appropriate algorithm to compensate for the resulting mass or momentum discrepancies can be then implemented.

Recent research has identified a number of appropriate numerical solvers for the shallow water equations applied to moving boundary problems. For example, non-oscillatory Galerkin FE solvers have been developed, such as the Streamwise Upwind/Petrov–Galerkin (SUPG) technique of Brookes and Hughes (1982). This method counteracts the oscillations that develop with Galerkin FE methods by adding diffusion to the solution, but only in a streamwise direction in order to minimise the impact of this additional diffusion on solution quality. Such methods reduce, but do not eliminate the negative depth problem and have been used successfully in a number of shallow water modelling studies (e.g. Bates *et al.*, 1998; Horritt, 2000). For FV methods, much attention has focused on the use of Riemann solvers (Van Leer, 1979; Roe, 1981; Toro, 1992) to evaluate the fluxes at the cell faces as such methods are able to capture wave propagation accurately (Bradford and Sanders, 2002). This method also has the advantage (Nicholas and Mitchell, 2003) that it is able to resolve the transition between subcritical and supercritical flows (e.g. in a dam break). Riemann solvers have been applied in fixed grid numerical models to simulate 1D and 2D dam-break flows (Alcrudo and Garcia-Navarro, 1993; Savic and Holly, 1993; Jha *et al.*, 1995) and wave run-up (Brocchini *et al.*, 2001), and have been modified to deal with situations involving variable topography, including fluvial flooding (Zhao *et al.*, 1994; Nujić, 1995; Tchamen and Kawahita, 1998; Beffa and Connell, 2001; Nicholas and Mitchell, 2003).

Next, one requires a scheme to identify and classify the partially wet cells correctly. For continuous topography, four types of element can be defined (Figure 6.1); fully wet, fully dry, partially wet (dam break-type) and partially wet (flooding type). For elements undergoing gradual flooding, wetting and drying occur from the bottom up and the regional water surface slope is typically small and its level lies within the height range of the partially wet element. By contrast, dam-break elements undergo rapid wetting from above or below as the leading edge of the flood wave moves up- or downslope. Such effects may not be solely confined to dam-break flows and can occur in fluvial flows where, for example, water flows into an enclosed depression on the floodplain. However, here we use the term 'dam-break element' to refer to this situation. Regional water slopes are therefore significantly non-horizontal and lie outside the height range of the element. A potential fifth type of element also exists, namely the particular case of a dam break floodwave arriving at an already partially flooded element. Of these only flooding-type elements may conclusively be identified as having spurious mass and momentum conservation properties within numerical schemes, yet dam break-type elements, share many distinguishing features. In particular, both contain one or more dry nodes and the interpolated free surface over the element is non-horizontal. However, for flooding-type elements it is intuitive that this results in mass and momentum conservation discrepancies, while for dam-break problems such a representation has at least the potential to be physically realistic.

Identification of partially wet elements in numerical models which use inclusion methods over continuous topography is further complicated by the fact that the numerical solution is initially applied to the partially wet zone and small positive water depths may be calculated at nodes which should otherwise be dry on the basis of the local water surface slope. This is a consequence of attempting to simulate a continuous water surface with a discrete numerical grid. Merely searching for elements where one or more nodes are dry may not, therefore, fully identify the set of cells to modify. To overcome this problem, previous wetting and drying schemes have used an arbitrarily assigned a small positive depth (King and Roig, 1988; Leclerc *et al.*, 1990), typically denoted ε, to define the wet/dry threshold. More sophisticated methods use the bed roughness length to assign the value for ε (Falconer and Chen, 1991), or use an analysis of the local water slope to overcome the need for an empirically assigned cut-off value (Hervouet and Janin, 1994). In the Hervouet and Janin (1994) scheme, a check is made to test whether the bed elevation at a given node is greater than the free surface elevation at any of the other nodes in the same element to which it belongs. Both these methods adequately identify all partially dry elements, but fail to distinguish between dam break and flooding types. As noted above, dam-break elements can satisfy both the above criteria, yet on the basis of the nodal hydraulic information available we cannot be certain that they require any mass or momentum correction. Rather, if we require a better representation of the wave front for dam break problems our only option is to refine the mesh in this vicinity. It is, however, important to distinguish between flooding type and dam break-type elements, in order that any correction scheme is applied only to the former class. A method to achieve this was presented by Bates and Hervouet (1999). Here the criterion of Hervouet and Janin (1994) was used to identify the complete set of flooding and dam-break elements. For these elements a further check was made to determine whether the regional or 'true' free surface lay within the height range of the element as defined by its minimum and maximum nodal elevations. In order to be able to apply the scheme to complex topography, assessment of the regional water surface level could not use an analysis of the free surface profile at adjacent nodes. Instead, Bates and Hervouet (1999) simply took for a given partially wet element the free surface height at its constituent node where the water depth was at a maximum as being the regional water level. If this free surface height is within the topographic range of the element as defined by its nodal bathymetry, then the element was classified as being of flooding type, if not then this was taken as indicative that a flood wave was arriving rapidly from above or below the element and as a consequence the element was classified as being of dam break type. This distinguishes between flooding and dam break situations for a linear FE model and also applies to more complex element shapes. Bates and Hervouet (1999) considered this as an improvement over previous identification methods, which were largely developed for relatively coarse FE mesh discretisations with simple topography and were thus confined to a narrow range of applications. Dam break problems were not studied with such algorithms and the simple topography meant that dam break-type elements were unlikely to occur in the flooding-type problems that were simulated. It should also be noted that both the ε threshold criteria and the more complex method of Bates and Hervouet (1999) could be used with models that exclude partially wet cells.

Once the partially wet cells have been identified they can be dealt with via exclusion or inclusion within the computational domain. In the latter case it is common for some correction to be applied to deal with the spurious hydraulic effects occurring over such elements. Numerous schemes have been suggested for such elements (see Balzano, 1998, for a detailed review) including:

1. modifications to the momentum conservation equations;
2. modifications to the mass conservation equation;
3. no correction to the controlling equations but with mesh adaptation at the shoreline using an interface tracking method.

Momentum equation corrections have been proposed by Hervouet and Janin (1994), Leclerc et al. (1990), Defina et al. (1994), Ip et al. (1998) and Defina (2000). For example, Hervouet and Janin (1994) suggest a method to treat the spurious water surface slopes that occur in partially wet elements in a continuous topography. Here such elements are included within the solution and it is recognised that the dominant driving term in the momentum equation for such elements will be:

$$\frac{\partial \mathbf{u}_d}{\partial t} = -g \frac{\partial Z_f}{\partial x} \qquad (6.6)$$

Other terms still act on such an element but will be either easy to calculate, e.g. friction, or be relatively unimportant, e.g. diffusion. The above water slope term may then be cancelled or replaced with a value more representative of the regional water surface slope to prevent the development of unrealistic velocities due to the spurious non-zero free surface gradient across the element. This is achieved by the analysis of free surface height at the node within the element where the water depth is at a maximum in a manner similar to the means used to distinguish between flooding-type and dam break-type elements. In the Leclerc et al. (1990) scheme, not only the free surface slope terms, but also any external source terms (**S** of equation 6.2) such as friction, Coriolis force or wind stress, are cancelled. In effect, frictional forces are subsumed within the turbulent viscosity parameter, ν_t, which introduces uniform smoothing over the element. This method is therefore similar to the Hervouet and Janin (1994) scheme but, in addition, requires careful choice of ν_t. In the Defina scheme (Defina et al., 1994; Defina, 2000), the friction term in the momentum equations is modified to take account of the additional form roughness generated by sub-grid topographic variations. This is achieved using steady state, uniform flow theory, assumptions about sub-grid topographic properties and an empirical relationship between these and the resulting energy loss.

Schemes which address mass conservation on partially wet elements include the model RMA-2 (King and Norton, 1978; MacArthur et al., 1990; Bates et al., 1992; Neilsen and Apelt, 2003), the model of Defina (Defina et al., 1994; Defina, 2000), the model of Ip et al. (Ip et al., 1998; Erturk et al., 2002) and the extension of the TELEMAC-2D model proposed by Bates and Hervouet (1999). In these near-identical methods a scaling coefficient, η, is defined which varies (Figure 6.3) from 0 to 1 for each element as it trends from fully dry (0) to fully wet (1). This coefficient (essentially equivalent to the porosity in groundwater modelling) is then used to scale the continuity equation to represent the true volume of water residing

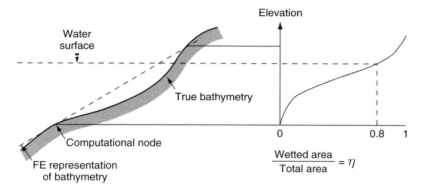

Figure 6.3 Definition sketch showing the significance of the scaling parameter, η (after Defina et al., 1994).

on that element at each time step. According to Defina *et al.* (1994) the continuity equation (6.1) thus becomes:

$$\eta \frac{\partial h}{\partial t} + h \nabla \mathbf{u}_d + \mathbf{u}_d \cdot \nabla h = 0 \qquad (6.7)$$

η therefore varies with water depth and the form of the relationship between η and h is dependent on the sub-grid topography (i.e. $\eta = f(Z_f | h)$). Such methods improve mass conservation properties, as by rescaling the volume of water on partially wet elements to correctly reflect the reduced volume of fluid, the correct physical quantity is conserved by the model. As a matter of fact, the volume of water in a domain Ω is no longer $\int_\Omega h \, d\Omega$, but rather $\int_\Omega \eta h \, d\Omega$. As η and h vary in time the mass of water at time t^n is:

$$\int_\Omega \eta^n h^n \, d\Omega \qquad (6.8)$$

Bates and Hervouet (1999) used these ideas to correct mass discrepancies for partially wet elements and implemented the Hervouet and Janin (1994) algorithm to correct the momentum equations. This model was then tested against a simple analytical solution for the run-up of a non-breaking wave on a 1×1 km planar beach (Figure 6.4). Manipulation of the water surface elevation at the seaward boundary of the model allowed a sinusoidal tide to be simulated entering the domain. By simplifying the beach topography to a planar slope and assuming a horizontal water surface, the position of the inundation front as it propagated across the domain and the water depth at any given point could be exactly known. Moreover, as the sub-grid topography was also planar, the value of η could be specified exactly. As an example, Figure 6.5 shows the comparison between a standard FE solution of the St Venant equations which includes the partially wet elements within the solution domain with the method proposed by Bates and Hervouet (1999) for water depths of 0.1, 0.05, 0.02 and 0.01 m at 3000 s into the simulation during the

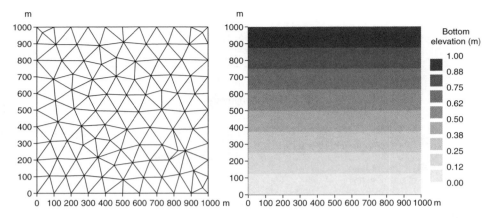

Figure 6.4 Finite element mesh and topography representing a planar beach generated as a simple analytical test case for wetting and drying problems. The cell resolution is approximately 100 m.

wetting phase. By employing their new technique Bates and Hervouet were able to improve the prediction of shallow water depth contours for moving boundary problems and also demonstrated that correct treatment of the near-boundary zone had a significant impact on the velocities predicted by the model over the whole domain. The tests also showed the Bates and Hervouet algorithm to be much less subject to unstable oscillations than a standard element inclusion method. Bates (2000) went on to show that sub-grid scaling of equation (6.1) reduced the need for excessive mesh refinement at the shoreline, and additionally that refinement could only reduce, but not eliminate, oscillations in the numerical solution.

This method is similar to the Volume of Fluid (VOF) approach (Ma *et al.*, 2002) which can be used to track both horizontal and vertical deforming boundaries in fixed grid 3D models. This tracks the volume fraction of water in three dimensions and therefore allows a grid with fixed vertical grid increments, thereby avoiding the problem of increasing cell height-to-length ratios as $h \to 0$. The position of the (horizontal or vertical) boundary is then extracted in a post-processing step by determining where the volume fraction is 0.5. However, as noted in section 6.1, avoidance of distorted elements for flows with horizontal extent 1–10 km and depths generally much less than 1–10 m still requires numerical grids that are horizontally very highly resolved and which are likely to incur a prohibitive computational cost. Neither has the VOF method been developed to consider the fact that for horizontal boundaries the volume fraction has a complex relationship with the sub-grid topography in each element.

Until recently a problem of schemes which are based on sub-grid parameters was that the topography at the sub-grid scale was rarely known and thus $\eta = f(Z_f | h)$ could not be readily parameterised. Instead, in testing applications of this theory researchers have frequently assumed η values based on experience or fractal scaling arguments (Bates *et al.*, 1992; Defina *et al.*, 1994). However, recent developments in topographic remote sensing such as airborne laser altimetry or LiDAR (Gomes-Pereira and Wicherson, 1999) have allowed accurate and fine spatial

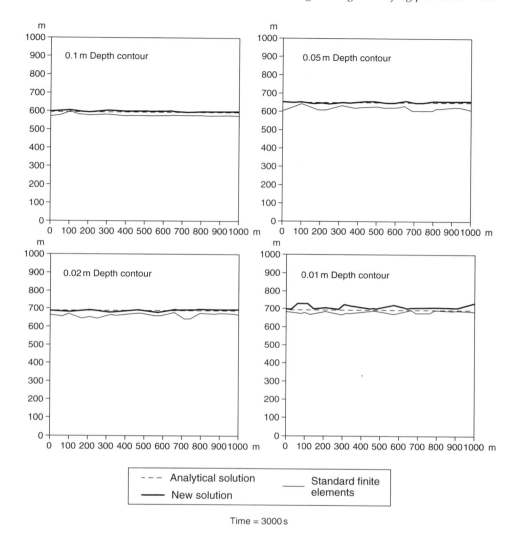

Figure 6.5 *Comparison of wetting phase shallow water depth contours predicted by standard finite elements and the new technique developed by Bates and Hervouet (1999). Each method is validated against the pseudo-analytical solution for the run up of a sinusoidal tide on the planar beach shown in Figure 6.4. At 3000 s into the simulation the analytical solution predicts that the 0.1, 0.05, 0.02 and 0.01 m contours should be positioned, respectively, at y = 600, 650, 680 and 690 m.*

resolution digital elevation data sets to be collected for fluvial and tidal floodplains (Chapter 11). Such data have spatial resolution of 1 m or less and are subject to a vertical rms error of only 10–15 cm. Given typical grid sizes of flood inundation models of the order 10–100 m (Bates *et al.*, 1998; Horritt and Bates, 2002), LiDAR potentially allows sub-grid-scale topography to be parameterised directly. This has been attempted by Bates and Hervouet (1999) and Bates *et al.* (2003) for beach run-up and fluvial flooding respectively with promising results. However, there is an urgent need for field data capable of testing such algorithms if their utility is to be demonstrated conclusively.

Lastly, as the size of the mass and momentum discrepancies resulting from the presence of partially wet cells in fixed grid models is dependent on the model grid scale, a further solution is to refine the model grid at the boundary. By reducing the size of partially wet elements one also minimises the problems generated. For transient problems where the shoreline moves during the simulation, this requires adaptive mesh refinement as the solution proceeds. While process gradients in nearshore regions may not justify high levels of refinement, this may be a way of dealing with the singularity that occurs as $h \to 0$. The computational cost may also be offset as a result of reduced grid resolution in other parts of the domain that might otherwise be over-refined in hand-generated meshes. The distinction here from fully deformable grids is that the domain extent remains fixed, and that partially wet cells are still incorporated in the solution, albeit in a way which attempts to minimise their impact. Adaptive gridding techniques have been used extensively in engineering disciplines such as aeronautics (for a review see Löhner, 2001, pp. 261–295) and whilst developed predominately for unstructured grid approaches, adaptation can also be undertaken for structured grids using multiblocks (Olsen and Kjellesvig, 1998) or quadtrees (Borthwick et al., 2001). Implementation of adaptive refinement thus requires: (1) that the user knows what the characteristics of an optimal mesh are; (2) the calculation of a spatially variable measure of mesh quality; and (3) an algorithm or strategy to refine or coarsen the mesh (Löhner, 2001, p. 262). Such a model for the shallow water equations is presented by Hubbard and Dodd (2002). Here adaptive mesh refinement is used without any special shoreline treatment and shown to be an acceptable method of treating wave run-up and overtopping prblems. However, as Lane et al. (Chapter 8) note, estimating mesh quality and defining the characteristics of an optimal mesh may be very difficult for real-world applications because of the complex relationship between the numerical solution, grid resolution and the representation of spatially heterogeneous topography and friction.

6.4.2 Fixed grid methods for discontinuous topography

Partially wet cells do not occur with discontinuous staircase topography as here entire elements are switched on/off depending on the computed water levels, and we assume instantaneous wetting of each element in a manner that conserves mass. However, Lynch and Gray (1980) note that momentum is still not conserved, and that the boundary should in fact move with the momentum of the fluid which will generally be small in shallow water flows. Lynch and Gray (1980) suggest that $\mathbf{u}_d \, \Delta t = (0.005 - 0.05) \Delta x$, where Δx is the grid size, which implies that it takes from 20 to 200 time steps for the boundary to move a distance Δx. Impulsive boundary motion, with potential consequences for model stability, is therefore also a feature of models based on the staircase topographic approximation. Moreover, instantaneous flooding of cells may also lead to an overestimation of the propagation speed of the wave front which will move with a velocity $\Delta x / \Delta t$. In other words, the front propagation velocity will be a function of the model grid scale unless the momentum balance at very small flow depths is treated carefully.

In models with discontinuous topography, definition of frontal cells cannot be made on the basis of one or more dry nodes as whole elements are either wet or dry. Instead a logical 'flag' needs to be assigned to each cell to indicate when it first becomes wet so that any correction can be applied. This also requires that an additional rule be implemented to identify when the frontal cell becomes fully wet so that the correction can be turned off. Far fewer methods of correcting for mass and momentum discrepancies have been published for models with discontinuous topography. One of the few available is described by Bradbrook et al. (2004) for the 2D storage cell model JFLOW. Instead of using a full solution of the shallow water equations, storage cell models simulate floodplain flows using an approximation to the 2D diffusive wave equation (see Bechteler et al., 1994; Bladé et al., 1994; Estrela and Quintas, 1994; Romanowicz et al., 1996; Bates and De Roo, 2000). A conservation equation is then solved for each grid cell, with fluxes between cells calculated analytically using Manning's equation. This has been shown (Horritt and Bates, 2001) to give near-identical results to numerical solutions of the 2D diffusive wave equation but at reduced computational cost. Storage cell codes generally employ a discontinuous topography and are therefore subject to the above problems in representing wave propagation accurately. Bradbrook et al. (2004) solve this problem by not allowing water to flow out of a cell until the cell is fully wet. Each cell has a property (w%) which represents the fraction of the cell that is wet. If the water cannot spread across the whole cell during a time step then the fraction of the grid cell that is wet is less than 1.0. The wetting parameter is updated each time step as the water travels across the cell, with outflow allowed once w% = 1.0.

6.5 Treating moving boundaries in advection–diffusion models

Whilst the previous discussion makes clear that considerable attention has been paid to moving boundary hydrodynamics in deformable and fixed grid models, the same cannot be said of coupled flow and transport problems where both a flow equation and an advection–diffusion equation are solved. Hardy et al. (2000a) note that the necessity to include dry areas within the model, in order to represent a dynamically moving inundation boundary, generates a number of difficulties that fixed grid FE solvers for fluvial sediment transport have, to date, failed to overcome. These include the generation of unstable terms in the advection–diffusion equation as the flow depth tends to zero and the artificial redirection of the transport flux in the vicinity of dynamically moving boundaries. In combination these problems lead to the development of highly diffuse solutions. This can be illustrated through consideration of a non-conservative formulation of the depth-averaged advection–diffusion equation:

$$\frac{\partial c}{\partial t} + \mathbf{u}_d \cdot \nabla c = \frac{1}{h} \nabla \cdot (h \mathbf{K} \nabla c) + \frac{S_m + c S_h}{h} \qquad (6.9)$$

where c is the tracer concentration [ML^{-3}]; $\mathbf{K}$ is the dispersion tensor [L^{2}T^{-1}] and S_m and S_h are source terms for the tracer mass and h respectively.

Equation (6.9) illustrates the problem of dealing with wetting and drying processes in numerical sediment transport models: as an element becomes dry, a division by zero is generated in both the diffusion and source terms. This problem still appears in other depth-averaged formulations of the advection–diffusion equation due to the link between concentration and mass per unit area:

$$c = \frac{M}{h} \qquad (6.10)$$

leading to physically unrealistic values for M (mass) as h (water depth) tends to zero when c (sediment concentration) is the conserved variable. A better solution to the development of a mass-conservative discretisation of the vertically integrated transport equation is to use the tracer mass, M, as the conserved variable. Hence the depth-averaged conservation equation would be:

$$\frac{\partial M}{\partial t} + \nabla \cdot (M\mathbf{u}_d) = \nabla \cdot (h\mathbf{K}\nabla c) + S_m \qquad (6.11)$$

However, the choice of M is not a good one for application to floodplains or tidal flats where the extent of the dry areas varies in time, as we are often interested more in the concentration, c, rather than the mass. Even choosing M as the conservative variable would still force us to compute equation (6.10) to obtain the concentration and again a singularity occurs. Thus far, only element exclusion methods have been used to solve this problem and schemes have been developed for FD (Falconer and Chen, 1991) and FE techniques (Moulin and Ben Slama, 1998; Hardy *et al.*, 2000a,b; Covelli *et al.*, 2002). A typical example is described by Hardy *et al.* (2000b) who define a function 'mask' for each element '*iel*' such that mask(*iel*) = 0 if the element is dry and mask(*iel*) = 1 if the element is wet. Using the FE method they then solve:

$$\sum_{iel} \int_{iel} \text{mask}(iel) \left[\frac{\partial c}{\partial t} + \mathbf{u} \cdot \nabla c \right] \varphi_i d\Omega$$

$$= \sum_{iel} \int_{iel} \text{mask}(iel) \left[\frac{1}{h} \nabla \cdot (\mathbf{K}\nabla c) + \frac{S_m + cS_h}{h} \right] \varphi_i d\Omega \qquad (6.12)$$

where φ_i is a Galerkin linear weighting function and Ω is the computational domain.

However, this still does not avoid all the difficulties associated with a numerical implementation of equation (6.9), as there is still the need to redirect transport in the presence of dynamic moving boundaries. As the flow depth tends to zero, the terms on the right-hand side in equation (6.9) will become large, leading to physically unrealistic values for the concentration, c, the dispersion, K, and any source term inputs. This highly diffuse solution in conjunction with artificially enhanced concentration levels typically manifests itself as a build-up of tracer mass in dry areas of the domain which may eventually lead to numerical instability. Clearly, careful choice of numerical procedure is required if such effects are to be avoided or minimised. Algorithms are thus required which damp spurious oscillations and preserve steep concentration gradients from wet (where there is a concentration) to dry areas (where the concentration is zero). Thus, similar to numerical hydrodynamic modelling in the presence of wetting and drying, the specification of an adequate method to deal with the advection term, $\mathbf{u}_d \cdot \nabla c$, is a key requirement (Ip *et al.*, 1998). Hardy *et al.* (2000b) solved this problem by implementing an SUPG-like weighting function for the advection term. The upwind coefficient was chosen to be the Courant number and the approach ensured stabilisation of the gradient term by adding artificial diffusion in the flow direction only. Spurious numerical oscillations were thus kept within their creation area and the build-up of tracer mass in areas of shallow depth was prevented or minimised. This is a promising approach; however, it is clear that much more work is required in this area, for example on sub-grid scaling approaches, if wetting and drying processes in coupled flow and transport models are to be comprehensively addressed.

6.6 Future research needs

This chapter has demonstrated the variety of approaches available to deal with wetting and drying processes in models of environmental hydraulics. It is also clear from the preceding discussion that all these approaches have strengths and weaknesses and that a definitive method of treating moving boundary problems is still lacking. All approaches discussed in some way compromise the representation of mass or momentum conservation to a greater or lesser extent, or involve some limiting assumption or additional computational cost. Continuing increases in computer power are also likely to change research in this area, and make mesh deformation approaches (which can fully treat the shoreline evolution) a more realistic proposition. However, this alone will not lead to a complete solution. In part this is a consequence of a lack of test data capable of discriminating between competing algorithms, particularly for real-world flows, and the acquisition of such data is probably the key research need in this area. A further requirement in the quest for improved models is also a better understanding of certain key processes. These include a need to understand more fully the circumstances in which 2D and 3D processes dominate the shoreline evolution and to acquire a better knowledge of the fluid dynamics of shallow water flow, particularly of the hydraulics of thin films of water over complex topography (cf. Balzano, 1998), the impact of infiltration losses to the subsurface and the interaction between shallow water flows and vegetation (Chapter 15).

Acknowledgements

The research reported in this chapter has been supported by a number of research grants including UK Natural Environment Research Council grant GR3 CO/030, UK Engineering and Physical Science Research Council grants GR/L95694 and GR/S17116 and the EU Framework 5 project 'Development of a European Flood Forecasting System (EFFS)'. Matt Horritt is supported by a UK Natural Environment Research Council Fellowship. Many thanks too go to Richard Hardy for his insightful review comments.

References

Akanbi, A.A. and Katopodes, N.D. 1988. Model for flood propagation on initially dry land. *Journal of Hydraulic Engineering*, American Society of Civil Engineers, **114**, 689–706.

Alcrudo, F. and Garcia-Navarro, P. 1993. A high-resolution Gudonov scheme in finite volumes for the 2D shallow water equations. *International Journal of Numerical Methods in Fluids*, **16**, 489–505.

Balzano, A. 1998. Evaluation of methods for numerical simulation of wetting and drying in shallow water flow models. *Coastal Engineering*, **34**, 83–107.

Bates, P.D. 2000. Development and testing of a sub-grid scale model for moving boundary hydrodynamic problems in shallow water. *Hydrological Processes*, **14**, 2073–2088.

Bates, P.D. and De Roo, A.P.J. 2000. A simple raster-based model for floodplain inundation. *Journal of Hydrology*, **236**, 54–77.

Bates, P.D. and Hervouet, J.-M. 1999. A new method for moving boundary hydrodynamic problems in shallow water. *Proceedings of the Royal Society of London, Series A*, **455**, 3107–3128.

Bates, P.D., Anderson, M.G., Baird, L., Walling, D.E. and Simm, D. 1992. Modelling floodplain flow with a two dimensional finite element scheme. *Earth Surface Processes and Landforms*, **17**, 575–588.

Bates, P.D., Stewart, M.D., Siggers, G.B., Smith, C.N., Hervouet, J.-M. and Sellin, R.H.J. 1998. Internal and external validation of a two-dimensional finite element model for river flood simulation. *Proceedings of the Institution of Civil Engineers, Water Maritime and Energy*, **130**, 127–141.

Bates, P.D., Marks, K.J. and Horritt, M.S. 2003. Optimal use of high-resolution topographic data in flood inundation models. *Hydrological Processes*, **17**, 537–557.

Bates, P.D., Horritt, M.S., Aronica, G. and Beven, K. 2004. Bayesian updating of flood inundation likelihoods conditioned on flood extent data. *Hydrological Processes*, **18**, 3347–3370.

Bechteler, W., Hartmaan, S. and Otto, A.J. 1994. Coupling of 2D and 1D models and integration into Geographic Information Systems (GIS). In White, W.R. and Watts, J. (eds), *Proceedings of the 2nd International Conference on River Flood Hydraulics*, John Wiley & Sons, Chichester, UK, 155–165.

Beckett, G., Mackenzie, J.A., Ramage, A. and Sloan, D.M. 2002. Computational solution of two-dimensional unsteady PDE's using moving mesh methods. *Journal of Computational Physics*, **182**, 478–495.

Beffa, C. and Connell, R.J. 2001. Two-dimensional flood plain flow – Part 1: Model description. *Journal of Hydraulic Engineering*, American Society of Civil Engineers, **6**, 397–405.

Behr, M. 2001. Stabilized space-time finite element formulations for free-surface flows. *Communications in Numerical Methods in Engineering*, **17**, 813–819.

Bladé, E., Gómez, M. and Dolz, J. 1994. Quasi-two dimensional modelling of flood routing in rivers and flood plains by means of storage cells. In Molinaro, P. and Natale, L. (eds), *Modelling of Flood Propagation Over Initially Dry Areas*, American Society of Civil Engineers, New York, 156–170.

Borthwick, A.G.L., Leon, S.C. and Jozsa, J. 2001. Adaptive quadtree model of shallow-flow hydrodynamics. *Journal of Hydraulic Research*, **39 (4)**, 413–424.

Bottasso, C.L. 1997. On the computation of the boundary integral of space-time deforming finite elements. *Communications in Numerical Methods in Engineering*, **13**, 53–59.

Bradbrook, K.F., Lane, S.N., Waller, S.G. and Bates, P.D. (2004). Two-dimensional diffusion wave modelling of flood inundation using a simplified channel representation. *International Journal of River Basin Management*, **2**, 211–223.

Bradford, S.F. and Sanders, B.F. 2002. Finite-volume model for shallow-water flooding or arbitrary topography. *Journal of Hydraulic Engineering*, American Society of Civil Engineers, **128 (3)**, 289–298.

Briggs, M.J., Synolakis, C.E., Harkins, G.S. and Green, D.R. 1995. Laboratory experiments of tsunami run-up on circular island. *Pure and Applied Geophysics*, **144 (3/4)**, 569–593.

Brocchini, M. and Peregrine, D.H. 1996. Integral properties of the swash zone and averaging. *Journal of Fluid Mechanics*, **317**, 241–273.

Brocchini, M., Bernetti, R., Mancinelli, A. and Albertini, G. 2001. An efficient solver for nearshore flows based on the WAF method. *Coastal Engineering*, **43**, 105–129.

Brookes, A.N. and Hughes, T.J.R. 1982. Streamline Upwind/Petrov Galerkin formulations for convection dominated flows with particular emphasis on the incompressible Navier–Stokes equations. *Computer Methods in Applied Mechanics and Engineering*, **32**, 199–259.

Carrier, G.F. and Greenspan, H.P. 1958. Water waves of finite amplitude on a sloping beach. *Journal of Fluid Mechanics*, **4**, 97–109.

Cha, K.S., Choi, J.W. and Park, C.G. 2002. Finite element analysis of fluid flows with moving boundary. *Korean Society of Mechanical Engineers International Journal*, **16 (5)**, 683–695.

Christian, C.D. and Palmer, G.N. 1997. A deforming finite element mesh for use in moving one-dimensional boundary wave problems. *International Journal of Numerical Methods in Fluids*, **25**, 407–420.

Covelli, P., Marsili-Libelli, S. and Pacini, G. 2002. SWAMP: A two-dimensional hydrodynamic and quality modeling platform for shallow waters. *Numerical Methods for Partial Differential Equations*, **18 (5)**, 663–687.

Defina, A. 2000. Two-dimensional shallow flow equations for partially dry areas. *Water Resources Research*, **36 (11)**, 3251–3264.

Defina, A., D'Alpaos, L. and Matticchio, B. 1994. A new set of equations for very shallow water and partially dry areas suitable to 2D numerical models. In Molinaro, P. and Natale, L. (eds), *Modelling Flood Propagation Over Initially Dry Areas*, American Society of Civil Engineers, New York, 72–81.

De Waal, J.P. and Van Der Meer, J.W. 1992. Wave run-up and overtopping on coastal structures. *Proceedings of the 23rd International Conference on Coastal Engineering*, American Society of Civil Engineers, New York, 1758–1771.

Erturk, S.N., Bilgili, A., Swift, M.R., Brown, W.S., Celikkol, B., Ip, J.T.C. and Lynch, D.R. 2002. Simulation of the Great Bay estuarine system: Tides with tidal flats wetting and drying. *Journal of Geophysical Research – Oceans*, **107 (C5)**, art. no. 3038.

Estrela, T. and Quintas, L. 1994. Use of GIS in the modelling of flows on floodplains. In White, H.R. and Watts, J. (eds), *Proceedings of the 2nd International Conference on River Flood Hydraulics*, John Wiley & Sons, Chichester, UK, 177–189.

Falconer, R.A. and Chen, Y. 1991. An improved representation of flooding and drying and wind stress effects in a two-dimensional numerical tidal model. *Proceedings of the Institution of Civil Engineers*, **91 (2)**, 659–678.

Feng, Y.T. and Perić, D. 2003. A spatially adaptive linear space-time finite element solution procedure for incompressible flows with moving domains. *International Journal of Numerical Methods in Fluids*, **43**, 1099–1106.

Gomes-Pereira, L.M. and Wicherson, R.J. 1999. Suitability of laser data for deriving geographical data: A case study in the context of management of fluvial zones. *Photogrammetry and Remote Sensing*, **54**, 105–114.

Gopalakrishnan, T. 1989. A moving boundary circulation model for regions with large tidal flats. *International Journal of Numerical Methods in Engineering*, **28**, 245–260.

Gopalakrishnan, T.C. and Tung, C.C. 1983. Numerical analysis of a moving boundary problem in coastal hydrodynamics. *International Journal for Numerical methods in Fluids*, **3 (2)**, 179–200.

Hardy, R.J., Bates, P.D. and Anderson, M.G. 1999. The importance of spatial resolution in hydraulic modelling of floodplain environments. *Journal of Hydrology*, **216**, 124–136.

Hardy, R.J., Bates, P.D. and Anderson, M.G. 2000a. Modelling suspended sediment deposition on a fluvial floodplain using a two-dimensional dynamic finite element model. *Journal of Hydrology*, **229**, 202–218.

Hardy, R.J., Bates, P.D. and Anderson, M.G. 2000b. Development of a reach scale two-dimensional finite element model for floodplain sediment deposition. *Proceedings of the Institution of Civil Engineers, Water Maritime and Energy*, **142**, 141–156.

Hervouet, J.-M. 2000. A high resolution 2-D dam-break model using parallelization. *Hydrological Processes*, **14 (13)**, 2211–2230.

Hervouet, J.-M. and Janin, J.-M. 1994. Finite element algorithms for modelling flood propagation. In Molinaro, P. and Natale, L. (eds), *Modelling Flood Propagation Over Initially Dry Areas*, American Society of Civil Engineers, New York, 102–113.

Hervouet, J.-M. and Van Haren, L. 1996. Recent advances in numerical methods for fluid flows. In Anderson, M.G., Walling, D.E. and Bates, P.D. (eds), *Floodplain Processes*, John Wiley & Sons, Chichester, UK, 183–214.

Hirt, C.W., Amsden, A.A. and Cook, J.L. 1974. An arbitrary Lagrangian Eulerian computing method for all flow speeds. *Journal of Computational Physics*, **14**, 227–253.

Horritt, M.S. 2000. Calibration of a two-dimensional finite element flood flow model using satellite radar imagery. *Water Resources Research*, **36**, 3279–3291.

Horritt, M.S. 2002. Evaluating wetting and drying algorithms for finite element models of shallow water flow. *International Journal for Numerical methods in Engineering*, **55 (7)**, 835–851.

Horritt, M.S. and Bates, P.D. 2001. Predicting floodplain inundation: Raster-based modelling versus the finite element approach. *Hydrological Processes*, **15**, 825–842.

Horritt, M.S. and Bates, P.D. 2002. Evaluation of 1-D and 2-D numerical models for predicting river flood inundation. *Journal of Hydrology*, **268**, 87–99.

Hubbard, M.E. and Dodd, N. 2002. A 2D numerical model of wave run-up and overtopping. *Coastal Engineering*, **47**, 1–26.

Ip, J.T.C., Lynch, D.R. and Friedrichs, C.T. 1998. Simulation of estuarine flooding and dewatering with application to Great Bay, New Hampshire. *Estuarine Coastal and Shelf Science*, **47 (2)**, 119–141.

Jamet, P. and Bonnerot, R. 1975. Numerical solution of the Eulerian equations of compressible flow by finite element method which follows the free boundary and interfaces. *Journal of Computational Physics*, **18**, 21–45.

Jha, A.K., Juichiro, J. and Ura, M. 1995. First- and second-order flux difference splitting schemes for dam break problems. *Journal of Hydraulic Engineering*, American Society of Civil Engineers, **121**, 877–884.

Johns, B., Dube, S.K., Sinha, P.C., Mohanty, U.C. and Roa, A.D. 1982. The simulation of a continuously deforming lateral boundary in problems involving the shallow water equations. *Computers and Fluids*, **10 (2)**, 105–116.

Kawahara, M. and Umetsu, T. 1986. Finite element method for moving boundary problems in river flows. *International Journal of Numerical Methods in Fluids*, **6**, 365–386.

King, I.P. and Norton, W.R. 1978. Recent applications of RMA's finite element models for two dimensional hydrodynamics and water quality. In Brebbia, C.A., Gray, W.G. and Pinder, G.F. (eds), *Proceedings of the Second International Conference on Finite Elements in Water Resources*, Pentech Press, London, 81–99.

King, I.P. and Roig, L. 1988. Two-dimensional finite element models for floodplains and tidal flats. In Niki, K. and Kawahara, M. (eds), *Proceedings of an International Conference on Computational Methods in Flow Analysis*, Okayama, Japan, 711–718.

Leclerc, M., Bellemare, J.-F., Dumas, G. and Dhatt, G. 1990. A finite element model of estuarine and river flows with moving boundaries. *Advances in Water Resources*, **13**, 158–168.

Lin, B. and Falconer, R.A. 1997. Three-dimensional layer-integrated modelling of estuarine flows with flooding and drying. *Estuarine, Coastal and Shelf Science*, **44 (6)**, 737–751.

Liu, P.L.F., Cho, Y.S., Briggs, M.J., Lu, U.K. and Synolakis, C.E. 1995. Run-up of solitary waves on a circular island. *Journal of Fluid Mechanics*, **302**, 259–285.

Löhner, R. 2001. *Applied CFD Techniques: An Introduction Based on Finite Element Methods*, John Wiley & Sons, Chichester, UK, 366pp.

Lynch, D.R. and Gray, W.G. 1980. Finite element simulation of flow deforming regions. *Journal of Computational Physics*, **36**, 135–153.

Ma, L., Ashworth, P.J., Best, J.L., Elliot, L., Ingham, D.B. and Whitcombe, L.J. 2002. Computational fluid dynamics and the physical modelling of an upland urban river. *Geomorphology*, **44 (3–4)**, 375–391.

MacArthur, R.C., Dexter, J.R., Smith, D.J. and King, I.P. 1990. Two dimensional finite element simulation of the flooding characteristics of Kawainui Marsh, Hawaii. *Proceedings of the 1990 National Hydraulic Engineering Conference*, American Society of Civil Engineers, New York (ISBN 0-87262-774-8), 664–669.

Marks, K. and Bates, P.D. 2000. Integration of high resolution topographic data with floodplain flow models. *Hydrological Processes*, **14**, 2109–2122.

Moulin, C. and Ben Slama, E. 1998. The two-dimensional transport module SUBIEF: Applications to sediment transport and water quality processes. *Hydrological Processes*, **12 (8)**, 1183–1195.

Neilsen, C. and Apelt, C. 2003. Parameters affecting the performance of wetting and drying in a two-dimensional finite element long wave model hydrodynamic model. *Journal of Hydraulic Engineering*, American Society of Civil Engineers, **129 (8)**, 628–636.

Nicholas, A.P. and Mitchell, C.A. 2003. Numerical simulation of overbank processes in topographically complex floodplain environments. *Hydrological Processes*, **17 (4)**, 727–746.

Nujić, M. 1995. Efficient implementation of non-oscillatory schemes for the computation of free surface flows. *Journal of Hydraulic Research*, **33**, 101–111.

Olsen, N.R.B. and Kjellesvig, H.M. 1998. Three-dimensional numerical flow modeling for estimation of maximum local scour depth. *Journal of Hydraulic Research*, **36 (4)**, 579–590.

Owen, M. 1980. Design of seawalls allowing for wave overtopping. *Technical Report no. EX924*, Hydraulics Research, Wallingford, UK.

Petera, J. and Nassehi, V. 1996. A new two-dimensional finite element model for the shallow water equations using a Lagrangian framework constructed along fluid particle trajectories. *International Journal of Numerical Methods in Engineering*, **39**, 4159–4182.

Quecedo, M. and Pastor, M. 2002. A reappraisal of Taylor-Galerkin algorithm for drying-wetting areas in shallow water computations. *International Journal for Numerical Methods in Fluids*, **38**, 515–531.

Roe, P.L. 1981. Approximate Riemann solvers, parameter vectors and difference schemes. *Journal of Computational Physics*, **43**, 357–372.

Roig, L.C. and Evans, R.A. 1993. Environmental modelling of coastal wetlands. *Proceedings of the 3rd International Conference on Estuarine and Coastal Modelling III*, American Society of Civil Engineers, New York, 522–535.

Romanowicz, R., Beven, K.J. and Tawn, J. 1996. Bayesian calibration of flood inundation models. In Anderson, M.G., Walling, D.E. and Bates, P.D. (eds), *Floodplain Processes*, John Wiley & Sons, Chichester, UK, 333–360.

Ryrie, S.C. 1983. Longshore motion generated on beaches by obliquely incident bores. *Journal of Fluid Mechanics*, **129**, 193–212.

Savic, I.J. and Holly, F.M. 1993. Dambreak flood waves computed by modified Gudonov method. *Journal of Hydraulic Research*, **31**, 187–204.

Shige-Eda, M. and Akiyama, J. 2003. Numerical and experimental study on two-dimensional flood flows with and without structures. *Journal of Hydraulic Engineering*, American Society of Civil Engineers, **129 (10)**, 817–821.

Simm, D.J. 1993. The deposition and storage of suspended sediment in contemporary floodplain systems: A case study of the River Culm, Devon. Unpublished PhD Thesis, University of Exeter, 347pp.

Soulaimani, A. and Saad, Y. 1998. An arbitrary Lagrangian–Eulerian finite element method for solving three-dimensional free surface flows. *Computer Methods in Applied Mechanics and Engineering*, **162**, 79–106.

Stoesser, T., Wilson, C.A.M.E. and Bates, P.D. and Dittrich, A. 2003. Application of a 3D numerical model to a river with vegetated floodplains. *Journal of Hydroinformatics*, **5**, 99–112.

Synolakis, C.E. 1987. The run-up of solitary waves. *Journal of Fluid Mechanics*, **185**, 523–555.

Tchamen, G.W. and Kawahita, R.A. 1998. Modelling wetting and drying effects over complex topography. *Hydrological Processes*, **12**, 1151–1183.

Tezduyar, T.E. 2001. Finite element methods for flow problems with moving boundaries and interfaces. *Archives of Computational Methods in Engineering*, **8 (2)**, 83–130.

Thacker, W.C. 1981. Some exact solutions to the non-linear shallow water equations. *Journal of Fluid Mechanics*, **107**, 499–508.

Toro, E.F. 1992. Riemann problems and the WAF method for solving the two-dimensional shallow water equations. *Philosophical Transactions of the Royal Society of London, Series A*, **338**, 43–68.

Van Leer, B. 1979. Towards the ultimate conservative difference scheme. V: A second order sequel to Gudonov's method. *Journal of Computational Physics*, **32**, 101–136.

Walling, D.E. and He, Q. 1996. Floodplains as suspended sediment sinks. In Anderson, M.G., Walling, D.E. and Bates, P.D. (eds), *Floodplain Processes*, John Wiley & Sons, Chichester, UK, 399–440.

Zhang, W. and Cundy, T.W. 1989. Modelling of two-dimensional overland flow. *Water Resources Research*, **25**, 2019–2035.

Zhao, D.H., Shen, H.W., Tabios, G.Q., Lai, J.S. and Tan, W.Y. 1994. A finite volume two-dimensional unsteady flow model for river basins. *Journal of Hydraulic Engineering*, American Society of Civil Engineers, **120**, 863–883.

Zheng, L.Y., Chen, C.S. and Liu, H.D. 2003. A modelling study of the Satilla River estuary, Georgia. I: Flooding-drying processes and water exchange over the salt marsh-estuary-shelf complex. *Estuaries*, **26 (3)**, 651–669.

7

Introduction to numerical methods for fluid flow

N.G. Wright

7.1 Introduction

This chapter addresses the question of how the equations that encapsulate the models of environmental systems can be converted to a form that is amenable to solution on a computer. It sets out the basic mathematical tools used to do this, drawing on the wider field of CFD in explaining the underlying numerical analysis required. Particular features that must be addressed in the application of CFD to environmental flows are outlined. The intention is to give readers enough information to decipher and interpret numerical modelling papers and to sound a cautionary note about the importance of attention to the detail of the application of CFD codes and the dangers of treating them as a black box. Inevitably, a certain amount of mathematical notation is used to explain the methods, which some readers may find daunting. However, it should be worth persevering with in order to gain an understanding of the methods

Numerical analysis is the branch of mathematics that deals with numerical solutions of mathematical problems. Most physical models are expressed as differential equations and their solution on a computer draws on the mathematical theories of the solution of differential equations. There are particular features of fluid flow that mean these techniques must be tailored for application here (e.g. flow direction, conservation of mass).

Whilst numerical analysis has been studied for over 100 years, the advent of the digital computer led to a dramatic increase in its applicability and hence research attention. Nowadays CFD is a viable commercial tool and a number of companies successfully market CFD software. What is now commonly recognised as CFD, began in the 1960s in order to study aeronautics and nuclear power systems. Since then CFD developments have increased its power and accuracy and also extended the fields of application. CFD is now an established part of design in aerospace, vehicle aerodynamics, nuclear design and safety, Chemical and Process Industries (CPI) and Heating, Ventilation and Air-Conditioning (HVAC). Consequently the international market for CFD software and services was estimated at over $300 million in 2000 (Löhner, 2001). The wide scope of CFD applications is reflected in the number of texts available. Those by Chung (2002) and Versteeg and Malalasekera (1995) represent good examples that an interested reader can consult. Chung is more comprehensive, but Versteeg and Malalasekera take a more practical approach. The application of CFD to environmental flows has begun to attract more attention in the past decade. This has led to articles in learned journals, but, as yet, few books that bring the different aspects together for the reader. This book begins to address this.

It is important to begin by clarifying the distinction between constructing a model of the physical situation and the subsequent solution of the model equations. This distinction is reflected in the question of verification and validation. The former is the question of whether a particular numerical solution is an accurate solution of the equations posed. This is a matter of numerical accuracy. Sensitivity checks must be carried out to ensure that the solution is an accurate one for the equations as posed. On the other hand, validation is the question of whether the models adequately represent the physical system and, once verification has been carried out, this can be done by comparing the results with physical measurements or theory. Sometimes coincidence can mean that a less numerically accurate solution can match measured data more closely than a more numerically accurate one. It is tempting to interpret this as meaning that the less accurate method is giving a "better" solution, but this is a false interpretation as the inaccuracies in the numerical solutions are just balancing out the modelling errors and this fortuitous agreement might lead to a completely wrong solution in another situation. It is necessary first to reduce errors in the numerical method and subsequently to identify modelling errors. This is discussed in more detail by Lane *et al.* (Chapter 8).

7.2 Methodology

The initial task, as mentioned above, is to convert the differential equations, which have continuously defined functions as solutions, to a set of algebraic equations that connect values at various discrete points that can be manipulated by a computer. This process is called discretisation. Various methods are used for this and the main three are described below.

7.2.1 Finite differences

Initial investigations into the solutions of Partial Differential Equations (PDEs) used the finite difference method and this was carried through into early CFD. The method is based on the use of a Taylor series to obtain expressions for values of a function surrounding a given point on a structured grid (Smith, 1978). These expressions are then combined to give an equation for the derivative at the given point in terms of those around it. A structured grid is one where neighbouring cells are addressed by using the convention of $(i-1, j)$, (i, j), etc. (Figure 7.1). In general, with structured grids, the grid lines are oriented in two or three directions for 2D or 3D problems, respectively. As an example, the following are three different approximations to the first derivative of the function of one variable, $T(x)$, at a point denoted by x_i:

Forward differencing

$$\frac{\partial T}{\partial x} = \frac{T_{i+1} - T_i}{\Delta x}$$

Backward differencing

$$\frac{\partial T}{\partial x} = \frac{T_i - T_{i-1}}{\Delta x}$$

Central differencing

$$\frac{\partial T}{\partial x} = \frac{T_{i+1} - T_{i-1}}{2\Delta x}$$

In order to demonstrate the finite difference method, consider the equation

$$\frac{d^2 T}{dx^2} - 2 = 0 \tag{7.1}$$

defined for x between 0 and 1 and with the boundary conditions that $T=0$ at $x=0$ and 1. Whilst this equation is much simpler than the equations for fluid flow, it serves to demonstrate the method and will be used for the methods presented later. In most cases an analytical solution is not available which is why a numerical one is sought, but in this case let us assume the particular solution (Figure 7.2):

$$T = x(x - 1) \tag{7.2}$$

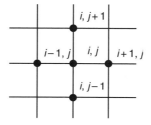

Figure 7.1 Finite difference grid.

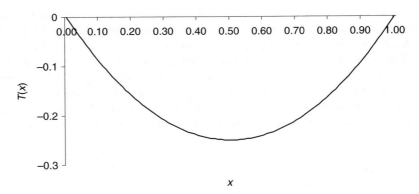

Figure 7.2 Analytical solution, T(x).

Considering three discrete values which we wish to find:

$$T_1 = T(0.0)$$
$$T_2 = T(0.5)$$
$$T_3 = T(1.0)$$

we need to derive a representation of the equation at $x = 0.5$. Using Taylor series

$$T_1 = T_2 - h\left(\frac{\partial T}{\partial x}\right)_2 + \frac{h^2}{2}\left(\frac{\partial^2 T}{\partial x^2}\right)_2 - \frac{h^3}{6}\left(\frac{\partial^3 T}{\partial x^3}\right)_2 + o(h^4)$$

$$T_3 = T_2 + h\left(\frac{\partial T}{\partial x}\right)_2 + \frac{h^2}{2}\left(\frac{\partial^2 T}{\partial x^2}\right)_2 + \frac{h^3}{6}\left(\frac{\partial^3 T}{\partial x^3}\right)_2 + o(h^4)$$

where h is the distance between the points which in this case is 0.5 and $o(h^4)$ denotes terms of order 4 and higher in h. Adding these two equations together and rearranging gives an expression for the second derivative:

$$\frac{\partial^2 T}{\partial x^2} = \frac{T_3 - 2T_2 + T_1}{h^2} + o(h^3)$$

Using this in the original equations with the boundary conditions gives:

$$\frac{0.0 - 2T_2 + 0.0}{0.5^2} - 2 = 0$$

which gives

$$T_2 = -\frac{1}{4}$$

This agrees with the value from the analytic solution.

The finite difference method is simple to calculate and implement for basic cases. However, it was found to have limitations in application to fluid flow problems where the geometries are irregular and the equations are nonlinear.

7.2.2 Finite elements

This method was originally developed for use in structural dynamics in the 1950s and developed for fluid dynamics later by Zienkiewicz and Cheung (1965). It sets up the equations so that a solution is found by minimising the global error. In contrast with the finite difference method defined earlier it is based on an unstructured grid which consists of polygons or polyhedra of arbitrary shape and size (Figure 7.3) which is attractive for dealing with complex geometries. The method sets up shape or basis functions in each element to approximate the local solution. A global function based on these is substituted into the governing PDEs. This equation is then integrated with weighting functions and the resulting error is minimised to give coefficients for the trial functions that represent an approximate solution.

To demonstrate the implementation of the finite element method we can again use equation 7.1. After splitting the interval from 0 to 1 into two equal elements (Figure 7.4) we assume a linear function for each element:

$$T^{(e)} = a_1 + a_2 x$$

where the superscript (e) denotes the element number – 1 or 2 in this case.

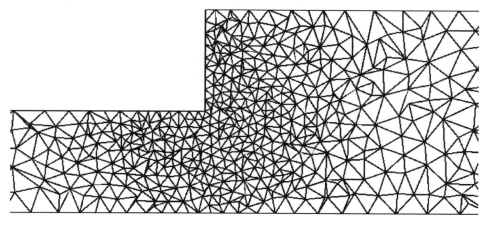

Figure 7.3 Unstructured tetrahedral mesh.

Figure 7.4 Finite elements.

For each element the values at each end are $T_1^{(e)}$ and $T_2^{(e)}$ so:

$$T_1^{(e)} = T^{(e)}(0) = a_1$$

$$T_2^{(e)} = T^{(e)}(h) = a_1 + a_2 h$$

We can use these to give values for a_1 and a_2:

$$a_1 = T_1^{(e)}$$
$$a_2 = \frac{T_2^{(e)} - T_1^{(e)}}{h}$$

So

$$T^{(e)} = \left(1 - \frac{x}{h}\right) T_1^{(e)} + \frac{x}{h} T_2^{(e)}$$

or

$$T^{(e)} = \sum_{N=1,2} \phi_N^{(e)} T_N^{(e)} s$$

where $\phi_N^{(e)}$ are called the trial functions:

$$\phi_1^{(e)} = 1 - \frac{x}{h}, \quad \phi_2^{(e)} = \frac{x}{h}$$

which are zero at one end of the element and one at the other.

Once these functions have been established the finite element method can proceed in different ways. The underlying concept is to combine these trial functions with test functions to approximate a solution and then to minimise the global error. For the example here, the Galerkin method will be shown for simplicity. Other methods are available and the Petrov–Galerkin method is widely used in fluid mechanics (Chung, 2002). In the Galerkin method the test functions are taken to be the same as the trial functions. A combination (inner product) of the residual of the local function with the test functions is taken, i.e.

$$\left(\phi_N^{(e)}(x), R^{(e)}\right) = \int_0^h \phi_N^{(e)}(x) \left(\frac{d^2\phi}{dx^2} - 2\right) dx = 0$$

Integrating by parts gives

$$\phi_N^{(e)} \frac{d\phi}{dx}\bigg|_0^h - \int_0^h \frac{d\phi_N^{(e)}(x)}{dx} \frac{dT^{(e)}(x)}{dx} dx - 2h \int_0^h \phi_N^{(e)}(x) dx = 0$$

For fixed value (Dirichlet) boundary conditions the first term is zero. So along, with substituting for $T^{(e)}(x)$, this gives

$$-\left[\int_0^h \frac{d\phi_N^{(e)}(x)}{dx}\frac{d\phi_M^{(e)}(x)}{dx}dx\right]u_M^{(e)} - 2\int_0^h \phi_N^{(e)}(x)dx = 0$$

Expanding gives

$$\begin{bmatrix} \int_0^h \frac{d\phi_1^{(e)}(x)}{dx}\frac{d\phi_1^{(e)}(x)}{dx}dx & \int_0^h \frac{d\phi_1^{(e)}(x)}{dx}\frac{d\phi_2^{(e)}(x)}{dx}dx \\ \int_0^h \frac{d\phi_2^{(e)}(x)}{dx}\frac{d\phi_1^{(e)}(x)}{dx}dx & \int_0^h \frac{d\phi_2^{(e)}(x)}{dx}\frac{d\phi_2^{(e)}(x)}{dx}dx \end{bmatrix} u_M^{(e)} = 2\begin{bmatrix} \int_0^h \phi_1^{(e)}(x)dx \\ \int_0^h \phi_2^{(e)}(x)dx \end{bmatrix}$$

which gives

$$\frac{1}{h}\begin{bmatrix} 1 & -1 \\ -1 & 1 \end{bmatrix} T_m^{(e)} = -h\begin{bmatrix} 1 \\ 1 \end{bmatrix}$$

This is for a general element. The matrix is termed the local stiffness matrix. For elements 1 and 2 we get

$$\frac{1}{h}\begin{bmatrix} 1 & -1 \\ -1 & 1 \end{bmatrix}\begin{bmatrix} T_1 \\ T_2 \end{bmatrix} = -h\begin{bmatrix} 1 \\ 1 \end{bmatrix}$$

$$\frac{1}{h}\begin{bmatrix} 1 & -1 \\ -1 & 1 \end{bmatrix}\begin{bmatrix} T_2 \\ T_3 \end{bmatrix} = -h\begin{bmatrix} 1 \\ 1 \end{bmatrix}$$

Combining these gives

$$\frac{1}{h}\begin{bmatrix} 1 & -1 & 0 \\ -1 & 2 & -1 \\ 0 & -1 & 1 \end{bmatrix}\begin{bmatrix} T_1 \\ T_2 \\ T_3 \end{bmatrix} = -h\begin{bmatrix} 1 \\ 2 \\ 1 \end{bmatrix}$$

where the matrix is now the global stiffness matrix. This is the final set of equations representing the finite element solution. For the example given earlier the equation for the global node 2 is

$$-T_1 + 2T_2 - T_3 = -2h^2$$

which is equivalent to the finite difference scheme, even if the method takes a bit longer! In cases that are more complex the finite element is no lengthier than the finite difference and generalises much better by virtue of the functional mathematical theory underlying it.

7.2.3 Finite volumes

The CFD researchers have developed and favoured an alternative method called the finite volume method. This technique is simpler and more numerically efficient than others and this has led to it being the most widely used in 3D CFD software.

The finite volume technique can use either structured or unstructured grids. The physical laws are integrated into each control volume to give an equation for each law in terms of values on the face of each control volume. These face values or fluxes are then calculated from adjacent values by interpolation. This method ensures that mass in conserved in the discrete form of the equations just as it is in the physical situation.

Consider a control volume as shown in Figure 7.5. We want an algebraic approximation to

$$\frac{d^2 T}{dx^2} - 2 = 0$$

Integrating this equation over the control volume gives

$$\int_{\Delta V} \left(\frac{d^2 T}{dx^2} - 2\right) dV = 0$$

or

$$\int_{-h/2}^{h/2} \int_{-h/2}^{h/2} \left(\frac{d^2 T}{dx^2} - 2\right) dx dy = 0$$

It is assumed that there is no variation in the y-direction to give

$$h \int_{-h/2}^{h/2} \left(\frac{d^2 T}{dx} - 2\right) dx = 0$$

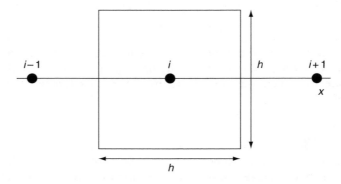

Figure 7.5 A control volume in the finite volume method.

Dividing by h and integrating gives

$$\left.\frac{dT}{dx}\right|_{-h/2}^{h/2} - 2x \left.\vphantom{\frac{dT}{dx}}\right|_{-h/2}^{h/2} = 0$$

This involves face values of $\frac{dT}{dx}$ that can be approximated in this example for the point $x = 1/2$ to give

$$\frac{T_3 - T_2}{h} - \frac{T_2 - T_1}{h} - 2\left[\frac{h}{2} - \frac{-h}{2}\right] = 0$$

$$T_3 - 2T_2 + T_1 - 2h^2 = 0$$

$$0 - 2T_2 + 0 = 2\left(\frac{1}{2}\right)^2$$

$$-2T_2 = \frac{1}{2}$$

$$T_2 = -\frac{1}{4}$$

This gives the same result as the finite difference and finite element methods for this simple case. However, for more complex cases this is rarely true, but the solution should be almost identical as the mesh is refined. The simplified example used here does not involve fluxes. The treatment of these is particularly appropriate with the finite volume and is a significant aspect of the scheme. Readers are referred elsewhere (Versteeg and Malalasekera, 1995) for a description of this.

Initially the finite volume method was applied only to structured grids, but as CFD developed there was a desire to apply it to geometries that did not conform to a rectilinear form. One solution is to deform the grid to fit the shape of the domain. This was successful, but cases arose where the grid was so deformed that it was difficult to obtain solutions to the problem.

A significant development in tackling complex geometries was a combination of the finite element and finite volume approaches. Researchers (Schneider and Raw, 1987) took the unstructured grids of finite elements, but applied the finite volume methodology. This approach now forms the basis for the main commercial 3D CFD packages (CFX, FLUENT, STAR-CD). It holds the advantages of ease of application to complex geometries (including natural geometries) and implicit conservation of physical quantities. Although the development of finite volume techniques proceeded relatively independently of the finite element approach, it can be shown that with a suitable choice of trial functions and weighting functions, a finite volume formulation can be developed through a finite element analysis showing that finite volume is in fact a subset of finite elements. The major difference is that finite volume methods minimise error locally whereas most general finite element methods reduce the error globally.

7.2.4 Lagrangian approaches

Most CFD is conducted using an Eulerian approach which is the one outlined earlier. This means that a fixed grid is used and the fluid flows across this. Velocities, pressures, etc., are defined at fixed points. In contrast to this a Lagrangian approach solves in a frame of reference that moves with the fluid. This can be particularly useful in cases where there is an interface between two fluids or for tracking the movement of particles.

7.3 Grids

Constructing a suitable mesh for a given geometry is often the most demanding part of a CFD simulation in terms of both operator time and expertise. A comprehensive discussion can be found in the book by Löhner (2001). Adequate details of the physical situation must be included in the mesh, but too great a level of detail will lead to a mesh that contains more cells than can be accommodated on the computer. So an appropriate spatial scale must be selected. Further, a mesh should be refined in certain key areas such as boundary layers, for example, where velocity gradients are highest.

In most modern CFD packages grid generation is automated to a large extent and the user may only need to select a global length scale for the mesh such as a maximum cell edge length. Once this is set, the software's own grid generator will set up a mesh. Even where automated grid generation is available this should only be viewed as a first stage in generating a solution. Once the problem has been set up and successfully solved, the solution and grid statistics relevant to each type of mesh should be examined to ascertain whether improvement is necessary. Changes to the grid can often make a significant difference to the solution (Figure 7.6). The following features are usually available in commercial softwares:

- Refinement of the grid in a specified area such as a boundary layer or the main channel in a floodplain.
- Specification of a specific type of cell in a boundary area. Hexahedral or prismatic meshes are generally accepted as giving more accurate solutions when the flow is aligned with one direction as it is in a boundary layer.

These types of mesh can be combined in various ways. For example, a hexahedral mesh on a boundary can be used in conjunction with a tetrahedral mesh in the rest of the domain. This would allow for accuracy in the boundary layer, but versatility in defining complex geometry away from the walls. With a structured mesh the domain can be separated into a series of contiguous, non-overlapping blocks each of which contains a structured grid. This "multiblock" technique retains the simplicity of structured grids, but allows for representation of more complex geometries than would be the case with a single block.

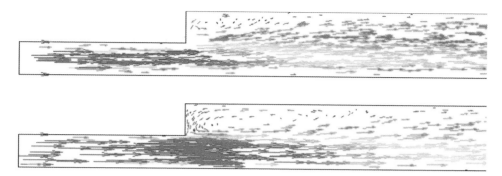

Figure 7.6 *Changes in solution with grid refinement. The lower velocity vectors are for a finer mesh than the upper ones. A clear difference between the size of the recirculation zone in each case.*

The construction of a grid always starts from a definition of the domain to be considered. This domain can be defined in two ways: by an analytic function (often just a plane) or by a set of points (such as a Digital Terrain Model or DTM). The latter is obviously often relevant in environmental flows. Careful consideration should be given to the balance between an accurate representation of the terrain and the complexity of the grid resulting from the large number of points that this would require. When fitting a grid to a DTM it should be borne in mind that using a denser DTM actually gives a different domain from a less dense DTM. If this is done as part of a grid independence study (section 7.6.1) it may appear that the grid refinement is not sufficient, whereas the change in the solution is due to the change in DTM resolution.

Increasingly, computer codes incorporate methods for automating the process of refining the grid. Instead of the user analysing the grid and results, this technique, known as adaptivity or adaption, uses a measure of the error or complexity of the solution to decide where to refine the grid. The criterion for refinement is often a gradient of the variables, but can be any function of the variables, e.g. velocity gradient, turbulence levels. A particular example would be for free surface flows where the grid can be refined wherever there is a change from water to air in order to increase the accuracy of the representation of the free surface. Adaption can also include the removal of cells where they are not required. More complex refinement strategies can be based on comparing solutions on two grids to give an estimation of truncation error (Smith, 1978).

7.4 Discretisation

Once a grid has been generated, the process of representing the PDEs can be carried out using one of the techniques introduced in section 7.2. Although the methodology is given there, it is worth reflecting on a few issues that can have a significant effect on the solution.

7.4.1 Truncation errors

Any discrete representation of a continuous solution will be an approximation and it is vital to have a measure of this error. Each methodology has its own representations, but these always involve a term based on the cell or element size. The evident fact, which is however worth restating, is that the smaller the elements used to represent the solution, the more accurate the numerical solution to the PDEs will be. The truncation error is a useful definition that is used to represent errors for finite difference methods (Smith, 1978; Löhner, 2001). It gives an estimate of error that is a function of grid size. Similar measures can be derived for finite elements and volumes, but only that for the finite difference method is highlighted here. The truncation error goes back to the use of Taylor series in deriving the finite difference method. The Taylor series representations of the discrete values such as

$$T_{i+1} = T_i + h\left(\frac{dT}{dx}\right)_i + \frac{h^2}{2}\left(\frac{d^2T}{dx^2}\right)_i + o(h^3)$$

are substituted into the finite difference expression (using forward differencing) where the overbar indicates an approximate value

$$\overline{\frac{dT}{dx}} = \frac{T_{i+1} - T_i}{h}$$

to give

$$\overline{\frac{dT}{dx}} = \frac{\left(T_i + h\left(\frac{dT}{dx}\right)_i + \frac{h^2}{2}\left(\frac{d^2T}{dx^2}\right)_i + o(h^3) - T_i\right)}{h}$$

$$= \left(\frac{dT}{dx}\right)_i + \frac{h}{2}\left(\frac{d^2T}{dx^2}\right)_i + o(h^2)$$

This indicates that the approximation is equivalent to the actual value plus a leading order term of the grid size multiplied by the second derivative at that position. This means that the error is reduced linearly as h is decreased and that the error will be higher in regions where there are rapid changes in gradients. If the same analysis is carried out for central differencing the following expression is obtained:

$$\overline{\frac{dT}{dx}} = \left(\frac{dT}{dx}\right)_i + h^2\left(\frac{d^3T}{dx^3}\right)_i + o(h^4)$$

From this it can be seen that the error in this case is proportional to the square of the grid size which in practice means that accuracy can be dramatically improved without having to resort to further grid refinement. Figure 7.7 demonstrates that use of a higher-order discretisation technique is a more cost-effective way of achieving accuracy than grid refinement. The derivative term in the error is now a third-order derivative and this is also significant. The second-order derivative in the truncation error of the forward differencing technique is similar to the diffusion term from the

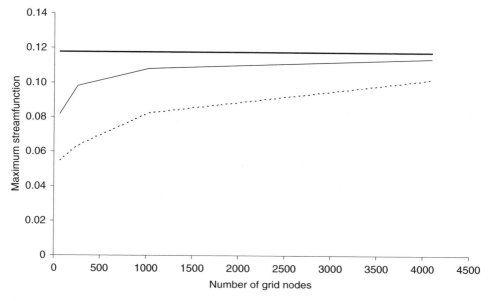

Figure 7.7 *Plot of streamfunction prediction for various numbers of grid nodes. The bold line is the benchmark solution. The dashed line is for a first-order discretisation and the solid line for a second-order discretisation. It can be seen that many fewer grid nodes are needed by the second-order scheme for a given level of accuracy.*

governing PDEs and the error introduces a false, numerical diffusion that can have a dramatic effect on the solution (Figure 7.8). Higher-order discretisations overcome this limitation by using information from more discrete nodes. Whilst this gives greater accuracy it does involve more calculations and these schemes can be more prone to instabilities in the solution process. However, in most cases the higher-order schemes give more accuracy without having to resort to the very fine grids that a low-order scheme would require.

It should be noted that although central differencing is more accurate than upwind differencing, it is not always stable for solutions of fluid flow problems. These difficulties will be discussed later.

7.4.2 Boundedness

In moving towards higher-order discretisation, researchers found that there were limitations due to lack of boundedness (Leonard, 1987). Any solution to a convection/diffusion problem such as found in fluid dynamics, in the absence of source terms, should not have a solution that is outside the bounds of the minimum and maximum values at the boundaries. This property is referred to as boundedness and is particularly important for environmental flows where transport of some scalar quantity is being studied. Upwind differencing possesses this property in all cases, but some higher-order schemes do not (Atias *et al.*, 1977; Leonard, 1979). Various schemes have been proposed that overcome this (i.e. SMART (Gaskell and Lau,

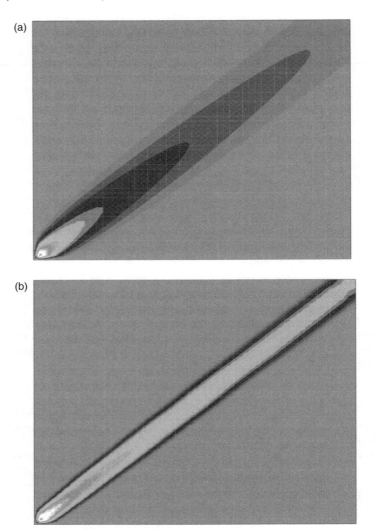

Figure 7.8 The effect of numerical diffusion: (a) First-order discretisation; and (b) Second-order discretisation. The contours represent the transport of a scalar in a flow from bottom left to top right. The false numerical diffusion can be seen in (a).

1988) and SHARP (Leonard, 1987, 1988)) and one or more of them are implemented in most CFD codes.

7.4.3 Conservation

The property of conservation is important in analysing fluid flow because conservation of mass is one of the fundamental laws. For a steady state simulation, conservation requires that the mass flowing into the domain is equal to that flowing out.

For an unsteady simulation, the inflow must equal the outflow plus any accumulation from sources within the domain.

Within a finite volume scheme, conservation is accomplished by ensuring that the flow out of one control volume face is equal to the flow into the next one. In a finite element method, where the integration is done over the sum of all the elements, this is not necessarily the case and so finite element solutions are not necessarily conservative. It can be argued that this lack of conservation is not significant as it should decrease as the grid is refined. However, the level of grid refinement required could be prohibitive. In view of this, particular care should be taken in situations where conservation is important such as flows with sediment transport and water quality indicators.

Conservation is obeyed if for two adjacent cells the flux through a shared face computed for one is equivalent to the flux computed for the other cell through the same face. More exactly, a discretisation scheme is conservative if there are no effective source terms in the algebraic analogue that do not appear in the governing PDE. If this is the case, then the algebraic equation mirrors the conservative property of the differential equation exactly. This can be stated mathematically as follows:

If $\overline{\frac{\partial \phi}{\partial x}}$ is the approximation to $\frac{\partial \phi}{\partial x}$ then the discretisation scheme is conservative if it is of the form:

$$\overline{\frac{\partial \phi}{\partial x}} = \frac{1}{h}[H(\phi_{i+\ell},\ldots,\phi_{i-\ell+1}) - H(\phi_{i+\ell-1},\ldots,\phi_{i-\ell})]$$

where H is a function of $2l$ arguments which must for consistency satisfy

$$H(\phi_i, \phi_i, \ldots, \phi_i) = \phi_i$$

7.5 Algebraic solution

Once the PDEs have been converted to the discrete form as described above we have a set of nonlinear algebraic equations. Given the nonlinearity and the size of grids currently in use (several million not being uncommon), the task of solving these is considerable. Further the form of the Navier–Stokes equations for incompressible flow means that we do not have an explicit equation for pressure although it is one of the dependent variables along with the velocity components.

The first technique to become widespread was the SIMPLE technique of Patankar and Spalding (1972). This is still popular today and is used in many research and commercial codes. Patankar and Spalding suggested using a pressure equation that was derived by combining the continuity and momentum equations. The procedure then solved for each variable in turn using "frozen" values of the other variables as follows:

1. Solve the momentum equation for u with assumed v, w and p values;
2. Solve the momentum equation for v with assumed u, w and p values;
3. Solve the momentum equation for w with assumed u, v and p values;
4. Solve the pressure correction equations with the latest u, v and w values;
5. Update the values of u, v, w and p to satisfy continuity.

This procedure is repeated until the required level of convergence is achieved.

During the 1980s, research was conducted (Schneider and Raw, 1987; Wright, 1988) on alternatives that solved for velocity and pressure simultaneously and removed the necessity for deriving a pressure equation. Coupled solvers have now been adopted for some commercial codes and deliver significant increases in computational efficiency and robustness. The efficiency comes about from the reduced number of iterations required which outweighs the increased computational cost of each iteration.

With the finite element technique, the equations are solved using either a direct or iterative method. The usual form of the former is the frontal or element-by-element method. This recognises that it is not necessary to assemble the full stiffness matrix from the local stiffness matrices for each element. The solution proceeds by solving for a particular element and any connected to it. In this way the solution progresses through the elements along a front.

As the finite element method was used to solve larger and larger problems with more and more elements, direct methods became inefficient. Iterative methods were then introduced which use much less computer storage. Popular ones are the conjugate gradient method and its generalisation, the generalized minimum residual method, GMRES (Saad and Schultz, 1986; Saad, 1996).

7.6 Some definitions: consistency, stability and convergence

Consistency
A numerical method is consistent (Chung, 2002) if it becomes the corresponding differential equations as the grid size and time step tend to zero.

Stability
This concept is often confused with convergence. Given that the governing equations are nonlinear, the techniques used to find solutions to the algebraic equations are iterative. A solution technique is stable (Chung, 2002) if the errors during the iterative process do not grow unboundedly as the iteration proceeds.

Convergence
A particular numerical solution is convergent (Chung, 2002) if the solution tends to a limiting value as the grid size is decreased. Once this limit is reached the solution is referred to as "grid independent" which reflects the fact that further refinement makes no difference to the solution. In many cases the level of grid refinement required is prohibitive. In this situation the finest grid possible should be compared with a coarser grid to establish how much the solution is changing by and in this way to obtain an estimate of how converged the solution is.

Consistency and stability imply convergence.

7.7 Boundary conditions

All fluid flow problems need to have values specified at the boundaries in order to find a solution. These are an important part of the problem definition and they must be chosen to reflect the physical situation and to be numerically realisable. Boundary conditions are usually categorised as Dirichlet or von Neumann (Smith, 1978) where Dirichlet conditions involve given values of the variables and von Neumann conditions involve given values of the derivatives of the variables.

One form of boundary that occurs commonly in environmental flows is an outlet boundary. This boundary is usually seen as having little significance to the main flow, but its implementation requires care. It has to be placed sufficiently far away from the main flow so that it can be assumed that the flow is uniform at the boundary and zero-derivative conditions can be applied. If it is applied too close it can have an effect on the solution that is unphysical or it can even prevent a solution from being found. Obviously it is preferable not to have it too far away as this can increase the size of the domain and require more grid cells. Sensitivity tests should be used to find an optimum position.

Some boundary conditions can extend into the domain and influence values at cells adjacent to a wall. An example of this is the Law of the Wall used with turbulence modelling. In this the value at the node adjacent to the wall is set to satisfy an assumed profile.

7.8 Solving the turbulence equations

The equations used to model the effect of turbulence on environmental flows are discussed by Sotiropoulos (Chapter 5). However, the equations pose particular challenges for numerical methods. For an RANS simulation it is important to ensure the boundedness property, else the value of turbulent kinetic energy may become negative which is physically wrong. These equations can be particularly "stiff" which means that the solution can change dramatically with small changes in the flow conditions. This makes iterative convergence very slow.

For Large Eddy Simulations (LES) the equations to be solved are relatively simple, but attention must be paid to grid quality and discretisation accuracy in order to ensure that turbulence is not generated by numerical error. Further the large number of grid cells used by LES requires efficient solution techniques such as multigrids.

7.9 Unsteady simulations

Early applications of CFD were mostly focused on steady state solutions. These have the benefit of being easy to solve in terms of complexity and computer requirements and they also produce fewer data. More and more, however, attention is being focused on unsteady simulations. This is especially pertinent for environmental flows such as atmospheric pollutant dispersion and fluvial flooding.

In these cases, in addition to the question of spatial discretisation, there is the question of temporal discretisation. The solution is advanced by discrete timesteps from one time level to another. This is usually denoted as from time level "n" to time level "$n+1$". The simplest time discretisation is backward differencing:

$$\left(\frac{dT}{dt}\right)^n_i = \frac{T_i^{n+1} - T_i^n}{\Delta t}$$

More accurate and stable are implicit schemes which use values at the new timestep, e.g. the Crank–Nicholson scheme which approximates:

$$\frac{\partial T}{\partial t} = \frac{\partial T^2}{\partial x^2}$$

with

$$\frac{T_i^{n+1} - T_i^n}{\Delta t} = \frac{1}{2}\left(T_{i+1}^{n+1} - 2T_i^{n+1} + T_{i+1}^{n+1}\right) + \frac{1}{2}\left(T_{i+1}^n - 2T_i^n + T_{i+1}^n\right)$$

The use of the next timestep as well as the current one to approximate the spatial derivative significantly increases stability which allows much larger values of the timestep to be used. Even more accuracy can be obtained by using the techniques for ordinary differential equations for the time derivatives, such as the Runge–Kutta method. This opens up the possibility of automatically controlling the timestep to deliver a given error level. In turn this can be used to balance the errors in the time and space derivatives through spatio-temporal error balancing. Given the large variations in spatial and temporal scales in environmental flows, locally refined timestepping (Crossley et al., 2003) can deliver significant advances in accuracy and reductions in solution times.

7.10 Simplification of the 3D equations for shallow water flows

In many situations it is possible to use assumptions about the fluid flow to simplify the Navier–Stokes equations. One example of this that is particularly pertinent to environmental flows is the case of the Shallow Water equations which can be used to describe the flow of water on the earth's surface, i.e. in rivers, estuaries and seas. In this case it is assumed that the lateral scales are large relative to the depth of the flow. From this it is deduced that the vertical pressure distribution is the same as that obtained in a stationary flow (often called a hydrostatic pressure distribution) and in turn this implies that the streamlines of the flows are almost parallel. The reduced set of equations can be considered in one or two dimensions. In the 1D case discharge and depth are related to each other and no account is taken of lateral or vertical variations in velocity. In the 2D case lateral variations are taken into account.

One-dimensional modelling is the mainstay of river management consultancy in many countries and softwares such as ISIS, MIKE11, SOBEK and HEC-RAS have been developed and successfully marketed for this work. These techniques use implicit methods such as the Preissmann or Abbott-Ionescu schemes to solve the equations (Cunge et al., 1980). The implicit nature of the solution techniques (Cunge et al., 1980) allows for large timesteps to be used. These methods are, in fact, not completely appropriate for these equations. Mathematical analysis of the equations would indicate that the use of characteristic-based methods (Smith, 1978; Cunge et al., 1980) or methods based on Riemann solvers (LeVeque, 1992; Toro, 1997) would be more appropriate. However, the traditional techniques have the advantage of allowing the straightforward incorporation of structures commonly found in rivers such as weirs, sluice gates, bridges, etc.

Whereas 1D modelling essentially represents the river as a straight, linear system, 2D modelling considers a grid in plan view. This grid can be Cartesian where the irregular geometry is represented by removing cells, boundary fitted or unstructured. Whichever technique is used, a set of cells are produced that cover the domain of interest. These are then solved using any of the three techniques outlined earlier: finite differences, finite elements and finite volumes. Finite differences can only be applied on the Cartesian grids and finite elements are usually only applied to unstructured grids. The finite volume method on the other hand is applied to all three. Many techniques exist for discretising the equations and for solving the subsequent algebraic equations: Alternating Direction Implicit (ADI) (or other forms of operator splitting) for Cartesian and boundary fitted grids (Falconer, 1980), fully implicit methods for unstructured methods, Riemann solvers (Sleigh *et al.*, 1998; Namin *et al.*, 2002) and functional iteration (Sleigh *et al.*, 1998) for unstructured grids. Finite element methods use the same techniques as in 3D solutions.

A number of commercial and academic codes are in use based on either the finite element or finite volume methodology. As mentioned earlier the finite element method does not conserve mass exactly and this must be borne in mind in Shallow Water modelling especially when considering the transport of sediment or water quality parameters.

Many codes have adopted the finite element approach due to its ability to deal with the geometric complexity usually found in river, estuarine and coastal areas. Examples are:

- TELEMAC-2D produced by EDF and marketed in the UK by HR Wallingford (Bates *et al.*, 1996);
- SMS produced by Brigham Young University based on codes from the USACE such as RMA2D (King and Norton, 1978);
- CCHE2D produced by NCCHE, University of Mississippi (Wang *et al.*, 1989).

Codes using the finite volume method have also been developed (Falconer, 1980) and applied successfully particularly for cases where conservation is important. In order to deal with geometrical complexity boundary fitted grids have been developed for use with finite volumes; examples are DIVAST (Falconer and Lin, 1997) and MIKE21C (DHI, 1998). In the past decade there has been significant development of unstructured finite volume codes (Anastasiou and Chan, 1997; Sleigh *et al.*, 1998; Olsen, 1999).

The issue of wetting and drying is a perennially difficult one for 2D models. As water levels drop, areas of the domain may become dry and the calculation procedure must remove these from the computation in a way that does not compromise mass conservation or computational stability. Most available codes can deal with this phenomenon, but they all compromise between accuracy and stability. This issue must be carefully examined in results from any 2D simulation where wetting and drying is significant.

The bathymetry of a river is represented in different ways for 1D and 2D models. In 1D a series of cross sections are used which represent bed heights at various distances along a line roughly perpendicular to the direction of flow. The distances between these cross sections should adequately represent the changes in cross section with shorter spacings in regions of rapid changes. However, in many cases the modeller is restricted by the choices made by the surveyor who relies more on ease of access and measurement. In 2D models the bathymetry is represented by bed heights at irregularly spaced (x, y) locations. Although these locations can be arbitrary, they are usually concentrated in the main channel and other areas with rapid changes in velocity or bed height.

Attention is increasingly being focused on model integration. In most modelling work we do not need a full 3D model; so the optimum approach is to construct a 1D model and to incorporate 2D or 3D models for specific areas where they are useful or necessary (Dhondia and Stelling, 2002). Additionally, it could be important to include a model of sub-surface or groundwater (Aradas *et al.*, 2002) in areas where soil saturation has a significant influence on flooding.

7.11 Guidance for CFD users

The rapid development of CFD has not brought it to the level of a "black box" analysis tool. Experience and understanding are still needed on the part of the operator in both setting up a problem and analysing its results. A number of efforts have been made to provide guidance and the recently produced European Research Community on Flow, Turbulence and Combustion (ERCOFTAC) Guidelines (Casey and Wintergerste, 2000) are an excellent example of these. Guidance is also available from reports such as that produced by the ASME (1993) and AIAA (1998). Lane and others have also discussed the more scientific questions of what is meant by validation and verification both in this text and elsewhere (Lane and Richards, 2001). Their arguments should be borne in mind by anyone applying CFD to environmental flows. An excellent book on verification and validation in CFD is that by Roache (1998).

7.12 The future

The future application of CFD in environmental flows will be driven by a number of factors:

- Increasing computer power continually expands the size of grids and length of runs that can be undertaken. Through this we will be able to include more and more physical effects in our models.
- New algorithms for turbulence modelling and roughness representation will improve our predictions of fluid flow.
- Remote sensing of input and output data for calibration will give better model definition and more exact methods for evaluating models.
- With increased computer power we will be able to carry out multiple runs for calibration and uncertainty estimation.

As the use of CFD for environmental flows expands, there will be a continuing need to educate users to use the technique intelligently rather than as a black box. This entails some understanding of the issues introduced in this chapter and the necessity of comparing results with reality. Again readers are directed to the guidelines in section 7.10.

References

AIAA (1998), *Guide for the Verification and Validation of Computational Fluid Dynamics Simulations*. AIAA.

Anastasiou, K. and C.T. Chan (1997), Solution of the 2D shallow water equations using the finite volume method on unstructured triangular meshes. *International Journal of Numerical Methods in Fluids*, **24**, 1225–1245.

Aradas, R.D., J.M. Wicks and N.G. Wright (2002), Integrated groundwater and surface water modelling for the río Salado master plan, Argentina, *Hydroinformatics 2002*. Cardiff, UK.

ASME (1993), Statement on the control of numerical accuracy. *Journal of Fluids Engineering*, **115**(3), 339–340.

Atias, M., M. Wolfstein and M. Israeli (1977), Efficiency of Navier–Stokes solvers. *AIAA Journal*, **15**(2), 263–266.

Bates, P.D., M.G. Anderson, D.A. Price, R.J. Hardy and C.N. Smith (1996), 'Analysis and development of hydraulic models for floodplain flows in floodplain processes', in *Floodplain Processes*, M. Anderson, D. Walling and P. Bates (eds). Chichester: Wiley.

Casey, M. and T. Wintergerste (2000), Best practice guidelines, *ERCOFTAC Special Interest Group on Quality and Trust in CFD*. ERCOFTAC Special Interest Group on Quality and Trust in CFD.

Chung, T.J. (2002), *Computational Fluid Dynamics*. Cambridge: CUP.

Crossley, A.J., N.G. Wright and C.D. Whitlow (2003), Local temporal adaptivity for modelling open channel flows. *ASCE Journal of Hydraulic Engineering*, **129**(6).

Cunge, J.A., F.M. Holly Jr and A. Verwey (1980), *Practical Aspects of Computational River Hydraulics*. London: Pitman.

DHI (1998), *MIKE21C: User Guide and Scientific Documentation*. Copenhagen: Danish Hydraulic Institute.

Dhondia, J.F. and G.S. Stelling (2002), Application of one dimensional – two dimensional integrated hydraulic model for flood simulation and damage assessment, *Hydroinformatics 2002*. Cardiff: IWA Publishing.

Falconer, R.A. (1980), *Numerical Modelling of Tidal Circulation in Harbours*. Proceedings of the Institution of Civil Engineers, Waterway, Port and Coastal Division, **106**, 32–33.

Falconer, R.A. and B. Lin (1997), Three-dimensional modelling of water quality in the Humber estuary. *Water Research, IAWQ*, **31**(5), 1092–1102.

Gaskell, P.H. and A.K.C. Lau (1988), Curvature compensated convective transport: Smart. *International Journal for Numerical Methods in Fluids*, **8**, 617–641.

King, I.P. and W.R. Norton (1978), Recent application of RMA's finite element models for two dimensional hydrodynamics and water quality. *Second International Conference on Finite Elements in Water Resources*. London: Pentech Press.

Lane, S.N. and K.S. Richards (2001), 'The "validation" of hydrodynamic models: Some critical perspectives', in *Model validation: Perspectives in Hydrological Science*, M.G. Anderson and P.D. Bates (eds). Chichester: John Wiley & Sons Ltd.

Leonard, B.P. (1979), A stable and accurate convective modelling procedure based on quadratic upstream interpolation. *Computational Methods in Applied Mechanics and Engineering*, **19**, 59–98.

Leonard, B.P. (1987), 'Locally modified QUICK scheme for highly convective 2-d and 3-d flows', in *Numerical Methods in Laminar and Turbulent Flow, Proceedings of the Fifth International Conference*. Swansea: Pineridge Press.

Leonard, B.P. (1988), Simple high-accuracy resolution program for convective modeling of discontinuities. *International Journal for Numerical Methods in Fluids*, **8**(10), 1291–1318.

LeVeque, R.J. (1992), *Numerical Methods for Conservation Laws*. Basel: Birkhauser Verlag.

Löhner, R. (2001), *Applied Computational Fluid Dynamics Techniques*. Chichester: John Wiley & Sons Ltd.

Namin, M.M., R.A. Falconer, M. Mohammadian, B. Lin and R. Kamalian (2002), Hydro-environmental modelling and analysis tool (HEMAT), a GUI-based 2DH hydrodynamic and contaminant transport model on unstructured triangular grids. *Fifth International Conference on Hydroinformatics*. Cardiff, Wales: IWA Publishing.

Olsen, N.R.B. (1999), Two-dimensional numerical modelling of flushing processes in water reservoirs. *Journal of Hydraulic Research*, **37**(1), 3–16.

Patankar, S. and D. Spalding (1972), A calculation procedure for heat, mass and momentum transfer in three-dimensional parabolic flows. *International Journal of Heat Mass Transfer*, **15**, 1787–1806.

Roache, P.J. (1998), *Verification and Validation in Computational Science and Engineering*. Albuquerque: Hermosa Publishers.

Saad, Y. (1996), *Iterative Methods for Sparse Linear Systems*. Boston: PWS Publishing.

Saad, Y. and M.H. Schultz (1986), GMRES: A generalized minimal residual algorithm for solving non-symmetric linear systems. *SIAM Journal of Science and Statistical Computing*, **7**, 856–869.

Schneider, G.E. and M.J. Raw (1987), Control-volume finite element method for heat transfer and fluid flow using co-located variables – 1 computational procedure. *Numerical Heat Transfer*, **11**, 363–390.

Sleigh, P.A., P.H. Gaskell, M. Berzins and N.G. Wright (1998), An unstructured finite-volume algorithm for predicting flow in rivers and estuaries. *Computers & Fluids*, **27**(4), 479–508.

Smith, G.D. (1978), *Numerical Solution of Partial Differential Equations: Finite Difference Methods*. Oxford University Press, Oxford, UK.

Toro, E.F. (1997), *Riemann Solvers and Numerical Methods for Fluid Dynamics*. Berlin: Springer Verlag.

Versteeg, H. and W. Malalasekera (1995), *An Introduction to Computational Fluid Dynamics – The Finite Volume Method*. London: Longman Group Ltd.

Wang, S.S.Y., V.V. Alonso, C.A. Brebbia, W.G. Gray and G.F. Pinder (1989), Finite elements in water resources. *Third International Conference on Finite Elements in Water Resources*. Mississippi, USA.

Wright, N.G. (1988), Multigrid solution of elliptic fluid flow problems, in Department of Mechanical Engineering. Unpublished PhD Thesis, University of Leeds.

Zienkiewicz, O. and Y. Cheung (1965), Finite elements in the solution of field problems. *The Engineer*, pp. 507–510.

8

A framework for model verification and validation of CFD schemes in natural open channel flows

S.N. Lane, R.J. Hardy, R.I. Ferguson and D.R. Parsons

8.1 Introduction

A number of disciplines that use Computational Fluid Dynamics (CFD) for the study of fluid problems have adopted a series of basic guidelines which a modeller must address in order for the results of the model to be deemed reliable. Most of the statements have originated from an editorial in the American Society of Mechanical Engineers *Journal of Fluid Engineering* (Roache *et al.*, 1986; Table 8.1). This required the quantification of the numerical accuracy of a numerical simulation prior to the work being deemed acceptable for publication within the journal. Since this original statement, several other journals have adopted similar explicit policies, including the American Society of Mechanical Engineers Editorial board (1993), the American Institute of Aeronautics and Astronautics (AIAA, 1994) and the *International Journal of Fluids Engineering* (Gresho and Taylor, 1994). Indeed, Leonard (1995), in his discussion on the policy statement on numerical accuracy for the *Journal of Fluids Engineering*, suggested that similar policies should be adopted by all journals in which numerical solution is used if sufficient publication quality is to be maintained. However, such a policy needs to be updated at least once every decade to accommodate software and hardware advances as well as the current interest in different

Table 8.1 The ASME Journal of Fluids Engineering *statement on the Control of Numerical Accuracy*

1. Authors must be precise in describing the numerical method used; this includes an assessment of the formal order of accuracy of the truncation error introduced by individual terms in the governing equations, such as diffusive terms, source terms, and most importantly, the convective terms. It is not enough to state, for example, that the method is based on a 'conservative finite volume formulation', giving then a reference to a general CFD textbook.
2. The numerical method used must be at least formally second-order accurate in space (based on a Taylor series expansion) for nodes in the interior of the computational grid. The computational expense of second, third, and higher-order methods are more expensive (per grid point) than first-order schemes, but the computational efficiency of these higher-order methods (accuracy per overall cost) is much greater. And, it has been demonstrated many times that, for first-order methods, the effect of numerical diffusion on the solution accuracy is devastating.
3. Methods using a blending or switching strategy between first and second-order methods (in particular, the well-known 'hybrid', 'power-law', and related exponential schemes) will be viewed as first-order methods, unless it can be demonstrated that their inherent numerical diffusion does not swamp or replace important modelled physical diffusion terms. A similar policy applies to methods invoking significant amounts of explicitly added artificial viscosity or diffusivity.
4. Solutions over a range of significantly different grid resolutions should be presented to demonstrate grid independence or grid-convergent results. This criterion specifically addresses the use of improved grid resolution to systematically evaluate truncation error and accuracy. The use of error estimates based on methods such as Richardson extrapolation or those techniques now used in adaptive grid methods, may also be used to demonstrate solution accuracy.
5. Stopping criteria for iterative calculations need to be precisely explained. Estimates must be given for the corresponding convergence error.
6. In time-dependent solutions, temporal accuracy must be demonstrated so that the spurious effects of phase error are shown to be limited. In particular, it should be demonstrated that unphysical oscillations due to numerical dispersion are significantly smaller in amplitude than captured short-wavelength (in time) features of the flow.
7. Clear statements defining the methods used to implement boundary and initial conditions must be presented. Typically, the overall accuracy of a simulation is strongly affected by the implementation and order of the boundary conditions. When appropriate, particular attention should be paid to the treatment of inflow and outflow boundary conditions.
8. In the presentation of an existing algorithm or code, all pertinent references or other publications must be cited in the paper, thus aiding the reader in evaluating the code and its method without the need to redefine details of the methods in the current paper. However, basic features of the code must be outlined according to item 1, above.
9. Comparison to appropriate analytical or well-established numerical benchmark solutions may be used to demonstrate accuracy for another class of problems. However, in general this does not demonstrate accuracy for another class of problems, especially if any adjustable parameters are involved, as in turbulence modelling.
10. Comparison with reliable experimental results is appropriate, provided experimental uncertainty is established. However, 'reasonable agreement' with experimental data alone will not be enough to justify a given single-grid calculation, especially if adjustable parameters are involved.

classes of flow simulations (Karniadakis, 1995). In this chapter, we develop those specified by the American Society of Mechanical Engineers (ASME) (1993) for the specific case of open channel flows. In doing so, we demonstrate that the ASME (1993) criteria are not sufficient for many practical river (and, by extensions, estuarine) flow problems, but they do provide a minimum framework around which to consider model verification and model validation.

8.2 Basic criteria for model verification

This section lists the basic criteria recommended by the ASME (1993) in relation to open channel flow problems. Where appropriate, additional comment is made.

Criterion 1: Standard of reporting; Criteria 2 and 3: Level of solution accuracy in space

1. Authors must be precise in describing the numerical method used; this includes an assessment of the formal order of accuracy of the truncation error introduced by individual terms in the governing equations, such as diffusive terms, source terms, and most importantly, the convective terms. It is not enough to state, for example, that the method is based on a 'conservative finite volume formulation', giving then a reference to a general CFD textbook.
2. The numerical method used must be at least formally second-order accurate in space (based on a Taylor series expansion) for nodes in the interior of the computational grid. The computational expense of second, third, and higher-order methods is higher (per grid point) than first-order schemes, but the computational efficiency of these higher-order methods (accuracy per overall cost) is much greater. And, it has been demonstrated many times that, for first-order methods, the effect of numerical diffusion on the solution accuracy is devastating.
3. Methods using a blending or switching strategy between first and second-order methods (in particular, the well-known 'hybrid', 'power-law', and related exponential schemes) will be viewed as first-order methods, unless it can be demonstrated that their inherent numerical diffusion does not swamp or replace important modelled physical diffusion terms. A similar policy applies to methods invoking significant amounts of explicitly added artificial viscosity or diffusivity.

Comments
The above criteria are clear guidelines that relate to good scientific practice in the choice of numerical scheme and the reporting of that scheme. Initially, item 3 requires additional comment. Blending or switching strategies have been used in applications of CFD to open channel flow problems (e.g. Bradbrook *et al.*, 2000; Ferguson *et al.*, 2003; Lane *et al.*, 2004). One of the reasons for doing this is that second-order methods can lead to problems in obtaining converged solutions with complex geometries (Shaw, 1992) which will be the norm in natural river channels, for example. This may be exacerbated by the common use of structured grids, so a move to unstructured grids and porosity-based methods (Chapter 10) could alleviate it. However, first-order methods may be inaccurate, especially if diffusion terms are dominant over convective terms. Thus, blending solutions may use a Peclet number

(a product of the velocity across a face and the ratio of the convective and diffusive terms evaluated across that face). Convergence problems may emerge in the form of numerical oscillations where the Peclet number is greater than two. In such situations, a hybrid scheme will use a first-order accurate scheme (e.g. upwind differencing). If it is less than two, then a second-order scheme (e.g. central differencing or quadratic upwind) is used. This is a practical compromise necessitated by the complex geometries that are now of interest. However, a more critical view of the ASME criterion would recognise that the order of a scheme will control how the solution accuracy decreases as mesh spacing is increased. In 2D schemes, and in certain cases in 3D schemes, a fine mesh with a first-order solution may yield 'adequately' accurate solutions that are still computationally efficient and provide better representation of the effects of finer-scale topographic variability. Here, we introduce the phrase 'adequately', and this emphasises that judgements over sufficient accuracy need some form of additional assessment. In the ASME criteria, this is dealt with under criterion 4 (mesh independence testing) but we go on to show that even this may only give a partial understanding of adequacy. This is one sense in which careful thought needs to be given to transferring the ASME statements into the potentially more complex geometries of natural river channels.

Criterion 4: Mesh independence testing

4. Solutions over a range of significantly different grid resolutions should be presented to demonstrate grid independence or grid-convergent results. This criterion specifically addresses the use of improved grid resolution to systematically evaluate truncation error and accuracy. The use of error estimates based on methods such as Richardson extrapolation, or those techniques now used in adaptive grid methods, may also be used to demonstrate solution accuracy.

Comments
When considering open channel flows, spatial discretisation is a recurring problem due to the complex topographic surfaces that are often found in natural environments and the fact that spatial variation in flow properties can occur across a range of spatial scales. A crucial aspect of spatial discretisation is the choice of mesh resolution. For many numerical modelling applications, the problem of specifying an optimum mesh resolution remains unbounded both because the correct solution is not known and because objective *a priori* rules for mesh construction do not exist (cf. Hardy *et al.*, 1999). However, one approach for assessing the choice of mesh resolution applies a Grid Convergence Index (GCI) (Roache, 1994, 1997, 1998). The appeal of this approach is that it provides an objective means for reporting the sensitivity of model solutions to numerical discretisation (Hardy *et al.*, 2003).

The GCI is an index of the uncertainty associated with the solution at a particular grid resolution (the mesh uncertainty), based on comparison with the solution at another resolution. It uses the theory of generalised Richardson extrapolation which assumes that, within a certain radius of convergence, the discrete solution for some flow variable converges monotonically at all points in the continuum as the grid spacing tends to zero. Once applied, an error estimate (the GCI value) is calculated which, if multiplied by a factor of safety, gives a conservative upper limit akin to a statistical confidence interval. The GCI can be applied for a single point on a numerical mesh, for an ensemble of points (e.g. in a cross section) or for an entire mesh, and should be determined for all variables of interest. A GCI value of zero for any one variable (at a point or for an ensemble) would indicate perfect mesh independence, but is unattainable if only because of numerical rounding errors. It is then up to the user to decide what is an acceptable level of mesh dependence. Of particular importance is the connection with solver order which is allowed for in the GCI calculation.

This technique would be regarded as a sufficient method for establishing mesh independence following the guidelines of the ASME (1993). However, we have found (Hardy *et al.*, 2003) that although the error bands on specific variables can be of acceptable numerical accuracy for use in predictive terms for certain variables (e.g. streamwise velocity), the same may not be true for variables which converge more slowly such as turbulent kinetic energy. Different variables converge at different rates depending on the processes dominating the flow, and so have different GCI values. In terms of velocity components, Hardy *et al.* (2003) found less acceptable GCI values for vertical velocity for a river confluence and for cross-stream velocity for a river meander. This sort of difference is likely to remain the case in studies of open channel flow, where shear may occur in a number of directions, and mesh resolution can have a major impact upon the extent to which the shear is properly resolved. However, it emphasises the need to be explicit about which parameter is the focus of the assessment. In practice, there are two possibilities: (i) the variable that is the most sensitive in the numerical scheme for that particular environment; or (ii) the variable that is of most physical interest. It may be necessary to identify the flow processes which control the system definition prior to the application of the scheme to deduce the convergence sequence and identify which variable to verify a scheme upon. This would imply that if a scheme is only acceptably verified on one parameter, the scheme can only be used to judge that parameter (or one below it on the convergence sequence). Undertaking some sort of mesh resolution test for one variable is not a sufficient demonstration of mesh resolution for all variables.

Criterion 5: Determination of solution convergence

5. Stopping criteria for iterative calculations need to be precisely explained. Estimates must be given for the corresponding convergence error.

Comments

A simulation is said to be converging when the residuals (or errors) in the equations decrease as the iterative solution proceeds. The solution is deemed to have converged when the sum of the absolute values of the residuals for that variable falls below a prespecified tolerance. The exact definition of the criterion, however, is somewhat ambiguous in fluvial applications with actual values rarely, if ever, defined. One approach is to set the tolerance for the mass and momentum flux residuals to 0.1% of the inlet flux. However, achieving this level of convergence for fluvial problems can be difficult, especially for high-resolution meshes. Indeed, especially where the bed topography and/or the associated flow field is complex, achieving this level of convergence may require good initial approximations. One approach is to apply the inlet boundary cross section to all downstream cross sections, as the fully developed flow profile likely to be specified at the inlet is likely to be closer to the true solution than a homogenous arbitrary value.

Furthermore, a possible reason why convergence is not being achieved, and the convergence criterion is not being met, is that too much under-relaxation is being applied. Relaxation is a technique that can be used to increase the convergence rate (or even achieve convergence when the problem would diverge) by slowing down the rate at which variables may change during the iteration procedure. There are two common techniques: linear relaxation and false time step relaxation. Linear relaxation specifies a multiplier, normally between 0 and 1, by which the solution can be changed. The lower the linear relaxation, the slower the permitted change in a value. If a conservation equation is being solved, false time step relaxation can be applied. In this case a source term is added to the finite volume equation for a given conserved variable. If the false time step is large, the rate of change of the solution is slower, thus increasing the number of iterations required for convergence. Under-relaxation reduces the possible amount of change (a low value for linear relaxation, a large time step for false time step), while over-relaxation is opposite. If too much relaxation is applied, perhaps unsystematically and to too many variables, then this may initiate a trend towards convergence, but actual convergence may not be ever achieved. Thus, relaxation should be used carefully.

Criterion 6: Solution accuracy in time

6. In time-dependent solutions, temporal accuracy must be demonstrated so that the spurious effects of phase error are shown to be limited. In particular, it should be demonstrated that unphysical oscillations due to numerical dispersion are significantly smaller in amplitude than captured short-wavelength (in time) features of the flow.

Comments

The time step chosen in the numerical simulation can have an important effect on the stability of the numerical scheme. For the case where conditions at the inlet are steady and the solution has no time dependence, the standard equation used to judge the stability of a solution is the Courant number (L):

$$L = u \frac{\Delta t}{\Delta x} \qquad (6.1)$$

where u is the largest velocity component in a given cell, Δt is the time step and Δx is the cell width in the direction of u. Equation (6.1) must be modified in 2D and 1D models of unsteady flow to include wave celerity. Thus, the Courant number defines the distance travelled through a cell per unit time. It provides a basic method for determining a time step through a simulation: as numerical solution involves the determination of gradients between two or more adjacent cells, if the scheme is explicit and first order (but note criterion 3), then the time step should be small enough that the change in one value of a variable should not propagate fully across an adjacent grid cell, and so escape re-evaluation on the adjacent grid cell. Thus, for explicit schemes, L has to be less than 1 for them to be stable. In implicit schemes, such as TELEMAC-2D, it is best for the value to be less than 1, although it may exceed 1 in localised cases (Hervouet, 1996). A discussion on Courant numbers for non-regular grids is given by Bates et al. (1995). A criterion may be implemented that allows time steps to vary within a solution. For instance, where the numerical gradients are steep, the time steps can be shortened to prevent unphysical oscillations due to numerical diffusion. Where gradients are low, time steps may be lengthened to save computational time.

A more complex problem arises when time-dependent calculations are considered. There are two typical cases which are applicable in open channel hydraulics where time-dependent calculations are applied: (1) where the Reynolds-averaged equations are used to deal with the large-scale unsteadiness of flow, such as where inlet discharge changes as a function of time (e.g. Bates et al., 1995); and (2) at smaller spatial scales, where the inherent unsteadiness of the flow is of interest (i.e. in relation to turbulence). Karniadakis (1995) notes that, with any time-dependent calculations, more emphasis should be put on the accuracy of these calculations. For a meaningful time-dependent simulation, a long-time integration will be required that extends over many time periods in relation to the processes that are of interest (e.g. vortex shedding cycles). Such time periods may be long, especially if low frequencies are present (Karniadakis, 1995). The same is true for turbulence models where the statistics need to be calculated over a very long time period. Estimation of the numerical error of time discretisation can be done in the same manner as with spatial errors (Roache, 1997). Roache (1997) argues that it is relatively straightforward to estimate the temporal error of the calculation by the application of an inexpensive temporal error estimator (Roache, 1994). The time error estimator uses the difference between the backward and forward time integration, implemented as an extrapolation, enabling an error estimate of accuracy for the time discretisation error for that time step. Usually, a user would be interested in the maximum over the spatial domain of the percentage deviation of the absolute value of the difference between time steps normalised by the range of values (Roache, 1997).

Criterion 7: Specification of boundary conditions

7. Clear statements defining the methods used to implement boundary and initial conditions must be presented. Typically, the overall accuracy of a simulation is strongly affected by the implementation and order of the boundary conditions. When appropriate, particular attention should be paid to the treatment of inflow and outflow boundary conditions.

Comments
The governing equations are represented as a series of partial differential equations that can be solved for a particular problem only when boundary conditions are specified for all the dependent variables. Hence, for open channel flow applications, boundary conditions will need to be specified for the mean velocity components, the mean scalar quantities and the turbulence quantities when transport equations are used for their calculation (ASCE, 1988). Alternatively, boundary conditions may be

specified as a relationship between a specified variable and an unspecified variable (e.g. turbulent quantities may be specified as a function of local flow velocity). However, the specification of the data necessary for the code to run is not so easy as one might expect (Roache, 1997). Moreover, the incorrect determination of these items can lead to unsuccessful model application, even if the code is entirely correct (Roache, 1997).

Boundary conditions need to be applied at the inflow and outflow boundaries, at all the solid boundaries and at the free surface. The inflow and outflow boundary conditions are problem-dependent (ASCE, 1988) and can either be constructed from experimental or field data (e.g. Hodskinson and Ferguson, 1998; Bradbrook *et al.*, 2001) or calculated based on hydraulic principles such as a fully developed flow profile (e.g. Bradbrook *et al.*, 2000). At the solid boundaries and at the free surface, more general rules may need to be adopted (ASCE, 1988). Indeed, many of the chapters in this book (e.g. Chapters 2, 10 and 13) explore the nature of these rules in relation to particular applications. For instance, at solid boundaries, the mean flow equations do not account for either viscous stresses or molecular diffusion terms, so the governing equations are not applicable in regions where these effects are important (e.g. in the viscous sub-layer) (ASCE, 1988). There are mean flow equations and turbulence models available to account for these effects, but since the gradients of most quantities are very steep in the viscous sub-layer, the numerical resolution of this region is very expensive and therefore not desirable. The practical approach is to bridge the viscous sub-layer to conditions at the wall (ASCE, 1988). This is the basis of the standard 'law of the wall', and leads to the need to specify a roughness height in 3D simulations of open channel flow. These issues are considered further in Nicholas (Chapter 13) and Lane and Ferguson (Chapter 10).

It is important to pay particular attention to model geometry when considering practical fluvial and estuarine problems. Most early applications of 3D CFD considered channels of simple cross-sectional shape (e.g. rectangular) and, in some cases, simple planform (e.g. straight, sinuous) (e.g. Leschziner and Rodi, 1981). In such situations, model geometry is well specified although meshing the geometry may still remain a challenge (e.g. meshing a tributary junction, with one tributary entering higher than the second, was difficult (Bradbrook *et al.*, 2001), even though both the channels had rectangular sections). The major development over the last 10 years has been application of 3D CFD to natural river channels (e.g. Hodskinson and Ferguson, 1998; Lane *et al.*, 1999, 2000; Nicholas and Smith, 1999; Bradbrook *et al.*, 2000; Booker *et al.*, 2001; Ferguson *et al.*, 2003). In these situations, there is a continuous variation in model geometry that relates to reach-scale and cross-section-scale channel morphology, as well as scales of variation associated with bedform and individual grain arrangements. Research has sought to generalise these geometries at a range of spatial scales (e.g. at the reach scale, through study of idealised representation of meanders (Hodskinson and Ferguson, 1998) and confluences (e.g. Bradbrook *et al.*, 2000, 2001); and at the grain scale, through study of water-worked gravel surfaces in straight channels (e.g. Lane *et al.*, 2004), in order to provide generic understanding of the effects of particular components of geometry upon flow processes. However, when models are to be applied to real river situations, containing the full range of scales of topographic variability, reproduction of field (or laboratory) measurements requires consideration of the full range of topographic scales. Model geometry takes on many of the characteristics of the boundary conditions discussed above as: (1) it can have a major, sometimes dominant, effect upon model predictions; (2) it can be very difficult to measure, especially at smaller spatial scales; and (3) it requires the introduction of various parameterisations through other boundary conditions (e.g. upscaling of roughness heights in the wall treatments applied to near-boundary cells). This creates two challenges. The first is the growing complexity of channel geometry that is being tackled, which makes creating suitable meshes for numerical solution a particular problem. Conventional structured grids (as used by Hodskinson and Ferguson, 1998; Lane *et al.*, 1999, 2000; Bradbrook *et al.*, 2000; Ferguson *et al.*, 2003) can lead to marked and rapid spatial variation and skewness in volume or element size, which can lead to numerical diffusion and instability. This is not really a boundary condition issue. However, as the ease with which model geometry can be measured (whether by remote sensing or field survey) improves, the dependence upon topographic parameterisation using boundary conditions may be reduced, but only if effective meshing methods, which lead to stable numerical solutions with minimal diffusion, can be developed. The second challenge is that whilst our ability to measure model geometry is improving, there still remain problems with measuring topography at the bedform and grain scales. Conventionally, this problem has been addressed through upscaling of roughness heights within wall treatments. However, Lane *et al.* (2004, in review) show that there are both theoretical reasons and empirical evidence that suggest that this will not provide a sufficient representation of the effects of sub-grid-scale topography upon flow processes in situations where the variability of bed topography is greater than c.5–10% of the flow depth (Chapter 10). This is primarily because roughness upscaling leads to an increase in magnitude of one of the sinks within the momentum equations in boundary-adjacent cells. It ignores the effects of sub-grid-scale topography upon blockage at the bed, which leads to an effective increase in the elevation at which velocity becomes zero, and which needs to be represented in the mass equations as well as the momentum equations. This discussion implies that particular attention has to be given to boundary conditions at the bed and adjacent to channel banks, but that the limit to conventional boundary condition treatments needs to be recognised, such that parameterisation is appropriate to the particular model application under consideration.

Criterion 8: Reporting of code

8. In the presentation of an existing algorithm or code, all pertinent references or other publications must be cited in the paper, thus aiding the reader in evaluating the code and its method without the need to redefine details of the methods in the current paper. However, basic features of the code must be outlined according to item 1, above.

Comments
This is an important criterion as many of the current applications of CFD to open channel flows (Hodskinson and Ferguson, 1998; Lane *et al.*, 1999, 2000; Bradbrook *et al.*, 2000, 2001; Morvan *et al.*, 2002) are making use of commercial codes that have not always been developed for application to open channels. Sometimes, the source codes themselves are directly modified (e.g. Bradbrook *et al.*, 2000; Lane *et al.*, 2004),

so that model functionality is extended. However, this criterion implies that it is important to make sure that: (1) the basis of the algorithms used in the code is adequately reported; and (2) any additional modifications made by the user are reported and justified explicitly.

(Optional) Criterion 9: Use of benchmark solutions

9. Comparison to appropriate analytical or well-established numerical benchmark solutions may be used to demonstrate accuracy for another class of problems. However, in general this does not demonstrate accuracy for another class of problems, especially if any adjustable parameters are involved, as in turbulence modelling.

Comments

One way of assessing a CFD result is by comparing a numerical solution based on the chosen algorithms with an analytical one, in order to demonstrate the range of conditions for which the model holds. This has been illustrated, for example, by Horritt (2000). Horritt considers the case of a channel that is annular in planform but rectangular in cross section, noting that this type of channel is a useful basis for study, as it would be expected to contain strongly rotational flows, and hence spatial variation in velocity magnitude and direction and free surface displacement. These are key characteristics for the effective representation of natural open channel flows. Comparison of a range of model solutions (with different meshes and numerical solvers) with the analytical solution allowed basic guidelines for mesh generation and numerical solution for this class of problem. Whilst Horritt (2000) demonstrates that some of these conclusions may be extended for the design of effective meshes for modelling flows in naturally meandering channels, it is also recognised that analytical solutions only provide a test for those cases defined by the assumptions made to allow analytical solutions to be reached (Oreskes *et al.*, 1994; Horritt, 2000). In essence, comparison with analytical solutions, as with other forms of model evaluation, should be viewed as one part of a much broader approach.

(Optional) Criterion 10: Comparison with experimental results

10. Comparison with reliable experimental results is appropriate, provided experimental uncertainty is established. However, 'reasonable agreement' with experimental data alone will not be enough to justify a given single-grid calculation, especially if adjustable parameters are involved.

Comments

This is the classic form of model evaluation, commonly called validation. One of the major issues that arises under this item is the conclusion that comparison with experimental results is appropriate but not required, unlike some of the other criteria. This is a surprising observation and one that is central to the difference between the views of modellers and those of experimentalists. At the outset, it is clearly impossible to demonstrate the truth of any proposition within a system, except in a closed system where all possible manifestations of the system are known

(Oreskes *et al.*, 1994; Lane and Richards, 2001). This is true of any study of any system, whether that is a numerical model, a laboratory experiment or a field survey (Lane, 2001): all involve some *a priori* research design which will cause questions to emerge over the difference between reality and the study's representation of reality (cf. Hacking, 1983). This means that the results of any study are of questionable validity, and validation is philosophically impossible (Popper, 1968; Konikow and Bredehoeft, 1992). This may lead to the conclusion that falsification is the only possible way forward. However, Lane and Richards (2001) argue that falsification is practically impossible as the complexity of the real world means that it is always possible to demonstrate that a model is false in some way, especially as we have to set the criteria that allow us to judge falsification, and to develop the model that is subject to that falsification. An alternative view is to explore the coherence of a model. This recognises that the ASME's eight required activities identified earlier are necessary, but not sufficient, steps in the process of evaluating the practical utility of a CFD scheme (Hardy *et al.*, 2003), something that Lane and Richards (2001) argue should be called 'model assessment', and which leads to decision as to whether or not, on balance, the evidence provided by a model can be believed. In section 8.3, we develop a set of criteria to add to the ASME list that are suitable for applications of open channel flow to CFD problems.

8.3 From verification to validation and searching for model coherence

8.3.1 Conventional validation

Conventional validation is based upon the premise that when a model *fails* to predict independent data adequately, *something* must be wrong (Luis and McLaughlin, 1992). However, when model predictions are correct, the model is not necessarily valid, as it is possible for an invalid model to provide an adequate representation of some aspects of reality. Bearing this in mind, model validation should consider (Flavelle, 1992): (i) the extent to which variability in the observations is explained by the predictions, which can vary from 0% (imprecise or poor fit) to 100% (precise or good fit); (ii) the extent to which predictions agree with observations, which can vary from perfect equality (accurate or unbiased) to perfect inequality (inaccurate or biased); and (iii) the extent to which the predictions provide sufficiently reliable information for them to be accepted when there are no check data (e.g. at non-measured locations within the model or in situations where boundary conditions or topographic parameters are different from those used when the check data were collected). The difference between model precision (i.e. (i)) and accuracy (i.e. (ii)) is important and needs further explanation (Figure 8.1). Figure 8.1a shows a situation where a plot of model predictions versus measurements has a line of best fit that lies almost on top of the line of 1:1 agreement. Figure 8.1b has the same situation, but the scatter of points about the line is different. Clearly, the model in Figure 8.1a is performing better than in Figure 8.1b, and this is represented as a measure of precision, with a greater proportion of the variability in measurements explained by the model in Figure 8.1a than in Figure 8.1b. In Figure 8.1c, there is the same level of explained variability as in Figure 8.1a, but strong deviation of the best fit line from the lineof 1:1 agreement. In this case, the range

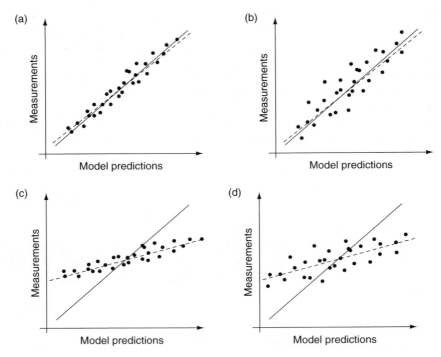

Figure 8.1 Four scenarios in the comparison of model predictions with an independent set of measurements. The solid line is the line of 1:1 agreement. The dashed line is the best fit line through the scatter of data points.

of model predictions is greater than that of measurements and there is a clear bias in model predictions: Figure 8.1a gives more accurate results than Figure 8.1c. In Figure 8.1d, there is both a lower level of precision than in Figure 8.1a as well as a higher level of bias. The description of both the precision and the accuracy of a model simulation is crucial to establishing how model predictions may be used. Good precision and good accuracy is the ideal case. If the precision is good, but there is bias, then it is likely that the patterns of model predictions are good, but their quantitative magnitudes are not. An empirical correction factor could be used in this instance. If the precision is poor, the model is less likely to be of use, even if the accuracy is good, as it implies substantial local uncertainty. At best, such a model may be used in general qualitative terms, coupled with the associated reporting of model uncertainty.

Lane and Richards (2001) review the issues surrounding the use of conventional validation in the hydrological sciences in general. In relation to CFD, this leads to a number of points. The first issue concerns how to describe the level of fit between model predictions and field observations and then how to determine what is an acceptable level of fit. The commonest form of model assessment is comparison of the measured and predicted vertical variation of velocity with elevation above the bed (e.g. Czernuszenko and Rylov, 2000; Meselhe and Sotiropolous, 2000; Sanjiv and Marelius, 2000; Sofialidis and Prinos, 2000; Lane et al., 2004). This is a logical comparison as one of the key uses of CFD is to provide estimates of shear at the bed in order to estimate sediment entrainment and if shear at the bed is not predicted correctly, then derived estimates of bed shear stress will not be correct. These qualitative assessments can give the appearance of a 'good' level of agreement. However, especially in the case of 3D applications, they are based upon a very small sample of model predictions, and quantitative assessment based upon principles of statistical inference is really required (Lane and Richards, 2001), if only so that the distinctions implicit in Figure 8.1 may be evaluated. This is especially the case as the model validation literature employs descriptors such as 'adequate' that are subjective and often poorly defined (Oreskes and Belitz, 2001). The only way in which levels of agreement can be formally defined will be through rigorous, modeller- and model-independent, quantitative assessment. This is why model precision and accuracy are important. The optimal way of doing such an assessment is some form of regression analysis, though for reasons explained later conventional ordinary least-squares (OLS) regression (e.g. Lane et al., 1999) may be inappropriate. The level of fit (or r^2) then gives model precision and the slope and intercept of the regression line give the accuracy. Unfortunately, this does not tell us what level of precision and what level of accuracy makes the model acceptable. With a large number of observations, a relatively low r^2 may be statistically significant, such that there is some signal in the validation data that can also be found in the model. However, a low r^2 may not be deemed sufficient even if it is statistically significant. Similarly, tests on the slope (and intercept) of the regression relationship are not to demonstrate that the model is accurate, but to check that the model is not inaccurate (Lane and Richards, 2001). With greater precision (i.e. more explained variance), it is easier to demonstrate that a model is giving us inaccurate predictions, and that there is bias in our model. As Flavelle (1992) notes, tests on the slope of a regression relationship only allow bias to be shown, and not lack of bias. Thus, whilst tests upon the slope of a relationship can be used to identify an inaccurate model, they cannot be used to demonstrate model accuracy. Hence, in terms of both precision and accuracy, what defines an acceptable level of agreement will remain a subjective decision (Flavelle, 1992; Jackson et al., 1992; Lane and Richards, 2001).

Second, it is important to be critical about what is being validated. A common observation in applications of CFD to natural river channels is that levels of explained variance of greater than 80% may be obtained for predictions of streamwise velocity. If the model is being run in a Cartesian framework, and streamwise flow direction changes with respect to the planform orientation of the grid (e.g. as in a meander) this may translate to high levels of explanation in both planform components of velocity. However, where the streamwise direction of the flow generally remains parallel to one mesh direction, vertical velocities typically have significantly lower levels of explanation, sometimes below 10% (Lane et al., in review). Thus, if the interest is in secondary circulation, a good level of agreement between predictions and observations of streamwise velocity should not be used to infer that secondary circulation is also well reproduced. The data chosen to validate a model should reflect what the model is to be used to predict. If a model has a number of sub-components (e.g. flow and sediment transport components), or a range of predictions (e.g. velocity components, turbulence parameters, bed shear stress), then where appropriate these sub-components should also be validated. For example, El-Hames and Richards (1998) illustrate a catchment-scale model in which sub-models are tested against a range of pre-existing model applications to simple test cases. A new hillslope runoff model was compared with an existing model (KININF, developed by Lima, 1992) on an application to a small experimental plot for which good rainfall-runoff data were available. It should be noted that this discussion implies that model performance does depend upon the data that are available for validation, and some data will provide a sterner test of a model than others. It implies a need to iterate carefully between model development and collection of laboratory or field validation data.

Similar issues occur in relation to spatial sampling. For instance, Lane et al. (2004) compared Acoustic Doppler Velocimeter (ADV) data, collected for a series of velocity profiles, with predictions from a model developed to provide an alternative method for representing complex topography in regular structured meshes. They found that errors in both vertical and streamwise velocities have an elevation dependence. If sampling is skewed towards bed measurements, then the model will be deemed to be performing more poorly than if they were skewed towards surface measurements. Again, this requires careful explanation of which areas in the model are of particular interest. It also requires recognition that error may be spatially variable and consideration of the spatial structure of the error field (e.g. Bradbrook et al., 1998).

Third, it is perhaps a peculiarly 'open channel flow' tradition that the ultimate judge of the success of a numerical model is seen to be independently acquired validation data, perhaps reflecting the significant progress in river research made using field observations and scaled laboratory models. The more general approach in the modelling community, as embodied in the 'acceptable but insufficient' label in ASME criterion 10, is not to require validation. One of the more powerful arguments for this is the difficulty of acquiring model validation data for the problem at hand, such that error may actually be due to: (1) problems with experimental design; (2) technological differences between the nature of predictions and measurements; and (3) contamination of measurements by processes that the model is not seeking to reproduce (Lane and Richards, 2001). Lane et al. (1999) demonstrate that difficulties with orienting a 2D electromagnetic current meter were at least partly responsible for the poor precision of model predictions reported by Lane et al. (1995). In relation to technological differences, in Lane et al. (2004), the vertical dependence of error was found to be related to differences between the extent of the ADV velocity measurement volume in the vertical (0.008 m) as opposed to model mesh resolution (0.002 m). Very strong shear at the bed meant that the model provided more reliable velocity predictions than the ADV measurements. Similar problems emerge when point measurements of floodplain velocity are made over a scale of a few centimetres when model elements cover many tens of metres. Contamination is a particular problem with field data when the geometrical variability that drives actual processes is not included in the model: for instance, in a model of urban flood inundation, a point measurement of velocity in a narrow corridor between two houses on sloping ground will bear no resemblance to model predictions of flow velocity where urban morphology is represented using upscaling of a roughness parameter. In theory, contamination ought to control model precision more than model accuracy: i.e. some aspects of the process are still reproduced in the model; there is just a high level of residual or unexplained variance. However, this assumes that the causes of low levels of precision are independent from controls upon model accuracy. This will not always be the case. For instance, roughness elements on the bed of a river may be a reason for poor levels of precision (Lane *et al.*, in review). However, research has shown that bed roughness is an important control upon turbulence anisotropy and hence secondary circulation (Nezu *et al.*, 1993). Thus, models with poor levels of precision may also be inaccurate, and contamination effects are not restricted to model precision.

Two comments follow from this. First, the existence of possible error in both model and measurements has implications for the type of statistics that should be used for model validation. In particular, the use of OLS regression to determine model accuracy (e.g. Lane *et al.*, 1999; Bradbrook *et al.*, 2000) needs to be reconsidered. OLS regression assumes error in only the dependent variable: the statistical model for observation i is $y_i = \alpha + \beta x_i + \varepsilon_i$. With perfectly precise measurements, it would be reasonable to use OLS regression of CFD predictions (as y) on field or laboratory measurements (as x). If any measurement error is suspected, OLS is inappropriate and will tend to underestimate the slope of the regression. OLS regression of field measurements on CFD predictions is always inappropriate since it assumes all error is in the measurements. If error is thought to be present in both the measurements and the simulation, and the relative amounts of error can be guessed, functional regression is appropriate. A special case of it, which assumes equal amounts of error, is reduced major axis regression. The weaker the correlation between measurements and predictions, the more the difference between the slope estimates using different forms of regression and the more important it is, therefore, to consider which choice is most appropriate.

Second, the term 'validation' needs to be applied carefully. If a model is deemed to have not predicted field or experimental data adequately, this does not lead to the conclusion that the model is wrong. Rather, there is a discrepancy between measurements and predictions in this particular situation that then has to be explained. This recognises that the measurements may be wrong, but also that the model may be right and it is the geometrical or boundary condition data used with the model that are wrong. Unfortunately, we now have a quandary as this argument also implies that a model *application* that is deemed adequate in one situation is not necessarily going to be adequate in other situations. Thus, whilst laboratory validation of a model has the appeal of reliability in terms of geometrical, boundary condition and validation data, this does not demonstrate that the predictions from a model used in a more complex field environment will also be valid. This is why field validation of CFD applications, and the subsequent development of CFD models in response to field application problems, is so crucial. Even then, there is little formal guidance as to when a successful validation of one application can be deemed transferable to a second application. At least one requirement of such a transfer must be a thorough comparison of the applications (boundary condition and geometrical characteristics) and the methods used to study those applications.

8.3.2 Model parameterisation and optimisation in CFD

Dangers of circularity can emerge where some demonstration of agreement between model predictions and reality is implied by looking at the results of a calibration test in which model predictions have been optimised against some sort of experimental data through adjusting parameters (cf. ASME criterion 10). This is what Oreskes *et al.* (1994) call 'forced empirical adequacy' and is common in situations where model sensitivity to a parameter change is high. It should be noted that in 2D and 3D CFD there are some dangers in transferring the conclusions regarding parameterisation from other modelling applications where parameter sensitivity can be exceptionally high (e.g. in certain hillslope hydrological models, Beven, 1989). In CFD applications, the role of parameters is partially subsumed by the influence of initial conditions and model geometric specification (e.g. Lane *et al.*, 2004). The extent to which this holds depends upon model dimensionality. As dimensionality increases, parameter number increases, but parameter dependence commonly decreases, such that opportunities for calibration are reduced. Thus, whilst 3D models may have many more parameters (e.g. associated with the turbulence model adopted or the nature of the boundary condition treatment) than 2D and 1D models, sensitivity to those parameters is much reduced. Lane *et al.* (1999) found much greater sensitivity to roughness treatment in a 2D depth-averaged solution than in a fully 3D solution, and this allowed more ready calibration of the 2D model to observed data. However, sensitivity to bed topography was significantly reduced in the 2D case (Lane and Richards, 1998), demonstrating an apparent shift from parameter sensitivity to geometric sensitivity as the dimensionality increases from 2D to 3D. It needs emphasising that there is a difference between these two model controls: many model parameters (e.g. roughness in 1D and 2D models) have an uncertain relationship to field-measurable parameters (e.g. grain size); geometric data, at least in theory, are directly measurable, even in complex gravel-bed rivers (e.g. Butler *et al.*, 2002) and in the absence of measurements may be inferred (e.g. Nicholas, 2001) from evidence that suggests a strong relationship between scale and surface variability (e.g. Nikora *et al.*, 1998; Butler *et al.*, 2001).

It is also the case that some of the adjustable parameters in 2D and 3D CFD applications have a much clearer physical meaning than their 1D counterparts. A good example of this is bed roughness. In a 1D model, the roughness parameter is an effective parameter, in that it represents the effects of turbulence, secondary circulation and boundary friction in a single adjustable parameter (e.g. Manning's n). Thus, the resemblance to data that can be measured in the field is poor, but it has proved to be an especially effective and highly sensitive (Romanowicz *et al.*, 1996) calibration parameter. In 3D CFD, roughness is a boundary condition, commonly included in the near-bed cells in order to drive a wall function that determines shear at the bed (or banks). As such, it represents the displacement of the height above the bed and into the flow at which velocity becomes zero. In theory, this is directly measurable from velocity profiles (i.e. hydraulic data), and so it seems to be just like Manning's n which can also be estimated from hydraulic variables (i.e. discharge, wetted perimeter and the slope of the energy line). Also like Manning's n, it could be used to calibrate the model. However, it differs in two important ways. First, it describes the effects of only one process: the height at which velocity becomes zero, which is controlled by the sub-grid-scale variability in surface elevations. Thus, it is easier to generate justifiable values of roughness height on the basis of the known characteristics of an application. Second, and related to this, sensitivity to roughness height in 3D models has been shown to be exceptionally low (Lane *et al.*, 2004;

Hardy *et al.*, in press), such that implausibly high values of roughness may be required to get sufficient shear at the bed. A good example of this is Booker *et al.* (2001) who found it necessary to calibrate bank roughness heights to implausibly high values in a 3D CFD application to represent the effects of vegetation. This is where more sophisticated treatments of vegetative roughness, as per Wilson *et al.* (Chapter 15), are required. In gravel-bed rivers, as discussed above, roughness height is a very poor calibration parameter as it represents only a sink in the momentum equations (Lane *et al.*, in review), rather than any effects on blockage through the mass conservation equations. It is easier to be critical of roughness height calibration because any roughness heights generated can be evaluated with respect to either known or estimated characteristics of sub-grid-scale topographic variability. This greater physical meaning means that blind model parameterisation is not well justified, especially where parameter estimates that have only a poor link to what is known about a particular application are determined. Even where the parameterisation is required because of inadequacies in available data, before parameterisation is adopted it is vital to be certain that the right components of the model are being parameterised and that parameterisation is not being adopted as a result of structural or meshing inadequacies in the overall model application.

8.3.3 Qualitative evaluation and model coherence

One of the difficulties identified earlier was the issue of what is a sufficient level of accuracy and this was noted to be a judgemental issue. One of the consequences of this is the need to view model evaluation as more than just giving primacy to observations over model predictions. Faith in model performance *is* partly informed by conventional validation, but also by a suite of other activities that support the conclusions reached. Lane and Richards (2001) note that this might include: (i) conceptual model assessment; (ii) sensitivity analysis; and (iii) visualisation.

Conceptual model assessment is based upon the premise that, for a model to have any chance of representing reality, it must include relevant processes within an adequate structure (Usunoff *et al.*, 1992). Thus, model evaluation should include a commentary on what processes are assumed to be operating in a particular model application and how these processes are represented within the model. The commentary might include: (i) assumptions regarding the spatial dimensionality of the problem; (ii) assumptions regarding the temporal scale of the problem; and (iii) assumptions regarding necessary processes for inclusion in the model. These three categories are commonly interlinked (Lane and Richards, 2001). For instance, if a depth-averaged solution is being used, attention should be given to the resultant dispersion terms associated with secondary circulation. Assumptions about (iii) in turn determine the spatial and temporal scales of model enquiry. Lane and Richards (2001) note that the commentary will have a strong grounding in existing knowledge and will demonstrate either that other modellers have used similar assumptions or that the assumptions being made in the model are an improvement upon those currently being made in the literature. The commentary should also provide an indication of where the model is likely to fail, and hence where it may be necessary to give particular attention to other aspects of evaluation. The importance of conceptual assessment must be emphasised: in situations where acquiring reliable validation data is a serious challenge, conceptual assessment provides a crucial way of demonstrating that process representation is sufficient.

Sensitivity analysis can perform at least five functions as part of a modelling exercise (Lane et al., 1994). It can: (i) determine that, in response to representative variation of model input parameter values and boundary conditions, conceptually realistic or justifiable model behaviour is experienced (e.g. Howes and Anderson, 1988; Lane et al., 1994); (ii) determine that the model is sufficiently sensitive to represent perceived behaviour in the real case (Howes and Anderson, 1988); (iii) identify those parameters to which the model is most sensitive (Young et al., 1971; McCuen, 1973; Beven, 1979) and which must therefore be given most attention in terms of data acquisition and quality; (iv) improve model representation through limiting the sensitivity of a particular model component (McCuen, 1973); and (v) assess the likely magnitude of error in a model prediction that arises from uncertainty about a particular parameter specification (e.g. McCuen, 1973; Lane et al., 1994). The level at which sensitivity analysis is applied may range from a narrow focus on one or more parameters to a much more fundamental comparison of entire schemes. For instance: (i) Horritt and Bates (2001) compared raster-based (simpler) and finite element (more complex) models of floodplain inundation as a means of assessing the degradation in model performance associated with adopting the simpler approach; (ii) Horritt and Bates (2002), with a similar rationale, compared 1D and 2D models of river flood inundation; and (iii) Lane et al. (1999) compared 2D and 3D predictions of flow velocity and bed shear stress for a shallow tributary confluence. These comparisons do not represent a form of model validation, as there is no use of independent data. Rather, they are a form of sensitivity analysis: in these examples, sensitivity to the simplicity of process representation contained within the models compared. When the results from two schemes agree (which, as with conventional validation criteria, requires a formal description of the level of agreement), then confidence can be placed in the performance of both models in relation to the particular application under study: there is an element of model coherence that suggests that the simpler scheme may be adopted. When they disagree, recourse to other datasets will be required as the more complex model may not necessarily result in better predictions, especially if its performance is limited by data availability.

In theory, the above discussion argues that sensitivity has much to offer the process of assessing that model behaviour is adequate. In practice, undertaking sensitivity analysis as part of a model validation strategy is complicated by a number of issues (Lane and Richards, 2001): (1) most models have many parameters, and hence many more possible parameter combinations, which is of particular concern if there is strong parameter interaction; (2) there can be strongly non-linear responses which make description of model behaviour and inter-comparison difficult to generalise into single statistical representations; (3) parameter values may need to be spatially distributed, which greatly increases the number of parameter combinations to be tried; and (4) model response may need to be evaluated in a spatially distributed way. Two other problems may be added to this list. First, especially in 3D CFD applications, and also in 2D applications using high-resolution meshes of large spatial extent, the number of factor combinations that can be simulated will be limited by the time it takes to run a simulation. Second, it is important to consider how to interpret the results from a sensitivity analysis and in particular what is making the model sensitive: the model, or the application? This matters because it has major implications for how the modeller must respond. If the model is the cause of the sensitivity, then effort must be placed upon making sure that the model is responding correctly and that the sensitivity is not being caused by problems such as numerical diffusion. If the application is the cause of the sensitivity, then the geometrical or boundary condition data that are causing the sensitivity will need to be evaluated and perhaps measured or managed in different ways.

The final form of model assessment is based around visualisation. This recognises that there will be a link, albeit a subjective one, between the phenomena that we see in the real world and those that we reproduce in the model. Especially with time-dependent 3D CFD simulations, measuring and handling data collected from many points within the model domain is a major headache. Animation may provide a feel for the extent to which the model reproduces temporally evolving, spatially distributed processes. This is a normal part of scientific validation, but one that presents difficulties given that conventional modes of presenting results for peer assessment in the scientific literature may not adequately capture those dynamics that are of interest (Lane and Richards, 2001) and 'Western Science' provides a world view in which we only believe what we can objectively measure.

In summary, conceptual assessment, sensitivity analysis and model visualisation provide alternative means of evaluating a model. This is best described as assessing a model's coherence. The rules associated with doing this may appear to introduce a strong subjectivity into the evaluation process. However, evaluation without recourse to quantification is an expected part of all scientific practice (e.g. in the refereeing of scientific papers or research grant applications), and there seem to be few arguments against broadening the definition of model evaluation to include such methods. The best description of this is establishing model coherence with respect to the range of resources that the scientist has: measurements, literature, observations, theories and visualisation.

8.4 Conclusion

What we seek to do in this conclusion is to identify a set of requirements of good practice for the evaluation of applications of CFD to open channel flow. The first premise is that these should build upon and not replace ASME criteria 1–9. This places CFD applications to open channel flow on an equal footing with other CFD application areas. The second premise follows Hardy *et al.* (2003): satisfying certain of the ASME criteria is unlikely to lead to a guarantee that model predictions are of sufficient use for practical purposes in real river channel systems; and that ASME criterion 10 needs to be replaced and to be developed in order to allow effective CFD application.

1. Applications of CFD to open channel flows should normally involve some form of model evaluation with respect to independently acquired data. In situations where evaluation is not undertaken, successful results from other model applications should not be used to infer validity unless a clear demonstration of the similarity of the applications is provided. This implies that laboratory evaluation of a numerical model is not a sufficient demonstration that the model can be used in a field context as the success of the model in a field context will be dependent

upon boundary condition and geometrical attributes that are not necessarily as readily and correctly determined.
2. Where experimental data are collected, modellers should report on both the precision and the accuracy based upon regression relationships which make defensible assumptions about possible error in measurements as well as predictions. These relationships should be subjected to formal statistical testing where appropriate. There should be a clear statement as to whether or not the identified levels of precision and accuracy are sufficient for the application under consideration. This should be recognised as a judgement, but stated explicitly rather than implicitly. It should specifically refer to which aspects of the model application are being reproduced sufficiently for practical interpretation purposes.
3. To support item 2, individual sub-components of the model (e.g. a sediment entrainment rule or a turbulence model) may also be evaluated. This is akin to the 'internal validation' recommended by modellers of catchment-scale hydrology and reach- to catchment-scale geomorphology (e.g. Hoey *et al.*, 2003). However, the need to do this is determined by the use to which the model will be put and by the number of sub-components of which the model is comprised.
4. Choices made in setting up the model may be justified by reference to any or all of empirical observations, established theoretical notions, and other model applications. This is post-positivist, because it replaces the assumed primacy of quantitative, empirical data with a series of more or less qualitative and theoretical evaluations (Brown, 1996).
5. Evaluation should include consideration of the spatial distribution of error as this is likely to identify either those parts of the model where predictions are least reliable or those areas where measurements are most likely to be in error. Such consideration is likely to be application-specific but might, for example, involve plotting error against distance from locations of strong shear (e.g. the bed, a shear layer), where errors are most likely to be spatially variable.
6. If model parameters are calibrated using data, evaluation of the results should, as far as possible, use independent data. This could involve evaluation using a different variable, or split samples of the same variable. However, it should be noted that the goodness-of-fit results from a calibration can still contain some elements of useful information: for example, a model optimised for a series of hydrographs may yield residual errors for individual hydrographs that in turn help in the identification of what is required to improve model performance. If only one data set is available, a split data test is recommended (e.g. random selection of a calibration dataset and a validation dataset from amongst the data available).
7. If model optimisation is undertaken, then a full justification of the parameter sets that result should be provided in relation to the model application. This reflects the stronger physical basis of many parameters in 2D and 3D models (e.g. roughness height), such that they have a better grounding in both fundamental theory and measurements than those in simpler models (e.g. Manning's n in a 1D model).
8. To understand the nature of the effects of model parameters, some form of sensitivity analysis may be presented. Given the computational demands of higher dimensionality CFD applications, and the clearer physical meaning of many model parameters, simple factor perturbation should be used to identify those parameters that warrant a more complex attention. If sensitivity appears to be low, then a more

detailed analysis might not be required. However, if it is required, then the causes of the sensitivity (i.e. application versus model) need to be carefully evaluated.
9. Visualisations of model predictions may be used as part of model evaluation, especially for 3D time-dependent solutions.

Acknowledgements

This paper is based upon research supported by the Natural Environment Research Council and grants GR3/9715, GR3/12588 and GR9/5059 in particular. Discussions with Keith Richards and members of the BGRG Working Group in CFD have been particularly important for this paper. An anonymous referee provided critical but constructive comments on an earlier draft of this paper.

References

AIAA, 1994. Editorial Policy statement on numerical accuracy and experimental uncertainty. *American Institute of Aeronautics and Astronautics*, **32**, 3.
ASCE, 1988. Turbulence modelling of surface water flow and transport: Part II. *Journal of Hydraulic Engineering*, **114**, 992–1014.
ASME, 1993. Statement upon the Control of Numerical Accuracy. *Journal of Fluids Engineering*, ASME, **115**, 339–340.
Bates, P.D., Anderson, M.G. and Hervouet, J.-M., 1995. An initial comparison of two 2-dimensional finite element codes for river flood simulation. *Proceedings of the Institution of Civil Engineers, Water, Maritime and Energy*, **112**, 238–248.
Beven, K.J., 1979. A sensitivity analysis of the Penman-Monteith actual evapotranspiration estimates. *Journal of Hydrology*, **44**, 169–190.
Beven, K.J., 1989. Changing ideas in hydrology – the case of physically-based models. *Journal of Hydrology*, **105**, 157–172.
Booker, D.J., Sear, D.A. and Payne, A.J., 2001. Modelling three-dimensional flow structures and patterns of boundary shear stress in a natural pool-riffle sequence. *Earth Surface Processes and Landforms*, **26**, 553–576.
Bradbrook, K.F., Biron, P., Lane, S.N., Richards, K.S. and Roy, A.G., 1998. Investigation of controls on secondary circulation and mixing processes in a simple confluence geometry using a three-dimensional numerical model. *Hydrological Processes*, **12**, 1371–1396.
Bradbrook, K.F., Lane, S.N. and Richards, K.S., 2000. Numerical simulation of time-averaged flow structure at river channel confluences. *Water Resources Research*, **36**, 2731–2746.
Bradbrook, K.F., Lane, S.N., Richards, K.S., Biron, P.M. and Roy, A.G., 2001. Flow structures and mixing at an asymmetrical open-channel confluence: A numerical study. *Journal of Hydraulic Engineering*, ASCE, **127**, 351–368.
Brown, H.I., 1996. The methodological roles of theory in science. In *The Scientific Nature of Geomorphology*, Rhoads, B.L. and Thorn, C.E. (eds), Chichester: John Wiley & Sons, 3–20.
Butler, J.B., Lane, S.N. and Chandler, J.H., 2001. Application of two-dimensional fractal analysis to the characterisation of gravel-bed river surface structure. *Mathematical Geology*, **33**, 301–330.
Butler, J.B., Lane, S.N., Chandler, J.H. and Porfiri, K., 2002. Through-water close-range digital photogrammetry in flume and field environments. *Photogrammetric Record*, **17**, 419–439.

Czernuszenko, W. and Rylov, A.A., 2000. A generalisation of Prandtl's model for 3D open channel flows. *Journal of Hydraulic Research*, **38**, 133–139.

El-Hames, A.S. and Richards, K.S., 1998. An integrated, physically-based model for arid region flash flood prediction capable of simulating dynamic transmission loss. *Hydrological Processes*, **12**, 1219–1232.

Ferguson, R.I., Parsons, D.R., Lane, S.N. and Hardy, R.J., 2003. Flow in meander bends with recirculation at the inner bank, *Water Resources Research*, **39**(11), 1322, DOI:10.029/2003WR001965.

Flavelle, P., 1992. A quantitative measure of model validation and its potential use for regulatory purposes. *Advances in Water Resources*, **15**, 5–13.

Gresho, P.M. and Taylor, C., 1994. Editorial. *International Journal of Numerical Methods in Fluids*, **19**, iii.

Hacking, I., 1983. *Representing and Intervening: Introductory Topics in the Philosophy of Natural Science*. New York: Cambridge University Press, 287pp.

Hardy, R.J., Bates, P.D. and Anderson, M.G., 1999. The importance of spatial resolution in hydraulic models for floodplain environments. *Journal of Hydrology*, **216**, 124–136.

Hardy, R.J., Lane, S.N., Ferguson, R.I. and Parsons, D.R., 2003. Assessing the credibility of a computational fluid dynamic code for open channel flows. *Hydrological Processes*, **17**, 1539–1560.

Hardy, R.J., Lane, S.N., Elliott, L. and Ingham, D.B., in press. Paper forthcoming in *Journal of Hydraulic Research*.

Hervouet, J.-M., 1996. TELEMAC-2D: Résolution des équations de Saint-Venant. In *Club des Utilisateurs de TELEMAC 2ème reunion [TELEMAC user's club 2nd meeting]* Janin, J.-M. (ed.), HE-42/95/077/A, EDF-DER.

Hodskinson, A. and Ferguson, R.I., 1998. Numerical modelling of separated flow in river bends: Model testing and experimental investigation of geometric controls on the extent of flow separation at the concave bank. *Hydrological Processes*, **12**, 1323–1338.

Hoey, T.B., Bishop, P. and Ferguson, R.I., 2003. Testing numerical models in geomorphology: How can we ensure critical use of model predictions? In *Prediction in Geomorphology*, Wilcock, P. and Iverson, R. (eds), American Geophysical Union research monograph series, 241–255.

Horritt, M.S., 2000. Development of physically based meshes for two-dimensional models of meandering channel flow. *International Journal of Numerical Methods in Engineering*, **47**, 2019–2037.

Horritt, M.S. and Bates, P.D., 2001. Predicting floodplain inundation: Raster-based modeling versus the finite-element approach. *Hydrological Processes*, **15**, 825–842.

Horritt, M.S. and Bates, P.D., 2002. Evaluation of a 1D and 2D numerical models for predicting river flood inundation. *Journal of Hydrology*, **268**, 87–99.

Howes, S. and Anderson, M.G., 1988. Computer simulation in geomorphology. In *Modelling Geomorphological Systems*, Anderson, M.G. (ed.), Chichester: John Wiley & Sons, 421–440.

Jackson, C.P., Lever, D.A. and Sumner, P.J., 1992. Validation of transport models for use in repository performance assessments: A view illustrated for INTRAVAL test case 1b. *Advances in Water Resources*, **15**, 33–45.

Karniadakis, G.E., 1995. Towards a numerical error bar in CFD. *Journal of Fluids Engineering*, **117**, 7–9.

Konikow, L.F. and Bredehoeft, J.D., 1992. Groundwater models cannot be validated. *Advances in Water Resources*, **15**, 75–83.

Lane, S.N., 2001. Constructive comments on D. Massey space-time, 'science' and the relationship between physical geography and human geography. *Transactions of the Institute of British Geographers*, **26**, 243–256.

Lane, S.N. and Richards, K.S., 1998. Two-dimensional modelling of flow processes in a multi-thread channel. *Hydrological Processes*, **12**, 1279–1298.

Lane, S.N. and Richards, K.S., 2001. The 'validation' of hydrodynamic models: Some critical perspectives. In *Model Validation for Hydrological and Hydraulic Research*, Bates, P.D. and Anderson, M.G. (eds), 413–438.

Lane, S.N., Richards, K.S. and Chandler, J.H., 1994. Distributed sensitivity analysis in modelling environmental systems. *Proceedings of the Royal Society, Series A*, **447**, 49–63.

Lane, S.N., Richards, K.S. and Chandler, J.H., 1995. Within reach spatial patterns of process and channel adjustment. In *Rivers*, Hickin, E.J. (ed.), Chichester: Wiley, 105–130.

Lane, S.N., Bradbrook, K.F., Richards, K.S., Biron, P.M. and Roy, A.G., 1999. The application of computational fluid dynamics to natural river channels: Three-dimensional versus two-dimensional approaches. *Geomorphology*, **29**, 1–20.

Lane, S.N., Bradbrook, K.F., Richards, K.S., Biron, P.M. and Roy, A.G., 2000. Secondary circulation in river channel confluences: Measurement myth or coherent flow structure? *Hydrological Processes*, **14**, 2047–2071.

Lane, S.N., Hardy, R.J., Elliott, L. and Ingham, D.B., 2004. Numerical modelling of flow processes over gravely-surfaces using structured grids and a numerical porosity treatment. *Water Resources Research*, W01302 JAN 8 2004.

Lane, S.N., Hardy, R.J., Keylock, C.J., Ferguson, R.I. and Parsons, D.R., in review. The implications of complex river bed structure for numerical modelling and the identification of channel-scale flow structures. Submitted to *Earth Surface Processes and Landforms*.

Leonard, B.P., 1995. Comments on the policy statement on numerical accuracy. *Journal of Fluids Engineering*, **117**, 5–6.

Leschziner, M. and Rodi, W., 1981. Calculation of strongly curved open channel flow. *Journal of the Hydraulics Division*, ASCE, **107**, 1111–1112.

Lima, J.T., 1992. Model KININF for overland flow on pervious surfaces. In *Overland Flow*, Parsons, A.J. and Abrahams, A.D. (eds), London: University College Press, 69–88.

Luis, S.J. and McLaughlin, D., 1992. A stochastic approach to model validation. *Advances in Water Resources*, **15**, 15–32.

McCuen, R.H., 1973. The role of sensitivity analysis in hydrologic modelling. *Journal of Hydrology*, **18**, 37–53.

Meselhe, E.A. and Sotiropolous, F., 2000. Three-dimensional numerical model for open-channels with free surface variations. *Journal of Hydraulic Research*, **38**, 115–121.

Morvan, H., Pender, G., Wright, N.G. and Ervine, D.A., 2002. Three-dimensional hydrodynamics of meandering compound channels. *Journal of Hydraulic Engineering*, **128**, 674–682.

Nezu, I., Tominaga, A. and Nakagawa, H., 1993. Field measurements of secondary currents in straight rivers. *Journal of Hydraulic Engineering*, **119**, 598–564.

Nicholas, A.P., 2001. Computational fluid dynamics modelling of boundary roughness in gravel-bed rivers: An investigation of the effects of random variability in bed elevation. *Earth Surface Processes and Landforms*, **26**, 345–362.

Nicholas, A.P. and Smith, G.H.S., 1999. Numerical simulation of three-dimensional flow hydraulics in a braided channel. *Hydrological Processes*, **13**, 913–929.

Nikora, V.L., Goring, D.G. and Biggs, B.J.F., 1998. On gravel-bed roughness characterisation. *Water Resources Research*, **34**, 517–527.

Oreskes, N. and Belitz, K., 2001. Philosophical issues in model assessment. In *Model Validation: Perspectives in Hydrological Science*, Anderson, M.G. and Bates, P.D. (eds), Chichester: John Wiley & Sons, 23–41.

Oreskes, N., Shrader-Frechette, K. and Belitz, K., 1994. Verification, validation and confirmation of numerical models in the earth sciences. *Science*, **263**, 641–646.

Popper, K.R., 1968. *The Logic of Scientific Discovery*. London: Hutchinson.

Roache, P.J., 1994. Perspective: A method for uniform reporting of grid refinement studies. *Journal of Fluids Engineering*, **116**, 405–413.

Roache, P.J., 1997. Quantification of uncertainty in computational fluid dynamics. *Annual Review of Fluid Mechanics*, **29**, 123–160.

Roache, P.J., 1998. Verification of codes and calculations. *AIAA*, **36**, 696–702.

Roache, P.J., Ghia, K.N. and White, F.M., 1986. Editorial policy statement on the control of numerical accuracy. *Journal of Fluids Engineering*, ASME, **108**, 2.

Romanowicz, R., Beven, K.J. and Tawn, J., 1996. Bayesian calibration of flood inundation models. In *Floodplain Processes*, M.G. Anderson, D.E. Walling and P.D. Bates (eds), Chichester: Wiley.

Sanjiv, S.K. and Marelius, F., 2000. Analysis of flow past submerged vanes. *Journal of Hydraulic Research*, **38**, 65–71.

Shaw, C.T., 1992. *Using Computational Fluid Dynamics*, New York: Prentice Hall.

Sofialidis, D. and Prinos, P., 2000. Turbulent flow in open channels with smooth and rough floodplains. *Journal of Hydraulic Research*, **37**, 615–640.

Usunoff, E., Carrera, J. and Mousavi, S.F., 1992. An approach to the design of experiments for discriminating among alternative conceptual models. *Advances in Water Resources*, **15**, 199–214.

Young, G.K., Tseng, M.T. and Taylor, R.S., 1971. Estuary water temperature sensitivity to meteorological conditions. *Water Resources Research*, **7**, 1173–1181.

9
Parameterisation, validation and uncertainty analysis of CFD models of fluvial and flood hydraulics in the natural environment

M.S. Horritt

9.1 Introduction

There is a large gulf between the use of CFD in industrial applications (one of its traditional uses) and in environmental scenarios. Whilst the fundamental processes being represented are identical (as described by the Navier–Stokes equations for example), and the problematic treatment of certain phenomena crops up in both cases (most notably turbulence), the application of CFD to the natural environment is hindered by the allied problems of data provision and uncertainty. For example, whilst engineering studies may investigate turbulence when modelling flow in a perfectly straight channel with well-defined dimensions and wall roughness, the problem is somewhat different for flow in a channel for which we have only a vague idea of the cross section at a small number of widely spaced locations, and a handful of samples of the bed material. The problem may become dominated by the uncertainty, to the extent that it is questionable whether CFD can be usefully applied at all in such circumstances. It is the interactions between CFD models, the data used to parameterise them (whether this be the specification of model-independent variables, geometry or boundary conditions) and the observations used to validate the approach that are discussed in this chapter.

Computational fluid dynamics is used in the environmental sciences, as in other disciplines, to fill in gaps in our knowledge: models are being used as a mapping between parameters (known) and predictions (unknown). For modelling to be useful, the parameters need to be in some way more easily observed than the predictions (otherwise it is easier to simply measure the quantities that the model is trying to predict). Thus the channel flow model using a small number of friction coefficients and channel cross sections to predict water surface elevations (Moussa and Bocquillon, 1996; Rutschmann and Hager, 1996) as a function of both space and time is an example of a worthwhile modelling exercise. At the other end of the complexity and scale spectra, the 3D turbulent flow model (Lane *et al.*, 1998, 1999; Bradbrook *et al.*, 2001) used to transform detailed measurements of bed form into flow and pressure fields is doing a useful job in predicting variables that would be impractical or impossible to measure in the field at a similar scale. These are both examples of the modelling methodology's ability to increase the dimensionality of the data provided, in the first case turning 1D measurements along the channel into depth and flow in 1D space and time, and in the second case using 2D topography to generate 3D flow fields. While such model applications are well posed in theory, we still require techniques to measure the parameters that drive model behaviour, and without these techniques the CFD models themselves are essentially useless in an environmental context. One commonly used indirect method of parameterisation deserves special note at this point: calibration. In this methodology, validation data along with an inverted model are used to derive model parameters in such a way that the model predictions best fit the observed validation data. This topic is discussed in its own right in section 9.2.2.

Once a CFD model has been parameterised and predictions obtained, some method of assessing its performance is required. This can be through either verification (using the terminology of Oberkampf and Trocano, 2002), where the predictions of a numerical model are compared with analytical solutions to the governing equations (e.g. Horritt, 2000a), or validation, where model predictions are compared with real-world observations. While the validation and verification processes are common to both the environmental and engineering sciences, the approaches are by necessity different. Series of detailed measurements using reproducible conditions are simply not practical in the field, so we are often left with sparse sets of inaccurate measurements as the best we can hope for. Again, an otherwise adequate model may be rendered useless if we have no way of determining whether it is performing satisfactorily. Thus the problem of validation is one of developing techniques for the measurement of variables in the field which are of sufficient accuracy and spatial resolution to actually prove useful in distinguishing the performance of different models, and thus establishing appropriate levels of process representation and numerical schemes for the problem being modelled.

Uncertainty exists in both parameterisation and validation data, and this uncertainty stems from two sources. First, measurements are subject to error, e.g. the finite vertical accuracy and precision of topographic measurements. While the non-linear behaviour of models may produce a complex response to such errors, they are at least (relatively) easy to quantify. Furthermore, research on error propagation in complex models has already been done in an engineering context (Chakraborty and Dey, 1996; Li and Ghanem, 1998; Lallemand *et al.*, 1999). A more insidious error arises from the spatial variability inherent to many spatial fields, from the necessarily limited spatial sampling undertaken by most measurement campaigns and from finite model resolution. A series of point measurements of ground elevation, for example, may hide a wealth of complex topographic variation in between the sampling points. In many cases this can be reduced by sampling at a higher resolution, but the fractal nature of many natural fields (Culling and Datko, 1987; Bates *et al.*, 1998a; Butler *et al.*, 2001) means that there will always be some structure at a scale finer than the sampling resolution. While it may be possible to define this variability in a statistical sense, non-linearity in model behaviours means their response to this variability is not easily characterised. Thus the problem of dealing with parameter uncertainty turns into one of dealing with sub-grid-scale variability. Uncertainty is also present in validation data, and this will tend to blur the validation problem: too much uncertainty and differently performing models will no longer be distinguishable. There may also be scaling problems similar to those encountered with parameterisation data. It may, for example, be extremely difficult to validate a model which predicts variables averaged over a computational element with point measurements, if there is sgnificant spatial variability in the quantity being measured. In this case we are trying to validate a model prediction with a measurement of the same quantity, but at such disparate spatial scales as to render the results incompatible.

The previous discussion has outlined the current issues in the parameterisation and validation of CFD models for environmental applications, and the related problem of uncertainty. In the following sections, techniques for addressing these issues are discussed, with emphasis being on the use of new technologies, followed by an examination of emergent techniques for uncertainty analysis. The discussion is based on the modelling of free surface water flows in fluvial and estuarine environments, which offers a good example of a field where considerable progress has been made in recent years in combining complex models with distributed parameterisation and validation data.

9.2 Model parameterisation

9.2.1 The measurement approach

The simplest way of obtaining parameterisation data is to go out and measure some. While ground-based surveys have long been used to parameterise models, recent advances in remote sensing techniques means we can now obtain data at a spatial extent and resolution unattainable by ground survey within reasonable cost constraints. There have, however, been some advances in ground-based survey in recent years, most notably with the advent of Global Positioning System (GPS) technology and its use in topographic mapping (Higgitt and Warburton, 1999; Connell *et al.*, 2001).

The two major parametric unknowns in modelling fluvial and estuarine flows are bathymetry and bed roughness. Both are in the form of spatially (and sometimes temporally) variable 2D fields, and are essential if recent advances in 2D and 3D numerical models are to be used to model natural processes. Previous approaches to quantifying these fields have generally been based on ground surveys, which due to the large areas required and the high resolution implied by the spatial heterogeneity are often prohibitively time-consuming and expensive. The 2D surficial nature of these fields (unlike, for example, hydraulic conductivity for groundwater flows) means that remote sensing techniques may be ideal tools for mapping topography and friction relatively quickly, cheaply and at a resolution commensurate with modern numerical models.

Aerial stereophotogrammetry has long been used to develop high resolution ($\sim$0.3 m horizontally), high precision ($\sim$10 cm vertically) Digital Elevation Models (DEMs) (Lane et al., 1994; Lane, 2000) for use in numerical modelling (Lane et al., 1999; Horritt, 2000b). Interferometric Synthetic Aperture Radar (INSAR) can be used to generate DEMs (Rocca et al., 2000), which may be used to generate bathymetry for hydraulic modelling, and although there are no examples of integrating this technique with CFD in the literature, it has been used for flood risk assessment (Sanders and Tabuchi, 2000). The potential to generate high-precision DEMs from airborne surveys of unvegetated environments ($\pm$5 cm for the intertidal zone of the Wadden See, Netherlands, reported in Wimmer et al., 2000) is clear, although satellite sensors may be too inaccurate ($\pm$4 m vertically for the ERS tandem mission in Rufino et al., 1998), and again vegetation is problematic (Slatton et al., 2001). Furthermore, as with other automated systems, it is not possible to exercise the level of control over DEM precision and resolution as in the photogrammetric case.

Airborne LiDAR (light detection and ranging) systems can provide DEMs with resolution and precision comparable to that from aerial stereophotogrammetry, but at a lower cost, since the post-acquisition processing can be more easily automated. This has led to the technique being adopted as the method of choice for generating high-resolution DEMs by, for example, the UK Environment Agency (EA), which has assessed the cost at £150 km^{-2}, compared with £1000 km^{-2} for stereophotogrammetry (EA, 1998). The system operates by measuring the time of flight of a laser pulse travelling between the platform, ground surface and back. Either the laser direction is fixed in relation to the aircraft, producing a profiling system as the laser footprint moves with the platform velocity (Ritchie, 1995), or it scans perpendicularly to the flight line (Flood and Gutelius, 1997) to give a 2D height map. Since the ground target may be diffuse due to the presence of microtopographic variation within the laser footprint (of the order of 10 cm for typical flying heights) and vegetation, the return pulse will be "smeared" into an elongated waveform. In most systems, the leading or trailing edge (or both) can be detected and hence the distance to the upper or lower height limits of the diffuse target measured. The distance between sensor and ground is combined with the platform position as measured by GPS (to either a local datum or a surveyed benchmark) and attitude from an Inertial Navigation System (INS) to give a target location in some (x, y, z) coordinate system. Height errors arising from instrument operation and GPS/INS positioning are typically 15 cm (Huising and Gomes-Pereira, 1998). This is the magnitude of error we would expect for flat, non-vegetated targets such as the intertidal zone (Lohani and Mason, 2001) and sparsely vegetated arid environments (Ritchie et al., 1994). We would expect greater errors for vegetated landcover types and steeper slopes, but the increase in error can be offset by using the last return of the laser pulse and image processing techniques, as in Cobby *et al.* (2001), where the topographic error under vegetation is limited to 17 cm when compared to ground-surveyed control points, except for the particularly difficult case of thickly wooded steep slopes where topography is effectively obscured by the dense canopy and is hard to interpolate. Topography derived from LiDAR surveys has been used to evaluate flood risk (Gomes-Pereira and Wicherson, 1999) and as input bathymetry for numerical models of flood flow (Horritt and Bates, 2001a). Figure 9.1 shows LiDAR data from a typical floodplain, which has been segmented into a topographic and vegetation signal (Cobby *et al.*, 2001).

LiDAR data can also be used to glean knowledge about vegetation cover, and this in turn may lead to new techniques for determining flow resistance, especially when this is dominated by vegetation, as it is on floodplains. The use of LiDAR to map vegetation is well established (Weltz *et al.*, 1994; Cobby *et al.*, 2001), and flume and theoretical studies of flow through vegetation have been carried out (Fathi-Maghadam and Kouwen, 1997; Nepf, 1999; Wu *et al.*, 1999; Kouwen and Fathi-Maghadam, 2000). It

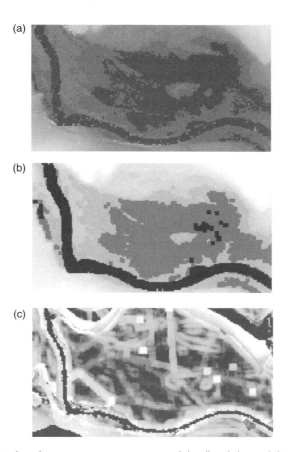

Figure 9.1 LiDAR data for a 450 m × 250 m area of the floodplain of the River Severn, UK: (a) raw LiDAR data sampled onto a 2.5 m grid; (b) extracted topography; and (c) extracted vegetation heights, with hedgerow structure and riparian vegetation clearly visible.

is still unclear whether vegetation height maps can be turned into useful distributed flow resistance values without more detailed biophysical parameters (Mason *et al.*, 2003). It may also be possible to derive more vegetation information from other sensors, such as the CASI multispectral scanner (Chen *et al.*, 1999), but this has yet to be attempted in the context of flow resistance modelling, and it may be that the best approach is to simply use LiDAR to classify vegetation (for example as grass, crops, hedges, etc.) and use a look-up table to derive maps of Manning's n. While the derivation of appropriate look-up values will still be problematic, with established values (Chow, 1959) strictly applicable only to 1D models, this approach still offers the merit of using remotely sensed data to determine spatial correlations in roughness classes, and hence reducing the number of free parameters in a model. The aerodynamic roughness for atmospheric flows has also been estimated using LiDAR data (Menenti and Ritchie, 1994), albeit with no assessment of the accuracy of this assessment.

Although conceptually straightforward, the measurement approach to model parameterisation may in practice yield inappropriate results. The first problem is one of reconciling the different scales of model and measurement resolution. With a sparse measurement data set (i.e. one of lower resolution than the model being parameterised), this is a question of interpolation. While the validity of modelling at a resolution higher than that of the available parameterisation data may be questionable, it is a valid approach where the spatial scale of processes being modelled is smaller than those of the parameter variations. The advent of high-resolution remotely sensed data sets generates the inverse problem, where a wealth of topographic data must be subsumed into a lower-resolution model, for example where modelling at the resolution of the parameterisation data is unfeasible due to computational constraints. Thus sensible elemental average values, which pay heed to hydraulically significant features, must be computed (Bates *et al.*, 2003). Data at a scale smaller than the model element can also be used to parameterise sub-grid process representation, for example in wetting and drying algorithms for moving boundary problems (Bates and Hervouet, 1999) and the treatment of small-scale bed features via a porosity approach in 3D models of gravel-bed rivers (Lane *et al.*, 2002). Thus small-scale features are not treated explicitly but are instead represented in a statistical manner.

A perhaps more serious difficulty is that the role of model parameters changes with scale, and hence their effective values are functions of model resolution and dimensionality. Manning's n values in 1D models are used to parameterise a different set of processes than in 2D models of flood flow. For example, Manning's n values in 1D models are used to model the effects of channel meandering, with channel sinuosities of >1.5 increasing Manning's n values by 30% (Chow, 1959; USGS, 1990). We would hope, however, that a 2D model would be able to represent some of the increase in energy loss processes explicitly, as the meandering channel planform is included in the model. Hence a 2D model of the same reach would have a different value for Manning's n in the channel than a 1D model, as the friction term would be representing different processes (Chapter 10). Similarly, a 2D model of meandering channel flow may not be able to represent secondary flows explicitly (Hsieh and Yang, 2003), and roughness parameters may have to be adjusted to compensate for the effects of these processes, which are explicitly handled in 3D models (Wilson *et al.*, 2003a,b). Model resolution will also affect the role of parameters; for example, in a 2D model of floodplain hydraulics, Manning's *n* values may be used to mimic the effects of sub-grid-scale topographic variation. A higher-resolution model will encompass smaller topographic variations within its smaller elements, and thus Manning's *n* values will be different.

9.2.2 The calibration approach

Of the two main parameters required to model fluvial and other environmental free surface flows, bathymetry and flow resistance, there has generally been more uncertainty associated with flow resistance. One only has to look at the ranges of values given in manuals of hydraulic design (e.g. Chow, 1959) to see that the specification of hydraulic roughnesses is a very inexact science, even if we do have a reasonable idea of the landcover on the floodplain and the channel bed material. The calibration approach avoids these difficulties by assuming very little *a priori* about friction coefficients, but choosing the values that produce a best fit between model predictions and observations. Calibration is thus an example of model inversion. Furthermore, it bypasses the difficulties of reconciling parameters effective at different model scales or dimensionalities with measured values.

The calibration methodology in hydrology originated in catchment modelling (Beven and Binley, 1992; Brath and Rosso, 1993; Titmarsh *et al.*, 1995; Lamb, 1999), where a small number of spatially and/or temporally "lumped" parameters are adjusted to maximise the fit between predicted and observed flood hydrographs, for example. The calibration problem can thus be thought of as a search through a parameter space of low dimensionality for an optimal set of model predictions. There may not be a simple relationship between the parameters found by a calibration process and the physical quantities they represent (e.g. soil hydraulic conductivity), as they can often be used to compensate for inadequate process representation, spatial heterogeneity or poor model numerics. The technique has been applied to numerical models of flood flow (Bates *et al.*, 1998b; Horritt, 2000b) by comparing model output with ground-based measurements of channel flow or maps of inundation extent from remote sensing (these sources of data are discussed in the section on validation data). Again model parameters can be used to compensate for inadequate process representation. This is illustrated in Figure 9.2 for three flood inundation models calibrated against inundation extent data derived from remote sensing (taken from Horritt and Bates, 2001b). The calibration responses are different for the three models, and this difference stems from the different process representations and numerical schemes used. This response of the parameters in the calibration process is double-edged: we may be able to produce better predictions if poorly represented processes are compensated for by different parameter values, but it makes model validation very difficult since model predictions can be improved just by better calibration rather than improved process representation.

Another drawback of the calibration approach is that it may become computationally intensive, especially if we choose to use spatially distributed parameters, effectively increasing the dimensionality of the parameter space. This can be offset by assuming that adjusting parameters in certain areas only affects model predictions in that area (Horritt, 2000b), or by limiting the spatial pattern of the parameters to

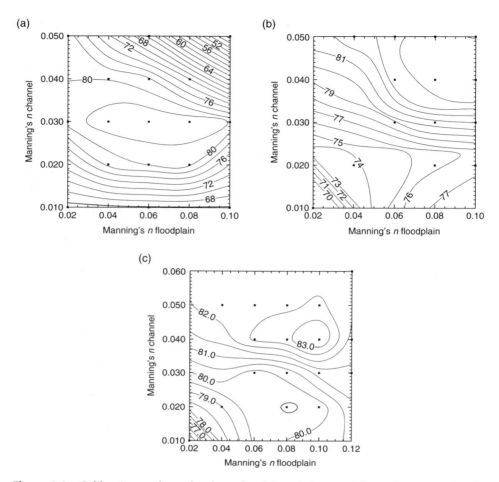

Figure 9.2 Calibration surfaces for three flood inundation models, with contour heights representing a measure of fit between predicted and observed inundation extent: (a) raster-based model using a kinematic wave approximation in the channel; (b) raster-based model using a diffusive wave approximation in the channel; and (c) 2D finite element model. The model performance is assessed against satellite imagery of flood extent from the ERS-1 SAR sensor, using the percentage of the domain predicted correctly as wet or dry (from Horritt and Bates, 2001b). (Reproduced by permission of John Wiley & Sons Ltd.)

a more manageable set of independent values (Refsgaard, 1997). In the context of flood modelling, this is most often implemented by assuming just two friction coefficients, one for the channel and one for the floodplain. A number of algorithms for searching a model's parameter space for the optimum calibration have also been developed in order to improve the efficiency of the calibration process. These range from the relatively simple simplex method (Nelder and Mead, 1965) to more complex approaches such as the SCE-UA (Shuffled Complex Evolution method developed at the University of Arizona) (Duan et al., 1992), genetic algorithms (Nditiru and

Daniell, 2001; Whigham and Crapper, 2001; Cheng *et al.*, 2002), and simulated annealing (Cooper *et al.*, 1997).

Crucial to the calibration process is the definition of a measure of fit between model predictions and observed data. This is essentially a subjective choice, as it depends on the modeller's view of what constitutes a good model. The choice also depends on the nature of the model predictions and observed data being compared (continuous or discrete, spatially or temporally distributed). Time series of discharge measurements may be compared with model predictions using the well known Nash-Sutcliffe efficiency:

$$R^2 = 1 - \frac{\sum (Q_{obs} - Q_{sim})^2}{\sum (Q_{obs} - \overline{Q})^2} \tag{9.1}$$

This is a measure of the mean square difference between simulated and observed discharges (Q_{sim} and Q_{obs}), scaled by the variance of the observed flow values. Similar measures can be defined using different methods of weighting different magnitudes of discharges, such as the Heteroscedastic Maximum Likelihood Estimator (HMLE) (Sorooshian, 1981) which allows the weighting of measurements to compensate for varying error magnitude. The difference in using these measures for parameter estimation in a calibration process will manifest itself in different optimal parameter sets. For example, if records of high flow are expected to be subject to large errors, these will be assigned a small weight accordingly, and hence calibrated model predictions will have larger discrepancies at high flows when compared with the observed data.

Binary pattern data information has also been used to calibrate distributed models, particularly remote sensing data (Grayson *et al.*, 2002). Maps of saturated areas have been used to parameterise the catchment model TOPMODEL (Franks *et al.*, 1998) in conjunction with discharge measurements. Flood extent data derived from satellite imagery have also been used to parameterise hydraulic models of inundation extent (Horritt, 2000b; Horritt and Bates, 2002). Again the parameterisation resulting from the calibration process depends on the choice of measure of fit between modelled and observed flood extent, but no work has been done on the effects of using different measures on the calibration process. The meteorological literature on comparing binary model predictions and observations (e.g. rain/no rain state) would, however, suggest that the choice of fit statistic has a significant effect on the assessment of model performance (Jolliffe and Stephenson, 2003). Difficulties may also arise as a result of the different scales of observed data and model predictions. When comparing model predictions of inundation at a coarse scale with higher-resolution observations, a significant improvement in model performance may be obtained by downscaling model results to the observation data scale. For example, model results for a 250 m inundation model have been improved by extrapolating the predicted water surface elevations and projecting them onto a 10 m DEM to give an inundation pattern at a scale commensurate with the satellite observations (Horritt and Bates, 2001a). This improves model results at the coarse scale to the equivalent of a 10 m resolution model.

9.3 Model validation

9.3.1 Field measurements

Field- and ground-based measurement is the traditional technique for validating CFD models of the natural environment. The use of stage and discharge (via rated section) data for model validation remains popular for many reasons. Data tend to be relatively inexpensive to collect, or are often available at zero cost to the scientist from water authorities. Predicted time series of stage and discharge are traditionally sought by flood-warning engineers, especially as it is relatively easy to transform a predicted stage into an approximate flood extent using a DEM. There is also a strong synergy between hydrological catchment models, for example predicting discharges through time at a few points in the catchment, and the data acquired. This is something of a chicken-and-egg about this situation, with models being developed to predict hydrographs because they are the only data available, and further data sources not being sought if hydrological models are used only to predict hydrographs. These point data have also been used to validate (and to calibrate) distributed hydraulic models (Bates *et al.*, 1998b).

Measurements of flow velocity have also been used to validate 3D models of channel flow. Nicholas and Sambrook-Smith (1999) compared model predictions of down- and cross-stream velocity with those measured by an Electromagnetic Current Meter (ECM), and found agreement with a correlation coefficient of $R^2 \sim 0.8$. Lane *et al.* (1999) found lower levels of correlation ($R^2 = 0.5$) for these two components when comparing model predictions with Acoustic Doppler Velocimetry (ADV) measurements, and a much poorer correlation for vertical velocity components ($R^2 = 0.25$). This is explained as being a result of errors in the bathymetry used to drive the model, which would be expected to affect vertical velocities more than the other components. This highlights a problem with all model validation studies: is the difference between predictions and observations a result of poor model process representation and numerics, or poor model parameterisation? A careful examination of error propagation through the model is required before any firm conclusions about process representation can be drawn. There is also the related problem of spatial scale. For fields that vary rapidly in space, errors can easily arise if two measurements are not located closely enough. Similarly, model element size will become an important factor, as predictions are essentially averages over areas which may be tens of metres across, whereas ADV measurements are at a scale of ~ 1 cm. This would indicate that in some cases we may be better off using the predicted free surface height to validate the models, as this tends to vary more smoothly than the velocity, and is not so strongly correlated with small-scale topographic features. The disadvantage is that we may only be testing large-scale flow behaviour, rather than the detailed flow structures that 3D models are meant to predict.

9.3.2 Remote sensing

Research has shown the potential of remotely sensed data to validate (and calibrate) hydraulic models (Bates *et al.*, 1997; Smith, 1997). Remote sensing gives a synoptic view of natural processes, often at a scale more commensurate with computational elements of larger-scale models (10–100 m) than field measurements (10–100 cm). Data also tend to be acquired at a number (often only one) of discrete points in time, thus directing modelling effort into capturing 2D distributed processes rather than dynamic behaviour.

The obvious candidates for remote sensing driven validation are 2D numerical models of flood flow which predict inundation extent. If the flooded area can be extracted from the imagery, it can be compared with predictions of inundation extent in either a distributed (Horritt, 2000b) or a lumped sense (Aronica et al., 2002). This approach really works only for out-of-bank flows where small changes in water surface height generate large changes in flood extent (due to low topographic gradients on the floodplain), although limited validation for in-bank flows covering or uncovering sandbanks has been done (Bates et al., 1997). Synthetic Aperture Radar (SAR) sensors are ideal for mapping the flooded area because as active microwave sensors they can detect open water beneath cloud and at night, and are therefore less susceptible to the vagaries of weather and illumination conditions than optical sensors, for example. SAR-derived flood extent maps have been used to good effect to validate and calibrate 2D models (Bates and De Roo, 2000; Horritt, 2000b; Horritt and Bates, 2001a,b), especially when parameterised with high-resolution elevation data. The efficacy of the technique is strongly dependent on the sensor parameters such as radar wavelength, polarisation and incidence angle (Figure 9.3).

Water surface elevation can also be used to validate numerical models, and these can be measured or inferred from a number of types of remotely sensed data. Inundation extent can be obtained by superimposing the waterline (i.e. the contour of zero depth) onto a DEM. Brakenridge et al. (1998) used this method and obtained vertical errors in the region of 1–3 m, most of this due to inaccuracies in the DEM

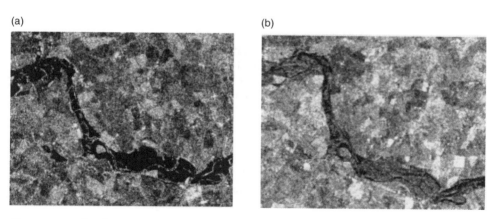

Figure 9.3 Flood mapping from different satellite SAR sensors for a 7 km × 7 km area of the River Severn floodplain. The flooded area is far more distinct in RADARSAT imagery ((a) – C-band, 5.6 cm wavelength, HH polarisation, 40° incidence angle) than in the ERS-2 imagery ((b) – C-band, 5.6 cm wavelength, VV polarisation, 23° incidence angle), where wind roughening of the water surface and low returns from dry areas are liable to induce misclassification. © Canadian Space Agency and European Space Agency.

derived from contour maps. Water levels can also be measured directly by satellite radar altimetry (Koblinksy *et al.*, 1993), with a vertical accuracy of 0.7 m. The main problem of the approach is the large size of the sensor's footprint (~5 km), which limits the technique to very wide rivers such as the Amazon. No use of the optical analogue to radar, LiDAR, to measure water surface elevations has so far been reported in the literature. Alsdorf *et al.* (2001) report a novel use of satellite SAR interferometry to measure water level changes over the Amazon floodplain. The technique gets around the usual phase incoherence over open water by using the double bounce reflections from emergent vegetation, and has detected changes in water surface elevations down to centimetre levels.

SAR interferometry can also be used to measure water surface currents (Graber *et al.*, 1996; Romeiser and Thompson, 2000; Frasier and Camps, 2001). The Doppler shift of signals scattered from surface capillary waves is measured, and hence the velocity of the scatterers inferred. This velocity is made up of the surface current and the wave velocities, and so the measurements need to be corrected for this wave velocity. The velocities produced will also need to be corrected if they are to be compared to, say, depth-averaged velocity predictions made by a model, using some knowledge of the surface velocity, flow depth and the vertical velocity profile (as in Moller *et al.*, 1998). The errors in the velocity fields may be too great to make a meaningful quantitative comparison with model results, but a qualitative comparison of flow structures may be possible (as in Thompson *et al.*, 1998).

9.4 Modelling and uncertainty

A common thread runs through the above discussions of parameterisation, calibration and validation data sets: all sources of data are subject to uncertainty. In applying CFD to the environment, uncertainty may well dominate the process and obscure any useful information about model process representation. To ensure we are drawing conclusions about real environmental processes, and not artefacts resulting from uncertainty, a rigorous analysis of error propagation through the model is required, along with a statistical comparison between uncertain model predictions and uncertain observations. Techniques for the study of error propagation are discussed below, where they are classified into Monte Carlo methods and stochastic methods.

9.4.1 Monte Carlo methods

Most CFD models are deterministic; they are designed to take a single, zero-uncertainty parameter set and transform it into a single, zero-uncertainty prediction set. The simplest way of reconciling this deterministic behaviour with the uncertain environment is to regard a deterministic simulation as a single realisation of a random process, and then build up the statistics of that process by performing multiple simulations. This is the Monte Carlo method.

The method has considerable advantages in the context of hydrological and hydraulic modelling, as it can implicitly capture the nonlinear behaviour of a deterministic model, along with the associated complex parameter interactions. An ensemble of deterministic simulations, mapping the parameter space to the model prediction space, can be easily inverted to give a mapping from model predictions back to the parameter space. Once a deterministic model has been developed (the usual first pass at modelling uncertainty), the technique may also be relatively easy to implement, as only a generator for the random parameter fields is required and if there is enough computing power to perform the simulations in a reasonable time. The implementation for 3D models of flow over uncertain topography, however, is more difficult if the computational mesh conforms to the bed form. The technique has proved popular in computing groundwater flows and solute transport (Jones, 1989, 1990; Marseguerra and Zio, 1998; Rong et al., 1998; Soutter and Musy, 1998; Bekesi and McConchie, 1999).

The Monte Carlo method has proved especially useful in hydrological (Beven and Binley, 1992) and hydraulic (Aronica et al., 1998, 2002; Romanowicz and Beven, 2003) modelling where parameters are poorly defined, due to the ease with which ensembles of simulations are inverted. In the simplest calibration procedure, some measure of fit between predictions and observations is calculated, and then the optimum parameter set can easily be found from the corresponding optimal measure of fit. This, of course, assumes that an optimal parameter set exists, which is not necessarily the case. There will almost certainly be uncertainty in the derived parameters (it will be impossible to define the parameters exactly with a calibration procedure based on uncertain data), and it may be that many disparate regions of the parameter space will produce equally good results when compared with the observations. The GLUE (Generalized Likelihood Uncertainty Estimation) procedure (Beven and Binley, 1992) instead aims to produce a distribution of parameters and thus addresses the problem of equifinality, where different parameters are equally good at reproducing the observed behaviour. The GLUE procedure takes an ensemble of simulations and weights each simulation according to a likelihood function, which describes how well model predictions fit the observations. The model predictions can be weighted with the same values to generate probabilistic predictions based on the calibrated parameters. A benefit of the technique is that the weighting by likelihood function can be cast in probabilistic terms as a form of Bayes' theorem (Beven and Freer, 2001a), and further observations can be used to update the parameter probability distributions. The technique has been applied to catchment hydrological models (Freer et al., 1996; Lamb et al., 1998; Beven and Freer, 2001b; Campling et al., 2002; Christiaens and Feyen, 2002) and 1D and 2D numerical models of flood hydraulics (Aronica et al., 1998, 2002; Romanowicz and Beven, 1998; Hankin et al., 2001), and can incorporate remotely sensed distributed data (Franks et al., 1998; Aronica et al., 2002). The procedure is illustrated for the case of inundation modelling in Figure 9.4, which shows a likelihood plot over the parameter space and the uncertain inundated area predicted by the GLUE procedure.

The main drawback of the GLUE procedure is its reliance on Monte Carlo simulations, and the associated computational burden. While this may not be relevant for a relatively simple model with few parameters, a complex model with distributed parameterisation will pose a considerable computational problem, since our Monte Carlo simulations will have to span a parameter space of high dimensionality, and each model realisation will take a long time to compute. We may be able to reduce this dimensionality either by identifying a smaller number of

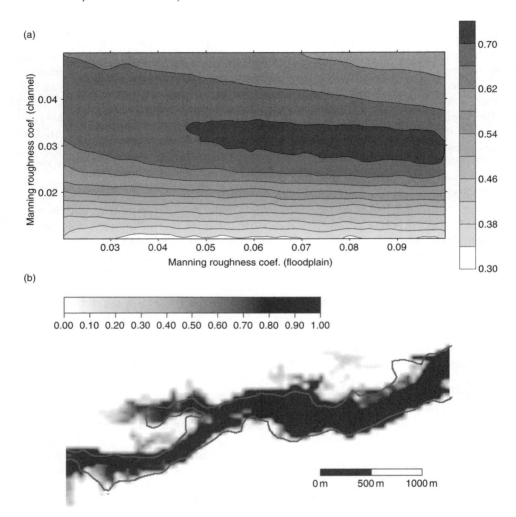

Figure 9.4 *Likelihood measure based on inundated area as a function of channel and floodplain friction (a). The likelihoods are treated as parameter probabilities and projected forwards to give an uncertain prediction of inundated area (b). The model is set up for an approximately 4 km reach of the River Thames, and inundated area observations are provided by ERS-1 satellite radar imagery (Aronica et al., 2002). (Reproduced by permission of John Wiley & Sons Ltd.)*

important parameters or parameter combinations, or by imposing arbitrary restrictions (e.g. limiting channel/floodplain friction to a single value, as in Aronica *et al.*, 2002). Monte Carlo simulations are also easy to parallelise (Beven and Binley, 1992), with each processor handling a single model realisation at a time, and the advent of relatively cheap parallel clusters based on PC technology has allowed Monte Carlo simulations to be performed more quickly in recent years.

9.4.2 Stochastic methods

The computational expense of Monte Carlo simulations for complex models with distributed parameterisations prompts a different approach to the problem. Rather than using a deterministic model, we may instead be able to formulate the model as a set of Stochastic Differential Equations (SDEs), differential equations with stochastic parameters. These may then be discretised using established methods (e.g. finite elements) and solved for the statistical properties (mean, covariance structure, higher-order moments) of the solutions directly, without the need for multiple simulations. An alternative approach is to discretise first and deal with uncertainty later, transforming the problem into a (possibly non-linear) system of equations with random coefficients. The limitation of the technique is that we can often only solve an approximate form of the SDEs, for example by expanding nonlinear terms using a perturbation approach, and we may have to assume something about the solution statistics (e.g. that they are Gaussian). First-order perturbation techniques have been used to model groundwater flows through random conductivity fields (Osnes and Langtangen, 1998) and with random recharge (Li and Graham, 1998), and a second-order expansion is used in Dainton *et al.* (1997). The results compare well with Monte Carlo simulations despite the approximations mentioned above.

Polynomial chaos (based on the work of Wiener, 1938) offers another approach to the uncertainty problem. A random process is viewed as being a function of a single parameter (the random parameter), analogous to a field varying in time, and the statistical properties of the process can be found by integration over this time-like parameter, according to ergodic theory. The solution to a SDE is written as an expansion of orthogonal functions of the random parameter, thus transforming the problem of measuring the statistics of an infinite number of random realisations into the problem of finding the properties of a countable set of functions (Ghanem and Red-Horse, 1999). Furthermore, this expansion can be simplified by truncation, when there is some physical justification such as the limited impact of spatial variation over short-length scales. The technique has been applied to modelling transport in heterogeneous media (Ghanem, 1998).

9.5 Conclusions

In this chapter we have discussed the challenges associated with the application of CFD to the natural environment, which have come to represent major research issues in themselves. The spatial heterogeneity of the natural environment, difference in scales between natural processes, model representation and observed validation and parameterisation data, and uncertainties in these data represent significant problems in the application of CFD models. These can be overcome by field measurement and remote sensing, but only partially, and there may always be a significant level of uncertainty to be dealt with. Thus the problem of uncertainty propagation through CFD models is one that must also be addressed before the worth of CFD techniques can be fully evaluated.

Acknowledgements

This work is supported by a NERC research fellowship (NER/I/S/2000/00847). LiDAR and SAR data were kindly provided by the EA. The author would like to thank the two referees for their helpful comments on this chapter.

References

Alsdorf, D.E., Smith, L.C. and Melack, J.M. 2001. Amazon floodplain water level changes measured with interferometric SIR-C radar. *IEEE Transactions on Geoscience and Remote Sensing*, 39(2), 423–431.

Aronica, G., Hankin, B. and Beven, K. 1998. Uncertainty and equifinality in calibrating distributed roughness coefficients in a flood propagation model with limited data. *Advances in Water Resources*, 22(4), 349–365.

Aronica, G., Bates, P.D. and Horritt, M.S. 2002. Assessing the uncertainty in distributed model predictions using observed binary pattern information within GLUE. *Hydrological Processes*, 16, 2001–2016.

Bates, P.D. and De Roo, A.P.J. 2000. A simple raster-based model for flood inundation simulation. *Journal of Hydrology*, 236, 54–77.

Bates, P.D. and Hervouet, J.-M. 1999. A new method for moving-boundary hydrodynamic problems in shallow water. *Proceeding of the Royal Society of London, Series A*, 455, 3107–3128.

Bates, P.D., Horritt, M.S., Smith, C.N. and Mason, D.C. 1997. Integrating remote sensing observations of flood hydrology and hydraulic modelling. *Hydrological Processes*, 11, 1777–1795.

Bates, P.D., Horritt, M.S. and Hervouet, J.-M. 1998a. Investigating two-dimensional, finite element predictions of floodplain inundation using fractal generated topography. *Hydrological Processes*, 12, 1257–1277.

Bates, P.D., Stewart, M.D., Siggers, G.B., Smith, C.N., Hervouet, J.-M. and Sellin, R.J.H. 1998b. Internal and external validation of a two-dimensional finite element code for river flood simulations. *Proceedings of the Institution of Civil Engineers, Water, Maritime and Energy*, 130, 127–141.

Bates, P.D., Marks, K.J. and Horritt, M.S. 2003. Optimal use of high-resolution topographic data in flood inundation models. *Hydrological Processes*, 17, 537–557.

Bekesi, G. and McConchie, J. 1999. Groundwater recharge modelling using the Monte Carlo technique, Manawatu region, New Zealand. *Journal of Hydrology*, 224(3–4), 137–148.

Beven, K. and Binley, A. 1992. The future of distributed models: Model calibration and uncertainty prediction. *Hydrological Processes*, 6, 279–298.

Beven, K. and Freer, J. 2001a. Equifinality, data assimilation, and uncertainty estimation in mechanistic modelling of complex environmental systems using the GLUE methodology. *Journal of Hydrology*, 249(1–4), 19–29.

Beven, K. and Freer, J. 2001b. A dynamic TOPMODEL. *Hydrological Processes*, 15(10), 1993–2001.

Bradbrook, K.F., Lane, S.N., Richards, K.S., Biron, P.M. and Roy, A.G. 2001. Flow structures and mixing at an asymmetrical open-channel confluence: A numerical study. *ASCE Journal of Hydraulic Engineering*, 127, 351–368.

Brakenridge, G.R., Tracy, B.T. and Knox, J.C. 1998. Orbital SAR remote sensing of a river flood wave. *International Journal of Remote Sensing*, 19(7), 1439–1445.

Brath, A. and Rosso, R. 1993. Adaptive calibration of a conceptual-model flash-flood forecasting. *Water Resources Research*, 29(8), 2561–2572.

Butler, J.B., Lane, S.N. and Chandler, J.H. 2001. Characterization of the structure of riverbed gravels using two-dimensional fractal analysis. *Mathematical Geology*, 33(3), 301–330.

Campling, P., Gobin, A., Beven, K. and Feyen, J. 2002. Rainfall–runoff modelling of a humid tropical catchment: The TOPMODEL approach. *Hydrological Processes*, 16(2), 231–253.

Chakraborty, S. and Dey, S.S. 1996. Stochastic finite element simulation of random structure on uncertain foundation under random loading. *International Journal of Mechanical Science*, 38(11), 1209–1218.

Chen, J.M., Leblanc, S.G., Miller, J.R., Freemantle, J., Loechel, S.E., Walthall, C.L., Innanen, K.A. and White, H.P. 1999. Compact Airborne Spectrographic Imager (CASI) used for mapping biophysical parameters of boreal forests. *Journal of Geophysical Research – Atmospheres*, 104(D22), 27945–27958.

Cheng, C.T., Ou, C.P. and Chau, K.W. 2002. Combining a fuzzy optimal model with a genetic algorithm to solve multi-objective rainfall-runoff model calibration. *Journal of Hydrology*, 168(1–4), 72–86.

Chow, V.T. 1959. *Open Channel Hydraulics*. McGraw-Hill, NY, 680pp.

Christiaens, K. and Feyen, J. 2002. Constraining soil hydraulic parameter and output uncertainty of the distributed hydrological MIKE SHE model using the GLUE framework. *Hydrological Processes*, 16(2), 373–391.

Cobby, D.M., Mason, D.C. and Davenport, I.J. 2001. Image processing of airborne scanning laser altimetry data for improved river flood modelling. *ISPRS Journal of Photogrammetry and Remote Sensing*, 56(2), 121–138.

Connell, R.J., Painter, D.J. and Beffa, C. 2001. Two-dimensional flood plain flow. II: Model validation. *Journal of Hydraulic Engineering*, 6(5), 406–415.

Cooper, V.A., Nguyen, V.T.V. and Nicell, J.A. 1997. Evaluation of global optimization methods for conceptual rainfall-runoff model calibration. *Water Sciences and Technology*, 36(5), 53–60.

Culling, W.E.H. and Datko, M. 1987. The fractal geometry of the soil-covered landscape. *Earth Surface Processes and Landforms*, 12, 369–385.

Dainton, M.P., Goldwater, M.H. and Nichols, N.K. 1997. Direct computation of stochastic flow in reservoirs with uncertain parameters. *Journal of Computational Physics*, 130, 203–216.

Duan, Q.Y., Sorooshian, S. and Gupta, V. 1992. Effective and efficient global optimization for conceptual rainfall-runoff models. *Water Resources Research*, 28(4), 1015–1031.

EA. 1998. Airborne LiDAR feasibility study. *Environment Agency*, Report TR E43.

Fathi-Maghadam, M. and Kouwen, N. 1997. Nonrigid, nonsubmerged, vegetative roughness on floodplains. *Journal of Hydraulic Engineering*, ASCE, 123(1), 51–57.

Flood, M. and Gutelius, W. 1997. Commercial implications of topographic terrain mapping using scanning airborne laser radar. *Photogrammetric Engineering and Remote Sensing*, 63(4), 327–366.

Franks, S.W., Gineste, P., Beven, K.J. and Merot, P. 1998. On constraining the predictions of a distributed model: The incorporation of fuzzy estimates of saturated areas into the calibration process. *Water Resources Research*, 34(4), 787–797.

Frasier, S.J. and Camps, A.J. 2001. Dual-beam interferometry for ocean surface current vector mapping. *IEEE Transactions on Geoscience and Remote Sensing*, 39(2), 401–414.

Freer, J., Beven, K. and Ambroise, B. 1996. Bayesian estimation of uncertainty in runoff prediction and the value of data: An application of the GLUE approach. *Water Resources Research*, 32(7), 2161–2173.

Ghanem, R. 1998. Probabilistic characterisation of transport in heterogeneous media. *Computer Methods in Applied Mechanics and Engineering*, 158, 199–220.

Ghanem, R. and Red-Horse, J. 1999. Propagation of probabilistic uncertainty in complex physical systems using a stochastic finite element approach. *Physica D*, 133, 137–144.

Gomes-Pereira, L.M. and Wicherson, R.J. 1999. Suitability of laser data for deriving geographical data: A case study in the context of management of fluvial zones. *Photogrammetry and Remote Sensing*, 54, 105–114.

Graber, H.C., Thompson, D.R. and Carande, R.E. 1996. Ocean surface features and currents measured with synthetic aperture radar interferometry and HF radar. *Journal of Geophysical Research – Oceans*, 101(C11), 25813–25832.

Grayson, R.B., Bloschl, G., Western, A.W. and McMahon, T.A. 2002. Advances in the use of observed spatial patterns of catchment hydrological response. *Advances in Water Resources*, 25(8–12), 1313–1334.

Hankin, B.G., Hardy, R., Kettle, H. and Beven, K.J. 2001. Using CFD in a GLUE framework to model the flow and dispersion characteristics of a natural fluvial dead zone. *Earth Surface Processes and Landforms*, 26(6), 667–687.

Higgitt, D.L. and Warburton, J. 1999. Applications of differential GPS in upland fluvial geomorphology. *Geomorphology*, 29(1–2), 121–134.

Horritt, M.S. 2000a. Development of physically based meshes for two-dimensional models of meandering channel flow. *International Journal for Numerical Methods in Engineering*, 47, 2019–2037.

Horritt, M.S. 2000b. Calibration and validation of a 2-dimensional finite element flood flow model using satellite radar imagery. *Water Resources Research*, 36(11), 3279–3291.

Horritt, M.S. and Bates, P.D. 2001a. Effects of spatial resolution on a raster based model of flood flow. *Journal of Hydrology*, 253, 239–249.

Horritt, M.S. and Bates, P.D. 2001b. Predicting floodplain inundation: Raster-based modelling versus the finite element approach. *Hydrological Processes*, 15, 825–842.

Horritt, M.S. and Bates, P.D. 2002. Evaluation of 1-D and 2-D models for predicting river flood inundation. *Journal of Hydrology*, 268, 87–99.

Hsieh, T.Y. and Yang, J.C. 2003. Investigation on the suitability of two-dimensional depth-averaged models for bend-flow simulation. *Journal of Hydraulic Engineering*, ASCE, 129(8), 597–612.

Huising, E.J. and Gomes-Pereira, L.M. 1998. Errors and accuracy estimates of laser data acquired by various laser scanning systems for topographic applications. *ISPRS Journal of Photogrammetry and Remote Sensing*, 53, 245–261.

Jolliffe, I.T. and Stephenson, D.B. 2003. *Forecast Verification: A Practitioner's Guide in Atmospheric Science*. Wiley & Sons Ltd, Chichester, 240pp.

Jones, L. 1989. Some results comparing Monte-Carlo simulation and 1st order Taylor-series approximation for steady groundwater-flow. *Stochastic Hydrology and Hydraulics*, 3(3), 179–190.

Jones, L. 1990. Explicit Monte-Carlo simulation head moment estimates for stochastic confined groundwater-flow. *Water Resources Research*, 26(6), 1145–1153.

Koblinsky, C.J., Clarke, R.T., Brenner, A.C. and Frey, H. 1993. Measurement of river level variations with satellite altimetry. *Water Resources Research*, 29(6), 1839–1848.

Kouwen, N. and Fathi-Maghadam, M. 2000. Friction factors for coniferous trees along rivers. *Journal of Hydraulic Engineering*, ASCE, 126(1), 732–740.

Lallemand, B., Cherki, A., Tison, T. and Level, P. 1999. Fuzzy modal finite element analysis of structures with imprecise material properties. *Journal of Sound and Vibration*, 220(2), 353–364.

Lamb, R. 1999. Calibration of a conceptual rainfall-runoff model for flood frequency estimation by continuous simulation. *Water Resources Research*, 35(10), 3103–3144.

Lamb, R., Beven, K. and Myrabo, S. 1998. Use of spatially distributed water table observations to constrain uncertainty in a rainfall-runoff model. *Advances in Water Resources*, 22(4), 305–317.

Lane, S.N. 2000. The measurement of river channel morphology using digital photogrammetry. *Photogrammetric Record*, 16(96), 937–957.

Lane, S.N., Chandler, J.H. and Richards, K.S. 1994. Developments in monitoring and terrain modelling small-scale river-bed topography. *Earth Surface Processes and Landforms*, 19, 349–368.

Lane, S.N., Bradbrook, K.F., Caudwell, S.W.B. and Richards, K.S. 1998. Mixing processes at river channel confluences: Field-informed numerical modelling. In Lee, J.H.W., Jayawardena, A.W. and Wang, Z.Y. (eds), *Environmental Hydraulics*. Balkema, Rotterdam, 345–350.

Lane, S.N., Bradbrook, K.F., Richards, K.S., Biron, P.M. and Roy, A.G. 1999. The application of computational fluid dynamics to natural river channels: Three-dimensional versus two-dimensional approaches. *Geomorphology*, 29, 1–20.

Lane, S.N., Hardy, R.J., Elliot, L. and Ingham, D.B. 2002. High-resolution numerical modelling of three-dimensional flows over complex river bed topography. *Hydrological Processes*, 16(11), 2261–2272.

Li, L. and Graham, W.D. 1998. Stochastic analysis of solute transport in heterogeneous aquifers subject to spatially random recharge. *Journal of Hydrology*, 206, 16–38.

Li, R. and Ghanem, R. 1998. Adaptive polynomial chaos expansions applied to statistics of extremes in nonlinear random vibration. *Probabilistic Engineering Mechanics*, 13(2), 125–136.

Lohani, B. and Mason, D.C. 2001. Application of airborne scanning laser altimetry to the study of tidal channel geomorphology. *ISPRS Journal of Photogrammetry and Remote Sensing*, 56(2), 100–120.

Marseguerra, M. and Zio, E. 1998. Monte Carlo simulation of the effects of human intrusion on groundwater contaminant transport. *Mathematics and Computers in Simulation*, 47(2–5), 361–369.

Mason, D.C., Cobby, D.M., Horritt, M.S. and Bates, P.D. 2003. Floodplain friction parameterisation in two-dimensional river flood models using vegetation heights derived from airborne scanning laser altimetry. *Hydrological Processes*, 17(9), 1711–1732.

Menenti, M. and Ritchie, J.C. 1994. Estimation of effective aerodynamic roughness of Walnut Gulch watershed with laser altimeter measurements. *Water Resources Research*, 30(5), 1329–1337.

Moller, D., Frasier, S.J., Porter, D.L. and McIntosh, R.E. 1998. Radar-derived interferometric surface currents and their relationship to subsurface current structure. *Journal of Geophysical Research – Oceans*, 103(C6), 12839–12852.

Moussa, R. and Bocquillon, C. 1996. Criteria for the choice of flood-routing methods in natural channels. *Journal of Hydrology*, 186, 1–30.

Nditiru, J.G. and Daniell, T.M. 2001. An improved genetic algorithm for rainfall-runoff model calibration and function optimization. *Mathematical and Computer Modelling*, 33, 695–706.

Nelder, J.A. and Mead, R. 1965. A simplex method for function minimization. *Computer Journal*, 7(4), 308–313.

Nepf, H.M. 1999. Drag, turbulence and diffusion in flow through emergent vegetation. *Water Resources Research*, 35(2), 479–489.

Nicholas, A.P. and Sambrook-Smith, G.H. 1999. Numerical simulation of three-dimensional flow hydraulics in a braided channel. *Hydrological Processes*, 13, 913–929.

Oberkampf, W.L. and Trocano, T.G. 2002. Verification and validation in computational fluid dynamics. *Progress in Aerospace Sciences*, 38, 209–272.

Osnes, H. and Langtangen, H.P. 1998. An efficient probabilistic finite element method for stochastic groundwater flow. *Advances in Water Resources*, 22(2), 185–195.

Refsgaard, J.C. 1997. Parameterisation, calibration and validation of distributed hydrological models. *Journal of Hydrology*, 198(1–4), 69–97.

Ritchie, J.C. 1995. Airborne laser altimeter measurements of landscape topography. *Remote Sensing of the Environment*, 53, 91–96.

Ritchie, J.C., Grissinger, E.H., Murphey, J.B. and Garbrecht, J.D. 1994. Measuring channel and gully cross-sections with an airborne laser altimeter. *Hydrological Processes*, 8(3), 237–243.

Rocca, F., Prati, C., Guarnieri, A.M. and Ferretti, A. 2000. SAR interferometry and its applications. *Surveys in Geophysics*, 21(2–3), 159–176.

Romanowicz, R. and Beven, K. 1998. Dynamic real-time prediction of flood inundation probabilities. *Hydrological Sciences Journal*, 43(2), 181–196.

Romanowicz, R. and Beven, K. 2003. Estimation of flood inundation probabilities as conditioned on event inundation maps. *Water Resources Research*, 39(3).

Romeiser, R. and Thompson, D.R. 2000. Numerical study on the along-track interferometric radar imaging mechanism of oceanic surface currents. *IEEE Transactions on Geoscience and Remote Sensing*, 38(1), 446–458.

Rong, Y., Wang, R.F. and Chou, R. 1998. Monte Carlo simulation for a groundwater mixing model in soil remediation of tetrachloroethylene. *Journal of Soil Contamination*, 7(1), 87–102.

Rufino, G., Moccia, A. and Esposito, S. 1998. DEM generation by means of ERS tandem data. *IEEE Transactions on Geoscience and Remote Sensing*, 36(6), 1905–1912.

Rutschmann, P. and Hager, W. 1996. Diffusion of floodwaves. *Journal of Hydrology*, 178, 19–32.

Sanders, R. and Tabuchi, S. 2000. Decision support system for flood risk analysis for the River Thames, United Kingdom. *Photogrammetric Engineering and Remote Sensing*, 66(10), 1185–1193.

Slatton, K.C., Crawford, M.M. and Evans, B.L. 2001. Fusing interferometric radar and laser altimeter data to estimate surface topography and vegetation heights. *IEEE Transactions on Geoscience and Remote Sensing*, 39(11), 2470–2482.

Smith, L.C. 1997. Satellite remote sensing of river inundation area, stage and discharge: A review. *Hydrological Processes*, 11(10), 1427–1439.

Sorooshian, S. 1981. Parameter-estimation of rainfall-runoff models with heteroscedastic streamflow errors – the non-informative data case. *Journal of Hydrology*, 52(1–2), 127–138.

Soutter, M. and Musy, A. 1998. Coupling 1D Monte-Carlo simulations and geostatistics to assess groundwater vulnerability to pesticide contamination on a regional scale. *Journal of Contaminant Hydrology*, 32(1–2), 25–39.

Thompson, K.R., Kelley, D.E., Sturley, D., Topliss, B. and Leal, R. 1998. Nearshore circulation and synthetic aperture radar: An exploratory study. *International Journal of Remote Sensing*, 19(6), 1161–1178.

Titmarsh, G.W., Cordery, I. and Pilgrim, D.H. 1995. Calibration procedures for rational and USSCS design flood methods. *Journal of Hydraulic Engineering*, ASCE, 121(1), 61–70.

USGS. 1990. Guide for selecting Manning's n roughness coefficients for natural channels and flood plains. *United Stated Geological Survey*, water supply survey paper 2339.

Weltz, M.A., Ritchie, J.C. and Fox, H.D. 1994. Comparison of laser and field measurements of vegetation height and canopy cover. *Water Resources Research*, 30(5), 1311–1319.

Whigham, P.A. and Crapper, P.F. 2001. Modelling rainfall-runoff using genetic programming. *Mathematical and Computer Modelling*, 33, 707–721.

Wiener, N. 1938. The homogeneous chaos. *American Journal of Mathematics*, 60, 897–936.

Wilson, C.A.M.E., Boxall, J.B., Guymer, I. and Olsen, N.R.B. 2003a. Validation of a three-dimensional numerical code in the simulation of pseudo-natural meandering flows. *Journal of Hydraulic Engineering*, ASCE, 129(10), 758–768.

Wilson, C.A.M.E., Stoesser, T., Olsen, N.R.B. and Bates, P.D. 2003b. Application and validation of numerical codes in the prediction of compound channel flows. *Proceedings of the Institution of Civil Engineers – Water and Maritime Engineering*, 156(2), 117–128.

Wimmer, C., Siegmund, R., Schwabisch, M. and Moreira, J. 2000. Generation of high precision DEMs of the Wadden Sea with airborne interferometric SAR. *IEEE Transactions on Geoscience and Remote Sensing*, 38(5), 2234–2245.

Wu, F-C., Shen, H.S. and Chou, Y-J. 1999. Variation of roughness coefficients for unsubmerged and submerged vegetation. *Journal of Hydraulic Engineering*, ASCE, 125(9), 934–942.

Part Two
Application Potential for Fluvial Studies

Part Two

10
Modelling reach-scale fluvial flows

S.N. Lane and R.I. Ferguson

10.1 Introduction

The aim of this chapter is to review the application of CFD to non-uniform flows in river channels. By reach scale, we mean situations where the domain length is longer than the channel width by a factor of at least 5 and up to 50 or more. What makes this type of application special and worthy of separate attention is that it involves channel geometries that are inevitably and specifically non-uniform, which has major implications both for process representation and for the way in which CFD is applied. Similarly, the reach scale is commonly the scale at which management activities are required, whether in relation to flow and sediment transport around structures such as bridge piers, in-channel sedimentation problems, bank erosion and channel migration, channel diversion, effluent discharge, flow abstraction or habitat enhancement. Thus, in addition to having complex geometries, reach-scale problems are also linked strongly to morphodynamics and habitat processes as well as the flow field *per se*. With this in mind, we begin this section by reviewing the use of 1D approaches for understanding reach-scale processes. This sets the context for a discussion of what processes need to be represented at the reach scale, and the methodological challenges that result. This is followed by a section that seeks to demonstrate what CFD can and cannot tell us in relation to classic aspects of reach-scale river behaviour. This links issues of process representation and CFD application to specific case studies to show what CFD can tell us, what it cannot tell us at present, and what it is unlikely to tell us however much research development is undertaken. Specific aspects of these issues are also discussed and exemplified in other chapters of this book (e.g. Chapters 4, 13 and 14).

10.2 One-dimensional approaches

One-dimensional approaches to reach-scale flow, morphodynamics and habitat problems are common, and likely to remain so in any situation where the reach of interest is longer than many tens of channel widths. This is because of computational and data availability limits to the application of 2D and 3D codes over long reaches. However, there are a number of situations in which 1D approaches can now be replaced by 2D and 3D approaches and in this section we seek to demonstrate what those situations are in order to constrain the range of situations in which 2D and 3D approaches might be deemed preferable.

10.2.1 Hydrodynamics

The starting point of 1D approaches is the treatment of the basic flow equations. Consider a reach of river that can be described by a cross section of area A, with an average velocity in that section of v. Thus, the discharge (Q) through that section is $Q = vA$. Both v and A can vary as a function of downstream distance (x). If the flow is steady ($\mathrm{d}Q/\mathrm{d}x = 0$), mass conservation gives

$$0 = -\frac{\mathrm{d}(vA)}{\mathrm{d}x} + i = -v\frac{\mathrm{d}A}{\mathrm{d}x} - A\frac{\mathrm{d}v}{\mathrm{d}x} + i \tag{10.1}$$

where i is the input from (or, if negative, loss to) storage per unit distance downstream. If the flow is unsteady, equation (10.1) becomes:

$$\frac{\partial A}{\partial t} = -v\frac{\partial A}{\partial x} - A\frac{\partial v}{\partial x} + i \tag{10.2}$$

The same analysis applies to momentum conservation. In principle, for an incompressible fluid, we state that the rate of change of momentum through time at a point will be a function of the spatial change of momentum plus sources (the driving forces):

$$\frac{\partial(Av)}{\partial t} = -\frac{\partial(Av^2)}{\partial x} + \text{sources} \tag{10.3}$$

The source terms are: (1) pressure gradients; (2) potential energy; and (3) friction that causes energy expenditure:

$$\frac{\partial(Av)}{\partial t} + \frac{\partial(Av^2)}{\partial x} = -Ag\frac{\partial h}{\partial x} + gA(S_\mathrm{o} - S_\mathrm{f}) \tag{10.4}$$

where h is mean flow depth, S_o is the bed slope of the channel (defining the potential energy term) and S_f is the friction slope (defining the friction term). If these equations are derived from the full 3D form of the Navier–Stokes equations (Chapter 2), then it becomes clear that the friction term in equation (10.4) is not just representing the effects of boundary resistance, but a whole set of other processes. These result in the extraction of momentum from the mean downstream flow (i.e. Av) and its transformation into flow components that are variable in the cross-stream and vertical directions (i.e. dispersion processes associated with secondary circulation) and through time (i.e. turbulence processes). The friction term is commonly defined under the assumption that the flow is locally uniform. This allows uniform flow equations to be used, such as the Darcy-Weisbach equation:

$$S_f = \frac{v^2 f}{8gR} \tag{10.5}$$

where R is the hydraulic radius and f is a 'friction parameter'. However, f is actually representing more than just friction such that, in most situations, its relationship to the actual surface roughness (as measured by local elevation variability, for instance) will be uncertain.

The above equations are commonly used for channel routing in relation to flood risk analysis and are the basis of a number of commercially available software packages for this purpose (e.g. HEC-RAS, ISIS, MIKE-11). There are good examples of their effective application to reach-scale situations, such as for modelling flow reversal in riffle–pool sequences (e.g. Richards, 1978; Keller and Florsheim, 1993; Carling and Wood, 1994). The main problems result when there is significant within-reach flow variability in either the vertical or the cross-stream direction. At the reach scale this is commonly associated with secondary circulation due to flow curvature, flow separation or anisotropic turbulence (e.g. Bradbrook et al., 2000a).

Judging when this is significant is a more difficult issue and commonly requires consideration of the geometry under consideration. For instance, for predicting the magnitude and timing of an out-of-bank flow, 1D models appear to be more than adequate provided proper attention is given to cross-section spacing and model calibration. However, where the interest is in within-reach flows, and the geometry is complex, at least a 2D or even a 3D treatment will be required. This has been illustrated effectively for the case of river meanders by Dietrich and Smith (1983). River meanders have strong secondary circulation associated with the interaction of bed morphology with water surface gradients and the associated pressure effects. If the flow field is considered in depth-averaged terms, both the cross-stream and downstream momentum equations have two topographic forcing terms as well as bottom shear stress terms, and Dietrich and Smith showed that these terms can be significant in natural channels. Although their research was based upon the analysis of force balance equations coupled to high-resolution field measurements, it implies that application of CFD to situations with topographic irregularity will need a 2D, if not a 3D, treatment.

10.2.2 Sediment transfer, morphodynamics and channel change

Reach-scale processes may involve significant sediment transport, potentially involving all size fractions. At the reach scale, washload may be transported through a reach in suspension: it is too fine to be deposited in the channel and its flux is limited by supply rather than local river transport capacity. This sediment transfer process would normally be represented by some form of advection–diffusion equation, which provides a generalised mathematical representation of both the advective and turbulent diffusive components of sediment transfer (there are direct analogies with turbulence modelling, described below, in this sense). Between the washload and the coarsest particles available for transport, sediment transfer will be governed by local transport capacity, whether moving along the bed or in suspension (normally only the finer sand). This transfer process matters as it controls erosion and deposition and hence channel change. It also controls the bed sedimentary environment, and hence may have major habitat implications. A hydraulic model can be combined with a sediment transport model to simulate such channel changes. 1D codes of this type (often called sediment routing models) perform all the calculations on a width-averaged basis with fixed banklines. 2D versions operate on a curvilinear spatial grid and in some cases allow for bank erosion with consequent remeshing of the grid. The basics of 1D sediment routing models are explained here. The extension to 2D is discussed in section 10.3.5.

Bed-material transport in such models is invariably represented using a continuum-mechanics approach, i.e. bulk properties of the flow are used to predict a bulk flux, rather than by considering the dynamics of individual particles. The sediment flux per unit width is normally represented as a nonlinear function of flow strength relative to a threshold for movement to occur. A well-known example is the Meyer-Peter and Müller equation

$$q_s^* = 8(\tau^* - \tau_c^*)^{1.5} \qquad (10.6)$$

where $q_s^* = q_s/(RgD_b^3)^{0.5}$; $\tau^* = \tau/(\rho RgD)$; q_s is the volume of transported sediment per unit width and time; D_b is a representative bed grain diameter; τ is bed shear stress; and R is the submerged specific gravity $(\rho_s - \rho)/\rho$ where ρ_s and ρ are sediment and water densities. Equation (10.6) may also explicitly include a correction factor for the effective shear stress: if not, it is assumed that shear stresses are corrected prior to application of the equation. The critical Shields stress τ_c^* is taken as constant, implying a critical shear stress $\tau_c \propto D$. Equations for the transport of individual size fractions $i = 1, 2, \ldots$ predict q_{si}/F_i from τ and τ_{ci}, where $q_{si} = q_s P_i$ is the flux of size class i and P_i and F_i are its volumetric proportions in the bedload and the bed, respectively. The critical stress τ_{ci} depends as much or more on D_b as on the representative diameter D_i of the individual size class, because of hiding and protrusion effects (Egiazaroff, 1965; Ashworth and Ferguson, 1989; Parker, 1990; Wilcock and Crowe, 2003).

Updating of bed elevation and Grain-Size Distribution (GSD) in morphodynamic models is based on conservation of sediment volume. The 1D form of the overall continuity equation (often termed the Exner equation) is

$$\frac{\partial z}{\partial t} = -\frac{1}{1-\lambda}\frac{\partial q_s}{\partial x} \qquad (10.7)$$

where z is bed surface elevation at time t and distance x downstream, and λ is bed porosity. Likely rates of aggradation and degradation are orders of magnitude lower than water velocities, so the hydraulic and sediment computations can be uncoupled in a simple alternating-sweep finite difference scheme using a downstream boundary condition for the flow and an upstream boundary condition (sediment supply) for transport.

The continuity equations for individual size fractions of the bed are normally formulated by assuming that bedload/bed exchange is restricted to a shallow active layer. The thickness L of this layer is usually set to the bedform amplitude for sand-bed rivers and a small multiple of D_b for gravel-bed rivers. Assuming constant porosity, the evolution of F_i at a point is described by:

$$(1-\lambda)\frac{\partial(LF_i)}{\partial t} = -\frac{\partial q_{si}}{\partial x} - (1-\lambda)E_i\left(\frac{\partial z}{\partial t} - \frac{\partial L}{\partial t}\right) \quad (10.8)$$

where the first term on the right represents deposition from (or entrainment to) bedload and the second term relates to the passive change of status of material near the base of the active layer as the latter moves up or down. E_i is the proportion of size i in this passively exchanged material. During degradation, E_i refers to previously inactive sediment, so the stratigraphy of the bed needs to be tracked by the model. During aggradation the natural assumption is $E_i = F_i$, but Hoey and Ferguson (1994) generalised this to $E_i = cF_i + (1-c)P_i$ (where $0 \leq c \leq 1$) to allow for the possibility that some deposited bedload infiltrates right through the active layer. Toro-Escobar et al. (1996) calculated that c was as low as 0.3 for one laboratory experiment, but most workers stick with $c = 1$ and thus $E_i = F_i$. Alternative versions of equation (10.8) can be obtained by eliminating $\partial z/\partial t$ using equation (10.7) and substituting $q_{si} = q_s P_i$; and the equation simplifies if L is assumed constant. Parker (1990) extended it to include size transformation through abrasion.

One-dimensional sediment routing models with a single grain size (e.g. HEC-6) have been used by engineers for decades to predict rates of scour and fill. The size-fraction continuity equation (10.8) was used by Hirano (1971) and Parker and Sutherland (1990) to model degradational armouring at a point. Multiple-fraction reach-scale models were developed in the late 1980s and early 1990s (e.g. Armanini and di Silvio, 1988; Rahuel et al., 1989; van Niekerk et al., 1992; Hoey and Ferguson, 1994). They have been used to simulate downstream fining in modern rivers (Hoey and Ferguson, 1994) and the rock record (Robinson and Slingerland, 1998), to retrodict the effects of meander rectification (Talbot and Lapointe, 2002), and to investigate the dispersion of sediment pulses (Cui et al., 2003). In most of these applications flow has been assumed steady, at some high 'dominant' discharge, but most 1D morphodynamic codes allow for hydrographs represented as a series of daily or other discharge values each of which is simulated using steady-flow equations.

The main limitation of 1D models of this kind is the width-averaging of the flow and transport calculations. This is perfectly adequate for canals or channelised rivers with trapezoidal cross sections, but means that 1D models can neither generate the bar–pool–riffle topography commonly found in natural rivers nor adequately simulate the associated local variation in flow and sediment transport conditions. Width-averaging can also lead to underestimation of bedload flux when 1D models are applied to highly non-uniform natural channels (Paola, 1996; Nicholas, 2000; Ferguson,

2003). The dependence of local transport rate on flow strength is nonlinear, especially in gravel-bed rivers near the threshold, so the extra flux in areas of relatively strong flow is liable to outweigh the reduction in flux in areas of weaker flow. However, the effect is weakened to the extent that the bed becomes coarser in areas of stronger flow. 2D models are better in principle for highly non-uniform reaches, and the more sophisticated ones have the additional advantage of allowing for changes in river width during aggradation, degradation and meander migration. They do, however, make much greater computational and data demands and raise other issues of process representation as discussed later (section 10.3.5).

A final issue that is particularly significant for multi-fraction models (whether 1D or 2D) is the specification of the incoming bed-material flux and its grain-size distribution. Hoey and Ferguson (1997) and Ferguson et al. (2001) have shown that 1D model response is sensitive to this boundary condition: overloading leads to aggradation, fining and increased flux whereas underloading leads to degradation, armouring and reduced flux. Reliable field measurements of bedload transport at a range of water discharges are rarely available, so assumptions usually have to be made. Often the safest one is to start the model domain at a section that appears to be stable and set supply equal to capacity, so that there is no change in bed elevation or grain-size distribution at the start of the reach.

10.2.3 Habitat

The available habitat in a river depends on several key physical parameters, notably flow velocity and depth, wetted perimeter, substrate, temperature and pH (e.g. Elso and Giller, 2001; Maddock et al., 2001; Chapter 16 in this book). These are often combined into some form of habitat score such as a Weighted Usable Area (WUA), as explained by Leclerc (Chapter 16). The best-known example of this is Physical Habitat Simulation System (PHABSIM) (e.g. Milhous et al., 1984) which is widely used to assess habitat suitability. However, the hydraulics component of PHABSIM is commonly simplified to a 1D treatment (section 10.2) and, even then, assumptions of uniform or gradually varied flow may be made (Milhous et al., 1989). Leclerc et al. (1995) note that this will not produce reliable results for less than $10\,\text{m}^2$, thus limiting its suitability in smaller streams or where spatial variation in habitat is high. This was confirmed by Ghanem et al. (1996) who obtained a more accurate representation of spatial patterns of flow velocity with a 2D model than with a 1D model. The result is severe over-estimation of the amount of available habitat (Crowder and Diplas, 2002). In addition, measurement of point velocities may provide insufficient habitat description as process gradients may also matter (Crowder and Diplas, 2002). In freshwater systems, a strong interaction between habitat composition/arrangement and habitat processes has been identified (Ward et al., 2001), which creates both spatial variation in habitat (diversity) and changes in habitat and its structure through time (dynamics). Ward et al. (2001) argue that the failure to recognise fully the diversity and dynamism of natural river corridors has hindered conceptual advances in river ecology and may have hindered river restoration. It follows that an awareness of the spatial assemblage of different habitat types, and how they change through time, is a crucial input to reach-scale assessment of habitat.

A good example of this is optimal brown trout habitat, which in general terms is characterised by (Raleigh et al., 1986) clear, cool to cold, water; a relatively silt-free rocky substrate in riffle areas; a 50–70% pool to 30–50% riffle habitat combination; availability of pools with sufficiently slow, deep water, and certainly areas of refuge where velocities never exceed $0.20 \, \text{m s}^{-1}$ (Vismara et al., 2001); well-vegetated, stable stream banks; abundant instream cover; and relatively stable annual water flow and temperature regimes. Unfortunately, within this species, things are somewhat more complicated. First, there is a life cycle dependence: abnormally low or high flows are particularly problematic for both egg deposition and fry emergence (Raleigh et al., 1986). For instance, redd destruction can result from a flood flow that leads to gravel movement (e.g. Gilvear et al., 2002). Second, a key habitat requirement is spatial diversity: fry prefer depths less than 0.20 m but older trout prefer water deeper than this (Kennedy and Strange, 1982): Third, whilst extreme events can be problematic for some stages of the life cycle, they are also important for the periodic removal of accumulated fine sediment (Gilvear et al., 2002), which can degrade spawning habitat: Raleigh et al. (1986) note that the optimal spawning conditions in gravels are less than 5% fines. Thus, some disturbance will be good for the system. Too much will undermine it, unless there are appropriate refugia for organisms in those stages of the life cycle that are sensitive to extreme flood events. Thus, assessing habitat without an assessment of the spatial arrangement of habitat suitability and its change through time is likely to lead to the highly conflicting results from research into *Salmo trutta* behaviour identified by Gilvear et al. (2002). It follows that 2D approaches (at least) will be required, along with an ability to simulate the transport of sediment which determines habitat suitability.

10.3 Two- and three-dimensional approaches: Process representation

The consideration of 1D treatments in section 10.2 has demonstrated that whilst there are many situations where a 1D treatment might be adequate, or even preferred, there are also situations where a 2D or 3D treatment is necessary. In this section, we seek to demonstrate the issues associated with process representation in reach-scale applications of CFD, including the different considerations for 3D rather than 2D modelling.

10.3.1 Reynolds-averaging, dimensionality and process representation

The full 3D form of the Navier–Stokes mass and momentum equations is given by Ingham and Ma (Chapter 2). Two modifications to these basic equations are commonly made in models used to estimate reach-scale river flows. First, the equations are Reynolds-averaged, with variables decomposed into time-averaged and time-varying components. This is necessary because at the Reynolds numbers and spatial scales typical of the reach scale, direct numerical simulation remains computationally unfeasible. Rather than modelling the time dependence explicitly, only the time-averaged components are determined. However, Reynolds-averaging results in the appearance of additional terms in the momentum equations known as the turbulent Reynolds stresses. These represent the transport of momentum by turbulence, which can never be zero in turbulent flow. Their importance is well established in reach-scale flows (e.g. in zones of strong shear, where turbulent motions may be generated). Unfortunately, the appearance of additional terms means the equations no longer close and it is necessary to introduce a turbulence model to represent the effects of turbulence upon the time-averaged flow properties. The nature of the turbulence model adopted does have important implications for the estimation of certain types of flow processes. This is especially the case for boundary cells, where certain assumptions about turbulence are made in order to determine necessary boundary conditions (section 10.3.3).

Second, the 3D form of the mass and momentum equations can be modified to a quasi-3D or a 2D depth-averaged form (Chapter 2). The quasi-3D modification involves assuming that, at the reach scale, flow is sufficiently shallow that the vertical scale can be assumed to be much smaller than the horizontal scale and the boundary layer extends throughout the flow depth. If this is the case, the w (vertical velocity) momentum equation is simplified to the hydrostatic pressure distribution. This simplification makes solution of the governing equations more straightforward. It should be noted that w is still determined while finding the solution, as it appears in the u and v momentum equations, as well as in the mass equations. Thus, in conceptual terms, this modification yields estimates of the w velocity that give a hydrostatic pressure distribution (i.e. one where the pressure is a function of water depth only). This means that the model will not represent properly any situation where there is a sudden change in pressure (such as an adverse pressure gradient). In these situations (section 10.4.4), the flow may separate from the bed, with a zone of negative dynamic pressure (i.e. deviation from hydrostatic pressure) forming in part of the flow.

Ingham and Ma (Chapter 2) also describe the case of 2D depth-averaging. This results in the introduction of two new groups of terms: bottom stresses and dispersion terms. The bottom stresses appear in the full 3D and the quasi-3D momentum equations, where they are commonly included in a wall treatment (section 10.3.3). In the depth-averaged case, their role takes on greater importance, and they have been widely expressed as a quadratic function of the depth-averaged velocity (e.g. Dietrich and Whiting, 1989; Nelson and Smith, 1989a,b; Shimizu and Itakura, 1989; Struiksma and Crosato, 1989; Shimizu *et al.*, 1990; Tingsanchali and Maheswaran, 1990; Yeh and Kennedy, 1993). This requires an acceptable description of the roughness coefficient (Lane, 1998) which, surprisingly given that this is concerned with a 2D flow, has often been specified in terms of a spatially uniform value of Manning's n (e.g. Shimizu and Itakura, 1989; Shimizu *et al.*, 1990; Tingsanchali and Maheswaran, 1990; Lane and Richards, 1998). Indeed, it is not surprising that friction is often used as one of the key calibration parameters in reach-scale hydraulic modelling in two dimension (e.g. Lane *et al.*, 1994; Chapter 9 in this book), just as it is in one dimension.

The dispersion terms are a product of vertical non-uniformity in the velocity field (which is inevitable because the velocities tend to fall to zero at the channel bottom) and represent deviations from the depth-averaged velocities within a vertical profile. Rodi (1980) and Rodi *et al.* (1981) note that the physical meaning of these dispersion terms is similar to the turbulent stress terms, in that both represent gradients of transport of momentum. From measurements of depth-averaged quantities it is usually not possible to distinguish between the turbulent and dispersion contributions to the momentum transport. As with the Reynolds shear stresses, no additional equations arise during the depth-averaging process and their determination requires either knowledge of the secondary flow field or a model of their effects on the mean flow properties. Field evidence has confirmed that dispersion terms can be important, associated with helical secondary flow due to either channel curvature (e.g. meander bends; Thorne and Hey, 1979) or the junction of two channels (Smith, 1974; Church and Jones, 1982; Ashmore *et al.*, 1992; Rhoads and Kenworthy, 1995, 1998). If no secondary flow – correction is made, depth-averaged models will tend to constrain the largest velocities to lie near the inside of channel bends (Bernard and Schneider, 1992) instead of predicting gradual velocity migration to the outside, as observed in the bends of many natural channels (e.g. Dietrich and Whiting, 1989). The most common treatment of these terms is semi-empirical or empirical where their effects on the mean transport of momentum (and hence the depth-averaged flow velocities) are modelled. The two main methods used to do this are reviewed in Lane (1998); see also Mosselman (Chapter 4). Both are based upon an analytical treatment of the effects of streamline curvature upon flow dispersion.

Even if given a justifiable treatment of bottom stresses and the dispersion terms, careful thought needs to be given to the acceptability of depth-averaging. It depends in the first place on the scale and aims of the study. If the reach length is many tens of channel widths and the main interest is in flood profiles or morphodynamics, not only is a 3D approach computationally very expensive but a 2D approach may be adequate and even advantageous, if undertaken carefully. For instance, numerical representation of changes in free surface elevation and hence in the lateral extent of flow is well developed (Chapter 11). This requires wetting and drying treatments which do not necessarily require remeshing of the computational domain. At present, 3D applications do not have this advantage and, whilst representation of free surface effects and their change through time is possible (e.g. using volume of fluid, see Chapter 2), it is computationally demanding and can lead to numerical problems (e.g. instability). Lane *et al.* (1999) compared a depth-averaged and a fully 3D solution and found little difference in predictive ability for horizontal velocity components of velocity, as judged by levels of explained variance with respect to check data. However, the depth-averaged predictions were biased in relation to the check data as compared with the 3D predictions, implying that additional calibration of the depth-averaged model was required to get the same level of agreement. This is where differences in the meaning of roughness terms between 2D and 3D models become important, as we explore in section 10.3.3. In particular, the dispersion term treatments reviewed in Lane (1998) will only represent aspects of secondary circulation that are embodied within them. Best (1987), Ashmore *et al.* (1992) and Biron *et al.* (1996) all describe the importance of topographic discordance as a secondary flow-generating mechanism, and treatments based upon planform curvature will not represent these effects. Similarly, these treatments will not represent secondary circulation that is caused by turbulent shear in the flow (e.g. Nezu *et al.*, 1993). The extent to which this matters at the reach scale needs debate: it is most probable that the topographic complexity of natural channels is such that mesoscale topographic forcing exceeds uncertainties associated with turbulence modelling.

What emerges from the above descriptions is that the ability to represent numerically those processes that might be deemed significant for reach-scale channel flows is sensitively dependent upon assumptions either associated with model choice or made during model development. The corollary of this is that what a particular model can predict will depend upon the assumptions made during model choice and development. This is illustrated in Table 10.1, which provides a very general commentary on issues of model dimensionality and turbulence representation in relation to different types of reach-scale flow phenomena. A number of points emerge. First,

Table 10.1 A basic summary of the relationship between process representation and model choice

Type of process	Issues associated with choice of model dimensionality and representation of turbulence
Secondary circulation processes in straight channels associated with cross-stream or downstream sediment sorting	Primarily driven by turbulence anisotropy. This means the required model must be three dimensional, with a turbulence model that does not assume isotropy (e.g. an algebraic stress model or a reynolds-stress model)
Planform streamline curvature such as in meanders and tributary junctions	Key process here is the effects of streamline curvature upon mass and momentum transfer. A three-dimensional treatment is ideal, but if the surface boundary condition specifies a rigid lid, some form of correction for mass conservation at the surface will be required. Simplification to a hydrostatic pressure distribution may be acceptable if the flow is hydrostatic throughout the channel. Two-dimensional depth-averaged models may be adequate for this problem, provided the interest is in the effects of secondary circulation upon the streamwise flow, and not in the streamwise flow components themselves. This will still require a correction to the mean flow properties due to the effects of streamline curvature, which may be achieved, for example, through a secondary circulation model
Flow separation such as is associated with a sudden increase in flow depth or channel width	The most commonly used k-ε turbulence model is unlikely to represent flow separation and a more sophisticated representation of turbulence is required. Extensive research has shown that even in the relatively simple flow separation associated with a backward-facing step, basic two-equation turbulence models do not reproduce separation lengths correctly. Some turbulence models (e.g. the RNG-type model) are modified to deal with this problem. Similarly, simplification of the full three-dimensional form of the Navier–Stokes equations through a hydrostatic pressure assumption will be inappropriate in this case and special attention may need to be given to the representation of the free surface boundary condition
Strong shear at tributary junctions associated with velocity differentials	One of the most complex problems to model, as the strong shear determines the nature of turbulence generation and dissipation which in turn controls the nature of the shear. Field and laboratory evidence suggests that a range of spatio-temporal scales of turbulence will be present. Requires either direct numerical simulation or large eddy simulation in order to get sufficient process representation

the type of model required, and the assumptions that can be made about process representation depend not only on the phenomenon under study, but also on which parameters in the system need to be determined. For instance, if the details of secondary circulation in a meander bend matter for erosion and deposition studies, simplification to a depth-averaged treatment, or even a quasi-3D treatment through a hydrostatic pressure assumption, may be inadequate. Second, some of the features that are of interest will require model treatments that are right at the forefront of research into CFD (e.g. advanced turbulence modelling). It is quite possible that some of the questions which fluvial geomorphologists wish to ask with respect to reach-scale flows cannot yet be answered using CFD. Finally, as process representation is considered, it is necessary to be particularly careful of the transfer of findings from channels with simple geometry to those of more complex geometry. Shankar et al. (2001) report on a simple study of a channel flow with a constriction. In turbulence terms, this is a challenge and they found that they needed variable vertical eddy diffusion to represent these flow processes properly. However, this does not imply that priority should be given to such a treatment over representation of other aspects of the problem. Both Lane et al. (1999) and Morvan et al. (2002) argue that correct representation of the channel geometry is likely to be of more value. This is not to imply that advanced turbulence modelling is not required, just that it should not be the prime area of research emphasis. Unfortunately, it also means that inter-comparison of turbulence models cannot be used to conclude that turbulence transport is a secondary influence (Morvan et al., 2002). Rather, it implies that our representation of turbulence may be insufficiently developed for its true effects to be identified.

10.3.2 The problem of discretisation

The above process representations have to be solved on some form of numerical mesh. This probably represents the greatest challenge to the application of CFD to reach-scale flows. On the one hand, if CFD can be successfully applied to the geometry of complex rivers, then it will overcome some of the main problems of relying upon the simplified geometries (e.g. trapezoidal section) of laboratory channels. However, research into the performance of numerical codes has emphasised, amongst other things, the need for careful investigation of discretisation. This includes the accuracy of discretisation (e.g. Wallis and Manson, 1997), convergence problems associated with fine grids in finite volume discretisations (e.g. Cornelius et al., 1999) and the spatial discretisation required for verification (Hardy et al., 2003). Thus, considering how best to discretise the domain of interest is a crucial part of the effective application of CFD at the river reach scale. In order to evaluate issues surrounding the discretisation of domains for complex geometries, this section begins with a very brief summary of the three broad options for discretisation that differ in quite fundamental ways: finite difference, finite volume and finite element. It should be emphasised that this description provides only a very brief summary of the issues associated with discretisation which is a complex issue and one that all CFD practitioners should pay serious attention to. Ferziger and Perić (1999) provide a thorough summary, and see also Wright (Chapter 7).

Finite difference schemes are based upon solution of the mass and momentum equations in differential form, which are approximated by a system of linear algebraic equations where the values of variables at the grid nodes are the unknowns (Ferziger and Perić, 1999). Each term of the partial differential equation at a particular node is replaced by a finite difference approximation. The differences may be evaluated in a backward, central or forward manner. Finite differences have two important properties. First, the continuous differential equations are expressed as a Taylor expansion which leads to first-, second-, third- and higher-order terms. If all of these derivatives are retained, the Taylor expansion is exact. However, the expansion is normally truncated. The effects of this truncation are controlled by: (1) the node (or grid spacing, with the size of the higher-order terms and hence the truncation error tending to zero as the grid) spacing tends to zero; and (2) the local rate of change of the variable under consideration where if this rate of change is high, the higher-order terms may be large. These reasons are why a crucial component of effective use of a finite difference scheme is grid refinement, either by increasing node density or through the strategic addition of nodes where local rates of variable change are high (e.g. in situations where there is strong shear). The convective components of the conservation equations can be treated as first-order derivatives. However, the diffusive components involve second-order derivatives. These may be dealt with by approximation of the first derivative twice, reflecting the geometrical principle that the second derivative is the slope of the line that is tangential to the curve representing the first derivative (Ferziger and Perić, 1999). Thus, in a finite difference scheme the most common approximation is a second-order central difference approximation.

Finite volume schemes use the integral form of the conservation equations rather than the differential form. The domain is divided into control volumes with a computational node at each volume centre, and integrals that apply to both the surface and the volume of the control. In a similar way to finite differences, an algebraic equation has to be determined for each control volume, which requires approximation of the surface and volume integrals using quadrature formulae (Ferziger and Perić, 1999). Surface integrals may be approximated using either (a) variable values at one or more locations on a cell face or (b) on the basis of nodal values at the cell centre. Interpolation is then required to transfer values from centres to faces. Volume integrals can be approximated using the product of the mean value in a volume and the volume of the control volume. No interpolation is required unless the mean value varies nonlinearly within the volume (Ferziger and Perić, 1999). Nonetheless, interpolation becomes a crucial step in a finite volume solution. This may be achieved through upwind methods (equivalent to forward or backward differencing), linear methods (equivalent to central differencing) or non-linear methods. In a finite volume solution, upwind schemes are only first-order accurate, with the second-order truncation error resembling a diffusive flux and which can become quite significant in situations where flow is oblique to the grid (Ferziger and Perić, 1999). Thus, use of upwind schemes require very fine grids. It is also possible to combine solution schemes (e.g. upwind and central difference). For instance, a Peclet number (a measure of the ratio of convective terms to diffusive terms) may be used to decide whether upwind or central differencing schemes should be used.

These finite difference and finite volume schemes assume a structured grid, that is a regular array of hexahedral grid cells which can be labelled $i = 1, 2, \ldots$ downstream, $j = 1, 2, \ldots$ laterally, and (in a 3D model) $k = 1, 2, \ldots$ vertically, with the same maximum values of j and k at all i. This essentially rectangular computational grid is transformed to fit the Cartesian coordinates of the channel planform and bed topography. Boundary Fitted Coordinates (BFCs) of this type are a natural choice for 2D models in which water depth is part of the solution. Commercial codes usually provide gridding tools to help the user interpolate a curvilinear grid between defined banklines. The numerical solution is simpler if the grid is locally orthogonal, as in DHI's MIKE21C code for example. BFCs have also been used in most attempts to model non-uniform reaches in three dimensions (e.g. Bradbrook et al., 1998, 2000a,b, 2001; Hodskinson and Ferguson, 1998; Lane et al., 1999, 2000; Nicholas and Smith, 1999; Nicholas, 2001). In general-purpose 3D CFD codes the grid is interpolated (sometimes in a linear rather than curvilinear way) between known cross sections. This can lead to significant problems, as noted by Lane et al. (1999). First, there is commonly significant spatial variation in water depth (e.g. along a channel cross section). With a structured mesh this can result in large variation in the average cell thickness. This may cause instability problems as the water depth becomes vanishingly small close to the channel margin. The marked spatial variation in grid size may confound attempts to achieve grid-independent solutions and also result in significant numerical diffusion. Second, it may also be problematic in relation to the sub-grid-scale wall treatment which has a range of minimum and maximum acceptable cell thicknesses (section 10.3.3). Third, in order to achieve mesh independence in the deeper parts of the flow, exceptionally large numbers of grid cells may be required throughout the flow domain. This leads to computational inefficiency as shallower areas are over-resolved. Fourth, if the rate of change of bottom geometry is rapid, it can lead to seriously skewed grid cells which not only enhance numerical diffusion, but may also lead to numerical instabilities. The same problem can occur (but commonly to a more significant degree) in relation to the effects of planform geometric variation, where sudden changes in-channel direction (e.g. in a tight meander bend) can lead to highly skewed grid cells; this is particularly serious if the gridding algorithm uses linear, rather than curvilinear, interpolation between specified sections. Fifth, fitting the computational grid to a Cartesian grid increases the probability that the primary flow direction is skewed with respect to the computational space, such that there can be high levels of numerical diffusion, even if the grid is not especially skewed. Finally, it is worth considering the effects of changing mesh resolution where there is a complex bed geometry. Lane et al. (1999) argue that as mesh resolution is increased there is the progressive introduction of geometric variability that is associated not with the bed topography, but with the way it has been sampled. The result is that it may be impossible to obtain a solution that is mesh-independent.

A number of solutions have been adopted to resolve these problems. First, it is common to limit the smallest water depths considered by specifying a minimum depth at the channel margin, normally 0.05 m or less. As the flow is generally very slow in these areas, this has a negligible effect upon model solution. Second, some of the real geometrical details may have to be sacrificed in order to achieve a stable solution with minimal diffusion. Third, connecting two (or more) single blocks together to create a multi-block solution can reduce mesh skewness in some situations (e.g. for a tributary junction; Bradbrook *et al.*, 2001). This is commonly called a block-structured grid (Ferziger and Perić, 1999). Finally, there are guidelines for profiling a mesh to identify locations that are leading to instability or high levels of numerical diffusion (e.g. Bradbrook, 1999). However, even with these, mesh construction for complex geometries is an exceptionally time-consuming process.

It is clear from the above that mesh definition for structured grids is a major issue. The two main alternatives available at present are based upon either unstructured grids or porosity treatments. Unstructured grids have a triangular, rather than quadrilateral, footprint which makes them easier to fit to an arbitrary boundary. This provides considerable flexibility for the study of reach-scale flows (Ferziger and Perić, 1999) and a direct link to finite element methods, although finite volume solutions can be developed for unstructured grids. Finite element solution methods are similar to finite volume methods in that they use volumes, but these are subject to weights that provide continuity across element boundaries. Finite volumes have proved popular in some applications to reach-scale river flows, especially very large scale floodplain flows solved using depth-averaged methods (e.g. Wan *et al.*, 2002; Caleffi *et al.*, 2003) and also in a large number of 3D applications (e.g. Apsley and Hu, 2003; Rameshwaran and Naden, 2003; Wilson *et al.*, 2003). Their main problem is that they require much more effort in relation to grid generation.

A simpler solution, and one of the most promising areas of current development, has yet to be applied at the reach scale. This uses structured schemes and represents bed geometry through blocking out cells using a numerical porosity term (Olsen and Stoksteth, 1995). Lane and Hardy (2002) and Lane *et al.* (2002) report on its application to high-resolution gravel topography and Lane *et al.* (2004) demonstrate a significant improvement in model predictive ability in quantitative terms when this approach is adopted. Use of the structured mesh facilitates proper grid-independence testing. It also permits the geometry to be changed without remeshing, thus allowing for bed erosion or deposition and for sensitivity analysis of how flow predictions depend on topographic representation. The extension of this method to the reach scale is something that would be particularly valuable, though it will have to await considerable increases in computing power if the spatial resolution is to remain as fine as in the cited pioneering studies.

10.3.3 Boundary conditions: Bed roughness

The third issue that merits consideration is the specification of boundary conditions. These need to be specified at the bed and (in a 3D model) at the free surface, as well as at the inlet(s) and outlet(s) within the model. The issue of how to specify the boundary condition on a river bed (and along its banks) is probably the most challenging of all boundary condition issues. A distinction needs to be made between 2D and 3D modelling as the meaning of the roughness parameter depends upon whether or not the 3D equations have been depth-averaged. In the case of a 3D model, the roughness parameter contributes directly to only the bottom grid cell, commonly as a control upon the elevation at which the velocity becomes zero within the grid cell (i.e. a roughness height). It has an explicit physical meaning and, at least in theory, is a measurable parameter. However, in two dimensions, the case is somewhat different. As the 3D equations are averaged over the flow depth, the bottom shear stress appears in the momentum equations explicitly, as a source term: i.e. it affects every cell, rather than being a condition that it affects only boundary cells. It is commonly assumed that the shear stress can be expressed as a square law of the depth-averaged velocity (U) using:

$$\tau_0 = k\rho U^2 \tag{10.9}$$

The parameter k is then expressed in terms of a roughness parameter, such as Chezy's C or Manning's n. For uniform flow it is readily found that equation (10.9) is equivalent to the Darcy-Weisbach equation (equation (10.5)) with $k = f/8$. Similarly, comparison with the Chezy and Manning equations gives:

$$k = \frac{g}{C^2} \tag{10.10}$$

$$k = \frac{gn^2}{h^{1/3}} \tag{10.11}$$

C (or f or n) can thus be regarded as an empirical parameter that determines the effective roughness required in order to get the correct relationship between shear stress and velocity. This has been widely studied in 1D models where these friction parameters are commonly back-calculated from measured hydraulic properties, estimated using prior experience or calibrated. Analogy with the 1D case is useful: in 1D models, roughness parameterisation is required in order to get the channel conveyance correct; in 2D models, it is required in order to get the depth-averaged velocity and local water depth correct. In both cases it is an *effective* roughness parameter that may show little resemblance to measured values of variables (e.g. bed grain size) that might in part control that roughness. Getting the roughness right is important in a depth-averaged morphodynamic model, since sediment transport and channel change will be computed using shear stress and equation (10.9) shows that results will depend on the specification of roughness. This is well illustrated by consideration of equations (10.9) and (10.11). The first-order uncertainty (σ_p) of parameter p due to uncertainty in variables q and r is given by:

$$\sigma_p = \sqrt{\left(\frac{\partial p}{\partial q}\sigma_q\right)^2 + \left(\frac{\partial p}{\partial r}\sigma_r\right)^2 + 2\frac{\partial p}{\partial q}\frac{\partial p}{\partial r}\sigma_{qr}} \tag{10.12}$$

Given that $|\sigma_{xy}| \leq \sigma_x \sigma_y$, we can write:

$$\sqrt{\left(\frac{\partial p}{\partial q}\sigma_q\right)^2 + \left(\frac{\partial p}{\partial r}\sigma_r\right)^2} \leq \sigma_p \leq \left|\frac{\partial p}{\partial q}\right|\sigma_q + \left|\frac{\partial p}{\partial r}\right|\sigma_r \tag{10.13}$$

Application of equation (10.9), with k set using equations (10.11)–(10.13) gives:

$$\sigma_{\tau_0} \leq \frac{2g\rho n\bar{U}}{h^{1/3}}(\bar{U}\sigma_n + n\sigma_U) \qquad (10.14)$$

Thus, the relative contribution of n and U to $\sigma_{\tau 0}$ is:

$$\frac{U\sigma_n}{n\sigma_U} \qquad (10.15)$$

Over a non-uniform river reach U is likely to have far greater spatial variability than n, so will be the dominant control over spatial variation in shear stress as defined by equation (10.9). Equation (10.14) shows that uncertainty in n feeds through to uncertainty in τ, both directly and via the dependence of estimated U on assumed n.

A more physically based alternative to the use of the traditional 1D friction factors to estimate k is to assume that the law of the wall holds throughout the full flow depth. The law of the wall is given by:

$$U = \frac{1}{\kappa}\sqrt{\frac{\tau_0}{\rho}}\ln\left(\frac{z}{z_0}\right) \qquad (10.16)$$

where κ is von Karman's constant; z is height above the bed; and z_0 is the height above the bed at which the velocity becomes zero. The depth-averaged value of velocity (U) is given by integration of equation (10.16) across the water depth (h) as:

$$U = \frac{1}{\kappa}\sqrt{\frac{\tau_0}{\rho}}\ln\left(\frac{h}{e \cdot z_0}\right) \qquad (10.17)$$

Thus:

$$\tau_0 = \left(\frac{\kappa}{\ln\left(\frac{h}{e \cdot z_0}\right)}\right)^2 \rho\bar{U}^2 \equiv k\rho\bar{U}^2 \qquad (10.18)$$

This gives the classic square law resistance velocity relationship assumed above. The derivation of equation (10.16) (e.g. Richards, 1982) is based upon the Boussinesq approximation in which shear stress is assumed to be proportional to a local strain rate (the vertical variation of velocity with elevation above the bed) and a constant of proportionality (the eddy viscosity). Research by Prandtl (1952) showed that the eddy viscosity is itself proportional to the local strain rate and a mixing length term related to Von Karman's constant and distance above the bed. This leads to equation (10.16), without any assumption of the relationship between shear stress and velocity, and is an important justification for the form of equation (10.9). Equation (10.17) also allows parameters like C (or n or f) to be expressed in terms of bed roughness height, which has a clearer physical meaning as the elevation above the bed at which the velocity becomes zero:

$$C = \frac{\sqrt{g}}{\kappa} \ln\left(\frac{h}{e \cdot z_0}\right) \qquad (10.19\text{a})$$

$$n = \frac{\kappa h^{1/6}}{\sqrt{g} \ln\left(\frac{h}{e \cdot z_0}\right)} \qquad (10.19\text{b})$$

Equations (10.19a) and (10.19b) have an important property: they demonstrate that the roughness parameters C, n and f depend upon local flow depth (h) as well as roughness height (z_0). That is, flow resistance depends on relative submergence (h/z_0), which can vary spatially and over time. The effect is small in most rivers but becomes significant for h/z_0 below about 0.01, i.e. in steep mountain channels or for riffles and bar tops in gravel-bed rivers. In the extreme case of boulder-bed streams with step-pool bed topography, n increases by an order of magnitude as depth decreases over a given bed (e.g. Lee and Ferguson, 2002). In shallow flows, therefore, the assumption of constant C (or n or f) may give an inaccurate representation of bottom stress terms in some parts of the domain. Lane and Richards (1998) used linear parameter perturbation to estimate the uncertainty in predicted depth-averaged velocities that might arise from uncertainties in n estimation. They specified the uncertainty in n as the standard deviation of n values estimated from distributed grid-by-number grain-size counts applied to a Strickler-type equation. This was applied to a spatially distributed estimate of parameter sensitivity to uncertainty. The uncertainties in velocity estimates were of a similar magnitude to that in the velocity estimates themselves, thus demonstrating the effects of uncertainty in n upon uncertainty in velocity estimates, which will propagate through to shear stress estimates. Some of the effects of n (and C or f) uncertainty can be reduced by specifying a roughness height, and applying it to relationships like those in equation (10.19) or directly into equation (10.18).

The question that follows is how to determine z_0. Following Nikuradse's 1930s work on roughened pipes, z_0 is equated with $k_s/30$ where the 'roughness height' k_s is associated in turn with a typical grain diameter of the bed. The problem is the definition of a 'typical' grain diameter for a poorly sorted mixture as commonly found in gravel-bed rivers. Back-calculation from stream-gauging data suggests k_s cannot be equated with D_{50} or D_{65} (as Nikuradse found) but is a few times D_{84} or several times D_{50} (e.g. Hey, 1979). This implies z_0 is around $0.1 D_{84}$, and this has been confirmed by researchers fitting the law of the wall to measured velocity profiles (Whiting and Dietrich, 1991; Ferguson and Ashworth, 1992). The need for k_s to be a multiple of even a coarse-tail grain size is thought to reflect the contribution of small-scale form drag on protruding clasts and pebble clusters, and a physically based calculation by Wiberg and Smith (1991) gave good agreement with the empirically determined multiplier. This is the sense in which the roughness height is an *effective* roughness parameter that results in the correct variation of vertical velocity with elevation above the river bed. The main problem to emphasise here is that this approach does assume that the law of the wall extends throughout the flow depth. Further, additional problems with the law of the wall also hold (e.g. in its derivation, it makes a zero-order turbulence closure in which the length scale is assumed to be a function of depth above the bed).

This issue of how to specify z_0 or k_s is equally relevant to 3D modelling. As noted earlier, the roughness boundary condition applies only to the bed- (or bank-) adjacent cells where it is common to apply some form of the law of the wall. Strictly, this is assumed to apply only within a wall region where shear within the fluid can be assumed to be the same as shear at the bed. The criterion

$$30 < z^+ < 300 \tag{10.20}$$

must be satisfied, where

$$z^+ = \frac{\sqrt{\frac{\tau_b}{\rho}}}{\nu} z_b \tag{10.21}$$

ν is the kinematic viscosity, and z_b is the distance of the centre of the wall-adjacent cell from the wall. In practice, equation (10.17) can be modified as it assumes local equilibrium of turbulence, which is violated under conditions of flow separation. Thus, Launder and Spalding (1974) recommend a non-equilibrium law of the wall in which shear velocity is replaced by the square root of the turbulent kinetic energy per unit mass (k) as the characteristic velocity scale:

$$\frac{u\sqrt{k}}{u_*^2} = \frac{\ln\left(0.25 c'_\mu \frac{\sqrt{k}}{u_*} \frac{z}{z_0}\right)}{0.25 c'_\mu \kappa} \tag{10.22}$$

where c'_μ is an empirically derived coefficient equal to 0.09 (Launder and Spalding, 1974). This sort of treatment aside, more of the fluid dynamics are represented explicitly in a 3D solution (e.g. through shear within a column of water), and model predictions are much less sensitive to roughness height specification than their 2D counterparts (Lane et al., 1999). This makes the approach less sensitive to specification of the exact form of the law of the wall. Indeed, where roughness is used as a calibration parameter in 3D models, it may have to be elevated to artificially high values before it has any effect upon model predictions.

In theory, specification of the roughness height ought to be more straightforward than in the 2D case: under the restriction upon cell thickness imposed by equation (10.21), the roughness height ought to relate to the topographic variability within the wall-adjacent cell. Thus, the specification of z_0 or k_s in terms of a grain diameter needs to reflect not just skin friction but also microtopography, that is sub-grid scale, with respect to the computational grid. This is similar to the 2D case in that some multiplier of roughness will be required. In theory, this could be downwards for very fine meshes, but the resolution of the topographic data used to fit the channel boundary to the mesh is often coarser than the grid itself, which therefore tends to be unrealistically smooth. Thus, roughness heights are commonly multiplied upwards, and ought to be both spatially variable and scale-dependent, in relation to both mesh resolution and the topographic content of the dataset used to describe the surface. As with the 2D case, roughness in a 3D model normally involves an effective roughness parameter. The roughness value required will depend not only upon local grain size and/or local topographic variability but also upon: (1) the scales of microtopographic variability contained within the topographic data; and (2) the way in which the mesh is used to represent the surface topography. Thus, field estimates of roughness height are not necessarily going to give the values of roughness height required in either a 2D or 3D model.

Three other issues need mentioning in relation to the 3D case. First, some studies have noted problems in terms of numerical stability and solution accuracy for flows characterised by high relative roughness (e.g. Nicholas and Smith, 1999) with these difficulties resulting from the existence of an upper limit of k_s for a given near-bed cell thickness (Nicholas, 2001). The implication of this is that the thickness of the near-bed cell limits the maximum shear velocity at the bed, so that near-bed velocities may be over-predicted in field situations that involve high relative roughness (Nicholas, 2001). Second, upscaling of the roughness height represents an increase in part of the sink within the momentum equations. It does not address how to set the reference height in the bed in the numerical mesh, which has a direct implication for the mass conservation equations. Lane *et al.* (2004) show that the effect of heterogeneous gravels upon the vertical variation of velocity is such that there needs to be an upward displacement in the numerical mesh in order to get the elevation of maximum shear and hence the downstream flux correctly predicted. A good example of the use of this is Wu *et al.* (2000) who specified their first mesh point outside the viscous sub-layer and above the bed roughness elements, and then upscaled k_s by a suitable amount. This approach will not represent the effects of spatial variability in flow components, especially near-bed cross-stream and vertical flow components. If these matter, there may be a logical limit to the use of roughness height as a means of parameterising sub-grid-scale topographic variability in CFD models of gravel-bed rivers (Lane *et al.*, 2004). Nicholas (Chapter 13) identifies a range of issues and methods that might be used to address this problem.

Third, in relatively deep narrow reaches, typical of small streams and those which are incising, the specification of bank roughness becomes important. This can be even more problematic than bed roughness, especially if the banks are vegetated. Wilson *et al.* (Chapter 15) discuss ways to represent vegetation effects. In the 3D reach-scale models already referred to in this chapter, bank resistance was commonly represented by the log law in cells touching the banks. Some of the studies used the same k_s value for banks and bed, but others found it necessary to use a higher value for the banks (much higher in the case of Booker *et al.* (2001), whose study reach has large woody debris along parts of the banks). This is the ultimate extreme of the problem of roughness parameterisation where highly inflated roughness heights may be required to achieve adequate process representation. Here, specific attention must be paid to the validity of the boundary condition treatments being used and it is quite probable that alternative approaches (e.g. Lane and Hardy, 2002) for parameterising vegetative resistance may be required if process representation is to be reliable.

Roughness merits one final point, relevant to both 2D and 3D applications. Correct estimation of the shear at the bed and the banks of a river is crucial if these models are to be used for sediment transport estimation. The strong interaction between roughness parameters and/or roughness heights and bed shear stress is implicit in the relationships described above and roughness values may exert a strong control on bed shear stress predictions. In comparing a 2D and 3D model of a river channel junction, Lane *et al.* (1999) found large differences between both the magnitudes and patterns of shear stress estimates which they attributed to the strong depth dependence of shear stress estimates in a 2D model as well as poor representation of the effects of vertical flow motions upon momentum transfer both towards and away from the river bed. The sensitivity (magnitude and patterns) of 2D models to changes in roughness parameters makes research into the representation of bed roughness in rivers of major importance, especially given the finding of Lane and Richards (1998) that sensitivity to roughness was greater than sensitivity to either turbulence or secondary circulation representation in a depth-averaged study of a river bifurcation.

10.3.4 Boundary conditions: Free surface and inlet

Ingham and Ma (Chapter 2) review the range of methods that might be adopted for the surface treatment. The key issue here is that a range of research into reach-scale river flows has shown that the spatial variation in water surface elevation is a key process driver (e.g. Dietrich and Smith, 1983; Rhoads and Kenworthy, 1995, 1998). Thus, surface treatments that allow for the representation of free surface effects are crucial. Ingham and Ma address both of the two main approaches that have been used in reach-scale applications to date: the porosity correction method, as used by Lane *et al.* (1999), Bradbrook *et al.* (2000a), and Ferguson *et al.* (2003); and the volume of fluid method as used by Ma *et al.* (2001). Both of these allow prediction of the probable water surface elevation map rather than its *a priori* specification, although there are now a range of reports of field measurements of water surface elevation at the reach scale (e.g. Chandler *et al.*, 1996; Biron *et al.*, 2002). This opens up the possibility of model validation with reference to water surface predictions (see also Chapter 9).

It is worth commenting here that the porosity correction method represents a solution to the problems of rigid lid treatments in 2D and 3D CFD applications, in which representation of surface pressure gradients (i.e. non-zero pressure terms acting on the lid) can be achieved through introducing additional source terms into the momentum equation (e.g. Bradbrook *et al.*, 2000a), but this leads to mass conservation errors close to the surface (regions of negative surface pressure where flow velocities are slower than they should be and positive surface pressure where velocities are higher than they should be). This is a similar effect to the situation that arises at the bed if the reference height is not set correctly. These methods are especially suited to steady flow. However, if the flow is unsteady, there is the possibility of change in both wetted area and stage, such that the volume occupied by the mesh needs to evolve. In three dimension, this can be an especially complex problem to address. A possible solution is to use an unsteady depth-averaged model to define the water surface elevation for use in a fully 3D solution (Wu *et al.*, 2000).

10.3.5 Extension to unsteady flows, sediment transport and channel change

The extension of 2D and 3D models to include unsteady flows, sediment transport and morphodynamics is an obvious challenge and has been the subject of recent research by several academic groups, engineering consultancies and national agencies.

We group these topics here because morphodynamic modelling introduces several of the same technical problems as unsteady flow modelling, as well as some different ones.

Dealing with gradually varied unsteady flow in a 2D model is straightforward, and proprietary hydrodynamic codes have long allowed users to specify inflow hydrographs. Bates *et al.* (Chapter 11) cover many of the issues relating to the treatment of unsteady flows in relation to floodplain investigation. In 3D CFD, solving the flow at a steady discharge is so slow with currently available computer power that modelling how the flow changes during a hydrograph is scarcely a practical proposition, other than by repeating a steady-flow simulation at different discharges. There have been few, if any, assessments of the effects of changes in discharge upon instream flow structures within a 3D CFD framework, although there are several for the 2D case (e.g. Heniche *et al.*, 2002; Caleffi *et al.*, 2003; Horritt, 2004). This may relate to a difficulty shared with studies of wetting and drying in floodplains: severe numerical instabilities can result as the volume of the domain evolves in response to changes in discharge, unless sophisticated treatments are adopted at the wetting front. In a 3D solution, this will involve lateral changes but, and more importantly for most instream flows, changes in the vertical.

Extension to sediment transport creates similar problems. Proprietary 2D codes have included sediment transport for several years but the extension to three dimension is still at the research frontier. The complexity of the problem from a computational point of view depends on whether modelling is restricted to sediment transport, includes bed erosion/deposition (necessitating remeshing vertically in a 3D approach) or includes both bed and bank change (necessitating remeshing horizontally in both 2D and 3D). Other issues that have to be tackled in both 2D and 3D codes include: (1) representation of heterogeneous bed material including sediment-sorting processes; (2) particle-size-dependent divergence between flow and sediment transport vectors; (3) a dynamic treatment of water surface changes in response to bed deformation, including wetting and drying due to both stage change and bank deformation; (4) representation of not just bed deformation, but also deforming banks; (5) the possible need for non-equilibrium sediment transport treatments (Guo and Jin, 1999); (6) specification of the correct sediment transport function (Vanoni, 1984; Guo and Jin, 1999); and (7) representation of the effects of turbulent transport upon sediment diffusion (Guo and Jin, 1999). The first four issues become especially notable as treatment moves beyond a width-averaged (1D) approach and these are addressed below.

Extending the 1D overall and fractional Exner equations (10.7) and (10.8), to two dimension requires only generalisation of the sediment divergence terms. The 2D forms of the equations can be solved by explicit or implicit methods. In implicit form, solution is based upon a time-marching solution initialised with the assumption that the GSD, active layer depth and sediment discharge should not change within a model time step. Provided that the GSD does evolve, and the solution is perpetuated with the same grain size for $t + dt$, the process converges when the predicted GSD is unchanged in subsequent computational time steps. As this is computationally expensive, the alternative is an explicit solution, which is implicitly stable but which assumes that within a computational time step changes in GSD and active layer depth do not have a significant effect upon sediment transport. This is essentially a first-order approximation of the Exner equation and so requires time steps to be sufficiently small. Application of the Exner equation involves GSD specification for a number of subsurface layers beneath the current active layer. Typically, these have to be dynamic which can lead to spurious vertical migration of sediment as the active layer is defined, if there are sudden discontinuities in sediment characteristics in the vertical. This can be avoided by making sure that the redefinition of the active layer involves small changes, which is also achieved by making sure that the numerical time step is not too great.

Application of the Exner equation provides a basic method for sediment accounting in terms of exchange between storage in the bed and transport. However, extension to two dimensions and three dimensions creates additional analytical issues. In a 1D model, sediment and water are assumed to move in one and the same direction but in 2D/3D this is not necessarily the case. Suspended material can be assumed to move in the direction of water flow in a 2D model (in a 3D model this may depend on height within the flow), but bedload is affected by any local bed slope transverse to the near-bed flow direction. The effect is stronger for coarser particles (the gravity force scales with D^3 but the fluid drag force only with D^2) so coarse particles tend to concentrate along thalwegs. Thus, divergence between sediment transport and flow direction must be incorporated into the model (e.g. Nelson and Smith, 1989a; Chapter 4 in this book).

In a 1D model, there is always flow over every cell but in a 2D model it is possible for initially submerged cells to emerge, and vice versa, in response either to local bed aggradation/degradation or to stage change caused by unsteady discharge or bed change elsewhere in the reach. A dynamic treatment of wetting and drying is therefore required. Thus, issues associated with modelling floodplain wetting and drying (Chapter 11) become relevant to sediment transport modelling. These problems have primarily been addressed in relation to erosion and deposition processes linked to the need to represent bank erosion. Recognition of lateral variation in flow conditions by going from one to two dimensions opens the possibility of modelling bank erosion as some function of near-bank conditions, but just how to do this is less obvious. It also leads to the need to remesh the grid once significant bank retreat has occurred, and to add a source term to the sediment continuity equation(s). There have been a growing number of attempts to develop reach-scale models of channel change based upon depth-averaged flow models. The simplest of these assume fixed banks. For instance, Kassem and Chaudhury (2002) used a depth-averaged, unsteady flow model, with a zero-order turbulence approximation and body-fitted coordinates and coupled this to the 2D sediment continuity equation. Their model included the divergence between flow direction and bed slope, but used an equilibrium treatment of sediment transport. Application to both 140° and 180° bends allowed simulation of both point bars and pools from an initially flat bed and these compared reasonably well in qualitative terms with previous experimental results. The use of fixed banks makes numerical simulation much more straightforward, especially if there is the potential for formation of complex channel patterns, as this can lead to highly skewed meshes, a particular problem if individual grid cells or elements become especially small, and the potential for numerical diffusion and/or instability. The problem is more acute in fully 3D solutions using structured grids, but it applies to all situations where there is active erosion and deposition of river banks. The result is commonly a simplification of the problem. For instance, Duan *et al.* (2001) address a wide range of meander types through assuming that the meandering channel must maintain a constant width. Their flow model is quasi-3D (i.e. it makes a hydrostatic pressure approximation), which will allow some representation of the effects of secondary circulation. The mesh deformation method was based upon requiring a constant channel width through time. For instance, consider a cross section with end points A and B. Their model determines the erosion rate at either A or B. If there is channel change, and A deposits (or erodes), then B will erode (or deposit) so as to maintain channel width. Within a time step, they then connect the centrelines of new sections. Finally, they moved cross sections so they remain equally spaced and orthogonal to channel centreline.

Nagata *et al.* (2000) go one step further by allowing for deforming banks and, most importantly, including a non-equilibrium sediment transport relation. The latter is especially important as bank erosion occurs intermittently at points when the bank collapses and the length scale of particle movement in the downstream direction is much greater than the rate of change of bottom topography close to the bank. Thus, the sediment transport treatment is based upon explicit treatment of sediment sources (a bank erosion failure algorithm and a sediment entrainment rule) coupled to a sediment routing algorithm, after entrainment, that explicitly recognises local geometric, inertial and flow field forces, and which is combined with a probabilistic sediment deposition rule. These process relationships are then time-integrated for a given time step using the mass conservation equation for each grid cell. The result is a Lagrangian-type treatment of sediment transport, in which entrainment at one point is coupled to determination of the trajectory of entrained material. This is applied for all locations within the model and then integrated across space to determine the net erosion and net deposition for each grid cell. They achieved an encouraging (qualitative) agreement between model predictions and corresponding laboratory experimental results.

As noted in section 10.3.1, a major issue for the effective representation of river channel processes is model dimensionality. The dynamics of erosion and deposition in river channels are commonly linked to issues of secondary circulation. For example, a classical explanation of scour in the outer part of meander bends is fluid downwelling which serves to transfer momentum towards the bed, and so encourage sediment transport and bed scour (e.g. Bathurst *et al.*, 1979). It is therefore likely that a full 3D treatment (e.g. Wu *et al.*, 2000; Olsen, 2003) may be necessary to achieve sufficient process representation, although the extent to which this is the case has yet to be demonstrated. Wu *et al.* (2000) used a full 3D model, with a basic two-equation turbulence model, to explore sediment transport in a 180° bend with a rectangular cross section. One of the major causes of instability in 3D solutions is the position of the free surface (as said earlier), which needs to evolve in response to sediment entrainment and deposition, as well as planform change. Thus, Wu *et al.* used a depth-averaged model (strictly speaking, valid only for gradually varied flow) to get a first approximation of the water surface for the 3D solution. The model dealt with the issue of how to represent bed roughness by specifying the first point above the bed outside the viscous sub-layer and above the bed roughness elements. The model included both a suspended sediment and a bedload treatment based upon a two-layer approach with dynamic exchange between them, and between them and the bed through both entrainment and deposition. Unlike the Nagata *et al.* (2000) treatment, the model assumes bedload is transported in the direction of the bed shear stress and no account is taken of either inertial or geometric forcing. This is clearly a limitation as research has shown (e.g. Dietrich and Smith, 1984) that divergence between the shear stress and sediment transport vectors is important in curved channels or where there is substantial topographic variability. Bed erosion and deposition was allowed through an adaptive curvilinear grid, but the channel was assumed to have fixed sidewalls. Wu *et al.* found a good qualitative agreement with experimental data for the same geometry (Odgaard and Bergs, 1988) and, as with other models, reach-scale geometry was reproduced quite well but not detailed surface variations.

10.4 Example applications

The aim of this section is to demonstrate how the application of CFD to reach-scale flow problems is enhancing our understanding of the physics, geomorphology and ecology of rivers at the reach scale. We address five aspects of reach-scale river behaviour. The first three relate to fundamental reach-scale 'river units': riffles and pools, meanders and confluences. The last two aspects address the dynamics of river behaviour, in relation to channel change, and links between reach-scale fluvial processes and habitat.

10.4.1 Riffles and pools

Most gravel-bed rivers contain pools at a spacing of a few to several channel widths, separated by bars over which the low-discharge flow is relatively shallow, steep and fast: a 'riffle'. Textbook diagrams tend to portray pools and riffles as 1D undulations in the long profile but the flow is usually two dimensional since bar fronts tend to be strongly oblique, facing left and right alternately. Riffles and pools are of interest to geomorphologists because of debates as to why they form and why they remain (sometimes with very little change) for long periods of time. The common theory invoked to explain their persistence is based upon the confusingly named 'velocity-reversal' hypothesis, in which as discharge rises, depth, surface slope and velocity increase faster in pools than riffles until there is a convergence or even reversal of the hydraulic inequalities between them, favourable to deposition on the riffles and scour in the pools (Keller, 1971). Connected to this is a strong feedback between riffle–pool development and local topographic forcing, where the riffle creates divergent flow which encourages particle deposition and the pool encourages convergent flow and hence particle erosion (Keller, 1972). Clifford (1992) proposed a mechanism for the initiation of riffle–pool sequences based upon the constraints imposed by river channel width upon the cross-stream extent, and hence the downstream length, of eddies. Once appropriate irregularities in the bed form at the small scale, the feedback identified above creates the possibility of positive feedback and riffle–pool growth. Riffle–pool systems may then stop increasing their total relief for a number of reasons, including the initiation of bank erosion and hence meandering, as well as maximum pool slopes above which particles can be evacuated from the pool.

Hydraulic modelling is an obvious line of approach to these questions. One-dimensional modelling has been reported by Richards (1978), Keller and Florsheim (1993), and Carling and Wood (1994); 2D by Miller (1994), Thompson *et al.* (1998), and Cao *et al.* (2003); and 3D by Booker *et al.* (2001) and Ma *et al.* (2001). The two 3D applications used similar CFD codes for this purpose (i.e. finite volume solution of the 3D dimensional Navier–Stokes equations), but applied to very different rivers. Booker *et al.* consider a sinuous channel with extensive and complex bank vegetation and a high relative relief (e.g. within cross-section variability approximately 50% of maximum water depth), but where the riffle–pool sequences are at a closer spacing than the channel inflexion points. Ma *et al.* consider an artificially straightened channel, with very low relative relief riffle–pool sequences (about 10% of bankfull flow depth). Booker *et al.* make some very important findings in relation to riffle–pool behaviour, primarily because of the high resolution of 3D data that their modelling approach can provide, as compared with field studies. First, they obtained evidence that supports the hypothesis of velocity reversal. Whilst depth increased over the riffle with increasing discharge, velocity fell slightly, resulting in a slight reduction of shear at the bed. In the pool, both velocity and shear stress increased with discharge. The effect of this was to reduce the spatial variability of near-bed hydraulics as discharge increased. This is a logical observation and reflects the progressive drowning out of topography at higher flows such that bed topography (or topographic forcing) reduces in importance as the topography is progressively submerged. This ties in with the observations of Ma *et al.* (2001) that in their study of extremely low relative relief riffle–pools, there was negligible topographic forcing of the bed and the flow was strongly aligned with the streamwise direction. Second, the different direction of velocity/shear stress change between riffles and pools demonstrates the danger of inferring patterns of bed shear stress from 1D hydraulic models based upon the mean uniform flow shear stress (e.g. Keller and Florsheim, 1993; Carling and Wood, 1994). This has a depth and water surface slope dependence which does not necessarily lead to the right changes in bed shear stress as discharge increases. Second, by being able to study a large number (5 riffles, 4 pools) of riffle–pool sequences simultaneously in the same model, it was possible both: (1) to explore differential behaviour of different riffle–pool couplets; and (2) following on from this, to incorporate explicitly the effects of inherited flow from each upstream pool on the next downstream riffle–pool sequence. This has been shown to be crucial to observed hydraulics. For instance, Thompson *et al.* (1996) describe how a constriction at a pool head results in jet-like flow into the pool itself, which has a major effect upon pool hydraulics (see also Cao *et al.*, 2003). Simultaneous study of a large number of couplets showed that not all riffle–pool sequences were associated with a velocity reversal, so that field measurements in a single riffle–pool sequence might lead to incorrect conclusions about the hypothesis. Of course, it might eventually prove possible to study a large number of riffle–pool sequences in the field, but to do this at a range of discharges would require a seriously complex and long measurement campaign. Thus, the numerical model, uncertainties in its predictions aside, allows a significant increase in process understanding and allows refinement of the types of field questions that need to be asked to confirm model results (e.g. a greater emphasis upon cross-comparison of adjacent riffle–pool sequences to identify under what conditions a velocity reversal occurs). The third important observation was that near-bed flows had an important effect upon maintenance of the riffle–pool sequence. They found that near-bed flow velocities suggested flow routing away from the deepest parts of the pools as the flows pass over the downstream slope of riffles and into pool heads. This would reduce sediment routing into pools, thus maintaining pool depth and not requiring increased erosion during velocity reversal to maintain the pool itself. This is important as Ma *et al.* (2001) found a strong streamwise orientation of bankfull flow, with relatively homogeneous near-bed flow velocities, that do not really support the idea of the velocity reversal which is required for pool maintenance. The Booker *et al.* work suggests that sediment routing at intermediate flows below bankfull, as confirmed by tracer experiments in the field, leads to pool maintenance by steering sediment away from accumulating in the pool.

10.4.2 Meander bends: Flow structures

Natural rivers are seldom straight at the reach scale of tens or hundreds of channel widths, and many turn first one way, then the other in a meandering pattern. A 3D helical motion is induced by the planform curvature of each bend, replacing the opposite-sense helix in the previous bend. The enhanced velocity and shear stress towards the outer bank at and past the apex of each bend causes bank erosion and meander migration unless the river is naturally confined or its banks are artificially protected. As the planform evolves the flow structure alters, but generally in a way which sustains the meandering process.

There have been many laboratory investigations of flow round bends with idealised geometry, and latterly many CFD simulations of such experiments. The first 3D computations, by Leschziner and Rodi (1979) and Demuren and Rodi (1986), were restricted to rectangular cross sections and made simplifying assumptions about the flow. A general finite volume treatment using BFCs was pioneered by Demuren (1993) and has become standard in subsequent work, though increasingly with some kind of free surface treatment (Meselhe and Sotiropolous, 2000). Overbank flow through (and over) a meandering reach has also been modelled and compared with lab measurements (Morvan *et al.*, 2002). The fit to experimental data in these studies (and several others that are not cited) has varied from reasonable to excellent depending on the variables of interest and the complexity of the situation being modelled. Spatial patterns in streamwise velocity, and broad vortex structures, are generally matched quite well but with appreciable errors in the details of turbulence, vertical velocity and bed shear stress patterns.

CFD modelling of *natural* meanders is much less well developed, probably because of their greater complexity and the lack of readily available, detailed and high-quality data sets with which to test models. The greater complexity of natural bends stems from the self-formed nature of their planform and bed topography. Most laboratory experiments have used bends of constant curvature (circular arcs) separated by straight segments; this square wave variation in curvature is used in some channelisation schemes, but natural bends almost always have continuously varying curvature, increasing to the apex of the bend and then decreasing. A sinusoidal variation in curvature gives a reasonable approximation of the planform of many bends, though highly sinuous bends may be asymmetric or have two curvature maxima. Natural meandering is also almost always associated with (and quite probably develops from) an alternate-bar pool–riffle sequence, with pools becoming locked alongside point bars at the bends (Seminara and Tubino, 1989) and riffles at the inflections between successive bends. Some laboratory experiments have used self-formed bed topography, but almost always within a symmetric cross section with rigid (often vertical) walls and often with a low aspect ratio. Curvature-induced helical motion occupies the full width in laboratory bends of symmetric cross section, but often only the outer part of the width in natural bends where flow is steered wholly outwards by the point bar even when it is submerged. As bends grow their maximum curvature also increases, and at radii below about twice the channel width it is common for flow to separate and recirculate, either just past a sharp inner-bank apex or just upstream of a sharp outer-bank apex. This complicates both the flow and the subsequent geomorphic evolution of the river. A final complication relates to boundary roughness: natural rivers have rough banks as well as rough beds, and in meanders there is strong spatial segregation of different-sized bed sediment so that roughness may vary spatially.

As far as we know the first published applications of 3D CFD to a natural meander bend were Hodskinson (1996) and Hodskinson and Ferguson (1998). These researchers were specifically interested in sharp bends with recirculation, and in the potential value of CFD for numerical experimentation on factors controlling the occurrence and extent of recirculation eddies. The computational approach was similar to Demuren (1993) except that the Renormalisation Group (RNG) version of the k-ε turbulence submodel was used in view of the strong shear and anisotropy associated with flow separation. Reasonable agreement was obtained with the limited test data that were obtained. Numerical experiments on outer-bank separation showed that its occurrence, and the size of the recirculation, depended less on centreline curvature than on outer-bank curvature (e.g. Figure 10.1a), the extent to which flow near the inner bank was blocked by a point bar, and the nature of the inflow as determined by the sense and magnitude of curvature in the next bend upstream.

More detailed follow-up work on flow in sharp bends has recently been reported by Ferguson *et al.* (2003) and Ferguson and Parsons (2004), with the focus now on the use of CFD to elucidate the flow structure. These authors reasoned that flow in natural rivers is seldom steady for long enough to build up as detailed a set of measurements as in a laboratory experiment, so it may be advantageous to collect data specifically to set up and validate a model and then use the model output to understand and illustrate the flow structure. This approach was applied to three bends, two with separation at the inner bank and one with separation at the outer bank. Bend geometry was represented by a boundary fitted grid, and the inflow to each bend at normal (i.e. fairly low) stage was measured carefully in order to set up 3D flow models. These were tested using distributed 3D velocity measurements using an Acoustic Doppler Velocimeter (ADV) in two of the bends. Agreement was very good for horizontal velocity components, and fair for vertical velocity, and the limits of recirculation were predicted to within one or two grid cells. The simulated flow fields in all three bends are dominated by a free stream close to one bank, separated by a shear layer from a recirculation eddy extending out from the opposite bank (Figure 10.1b). A simulation of bankfull conditions in one bend showed that the

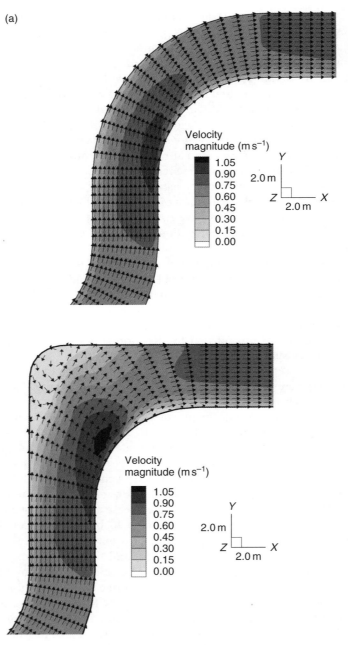

Figure 10.1 Surface velocities in meander bends with recirculation, as simulated by 3D CFD: (a) numerical experiment to show outer-bank separation depends on curvature of outer bank, not centreline; and (b) simulations of natural bends with recirculation at inner (lower map) or outer bank (upper map). Maps in (a) are recalculated after Hodskinson and Ferguson (1998); those in (b) are redrawn from results presented in Ferguson et al. (2003) and Ferguson and Parsons (2004). Maps show velocity magnitude by shading, direction by unit arrows.

(b)

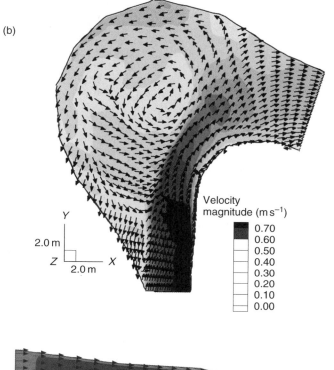

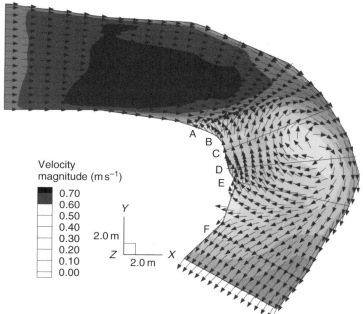

Figure 10.1 (Continued).

recirculation survived but was rather smaller. The lateral compression of the free stream as it passes alongside the recirculation might be expected to strengthen the helical motion induced by bend curvature, but the simulations showed this is more than offset by vertical expansion from inflow riffle to bend pool: in contrast to what is normally found in meander bends, the helical motion decays rapidly and has disappeared by the start of the next bend. In the bends with inner-bank separation the fastest current is close to the outer bank, and downwells there in the usual way, but the strongest attack on the bank is further upstream than usual. In the bend with outer-bank separation the strongest current is close to the inner bank upstream of the apex, and the main downwelling is in the mid-channel shear layer. The ability to simulate such complicated flows holds out promise of a better understanding of the mechanisms of dispersion and also of the implications of flow separation for future meander migration.

10.4.3 Meander bends: Erosion, deposition and bend migration

Whiting and Dietrich (1993) reported a set of flume experiments involving weakly meandering channels with fixed banks but mobile beds. Figure 10.2 shows a series of plots of bar migration through one of these channels, but based upon numerical modelling. It shows a 10° meander bend, with a wavelength of $8w$ where w is the channel width, and a steady discharge of $1.47 \, \text{L s}^{-1}$. The bed material was homogeneous sand with a uniform grain size (0.67 mm). Figure 10.2 shows a characteristic form of behaviour. At $t = 120$ s, the bed has washed out, and there is little relief in the channel. On the true left, there is a small bar in the upstream bend inflection and a small pool just downstream of the bed apex. This is mirrored on the true right of the channel, with a small pool opposite the inflection point bar and a small bar opposite the apex pool. During the next 1200 s, there is progressive growth of both the bars and the pools. This is rapid for the first 600 s, then much slower. Accompanying this bar-form growth, there is steady extension of the true left of the point bar into the bend apex, along the outer bank. The bar becomes progressively more asymmetric, with a bar front that extends further downstream away from the true left bank. This is most notable around 720 s, after which the bar starts to grow more rapidly along the true left bank itself. The true right pool that forms at the upstream inflection point reaches its maximum scour slightly before the maximum height of the bar (at about 1080 s). The front edge of the pool migrates slowly around the inner bank on the true right. The trailing edge of the pool migrates quite rapidly at first, then more slowly. The net result, by 1200 s, is that the maximum elevation of the bar is just upstream of the outer-bank apex, on the true right, and the maximum depth of the pool, in relative terms, is slightly further upstream on the inner bank. After 1200 s, both the bar and the pool start to diminish. This is slow at first but, of note, it is connected to continued migration of the bar front downstream from the outer bank of the meander. As soon as the maximum bar elevation is downstream of the apex (1560 s), the bar and pool relief starts to fall more rapidly until, by 2280 s, the channel is almost a flat bed. This is extremely brief, with rapid formation and extension of a new bar in the true left bank from 2340 s onwards. Figure 10.3a shows the relationship between bedform relief and bar and pool position in more detail. This shows how there is a continual migration of the bar–pool system and associated with the

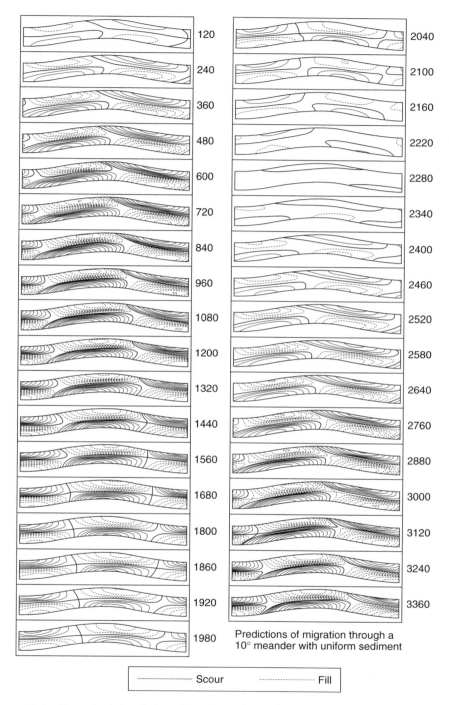

Figure 10.2 Numerical simulation of bar formation and migration in a weakly meandering flume channel with fixed sidewalls and uniform bed sediment.

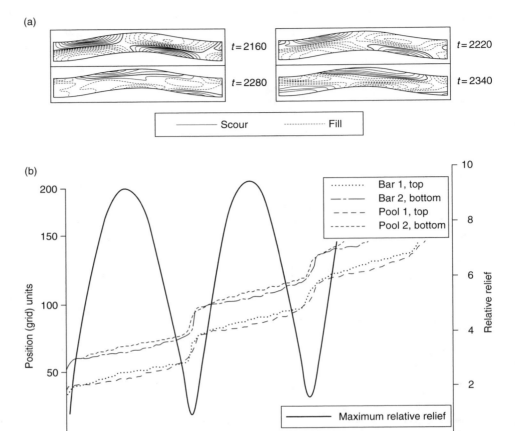

Figure 10.3 Summary data of the simulations in Figure 10.2 to show the relationship between rapid bed migration and bar-form development (a) and more temporal detail of the patterns shown in (b).

growth of maximum relative relief. However, whilst this migration continues at a steady rate, the maximum relative relief starts to fall at the point noted above where the maximum elevation of the bar is just upstream of the outer-bank apex. As the maximum relative relief reaches its lowest values, the pool and the bar both migrate more rapidly. This demonstrates that the migration of bar–pool units is self-regulatory, and is in accordance with the observations of Whiting and Dietrich (1993). If we explore the period from 2160 to 2340 s in more detail (Figure 10.3b), we get a clearer understanding of these dynamics. The bar–pool system is rapidly washed out between 2160 and 2280 s. During this period, the pool begins to migrate around the outer bank, slowly at first, but then very rapidly once the bar relief is at a minimum.

Although the bar is destroyed, the pool is never destroyed, and as soon as it has migrated to downstream of the apex the migration rate slows down and the relief starts to increase once again.

All the bend-flow simulations mentioned so far were of flow in single bends, either in a rigid channel or at a particular time in an evolving natural channel. Until recently this CFD-led strand of research has been quite separate from another, application-led, strand concerned with modelling the evolution of long reaches of meandering river. However, these two areas of research are starting to converge: bend-scale studies have delivered an increased appreciation of the nature of erosion and deposition processes that lead to bend migration, whilst larger-scale studies are starting to benefit from computational and measurement advances that allow a more realistic representation of those processes over many multiples of channel width.

Plausible simulations of long-term meander kinematics have been obtained using 1D models for some time. Commonly, these treat the meandering channel as a zigzag line whose nodes shift laterally at a rate controlled by local and upstream curvature via some implicit representation of coupled 2D features of the flow and bed topography (e.g. Johannesson and Parker, 1989; Howard, 1992; Sun *et al.*, 1996; Lancaster and Bras, 2002). The simplifying assumptions differ slightly between authors, but include depth-averaging of the flow, treating the helical motion as a linearised perturbation of the primary flow, representing cross-sectional asymmetry by a constant transverse slope, and assuming bank erosion is directly proportional to near-bank excess velocity or shear stress. Local variations in bank strength can be represented by altering the constant in this last proportionality. Models of this type can generate characteristic phenomena that have been detected in observational studies: the asymmetry of sinuous bends, the development of new lobes in very long bends, and the distortion of initially regular meander trains when floodplain erodibility is spatially variable because of the traces of old channel fills and oxbows. This suggests that even quite simple models can help us understand the complexity of natural meandering when viewed over extended time and space scales. However, the authors of these models have not claimed that they are operational tools for forecasting in detail the evolution of specific bends. For this to become possible, the simplifying assumptions need to be tested and, if unsatisfactory, replaced by more sophisticated alternatives. The development of 2D bend models by other researchers is starting to permit this testing and refinement. In relation to meander migration, adequate representation of bank erosion remains the primary challenge. It is the subject of much ongoing research as regards how the process is modelled (e.g. Darby *et al.*, 2002), its possible intermittency even in steady-flow conditions (e.g. Nagata *et al.*, 2000), and the treatment of sediment continuity and source terms at the foot of the bank (Nagata *et al.*, 2000; Duan *et al.*, 2001; Darby *et al.*, 2002). All three of these papers obtained encouraging results in comparisons with laboratory data from curved channels with noncohesive banks, though as previously noted the model of Duan *et al.* (2001) is restricted to the case where channel width does not alter as one bank erodes. Darby *et al.* (2002) also simulated a natural gravel-bed meandering reach (Goodwin Creek in the southern US); aspects of the results were again encouraging but pool depths were underestimated and consequently so was bank retreat.

This brief survey makes it clear that whilst there have been some exciting developments in modelling flow in meander bends and its consequences for sediment transport and channel change, a great deal remains to be done. Alternative sub-grid-scale parameterisations (of turbulence and boundary roughness) in 3D flow models have not been thoroughly investigated, and nor have the endless combinations of flow, transport and bank-erosion sub-models that are possible in morphodynamic modelling. There will doubtless be a debate between those who see a full 3D CFD treatment of the flow as the only secure foundation for modelling sediment transport and channel change, and those who see a 2D flow treatment as adequate for these purposes given the approximations that are involved once erosion and transport are tackled. But what is also clear is the huge potential for better understanding of meandering rivers through use of CFD techniques.

10.4.4 Confluences

Open channel confluences (also known as tributary junctions or river junctions) have represented a major challenge for 3D applications of CFD. They must include (Bradbrook *et al.*, 2000a): (1) reach-scale pressure gradient forces associated with realignment of one or both channels into the post-confluence channel, commonly accompanied by a change in width–depth ratio; (2) topographic steering associated with the complex morphologies commonly observed at tributary junctions (e.g. Rhoads and Kenworthy, 1995, 1998; Rhoads, 1996; Rhoads and Sukhodolov, 2001), which may include a deep scour hole (e.g. Best and Ashworth, 1997) and point bars attached to one or both of the banks of the post-confluence channel; (3) zones of flow separation, both in planform and over the lee of avalanche faces where there is tributary discordance (e.g. Biron *et al.*, 1996); and (4) strong shear, both between the tributary flows themselves (e.g. Biron *et al.*, 1996) and between zones of planform flow separation and the tributaries. The result will be flow structures that are strongly three dimensional, highly unsteady and which will have important implications for mixing processes and the transport of both fine and coarse sediment. In order to represent these flow processes, the numerical model adopted will need to include a number of processes. First, a full solution of the 3D Navier–Stokes equations will be required as it is possible that the flow is not hydrostatic (precluding simplification of the z momentum equation to $p = -\rho g$) and probable that existing treatments of secondary circulation in depth-averaged models (e.g. Lane and Richards, 1998) are unlikely to represent the rapid evolution of flow structures associated with streamline curvature and topographic forcing at the bed. Second, it is possible that there are quite considerable lateral and downstream gradients of water surface elevation (e.g. Biron *et al.*, 2002). Thus, the model must be capable of representing spatial variation in free surface elevation either explicitly or implicitly through a rigid lid approximation combined with appropriate correction of the mass conservation equations. Third, the potential complexity of planform and bed geometry will require very careful mesh design and close attention to be given to grid independence. With present technology, this almost certainly means use of BFCs as well as close consideration of the extent to which the generated meshes result in significant numerical diffusion. Fourth, the representation of both flow separation and turbulence will mean that the basic two-equation k-ε turbulence model is unlikely to predict Reynolds-averaged flow properties correctly. Finally, the extent to which prediction of Reynolds-averaged flow properties is acceptable needs careful consideration. There is growing realisation that the predictions of flux (e.g. of neutrally buoyant particles) from Reynolds-averaged predictions differ from time-dependent predictions (Lane et al., 2000). It is probable that both large eddy simulation and direct numerical simulation will be required.

Some of the earliest attempts at confluence modelling were undertaken by Weerakoon and Tamai (1989) and Weerakoon et al. (1991). Weerakoon and Tamai (1989) imposed a rigid lid solution, adopted a two-equation k-ε type turbulence model and used a parabolic treatment. The focus of this work was simple (i.e. rectangular and trapezoidal) channels. Use of a parabolic treatment precluded consideration of flow recirculation, but a training levee between the tributary and the main flow made this assumption acceptable. Weerakoon et al. (1991) used a fully elliptic treatment in a study of an asymmetrical confluence with a 60° junction angle between the tributary and the main flow. In a qualitative comparison, they found a reasonable agreement between model predictions and laboratory results, although the predicted recirculation zone length, in the downstream direction, was approximately 30% too short. This was attributed to numerical diffusion associated with grid discretisation, the use of the basic form of the k-ε type turbulence model and the treatment of the water surface as a rigid lid without appropriate correction of the mass conservation equation.

This work was developed by Bradbrook et al. (1998, 2001) who also used idealised geometries comprising channels of rectangular section to study the flow structures associated with asymmetrical junctions. This work extended that of Weerakoon et al. (1991) through: (1) use of a modified form of the k-ε type turbulence model to deal with situations with strong strain; (2) a porosity-based mass conservation correction coupled with a rigid lid treatment to represent the free surface (Bradbrook et al., 2000a); and (3) extensive tests on mesh dependence and sensitivity to mesh design, to assess the extent to which the solution had significant numerical diffusion. As with some of the meander work described above, the turbulence model modification was based upon RNG theory in which an extra term is introduced to represent the way in which high mean strain rates can increase turbulent dissipation. In the k-ε formulation, increasing turbulent dissipation results in a local decrease in eddy viscosity and hence the rate of extraction of momentum from the mean flow. This has the effect of increasing the size of computed recirculation zones.

The main focus of the Bradbrook et al. work was an assessment of the interaction between a set of four key controls that will influence the magnitude of streamline curvature and the degree of topographic forcing: the velocity ratio between the two tributaries; the junction angle between the two tributaries; the level of confluence asymmetry; and the degree of discordance between each tributary and the post-confluence channel. The simplest configuration studied (Bradbrook et al., 1998) was a parallel confluence, based on the laboratory study of Best and Roy (1991). This involved a junction angle of 0° and varying degrees of bed discordance between each tributary and the post-confluence channel (Figure 10.4a). The more conventionally studied discharge or momentum ratio was disaggregated into depth and velocity ratios. Negligible secondary circulation resulted with a depth ratio and a velocity ratio of 1. Mixing of flows from the two channels was slow, driven only by the

252 Computational Fluid Dynamics

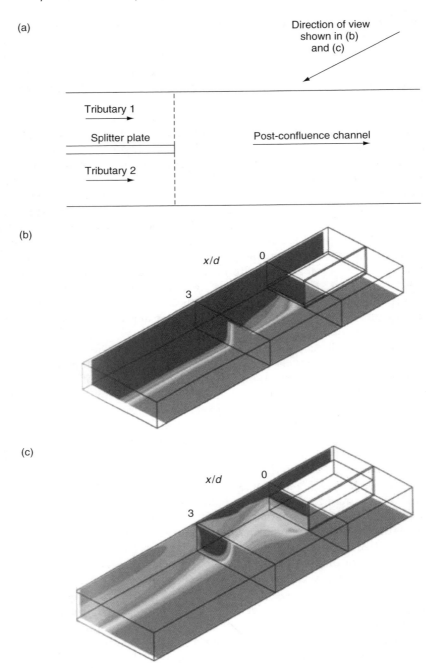

Figure 10.4 *The zero junction angle (or parallel) confluence experiments (a) showing fluid mixing with a negligible depth ratio 0.9 (b) and a larger depth ratio 0.5 (c). In (b) and (c), x/d refers to the distance downstream from the step (x) divided by the step height (d), in the 0.5-depth ratio case.*

low-momentum fluid that formed downstream of the splitter plate due to velocity gradients between each tributary and the splitter plate. Introduction of a small depth ratio (0.9, created by elevating tributary 2 to reduce its flow depth by 10%) resulted in cross-stream transfer of fluid from the deeper channel into the separation zone that formed behind the step (Figure 10.4b), and hence secondary circulation. As the step height increased, this transfer increased (Figure 10.4c). The associated secondary circulations were stronger when the section-averaged velocity in tributary 2 was greater than that in tributary 1. Essentially, this was because increasing the momentum of the shallower tributary increased the separation length downstream from the associated step, resulting in greater cross-stream transfer of flow from the deeper channel. It was shown that depth and velocity ratios interact to control the cross-stream gradient of pressure at the bed and the vertical pressure gradient, the former controlling the magnitude of the cross-stream velocities and the latter the relative depth of flow over which those lateral velocities could exist (Bradbrook et al., 1998). In turn, this determines the strength of both mixing and secondary circulation processes.

The introduction of a junction angle, whilst retaining tributary symmetry (Figure 10.5, symmetrical), resulted in more intense secondary circulation, with the intensity increasing as junction angle increased (Bradbrook, 1999). This was related to the increased hydrostatic pressure at the stagnation point that forms at the upstream junction corner, which was in turn due to water surface super-elevation and was associated with twin back-to-back helical cells (Figure 10.5a, symmetrical). Inertial effects were important, with the surface flow responding more slowly than the bed flow, and this resulted in strong sensitivity to velocity ratio. For example, an asymmetry in secondary circulation may result in a confluence with a symmetrical planform due to velocity ratio effects. Hence, the relative discharges of the two tributaries, especially where discharges change in response to differential sub-basin hydrological response, are important. Extension to the asymmetric case (Figure 10.5, asymmetrical) showed that the zone of water surface super-elevation formed initially in the central confluence zone but then migrated towards the channel sidewall opposite the tributary. The result was translation of twin back-to-back cells into a single cell over a short-length scale, with strong upwelling into the separation zone that formed at the downstream junction corner on the tributary side of the channel (Figure 10.5a, asymmetrical). The additional inclusion of bed discordance (Bradbrook et al., 2001) resulted in much stronger secondary circulation, with fluid drawn into the lee of the step and leading to strong upwelling.

These results could have been obtained using an intensively designed series of scaled laboratory experiments. However, the main appeal of using a numerical model in this situation is that it is possible to understand the reasons behind the predicted flow structures. In particular, it is possible to consider the spatial patterns of deviation from hydrostatic pressure (conventionally known as the dynamic pressure). Figure 10.5d (from Bradbrook et al., 2000a) shows the spatial patterns of dynamic pressure for the symmetrical and asymmetrical cases, and without bed discordance. This shows that even in the absence of bed discordance, there can be strong spatial variation in pressure at the bed. In the symmetrical case, zones of negative dynamic pressure form symmetrically on the downstream junction corners at the bed (Figure 10.5d, symmetric) and correspond to zones of cross-stream fluid

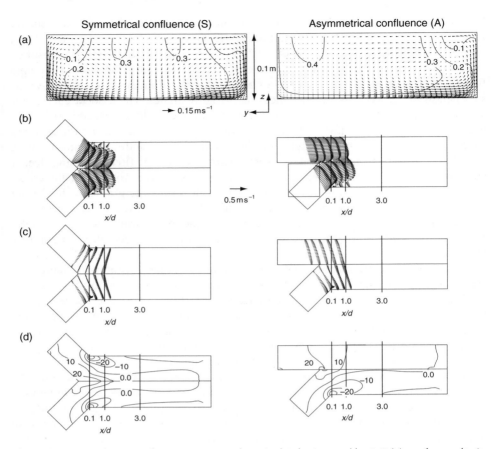

Figure 10.5 Predictions of cross-stream and vertical velocity at x/d = 3.0 (a); surface velocity (b); bed velocity (c); and bed pressure expressed as deviation from hydrostatic (d). Reproduced from Bradbrook et al. (2000a). The symmetrical confluence involves a 90° junction angle between the tributaries and the asymmetrical confluence a 45° junction angle between one tributary and the post-confluence channel.

entrainment, strong fluid upwelling (Figure 10.5a, symmetric) and strong flow separation at the surface (Figure 10.5c, symmetric). In the asymmetric case, the zone of negative dynamic pressure forms only on the true right of the channel. Bradbrook et al. (2001) showed that introduction of discordance resulted in a significant increase in the absolute magnitude of the negative dynamic pressure, which moved to below the middle of the tributary entry point and extended much further downstream. This explains the intense secondary circulation observed in field confluences in the presence of asymmetry (Biron et al., 1996). The numerical results showed that, in the discordant case, the magnitude of water surface super-elevation was decreased, implying the reduced importance of streamline curvature, and following from the reduction in momentum due to having a shallower tributary and no increase in inlet velocities. This demonstrates how topographic discordance can significantly increase the intensity of secondary circulation even where the contribution from streamline curvature is reduced.

The above simulations have added significantly to our understanding of the controls upon secondary circulation in the type of idealised channel geometries that have commonly been investigated in laboratory experiments. The extension of CFD to natural river channel confluences is reported in Lane *et al.* (1999, 2000) and Bradbrook *et al.* (2000a,b). A number of important results have emerged from this work. First, in methodological terms, the work has demonstrated the need to be cautious in inferring patterns of secondary circulation from field data without fully 3D flow measurements (Lane *et al.*, 2000). In order to define secondary circulation, it is necessary to specify the direction of primary velocity and then to locate cross-sections orthogonal to this. Lane *et al.* (2000) show the extent to which different methods of section rotation reveal different patterns of secondary circulation. Second, application of the model to the confluence of the Kaskaskia River and the Copper Slough, studied by Rhoads and Kenworthy (1995, 1998) and Rhoads (1996), confirmed that the helical circulation that formed resembled two back-to-back meanders. The extent to which this is the case will depend upon the level of tributary symmetry in the confluence (compare Figure 10.5a symmetrical and 10.5a asymmetrical) as judged in relation to streamline curvature and momentum ratio. The water surface elevation difference across a section normal to the direction of streamline curvature (ΔE_{ni}) will depend upon the centrifugal acceleration, and for tributary i is given by:

$$\Delta E_{ni} = \frac{\overline{U_{si}}^2}{g} \cdot \frac{w_i}{R_i} \quad (10.23)$$

where $\overline{U_{si}}$ is the streamwise section-averaged velocity in tributary i, w_i is the width of tributary i, g is the acceleration due to gravity and R_i is the radius of curvature of tributary i. This provides a measure of the relative strength of two tributaries in relation to planform curvature, channel width and a measure of flow inertia:

$$\frac{\Delta E_{n1}}{\Delta E_{n2}} = \frac{\overline{U_{s1}}^2}{\overline{U_{s2}}^2} \cdot \frac{w_1}{w_2} \cdot \frac{R_2}{R_1} \quad (10.24)$$

The results from the numerical modelling of both the field case and the idealised channel showed that if the ratio in equation (10.24) was greater than 1, then tributary 1 dominated the post-confluence zone, forcing a change in the direction of streamline curvature of tributary 2. If it was less than 1, the opposite held. It was also noted that this had important geomorphological implications. First, the zone of water surface super-elevation and its associated scour hole migrated with the streamlines associated with the stronger tributary towards either the river bank opposite tributary 1 where $\Delta E_{n1} > \Delta E_{n2}$, or tributary 2 where $\Delta E_{n2} > \Delta E_{n1}$. Second, the timescales of adjustment of U, w and R are different, with U responding to changes in the sub-catchment hydrology of individual tributaries, possibly associated with individual storm events, but w and R adjusting over much longer timescales. Thus, the hydrodynamics of the confluence within a storm event may respond to differences in

256 Computational Fluid Dynamics

tributary response in relation to a given combination of tributary widths and radii of curvature.

The numerical simulations have also allowed exploration of the role of topographic forcing in addition to that associated with streamline curvature. Bradbrook et al. (2000a) compared model predictions with the true bed topography and with the scour hole artificially filled in (Figure 10.6). Whilst the physical meaning of this simulation should be questioned (the flow and topography will not necessarily be in equilibrium and so the simulation does not say anything about the effects of flow processes upon scour hole generation), it is clear that the scour hole is a major factor in causing downwelling, in addition to that caused by streamline curvature. Making sense of this observation is again permitted by consideration of the dynamic pressure

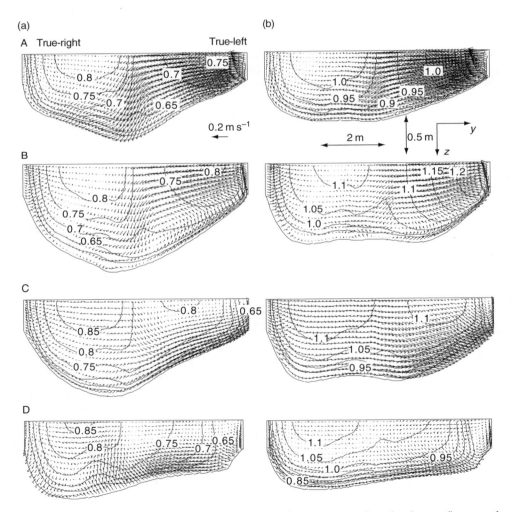

Figure 10.6 Model predictions of downstream and cross-stream flow for the confluence of Kaskaskia River and Copper Slough with the scour hole (a) and with the scour hole filled in (b).

field. This showed that above the scour hole there was a zone of negative dynamic pressure (i.e. where the pressure was lower than would be expected due to hydrostatic effects alone) even though the scour hole slope was too shallow for there to be flow separation. This was coincident with the zone of maximum downwelling into the scour hole. This emphasises the need to consider the effects of flow depth expansion into a scour hole as a continuum from the flat bed case, where negative dynamic pressure can only be associated with water surface super-elevation and flow separation effects through weak entry slopes (where there will be negative dynamic pressure and weak downwelling but no flow separation) to the most extreme case (where the flow separates from the bed, there is strong negative dynamic pressure and downwelling to a reattachment point further down the confluence, and the tributary is truly discordant).

Most recently, this work has been extended to deal with one of the major concerns over the application of CFD to natural river channels: the implications of bed roughness. As already discussed, the common approach used in the confluence modelling to date (e.g. Lane et al., 1999; Bradbrook et al., 2000a) has involved increasing the roughness height to represent the effects of those topographic scales between the grain scale and the model mesh resolution that are not explicitly incorporated into the model's topography. However, research by Lane et al. (in review) has shown that representation of these sub-grid scales explicitly as topography can result in significant vertical and cross-stream flows that are not present when the topography is represented through scaling of the roughness term. The extent to which this is a problem depends upon the ratio of the water depth to some measure of topographic variability (e.g. the D_{84}) and so it becomes more of a problem in channels with coarser grain sizes. As these effects are noted in the vertical and cross-stream components of flow, obvious questions emerge as to how they interact with secondary circulation associated with larger scales of planform and bed topographic forcing. As a first attempt to explore this issue, Lane et al. (in review) compared numerical results for simulations of a parallel confluence with one tributary entering at a higher elevation than the other. The comparison involved a smoothed bed with a scaled roughness height (mirroring the laboratory experiment of Best and Roy [1991] and the numerical experiments of Bradbrook et al. [1998]) and a gravel bed with explicit incorporation of gravel grains and bedforms measured using digital photogrammetry and represented using a porosity treatment (Lane and Hardy, 2002; Lane et al., 2004). These results showed that explicit representation of the gravel topography in the post-confluence channel reduced the cross-stream transfer of fluid into the zone of negative dynamic pressure which forms in the lee of the shallower tributary. However, the rate of mixing between the two tributaries was substantially increased because: (1) the tributary inlet had greater flow field variability as a result of the rough bed; and (2) topographic variability at the bed created a higher spatial resolution variability in the flow field that reduced the downstream advection of secondary circulation associated with the tributary discordance. It appears from these results that gravel beds enhance the rate of mixing at tributary junctions.

The main weakness associated with the modelling described above is its dependence upon Reynolds-averaged solutions with an RNG-type k-ε turbulence model. Bradbrook et al. (2000b) used large eddy simulation in order to produce time-dependent model simulations with much less dependence upon parameterisation of turbulence effects using a turbulence model. In the first instance, this was applied to a parallel confluence with bed discordance, identical to the Best and Roy (1991) and Bradbrook *et al.* (1998) simulations described above. This demonstrated that the zone of negative dynamic pressure in the lee of the shallow tributary resulted in the periodic draw of fluid across from the deeper tributary. This results in the increase of dynamic pressure until a critical point of stability was reached, at which point an eddy was shed. This eddy had strong streamwise vorticity and was advected down the tributary side of the post-confluence channel. Once the eddy was shed, the process of cross-stream transfer of fluid into the lee of the shallow tributary began again. Whilst the process was found to be periodic, the periodicities were irregular in terms of both magnitude and duration. Large eddy simulation has also been applied to natural channels (e.g. Bradbrook *et al.*, 2000b). Whilst problems of mesh resolution and design have precluded representation of higher frequency velocity fluctuations thus far, it has been possible to reproduce the periodic migration and deformation of the shear layer reported by Biron *et al.* (1996).

10.4.5 Investigation of habitat at the reach scale

With the growing application of CFD at the reach scale, there has been an increasing number of applications of CFD to habitat modelling (Chapter 16). Adopting a 2D approach is especially beneficial here as fish are migratory organisms and hence they can respond as their habitat demands change as a function of either life cycle or habitat (e.g. erosion/deposition). One-dimensional approaches do not represent the substantial within-reach variability in hydraulic variables. Leclerc *et al.* (1995, 1996) used a finite element solution of the depth-averaged flow equations, including a wetting and drying treatment, with a Habitat Suitability Index (HSI), to explore habitat changes on the Moisie River, Quebec (Leclerc *et al.*, 1995) and the Ashuapmushuan River, Quebec (Leclerc *et al.*, 1996). The HSI was derived from observed vertically averaged velocity, depth and substrate characteristics for Atlantic salmon fry and parr, using multivariate statistical techniques. The HSIs were driven by the hydraulic model predictions and were used to determine how the percentage usable area varied as a function of discharge. The advantage of this approach is that it recognises that as some habitat becomes less suitable as flow depth and/or velocity locally rises, other habitats become more suitable (e.g. if they were originally dry but become wet). This showed that there was a rapid increase in the percentage usable area for both parr and fry habitat up to $50 \, \text{m}^3 \, \text{s}^{-1}$. Above this, it remained relatively constant for fry, but rose more slowly for parr. Most importantly, they noted that the spatial scale of model predictions was more similar to that of known ecological function: salmonids defend territories that rarely exceed $4 \, \text{m}^2$. In a similar study Tiffan *et al.* (2002) used a depth-averaged model to simulate flow depths and velocities at 36 steady state discharges. The biological model was based upon multivariate logistic regression in which the probability of subyearling fall Chinook salmon presence was predicted from physical habitat parameters predicted by the model. The results showed that estimates of rearing habitat decreased as flows increased and that estimates of the area that fish could become stranded in initially rose, but then fell. However, they noted that even the $16 \, \text{m}^2$ resolution adopted was too coarse to characterise adequately the habitat needs of the subyearling fish. The biological model in the Tiffan *et al.* study was based upon a Habitat Probabilistic Index (HPI) model and Guay *et al.* (2000) used a similar approach, but compared it with HSI approaches. They found that the HPI produced better results than the HSI, and they noted that this may be because of the multivariate nature of the HPI approach in which predictions from the hydraulic models are considered simultaneously rather than independently, thus dealing with a common criticism of approaches like PHABSIM.

Crowder and Diplas (2000a) extend these approaches to the determination of energy gradients and 'velocity shelters' or refugia in gravel-bed streams. They recognise that point predictions of hydraulic variables may not always provide a sufficient representation of habitat as spatial variation in those variables is also a crucial fish (e.g. Hayes and Jowett, 1994) and macro-invertebrate (e.g. Lancaster and Hildrew, 1993) requirement. Crowder and Diplas undertook a higher-resolution modelling study, which included mesoscale topographic features (e.g. boulders) and which generated predictions for input into spatial habitat metrics. These seek to capture the physical habitat difference between two locations with the same velocity but different surrounding velocities in terms of velocity and kinetic energy gradients. For instance, by scaling the spatial change in kinetic energy between two points by the kinetic energy at the point with the smaller velocity, they were able to derive a metric that represents the kinetic energy that must be spent by an organism in order to move from the point of lower velocity to the point of higher velocity. As these are based upon gradients, the determined metrics will depend upon the spatial scale over which calculations are made, and this will need to be evaluated in relation to the spatial resolution of the mesh used in the model in relation to the behavioural aspects of the organism being considered. The research showed that the presence of boulders resulted in a substantially more complex spatial metric which provides a greater habitat range for fish. Linking this to observed fish behaviour (Crowder and Diplas, 2002) confirmed that boulders enhanced the potential availability of the right habitat.

10.5 Conclusion: Future research needs

The last 20 years have seen a major improvement in our representation of flow processes and sediment transport in terms of reach-scale application. Initial attempts to model natural channel geometries in two dimensions in the late 1980s were extended to three dimensions in the 1990s. In terms of research needs for the next 10 years, they can be divided into two broad areas. The first relates to continued methodological development. The progressive improvement in computer power and field measurement capability means that numerical models of reach-scale fluvial processes can now be developed for much larger spatial scales (e.g. in 2D for wide rivers, with complex channel patterns, for lengths up to about 100 channel widths; in 3D, for narrower rivers, for up to about 10 channel widths). Indeed, it could be argued that a single meander bend or tributary confluence is not really a reach-scale study. This increase in scale is potentially exciting as it will reduce the dependence upon explicit specification of initial conditions as the distance between the reach of interest and the upstream and downstream inlet conditions can be increased. In substantive terms, it will allow study of the downstream propagation of flow structures through many multiples of river channel width, and hence a much better understanding of the spatial scale at which flow structures form and evolve.

This improvement aside, a number of research developments are still required. First, whilst computer power and data collection capabilities have allowed scaling up, it will remain the case that the required complexity of flow field representation (including the choice between 2D and 3D approaches) will vary between applications, and perhaps even spatially within a single application. Numerical modelling of floodplain flow inundation has demonstrated that coupling 1D flood wave routing in the river channel to 2D routing across the floodplain can provide a highly effective representation of the flooding process (e.g. Bates and De Roo, 2000). The same nesting of models with varying process representation may be one of the most efficient means of capturing sufficient process representation for reach-scale flow modelling. For instance, in a current field and laboratory modelling project at the confluence of the Rivers Paraguay and Parana in Argentina, the width of the confluence (c.2 km in places) precludes a 3D modelling approach. Initial assessment of the width–depth ratio (greater than 200) might suggest that a 3D approach is not necessary. However, acoustic Doppler current profiler measurements, as well as field experience, suggest that the flow is very strongly three dimensional in places. A 3D treatment of the entire system is not possible and so we are exploring nesting a 3D model within a 2D model.

Second, the discretisation of complex channel geometries in reach-scale CFD remains a major challenge. This is especially the case for 3D applications, but it may occasionally apply to 2D applications, especially where wetting and drying processes are involved. The porosity method described earlier (Lane and Hardy, 2002; Lane et al., 2004) needs to be extended and modified for reach-scale application as this will allow representation of complex bed geometries without recourse to BFCs and the associated problems of numerical diffusion and instability.

Third, there remain issues associated with boundary condition specification, especially in relation to data availability. In 3D applications this is especially the case in relation to sub-grid-scale topographic variability. In the majority of situations, such data will not be available and, as Lane et al. (in review) show, failure to represent the bed surface adequately may lead to a serious degradation of model predictions of flow over irregular gravel beds. One of the ways forward may be to start to reconstruct sub-grid-scale surface variability. Use of a standard deviation of bed sizes to do this is not necessarily appropriate due to autocorrelation in the elevation surface. However, fractal-based or other stochastic characterisations of the surface structure of both sand-bedded (e.g. Robert and Richards, 1988) and gravel-bedded (e.g. Butler et al., 2001; Chapter 13 in this book) channels may provide a means of representing the effects of sub-grid-scale variability upon flow structure, if not the detail, in a manner that is much more effective than multiplication upwards of roughness heights.

Fourth, and perhaps most importantly, there remains the major challenge of effective coupling of flow models to sediment transport models and channel change treatments. Of particular importance is the ability to model the interaction between planform, flow and channel migration. Whether or not this needs a 2D, quasi-3D or fully 3D treatment of flow has yet to be fully established. Research is also needed on the most efficient methods of remeshing after channel change, and on the possibility of avoiding the need for this by using the porosity method in a regular grid.

These methodological developments will enhance our reach-scale modelling capability. However, the sophistication of reach-scale modelling is such that considerable opportunity for application of reach-scale models to substantive problems is emerging. The above discussion has provided specific examples of the use of CFD in two dimension and three dimension to understand flow processes in riffle–pool sequences, confluences and meanders. There are now examples of application of 2D models to entire braided river systems. These studies have combined detailed consideration of particular field case studies with more generic application to generalised cases (sometimes referred to as laboratory channels). This iteration between field application and generalised cases has resulted in major improvements in our understanding of the interaction between complex geometries and flow structures. The same attention has yet to be given to sediment transport and its implications for channel change, and also to aspects of river channel habitat.

Many of the methodological issues discussed above have implications for our ability to predict not just flow but also sediment transport and habitat availability. For instance, Crowder and Diplas (2002) note that model resolution improves when more comprehensive topographic data is incorporated and that if the flow patterns generated by certain topographic features (e.g. exposed boulders) lead to biologically meaningful habitat characteristics, the model (and the data used to derive the model) needs to represent them. This requires not only an ability to measure the necessary topographic data in the field, but also an ability to represent that topography in the numerical mesh, something that few studies have addressed (Crowder and Diplas, 2000b). This feeds back into data needs in relation to the quality and density of information that is required to provide the right resolution of model predictions for habitat representation. Thus, the substantive work that needs to be done ranges in both scale and purpose. In terms of scale, this ranges from exploration of the nature of the multiplier of roughness height using the porosity method and high-resolution 3D microtopographic data through to reach-scale identification of habitat suitability over many multiples of channel width. In purpose, it ranges from conventional assessment of river channel stability problems in relation to channel migration through to advanced assessment of habitat suitability in relation to different flow and sediment delivery conditions.

The final comment to make is the issue of practicality. Despite advances in computational power and data handling ability, there is a significant cost (in terms of computer time, data requirements and model learning curve) when one moves from one to two dimensions and again from two to three dimensions. There are clearly situations (and scales of analysis) where higher-dimensionality models are preferable, but the fact remains that at any given time the available computer power will allow lower-dimensionality models to be applied to longer river reaches. However, choosing the right dimensionality does require careful consideration of the problem being investigated. This choice will impact upon the associated data requirements (which may in turn restrict the model dimensionality that can be adopted) as well as the model parameterisation and calibration that are required. Ultimately, this will define the limits of model applicability. Now that higher-dimensionality models can be applied more readily to reach-scale flow and sediment transfer problems it will become increasingly important to evaluate these limits such that model suitability, rather than cost-effectiveness alone, informs model choice.

Acknowledgements

This chapter is based upon research supported by the Natural Environment Research Council through grants GR3/9407, GR3/9715, GR3/12588, GR9/5059, and NER/B/S/2002/00354. Dan Parsons supplied Figure 10.1 and Anthony Reeves undertook the meander bed-form modelling. The work has benefited from discussions amongst the British Geomorphological Research Group fixed-term working party on computational fluid dynamics applications in geomorphology, and also from discussions and research undertaken with Pascale Biron, Kate Bradbrook, Mike Church, Richard Hardy, Trevor Hoey, Dan Parsons, Anthony Reeves, Keith Richards and André Roy. Keith Richards provided valuable comments on an earlier draft of this chapter.

References

Apsley, D. and Hu, W., 2003. CFD simulation of two- and three-dimensional free-surface flow. *International Journal for Numerical Methods in Fluids*, **42**, 465–491.

Armanini, A. and di Silvio, G., 1988. A one-dimensional model for the transport of a sediment mixture in non-equilibrium conditions. *Journal of Hydraulics Research*, **26**, 275–292.

Ashmore, P., Ferguson, R.I. Prestegaard, K., Ashworth, P. and Paola, C., 1992. Secondary flow in anabranch confluences of a braided, gravel-bed stream. *Earth Surface Processes and Landforms*, **17**, 299–311.

Ashworth, P.J. and Ferguson, R.I., 1989. Size-selective entrainment of bedload in gravel-bed streams. *Water Resources Research*, **25**, 627–634.

Bates, P.D. and De Roo, A.P.J., 2000. A simple raster-based model for flood inundation simulation. *Journal of Hydrology*, **236**, 54–57.

Bathurst, J.C., Thorne, C.R. and Hey, R.D., 1979. Secondary flow and shear stress at river bends. *Journal of the Hydraulics Division*, ASCE, **105**, 1277–1295.

Bernard, R.S. and Schneider, M.L., 1992. Depth-averaged numerical modelling for curved channels. *Technical Report HL-92-9*, US Army Corps Engineers, Waterways Experiment Research Station, Vicksburg, Mississippi, US.

Best, J.L., 1987. Flow dynamics at river channel confluences: Implications for sediment transport and bed morphology. In Etheridge, F.G., Flores, R.M. and Harvey, M.D. (eds), *Recent Developments in Fluvial Sedimentology*, SEPM Special Publication, **39**, 27–35.

Best, J.L. and Ashworth, P.J., 1997. Scour in large braided rivers, and the recognition of sequence stratigraphic boundaries. *Nature*, **387**, 275–277.

Best, J.L. and Roy, A.G., 1991. Mixing-layer distortion at the confluence of channels of different depth. *Nature*, **350**, 411–413.

Biron, P.M., Best, J.L. and Roy, A.G., 1996. Effects of bed discordance on flow dynamics at river channel confluences. *Journal of Hydraulic Engineering*, ASCE, **122**, 676–682.

Biron, P.M., Richer, A., Kirkbride, A.D., Roy, A.G. and Han, S., 2002. Spatial patterns of water surface topography at a river confluence. *Earth Surface Processes and Landforms*, **27**, 913–928.

Booker, D.J., Sear, D.A. and Payne, A.J., 2001. Modelling three-dimensional flow structures and patterns of boundary shear stress in a natural pool-riffle sequence. *Earth Surface Processes and Landforms*, **26**, 553–576.

Bradbrook, K.F., 1999. *Numerical, Field and Laboratory Studies of Three-Dimensional Flow Structures at River Channel Confluences*. Thesis submitted in partial fulfilment of the requirements for award of the PhD degree, University of Cambridge, 362pp.

Bradbrook, K.F., Biron, P., Lane, S.N., Richards, K.S. and Roy, A.G., 1998. Investigation of controls on secondary circulation and mixing processes in a simple confluence geometry using a three-dimensional numerical model. *Hydrological Processes*, **12**, 1371–1396.

Bradbrook, K.F., Lane, S.N. and Richards, K.S., 2000a. Numerical simulation of time-averaged flow structure at river channel confluences. *Water Resources Research*, **36**, 2731–2746.

Bradbrook, K.F., Lane, S.N., Richards, K.S., Biron, P.M. and Roy, A.G., 2000b. Large Eddy Simulation of periodic flow characteristics at river channel confluences. *Journal of Hydraulic Research*, **38**, 207–216.

Bradbrook, K.F., Lane, S.N., Richards, K.S., Biron, P.M. and Roy, A.G., 2001. Flow structures and mixing at an asymmetrical open-channel confluence: A numerical study. *Journal of Hydraulic Engineering*, ASCE, **127**, 351–368.

Butler, J.B., Lane, S.N. and Chandler, J.H., 2001. Application of two-dimensional fractal analysis to the characterisation of gravel-bed river surface structure. *Mathematical Geology*, **33**, 301–330.

Caleffi, V., Valiani, A. and Zanni, A., 2003. Finite volume method for simulating extreme flood events in natural channels. *Journal of Hydraulic Research*, **41**, 167–177.

Cao, Z., Carling, P. and Oakey, R., 2003. Flow reversal over a natural pool-riffle sequence: A computational study. *Earth Surface Processes and Landforms*, **28**, 689–705.

Carling, P.A. and Wood, N., 1994. Simulation of flow over pool-riffle topography: A consideration of the velocity reversal hypothesis. *Earth Surface Processes and Landforms*, **19**, 319–322.

Chandler, J.H., Lane, S.N. and Richards, K.S., 1996. The determination of water surface morphology at river channel confluences using automated digital photogrammetry and their consequent use in numerical flow modelling. *International Archives of Photogrammetry and Remote Sensing*, **XXXI**(B7), 99–104.

Church, M.A. and Jones, D., 1982. Channel bars in gravel-bed rivers. In Hey, R.D., Bathurst, J.C. and Thorne, C.R. (eds), *Gravel-Bed Rivers*, Wiley, Chichester, 291–338.

Clifford, N.J., 1992. Formation of riffle pool sequences – field evidence for an autogenic process. *Sedimentary Geology*, **85**, 39–51.

Cornelius, C., Volgmann, W. and Stoff, H., 1999. Calculation of three-dimensional turbulent flow with a finite volume multigrid method. *International Journal of Numerical Methods in Fluids*, **31**, 703–720.

Crowder, D.W. and Diplas, P., 2000a. Evaluating spatially explicit metrics of stream energy gradients using hydrodynamic model simulations. *Canadian Journal of Fisheries and Aquatic Science*, **57**, 1497–1507.

Crowder, D.W. and Diplas, P., 2000b. Using two-dimensional hydraulic models at scales of ecological importance. *Journal of Hydrology*, **230**, 172–191.

Crowder, D.W. and Diplas, P., 2002. Assessing changes in watershed flow models with spatially explicit hydraulic models. *Journal of the American Water Resources Association*, **38**, 397–408.

Cui, Y., Parker, G., Pizzuto, J.E. and Lisle, T.E., 2003. Sediment pulses in mountain rivers, Part II: Comparison between experiments and numerical predictions. *Water Resources Research*, **39**(9), 1240.

Darby, S.E., Alabyan, A.M. and Van de Wiel, M.J., 2002. Numerical simulation of bank erosion and channel migration in meandering rivers. *Water Resources Research*, **38**(9), 1163, DOI:10.1029/2001WR000602.

Demuren, A., 1993. A numerical model for flow in meandering channels with natural bed topography. *Water Resources Research*, **29**, 1269–1277.

Demuren, A.O. and Rodi, W., 1986. Calculation of flow and pollutant dispersion in meandering channels. *Journal of Fluid Mechanics*, **172**, 63–92.

Dietrich, W.E. and Smith, J.D., 1983. Influence of the point bar on flow through curved channels. *Water Resources Research*, **19**, 1173–1192.

Dietrich, W.E. and Smith, J.D., 1984. Bed load transport in a river meander. *Water Resources Research*, **20**, 1355–1380.

Dietrich, W.E. and Whiting, P.J., 1989. Boundary shear stress and sediment transport in river meanders of sand and gravel. In Ikeda, S. and Parker, G. (eds), *River Meandering*, AGU Water Resources Monograph, **12**, 1–50.

Duan, J.G., Wang, S.S.Y. and Jia, Y., 2001. The application of the enhanced CCHE2D model to study the alluvial channel migration process. *Journal of Hydraulic Research*, **39**, 469–480.

Egiazaroff, I.V., 1965. Calculation of nonuniform sediment concentrations. *Journal of Hydraulics Division*, ASCE, **91**, 225–247.

Elso, J.I. and Giller, P.S., 2001. Physical characteristics influencing the utilization of pools by brown trout in an afforested catchment in Southern Ireland. *Journal of Fish Biology*, **58**, 201–221.

Ferguson, R.I., 2003. The missing dimension: Effects of lateral variation on 1-D calculations of fluvial bedload transport. *Geomorphology*, **56**, 1–14.

Ferguson, R.I. and Ashworth, P.J., 1992. Spatial patterns of bedload transport and channel change in braided and near-braided rivers. In Billi, P., Hey, R.D., Thorne, C.R. and Tacconi, P. (eds), *Dynamics of Gravel-Bed Rivers*, Wiley, Chichester, 477–496.

Ferguson, R.I. and Parsons, D.R., 2004. Flow structures in meander bends with recirculation. In Jirka, G. and Uijttewaal, W. (eds), *Shallow Flows*, Balkema, Leiden, 325–331.

Ferguson, R.I., Church, M. and Weatherly, H., 2001. Fluvial aggradation in Vedder river: Testing a one-dimensional sedimentation model. *Water Resources Research*, **37**, 3331–3347.

Ferguson, R.I., Parsons, D.R., Lane, S.N. and Hardy, R.J., 2003. Flow in meander bends with recirculation at the inner bank. *Water Resources Research*, **39**, 2003WR001965.

Ferziger, J.H. and Perić, M., 1999. *Computational Methods for Fluid Dynamics*, 2nd edition, Springer, Berlin, 389pp.

Ghanem, A., Steffler, P. and Hicks, F., 1996. Two-dimensional hydraulic simulation of physical habitat conditions in flowing streams. *Regulated Rivers: Research and Management*, **12**, 185–200.

Gilvear, D.J., Heal, K.V. and Stephen, A., 2002. Hydrology and the ecological quality of Scottish river ecosystems. *Science of the Total Environment*, **294**, 131–159.

Guay, J.C., Boisclair, D., Rioux, D., Leclerc, M., Lapointe, M. and Legendre, P., 2000. Development and validation of numerical habitat models for juveniles of Atlantic salmon (*Salmo salar*). *Canadian Journal of Fisheries and Aquatic Science*, **57**, 2065–2075.

Guo, Q-C. and Jin, Y-C., 1999. Modelling sediment transport using depth-averaged and moment equations. *Journal of Hydraulic Engineering*, **125**, 1262–1269.

Hardy, R.J., Lane, S.N., Ferguson, R.I. and Parsons, D., 2003. Assessing the credibility of a series of computational fluid dynamic (CFD) simulations of open channel flow. *Hydrological Processes*, **17**, 1539–1560.

Hayes, J.W. and Jowett, I.G., 1994. Microhabitat models of large drift-feeding brown trout in three New Zealand rivers. *North American Journal of Fisheries Management*, **14**, 710–725.

Heniche, M., Secretan, Y., Boudreau, P. and Leclerc, M., 2002. Dynamic tracking of flow boundaries in rivers with respect to discharge. *Journal of Hydraulic Research*, **40**, 589–602.

Hey, R.D., 1979. Flow resistance in gravel-bed rivers. *Journal of the Hydraulics Division*, ASCE, **105**, 365–379.

Hirano, M., 1971. River bed degradation with armouring. *Transactions of the Japanese Society of Civil Engineers*, **3**, 194–195.

Hodskinson, A., 1996. Computational fluid dynamics as a tool for investigating separated flow in river bends. *Earth Surface Processes and Landforms*, **21**, 993–1000.

Hodskinson, A. and Ferguson, R.I., 1998. Numerical modelling of separated flow in river bends: Model testing and experimental investigation of geometric controls on the extent of flow separation at the concave bank. *Hydrological Processes*, **12**, 1323–1338.

Hoey, T.B. and Ferguson, R.I., 1994. Numerical modelling of downstream fining by selective transport in gravel-bed rivers: Model development and illustration. *Water Resources Research*, **30**, 2251–2260.

Hoey, T.B. and Ferguson, R.I., 1997. Controls of strength and rate of downstream fining above a river base level. *Water Resources Research*, **33**, 2601–2608.

Horritt, M., 2004. Development and testing of a simple 2D finite volume model of sub-critical shallow water flow. *International Journal for Numerical Methods in Fluids*, **44**, 1231–1255.

Howard, A.D., 1992. Modelling channel migration and floodplain construction. In Carling and Petts (eds), *Lowland Floodplain Rivers*, Wiley, Chichester, 1–41.

Johannesson, H. and Parker, G., 1989. Linear theory of river meanders. In Ikeda, S. and Parker, G. (eds), *River Meandering*, AGU Geophysical Monograph, 181–214.

Kassem, A.A. and Chaudhury, M.H., 2002. Numerical modelling of bed evolution in channel bends. *Journal of Hydraulic Engineering*, **128**, 507–514.

Keller, E.A., 1971. A real sorting of bed material: The hypothesis of velocity-reversal. *Geological Society of America Bulletin*, **83**, 915–918.

Keller, E.A., 1972. Development of alluvial streams, a five stage model. *Geological Society of America Bulletin*, **83**, 1531–1536.

Keller, E.A. and Florsheim, J.L., 1993. Velocity reversal hypothesis: A model approach. *Earth Surface Processes and Landforms*, **18**, 733–740.

Kennedy, G.J.A. and Strange, C.D., 1982. The distribution of salmonids in upland streams in relation to depth and gradient. *Journal of Fish Biology*, **20**, 579–591.

Lancaster, J. and Hildrew, A.G., 1993. Flow refugia and the microdistribution of lotic macroinvertebrates. *Journal of the North American Benthological Society*, **12**, 385–393.

Lancaster, S.T. and Bras, R.L., 2002. A simple model of river meandering and its comparison to natural channels. *Hydrological Processes*, **16**, 1–26.

Lane, S.N., 1998. Hydraulic modelling in hydrology and geomorphology: A review of high resolution approaches. *Hydrological Processes*, **12**, 1131–1150.

Lane, S.N. and Hardy, R.J., 2002. Porous rivers: A new way of conceptualising and modelling river and floodplain flows? In Ingham, D.B. and Pop, I. (eds), *Transport Phenomena in Porous Media*, **2**, Elsevier, Pergamon.

Lane, S.N. and Richards, K.S., 1998. Two-dimensional modelling of flow processes in a multi-thread channel. *Hydrological Processes*, **12**, 1279–1298.

Lane, S.N., Richards, K.S. and Chandler, J.H., 1994. Distributed sensitivity analysis in modelling environmental systems. *Proceedings of the Royal Society, Series A*, **447**, 49–63.

Lane, S.N., Bradbrook, K.F., Richards, K.S., Biron, P.M. and Roy, A.G., 1999. The application of computational fluid dynamics to natural river channels: three-dimensional versus two-dimensional approaches. *Geomorphology*, **29**, 1–20.

Lane, S.N., Bradbrook, K.F., Richards, K.S., Biron, P.M. and Roy, A.G., 2000. Secondary circulation in river channel confluences: Measurement myth or coherent flow structure? *Hydrological Processes*, **14**, 2047–2071.

Lane, S.N., Hardy, R.J., Elliott, L. and Ingham, D.B., 2002. High resolution numerical modelling of three-dimensional flows over complex river bed topography. *Hydrological Processes*, **16**, 2261–2272.

Lane, S.N., Hardy, R.J., Elliott, L. and Ingham, D.B., 2004. Numerical modelling of flow processes over gravely-surfaces using structured grids and a numerical porosity treatment. *Water Resources Research*, W01302 JAN 8 2004.

Lane, S.N., Hardy, R.J., Keylock, C.J., Ferguson, R.I. and Parsons, D.R., in review. The implications of complex river bed structure for numerical modelling and the identification of channel-scale flow structures. *Earth Surface Processes and Landforms*.

Launder, B.E. and Spalding, D.B., 1974. The numerical computation of turbulent flows. *Computer Methods in Applied Mechanics and Engineering*, **3**, 269–289.

Leclerc, M., Boudreault, A., Bechara, J.A. and Corfa, G., 1995. Two-dimensional hydrodynamic modelling: A neglected tool in the Instream Flow Incremental Methodology. *Transactions of the American Fisheries Society*, **124**, 645–662.

Leclerc, M., Boudreault, A., Bechara, J.A. and Belzile, L., 1996. Numerical method for modelling spawning habitat dynamics of landlocked salmon, *Salmo salar*. *Regulated Rivers: Research and Management*, **12**, 273–285.

Lee, A.J. and Ferguson, R.I., 2002. Velocity and flow resistance in step-pool streams. *Geomorphology*, **46**, 59–71.

Leschziner, M.A. and Rodi, W., 1979. Calculation of strongly curved open channel flow. *Journal of Hydraulic Engineering*, **105**, 1297–1314.

Ma, L., Ashworth, P.J., Best, J.L., Elliott, L., Ingham, D.B. and Whitcombe, L.J., 2001. Computational fluid dynamics and the physical modelling of an upland river. *Geomorphology*, **44**, 375–391.

Maddock, I.P., Bickerton, M.A. and Spence, R., 2001. Reallocation of compensation releases to restore river flows and improve instream habitat availability in the Upper Derwent catchment, Derbyshire, UK. *Regulated Rivers*, **17**, 417–441.

Meselhe, E.A. and Sotiropolous, F., 2000. Three-dimensional numerical model for openchannels with free surface variations. *Journal of Hydraulic Research*, **38**, 115–121.

Milhous, R., Updike, M. and Snyder, D., 1989. PHABSIM system reference manual: Version 2. *US Fish and Wildlife Service*, FWS/OBS 89/16 (Instream Flow Information Paper 26).

Milhous, R.T., Wegner, D.L. and Waddle, T., 1984. Users guide to the physical habitat simulation system. *US Fish and Wildlife Service Biological Services Program*, FWS/OBS-81/43.

Miller, A.J., 1994. Debris-fan constrictions and flood hydraulics in river canyons: Some implications from two-dimensional modelling. *Earth Surface Processes and Landforms*, **19**, 681–697.

Morvan, H., Pender, G., Wright, N.G. and Ervine, D.A., 2002. Three-dimensional hydrodynamics of meandering compound channels. *Journal of Hydraulic Engineering*, **128**, 674–682.

Nagata, N., Hosoda, T. and Muramoto, Y., 2000. Numerical analysis of river channel processes with bank erosion. *Journal of Hydraulic Engineering*, **126**, 243–252.

Nelson, J.M. and Smith, J.D., 1989a. Evolution and stability of erodible channel beds. In Ikeda, S. and Parker, G. (eds), *River Meandering*, AGU Geophysical Monograph, 321–378.

Nelson, J.M. and Smith, J.D., 1989b. Flow in meandering channels with natural topography. In Ikeda, S. and Parker, G. (eds), *River Meandering*, AGU Geophysical Monograph, 69–102.

Nezu, I., Tominaga, A. and Nakagawa, H., 1993. Field-measurements of secondary currents in straight rivers. *Journal of Hydraulic Engineering*, **119**, 598–564.

Nicholas, A.P., 2000. Modelling bedload yield in braided gravel bed rivers. *Geomorphology*, **36**, 89–106.

Nicholas, A.P., 2001. Computational fluid dynamics modelling of boundary roughness in gravel-bed rivers: An investigation of the effects of random variability in bed elevation. *Earth Surface Processes and Landforms*, **26**, 345–362.

Nicholas, A.P. and Smith, G.H.S., 1999. Numerical simulation of three-dimensional flow hydraulics in a braided channel. *Hydrological Processes*, **13**, 913–929.

Odgaard, A.J. and Bergs, M.A., 1988. Flow processes in a curved alluvial channel. *Water Resources Research*, **24**, 45–56.

Olsen, N.R.B., 2003. Three-dimensional CFD modeling of self-forming meandering channel. *Journal of Hydraulic Engineering*, **129**, 366–372.

Olsen, N.R.B. and Stoksteth, S., 1995. Three-dimensional numerical modelling of water flow in a river with large bed roughness. *Journal of Hydraulic Research*, **33**, 571–581.

Paola, C., 1996. Incoherent structure: Turbulence as a metaphor for stream braiding. In Ashworth, P.J. Bennett, S., Best, J.L., and McLelland, S.M (eds), *Coherent Flow Structures in Open Channels*, Wiley, Chichester, 705–723.

Parker, G., 1990. Surface-based bedload transport relationship for gravel rivers. *Journal of Hydraulic Research*, **28**, 417–436.

Parker, G. and Sutherland, A.J., 1990. Fluvial armor. *Journal of Hydraulic Research*, **28**, 529–543.

Prandtl, L., 1952. *Essentials of Fluid Dynamics*. Blackie & Sons, London.

Rahuel, J.L., Holly, F.M., Chollet, J.P., Belleudy, P.J. and Yang, G., 1989. Modelling of riverbed evolution for bedload sediment mixtures. *Journal of Hydraulic Engineering*, **115**, 1521–1542.

Raleigh, R.F., Zuckerman, L.D. and Nelson, P.C., 1986. Habitat suitability index models and instream flow suitability curves: Brown trout revised. *US Fish and Wildlife Service Biological Report*, **82**(10.124).

Rameshwaran, P. and Naden, P.S., 2003. Three-dimensional numerical simulation of compound channel flows. *Journal of Hydraulic Engineering*, **129**, 645–652.

Rhoads, B.L., 1996. Mean structure of transport-effective flows at an asymmetrical confluence when the main stream is dominant. In Ashworth, P.J., Bennett, S., Best, J.L. and McLelland, S.M. (eds), *Coherent Flow Structures in Open Channels*, Wiley, Chichester, 459–490.

Rhoads, B.L. and Kenworthy, S.T., 1995. Flow structure at an asymmetrical stream confluence. *Geomorphology*, **11**, 273–293.

Rhoads, B.L. and Kenworthy, S.T., 1998. Time-averaged flow structure in the central region of a stream confluence. *Earth Surface Processes and Landforms*, **23**, 171–191.

Rhoads, B.L. and Sukhodolov, A.N., 2001. Field investigation of three-dimensional flow structure at stream confluences: 1. Thermal mixing and time-averaged velocities. *Water Resources Research*, **37**, 2393–2410.

Richards, K.S., 1978. Simulation of flow geometry in a riffle-pool stream. *Earth Surface Processes*, **3**, 345–354.

Richards, K.S., 1982. *Rivers: Form and Process in Alluvial Channels*. Methuen, London.

Robert. A. and Richards, K.S., 1988. On the modelling of sand bedforms using the semi-variogram. *Earth Surface Processes and Landforms*, **13**, 459–473.

Robinson, R.A.J. and Slingerland, R.L., 1998. Origin of fluvial grain-size trends in a foreland basin: Pocono Formation of the central Appalachian basin. *Journal of Sedimentary Research*, **68**, 473–486.

Rodi, W., 1980. *Turbulence Models and Their Application in Hydraulics*. IAHR, Delft, 104pp.

Rodi, W., Pavlovic, R.N. and Srivatsa, S.K., 1981. Prediction of flow and pollutant spreading in rivers. In Fischer, H.B. (ed.), *Transport Models for Inland and Coastal Waters*, Academic Press, New York, 542pp.

Seminara, G. and Tubino, M., 1989, Alternate bars and meandering: Free, forced, and mixed interactions. In Ikeda, S. and Parker, G. (eds), *River Meandering*, AGU Geophysical Monograph, 267–320.

Shankar, N.J., Chan, E.S. and Zhang, Q.Y., 2001. Three-dimensional numerical simulation for an open channel flow with a constriction. *Journal of Hydraulic Research*, **39**, 187–201.

Shimizu, Y. and Itakura, T., 1989. Calculation of bed variation in alluvial channels. *Journal of Hydraulic Engineering*, ASCE, **115**, 367–384.

Shimizu, Y., Yamaguchi, H. and Itakura, T., 1990. Three-dimensional computation of flow and bed deformation. *Journal of Hydraulic Engineering*, ASCE, **116**, 1090–1106.

Smith, N.D., 1974. Sedimentology and bar formation in the Upper Kicking Horse River, a braided outwash stream. *Journal of Geology*, **82**, 205–223.

Struiksma, N. and Crosato, A., 1989. Analysis of a two-dimensional bed topography model for rivers. In Parker, G. (ed.), *River Meandering*, American Geophysical Union Monograph, 153–180.

Sun, T., Meakin, P. and Jossang, T., 1996. A simulation model for meandering rivers. *Water Resources Research*, **32**, 2937–2954.

Talbot, T. and Lapointe, M., 2002. Numerical modeling of gravel bed river response to meander straightening: The coupling between the evolution of bed pavement and long profile. *Water Resources Research*, **38**, DOI:10.1029/2001WR000330.

Thompson, D.M., Wohl, E.E. and Jarrett, R.D., 1996. A revised velocity-reversal and sediment-sorting model for a high-gradient, pool-riffle stream. *Physical Geography*, **17**, 142–156.

Thompson, D.M., Nelson, J.M. and Wohl, E.E., 1998. Interactions between pool geometry and hydraulics. *Water Resources Research*, **34**, 3673–3681.

Thorne, C.R. and Hey, R.D., 1979. Direct measurements of secondary currents at a river inflexion point. *Nature*, **280**, 226–228.

Tiffan, K.F., Garland, R.D. and Rondorf, D.W., 2002. Quantifying flow-dependent changes in subyearling fall Chinook Salmon rearing habitat using two-dimensional spatially explicit modelling. *North American Journal of Fisheries Management*, **22**, 713–726.

Tingsanchali, T. and Maheswaran, S., 1990. Two-dimensional depth-averaged flow computation near groynes. *Journal of Hydraulic Engineering*, ASCE, **116**, 103–125.

Toro-Escobar, C.M., Parker, G. and Paola, C., 1996. Transfer function for the deposition of poorly sorted gravel in response to streambed aggradation. *Journal of Hydraulic Research*, **34**, 35–53

Van Niekerk, A., Vogel, K.R., Slingerland, R.L. and Bridge, J.S., 1992. Routing of heterogeneous sediments over movable bed: Model development. *Journal of Hydraulic Engineering*, **118**, 246–261.

Vanoni, V.A., 1984. Fifty years of sedimentation. *Journal of Hydraulic Engineering*, **110**, 1021–1057.

Vismara, R., Azzellino A., Bosi, R., Crosa, G. and Gentili, G., 2001. Habitat suitability curves for brown trout (*Salmo trutta fario* L.) in the River Adda, Northern Italy: Comparing univariate and multivariate approaches. *Regulated Rivers*, **17**, 37–50.

Wallis, S.G. and Manson, J.R., 1997. Accurate numerical simulation of advection using large time steps. *International Journal of Numerical Methods in Fluids*, **24**, 127–139.

Wan, Q., Wan, H.T., Zhou, C.H. and Wu, Y.X., 2002. Simulating the hydraulic characteristics of the lower Yellow River by the finite-volume technique. *Hydrological Processes*, **16**, 2767–2779.

Ward, J.V., Malard, F. and Tockner, K., 2001. Landscape ecology: A framework for integrating pattern and process in river corridors. *Landscape Ecology*, **17**, 35–45.

Weerakoon, S.B. and Tamai, N., 1989. Three-dimensional calculation of flow in river confluences using boundary fitted co-ordinates. *Journal of Hydroscience and Hydraulic Engineering*, **7**, 51–62.

Weerakoon, S.B., Kawahara, Y. and Tamai, N., 1991. Three-dimensional flow structure in channel confluences of rectangular section. *Proceedings of the 25th I.A.H.R. Congress A*, IAHR, Madrid, 373–380.

Whiting, P.J. and Dietrich, W.E., 1991. Convective accelerations and boundary shear stress over a channel bar. *Water Resources Research*, **27**, 783–796.

Whiting, P.J. and Dietrich, W.E., 1993. Experimental studies of bed topography and flow patterns in large-amplitude meanders, 1. Observations. *Water Resources Research*, **29**, 3605–3614.

Wiberg, P.L. and Smith, J.D., 1991. Velocity distribution and bed roughness in high-gradient streams. *Water Resources Research*, **27**, 825–838.

Wilcock, P.R. and Crowe, J.C., 2003. A surface-based transport model for sand and gravel. *Journal of Hydraulic Engineering*, **129**, 120–128.

Wilson, C.A.M.E., Boxall, J.B., Guymer, I. and Olsen, N.R.B., 2003. Validation of a three-dimensional numerical code in the simulation of pseudo-natural meandering flows. *Journal of Hydraulic Engineering*, **129**, 758–768.

Wu, W., Rodi, W. and Wenka, T., 2000. 3D numerical modelling of flow and sediment transport in open channels. *Journal of Hydraulic Engineering*, **126**, 4–15.

Yeh, K-C. and Kennedy, J.F., 1993. Moment model of non-uniform channel-bend flow; 1 Fixed Bed. *Journal of Hydraulic Engineering*, ASCE, **119**, 776–795.

11

Numerical modelling of floodplain flow

P.D. Bates, M.S. Horritt, N.M. Hunter, D. Mason and D. Cobby

11.1 Introduction

Numerical modelling of floodplain flow is undertaken for two main reasons. First, one may wish to use such models as an alternative to laboratory experiments in order to improve process understanding. Second, one may wish to obtain predictions of quantities relevant to the management of floodplain systems, such as discharge, water surface elevation, inundation extent and flow velocity. In each case, the model is a symbolic representation, either in mathematical notation or computer code, of the processes that we perceive are relevant to the problem in hand. The model therefore embodies a hypothesis that can be tested through comparison to analytical solutions, scale models or field data. For the first class of application, physical realism of the simulated processes is often emphasized, whereas for the second, computational constraints and the potential need to evaluate model uncertainty may influence the class of model selected. Whilst in reality we know that compound channel flows will be fully turbulent over a wide range of space scales and unsteady in time, simulation of such complexity may be computationally prohibitive. Moreover, the processes perceived by modellers to be relevant to the accurate simulation of floodplain flow for a particular purpose are likely to comprise only a subset of the known physical mechanisms. The key step in selecting an appropriate numerical modelling framework for floodplain flows is therefore to identify those processes that are relevant to a particular modelling problem and decide how these can be discretized and parameterized in the most computationally efficient manner. In the next section (11.2), flow processes in compound channels will be discussed along with the simplifying assumptions frequently made in floodplain flow modelling. Examples of numerical modelling tools that embody these simplifying assumptions and appropriate case studies will then be discussed in section 11.3, followed by a discussion of parameterization, validation and uncertainty analysis requirements for floodplain flow modelling in section 11.4. The chapter will conclude with a discussion of future possibilities and research needs (section 11.5).

11.2 Flow processes in compound channels

Most lowland rivers consist of a main channel with one or two adjacent floodplain areas. During the passage of a flood wave, the flow depth may exceed bankfull height, and water may rapidly extend and retreat over the low-lying and flat floodplains. Floodplains therefore act either as storage areas for flood water or provide an additional route for flow conveyance. In fluid dynamics terms, a flood is a long, low-amplitude wave passing through a compound channel of complex geometry. In the largest basins such waves may be up to $\sim 10^3$ km in length or greater but with amplitude of only $\sim 10^1$ m, and may take several months to traverse the whole system. Flood waves are translated downstream with speed or celerity, c [LT^{-1}], and attenuated by frictional losses such that in downstream sections the hydrograph is flattened out (Figure 11.1). Wave speeds can be shown (NERC, 1975) to vary with discharge such that maximum wave speed occurs at approximately two-thirds bankfull capacity (Knight and Shiono, 1996). Typical observed values for c reported by NERC (1975) and Bates *et al.* (1998) for UK rivers with widths in the range 10–50 m are in the ranges 0.5–1.8 m s^{-1} and 0.3–0.67 m s^{-1}, respectively.

Below the scale of the flood wave itself, other significant in-channel processes can be identified, each with a characteristic length scale. These include: (1) shear layers forming at the junction between the main flow and slower moving 'dead zones' at the scale of the channel planform (Hankin *et al.*, 2001; Chapter 10 in this book); (2) secondary circulations at the scale of the channel cross section (Bridge and Gabel,

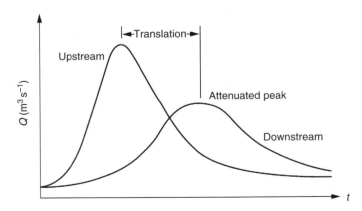

Figure 11.1 Translation and attenuation of a flood wave between two gauging stations.

1992; Nezu et al., 1993); and (3) turbulent eddies. The latter range from heterogeneous structures at the scale of roughness elements and obstructions on the bed (Ashworth et al., 1996; McLelland et al., 1999; Shvidchenko and Pender, 2001) down through the turbulent energy cascade (Hervouet and Van Haren, 1996) to the Kolmogorov length scale, η [L], where turbulent kinetic energy is finally dissipated. In typical channel flows these smallest eddies may be only $\sim 10^{-2}$ mm across (Hervouet and Van Haren, 1996) and are highly transient.

When flow depth exceeds bankfull height to produce a compound-flow cross-section, all the above processes will continue to exist, but in addition, a number of new physical mechanisms will begin to operate. Principally, these are: (1) momentum exchange between fast moving water in the channel and slower moving water on the floodplain (Knight and Shiono, 1996); and (2) the interaction between meandering channel flow and flow on the floodplain (Sellin and Willetts, 1996). Channel–floodplain momentum exchange occurs across a shear layer created at the interface between channel and floodplain which is manifest (Figure 11.2) as a series of vortices with vertically aligned axes (Sellin, 1964; Fukuoka and Fujita, 1989; Shiono and Knight, 1991). Knight and Shiono (1996) draw attention to the 3D structure of this interface zone, whilst Ervine and Baird (1982) conclude that failure to account for the resulting momentum exchange can lead to errors of up to ±25% in the discharge calculated using uniform flow formulae such as the Manning and Chezy equations. Further vigorous momentum exchange occurs during out-of-bank flow in meandering compound channels (see Sellin and Willetts, 1996, for a discussion). Here water spills from the downstream apex of channel bends and flows over meander loops before interacting with

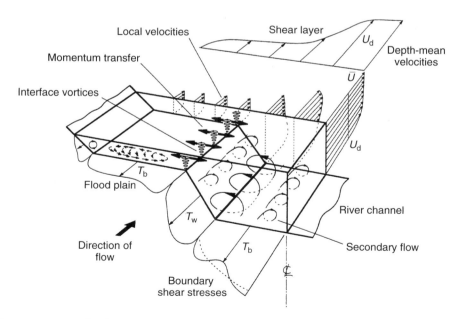

Figure 11.2 Hydraulic processes associated with overbank flow in a straight compound channel (after Shiono and Knight, 1991). (Reproduced by permission of Cambridge University Press.)

channel flow in the next meander (Figure 11.3). These 3D interactions modify secondary circulations within the channel and represent an additional energy loss in the near channel area. Floodplain flows beyond the meander belt will not be subject to such energy losses and this region may provide a route for more rapid flow conveyance. The impact of these additional energy losses will be at a maximum at some shallow overbank stage, when the interaction between main channel and floodplain is at its greatest (Knight and Shiono, 1996), before slowly decreasing as depth increases and the whole floodplain and valley floor begins to behave as a single channel unit.

Away from the near channel zone, water movement on the floodplain may be more accurately described as a typical shallow water flow (i.e. one where the width:depth ratio exceeds 10:1) as the horizontal extent may be large (up to several kilometres) compared to the depth (usually less than 10 m). Such shallow water flows over low-lying topography are characterized by rapid extension and retreat of the inundation front over considerable distances, potentially with distinct processes occurring during the wetting and drying phases (see Nicholas and Mitchell, 2003). Correct treatment of this moving boundary problem is therefore important both to capture adequately the shallow water energy losses (which may be high due to large relative roughness) and because flood extent is a common prediction requirement from hydraulic codes.

Flow interactions with micro-topography (see Walling *et al.*, 1986), vegetation (Lopez and Garcia, 2001) and structures (Meselhe *et al.*, 2000) may all be important, thereby giving a complex modelling problem. In particular, where the floodplain acts as a route for flow conveyance rather than just as storage, energy losses are dominated by vegetative resistance. Yet despite a small number of pioneering studies (see for example Kouwen, 1988; Nepf, 1999; Ghisalberti and Nepf, 2002; Wilson and Horritt, 2002), the interaction between plant form, plant biomechanics, energy loss

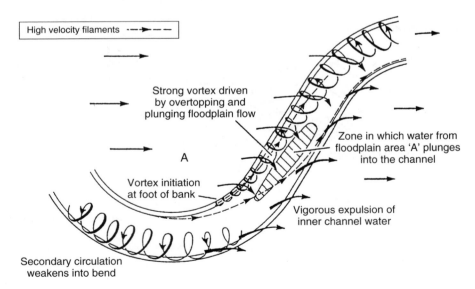

Figure 11.3 *Representation of principal flow structures occurring during overbank flow in a meandering compound channel (after Sellin and Willetts, 1996). (Published by John Wiley & Sons Ltd.)*

and turbulence generation is at present relatively poorly understood (Chapter 15). Moreover, many numerical models of floodplain flow assume that the channel bed is fixed over the course of the event, and for very large floods this may not be the case as embankment failure or geomorphic change may considerably affect the flow field.

Whilst overbank flow in compound channels is clearly a 2D process, many practical floodplain management questions require the prediction of water levels only at particular points of interest. In such cases, the modeller is primarily concerned with the downstream routing of flow through a compound cross section, and may be less concerned to represent floodplain flow and storage accurately. Here, the flow processes of interest are one dimensional in the down-valley direction and 1D models may therefore be used to represent such flows. Whilst this is often considered a gross simplification of the flow field (see Knight and Shiono, 1996), one can justify the approach by assuming that the additional approximations involved in continuing to treating out-of-bank flow as if it were one dimensional are small compared to other uncertainties (for a discussion see Ali and Goodwin, 2002). Alternatively, one can attempt to correct 1D flow routing methods to account for the additional energy losses and/or mass transfers (see Knight and Shiono, 1996) or develop hybrid schemes that combine 1D modelling for channel flows with a 2D treatment of the floodplain (see Bates and De Roo, 2000).

Lastly, whilst typical hydraulic models do not consider water exchanges with the surrounding catchment, for whole catchment modelling or flood inundation simulation over long river reaches, such exchanges may, at particular times, become important (e.g. Stewart et al., 1999; Woessner, 2000). Such processes include direct precipitation or runoff to the floodplain surface (e.g. Mertes, 1997), evapo-transpiration losses, the so-called bank-storage effect (Pinder and Sauer, 1971; Squillace, 1996) resulting from interactions between the river water and alluvial groundwaters contained within the hyporheic zone (Stanford and Ward, 1988; Castro and Hornberger, 1991; Wroblicky et al., 1998), subsurface contributions to the floodplain groundwater from adjacent hillslopes (e.g. Bates et al., 2000; Burt et al., 2002) and flows along preferential flow paths, such as relict channel gravels, within the flood plain alluvium (e.g. Haycock and Burt, 1993; Poole et al., 2002). Over particular reaches and in particular environments, integration of some or all of these processes with flood routing models may be required and necessitate complex modelling structures (e.g. Kohane and Welz, 1994; Stewart et al., 1999) which may be difficult to fully parameterize.

11.3 Numerical modelling tools

Hydraulic models for compound channel flow can be classified according to the number of dimensions in which they represent flow processes, and the above discussion indicates that for particular problems a one-, two- or even 3D model may be most appropriate. These models are therefore discussed in the following section.

All such tools are ultimately derived from the 3D Navier–Stokes momentum equation that for an incompressible fluid of constant density can be expressed in Cartesian vector notation as:

$$\rho \frac{D\mathbf{u}}{Dt} = -\nabla p + \mu \nabla^2 \mathbf{u} + F \qquad (11.1)$$

where ρ is the fluid density [ML^{-3}]; **u** is the velocity [LT^{-1}]; t is the time [T], p is the pressure [ML^{-1}T^{-2}]; μ is the viscosity [ML^{-1}T^{-2}] and F is the set of source terms (for example gravity, Coriolis and friction) to be included in the specification of a particular problem.

This when combined with the equation of continuity:

$$\nabla \cdot \mathbf{u} = 0 \qquad (11.2)$$

gives a system of equations that can be solved to yield the 3D velocity vector $\mathbf{u} = (u\ v\ w)$, where u, v and w are the three components of **u** in the x-, y- and z-directions, respectively, and pressure, p, for a given point in time and space. In free surface models the pressure is typically replaced with the flow depth, h [L]. For a more detailed discussion of the Navier–Stokes equations and derived flow equations relevant to fluvial flow modelling the reader is referred to Ingham and Ma (Chapter 2).

Theoretically, equations (11.1) and (11.2) or their lower dimensional derivates can be used, with appropriate boundary conditions, to fully describe any open channel flow. However, direct application of these equations to turbulent flows in compound channels requires a sufficiently resolved model discretization capable of capturing the smallest time and length scales of turbulent motion. For studies of compound channel flow, for both research and predictive purposes, we are often not interested in the details of the instantaneous velocity field at all scales but, rather, the mean flow properties at some larger scale. Reynolds-averaging is therefore employed to split each variable in equations (11.1) and (11.2) (for example **u**) into two components: a mean value ($\bar{\mathbf{u}}$) and a random variation about it ($\mathbf{u}'$) (Reynolds, 1895). We assume that the random variations about the mean are normally distributed and that over a sufficient time period these cancel to zero. We can then replace **u** in equations (11.1) and (11.2) with ($\bar{\mathbf{u}}$) and obtain the Reynolds-averaged Navier–Stokes equations (RANS). This does, however, lead to the introduction of new terms in equation (11.1) representing the shear stress on the mean flow due to turbulence. As these depend on the instantaneous velocity fluctuations their existence is a problem for a time-averaged model as these values are unknown. To close the RANS equations we need to provide values for the Reynolds stresses by introducing some model of turbulence. For a detailed discussion of Reynolds-averaging, Reynolds stresses and choice of turbulence closure model, the reader is referred to Rodi (1993), Rodi *et al.* (1997) and Sotiropoulos (Chapter 5).

11.3.1 Three-dimensional models

As noted in section 11.2, flow processes in compound channels are fully three dimensional, and the 3D aspects of the flow may dominate in the near-channel region. Representation of such processes therefore demands a 3D treatment based on the solution of the Reynolds-averaged form of equations (11.1) and (11.2) using some appropriate numerical technique such as finite differences, finite elements or finite volumes (for a more detailed discussion of these methods see Chapter 7). For problems such as sediment transport and flow–vegetation interaction, where this dimensionality of representation is important, a number of commercial and research-use modelling tools are available. The key decisions for the modeller lie in the construction of an appropriate discretization so that the relevant flow structures can be represented and the choice of turbulence closure scheme to be employed. These decisions need to be balanced against computational cost and this leads to a trade off between cost and the size of either the cell or the element size (and hence the size of domain) or the realism with which turbulent processes can be represented. A typical application is described by Stoesser *et al.* (2003), who report the application of a 3D RANS model to a 3.5-km channel–floodplain reach of the River Rhine in South-West Germany. The governing equations were solved for a general 3D geometry, discretized by the finite volume method on curvilinear co-ordinates. The study used a relatively high-resolution mesh of 198 144 cells comprising $258 \times 64 \times 12$ cells in the streamwise, cross-stream-wise and vertical directions respectively (Figure 11.4) and giving an approximate cell size of $13\,\text{m} \times 3\,\text{m} \times 0.5\,\text{m}$. Correct grid spacing in the vertical was obtained through the σ-transform (see Hervouet and Van Haren, 1996, p. 206) to give a normalized thickness of vertical layers irrespective of bottom topographic variations. The model successfully reproduced reach-scale floodplain velocity observations determined from dye tracing experiments taken during an approximately 100-year recurrence interval flow with discharge of $\sim 3600\,\text{m}^3\,\text{s}^{-1}$ and water levels for an independent verification event when discharge equalled $\sim 2400\,\text{m}^3\,\text{s}^{-1}$. The study provided evidence of the

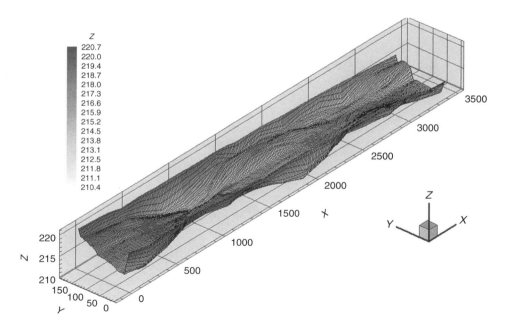

Figure 11.4 Three-dimensional grid used to simulate a 3.5-km reach of the River Rhine, South-West Germany by Stoesser et al. (2003). (Published by IWA Publishing.)

computational feasibility of a 3D approach for river reaches of this scale for steady state flows and demonstrated the range of hydraulic quantities that can be calculated (Figure 11.5).

Higher-order turbulence modelling approaches for 3D flow in compound channels have also been attempted and include algebraic stress (Shao *et al.*, 2003) and large eddy simulation (Thomas and Williams, 1995) schemes. However, these approaches incur large computational costs and have yet to be applied to anything other than experimental channels of regular geometry. Full field-scale models of complex natural geometry will no doubt become possible in the future, but may be over-specified for the problem in hand given the quality of the available validation data.

As well as possible computational constraints, 3D approaches which use either a deformable mesh (Feng and Perić, 2003) or the σ-transform (Stoesser *et al.*, 2003) to discretize the grid in the vertical may suffer from stability problems during dynamic shallow water flows as cell height–length ratios may become highly distorted as water depth, h, approaches zero. Whilst the volume of fluid method (Ma *et al.*, 2002) can be used to track the horizontal moving boundary in 3D models whilst retaining fixed vertical grid increments (Chapter 6), avoidance of distorted elements for flows with horizontal extent 100–10 000 m and depths generally much less than 1–10 m still requires numerical grids that are horizontally very highly resolved. Such grids are likely to incur a prohibitive computational cost. Thus, to date, most 3D numerical models of compound channel flow have not been applied to problems with significant changes in domain extent over low-gradient floodplains. Neither may such approaches be necessary, as for many

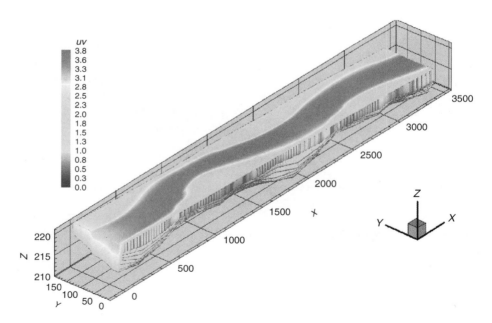

Figure 11.5 *Scalar product of u and v velocities in m s^{-1} predicted by Stoesser et al. (2003) for the 3.5-km reach of the River Rhine shown in Figure 11.4. (Published by IWA Publishing.) (See Plate 1.)*

scales of compound channel flows the shallow water approximation may be adequate. For these reasons, dynamically varying flows in compound channels have, to date, typically been treated with 2D models and these are therefore discussed in section 11.3.2.

11.3.2 Two-dimensional models

Two-dimensional approaches typically involve integration of the Navier–Stokes equations (11.1) and (11.2) over the flow depth to produce depth-averaged values of the velocity. Depending on the precise approximations used this can result in a number of different equation sets such as the St Venant equations, which assume a hydrostatic pressure distribution, or the Boussinesq equations, which do not (see Hervouet and Van Haren, 1996). For example, the St Venant equations are given in non-conservative form as:

Continuity equation

$$\frac{\partial h}{\partial t} + \vec{u_d} \cdot \overrightarrow{\text{grad}}(h) + h \text{div}(\vec{u_d}) = 0 \qquad (11.3)$$

Momentum equations

$$\frac{\partial u_d}{\partial t} + \vec{u_d} \cdot \overrightarrow{\text{grad}}(u_d) + g\frac{\partial h}{\partial x} - \text{div}(\nu_t \cdot \overrightarrow{\text{grad}}(u_d)) = S_x - g\frac{\partial Z_f}{\partial x} \qquad (11.4)$$

$$\frac{\partial v_d}{\partial t} + \vec{u_d} \cdot \overrightarrow{\text{grad}}(v_d) + g\frac{\partial h}{\partial y} - \text{div}(\nu_t \cdot \overrightarrow{\text{grad}}(v_d)) = S_y - g\frac{\partial Z_f}{\partial y} \qquad (11.5)$$

where u_d, v_d are the depth-averaged velocity components [LT^{-1}] in the x and y cartesian directions [L]; Z_f is the bed elevation [L]; ν_t is the kinematic turbulent viscosity [L^{2}T^{-1}]; S_x, S_y are the source terms (friction, Coriolis force and wind stress) and g is the gravitational acceleration [LT^{-2}].

These can then be solved using some appropriate numerical procedure (Chapter 7) and turbulence closure (Chapter 5) to obtain predictions of the water depth, h, and the two components of the depth-averaged velocity, u_d and v_d. The shallow water equations are most often applied to flows that have a large areal extent compared to their depth and where there are large lateral variations in the velocity field. They are thus well suited to the computation of overbank flood flows in compound channels, tides, tsunamis or even dam breaks (Hervouet and Van Haren, 1996). Whilst 2D models cannot fully represent the complex flow processes in the near channel region of a compound channel, they will still capture certain aspects of these processes. Here, the modeller assumes that this is sufficient to reproduce the particular flow features that are of interest for the problem in hand at a given scale. Moreover, 2D schemes can also more easily represent moving boundary effects and may therefore be of more use for simulating problems where inundation extent changes dynamically through time (Chapter 6).

Whether one can conclusively discriminate between 2D and 3D approaches will also depend, to an extent, on the type and quality of validation data available. With flume data such discrimination may be possible (e.g. Wilson *et al.*, 2003), but field data often consist only of bulk flows from gauging stations or at best limited amounts of inundation extent or point state variable data. Given that all validation data will be subject to some error, it may be easy to calibrate many different hydraulic model structures to provide an acceptable match. In practice, therefore, one may not have conclusive evidence that necessitates the use of a 3D code, particularly if the quantities one wants to predict can also be obtained from a lower-dimensional scheme.

Numerous classes of 2D approaches have been developed, and these include storage cell codes (section 11.3.4), full solutions of the 2D St Venant or shallow water equations (Gee *et al.*, 1990; Feldhaus *et al.*, 1992; Bates *et al.*, 1998; Nicholas and Mitchell, 2003) and simplified shallow water models (Molinaro *et al.*, 1994) where certain terms, such as inertia, are omitted from the controlling equations. These equations can all can be discretized using structured (Nicholas and Walling, 1997; Bates and De Roo, 2000; Nicholas and Mitchell, 2003) or unstructured grids (Hervouet and Van Haren, 1996; Sen, 2002), and more complex schemes even allow the grid to deform to follow the moving inundation front through time (Lynch and Gray, 1980; Kawahara and Umetsu, 1986; Benkhaldoun and Monthe, 1994). Fixed grid approaches require additional algorithms to treat spurious mass and momentum conservation effects that occur in partially wet cells at the flow field boundary (King and Roig, 1988; Leclerc *et al.*, 1990; Defina *et al.*, 1994, Ip *et al.*, 1998; Tchamen and Kawahita, 1998; Bates and Hervouet, 1999; Defina, 2000). Given the computational cost of re-meshing and potential problems in maintaining numerical stability and grid independence, fixed grid approaches have been preferred to date. This also has the advantage that it allows use of a more resolved spatial discretization and thus a greater degree of topographic complexity can be included in the model. This is clearly critical, as topography is a major control on the inundation process (e.g. Nicholas and Walling, 1997). For a full review of approaches to moving boundary problems see Bates and Horritt (Chapter 6).

A number of typical 2D model applications are reviewed by Bates *et al.* (1998). Here finite element solutions of equations (11.3)–(11.5) were developed for five river–floodplain reaches varying in length from 0.5 to 60-km discretized as unstructured grids. The models were calibrated to replicate wave speeds through each reach and were validated against a variety of independent level and discharge data. A typical finite element mesh is shown in Figure 11.6 for a 60-km reach of the River Severn, West-Central England, UK, between the gauging stations at Montford Bridge and Buildwas.

11.3.3 One-dimensional models

Most simply, for flow routing problems, floodplain flow can be treated as one dimensional in the down-valley direction. An equation for 1D channel flow can be derived by considering mass and momentum conservation between two cross sections Δx apart. This yields the well known 1D St Venant or shallow water equations:

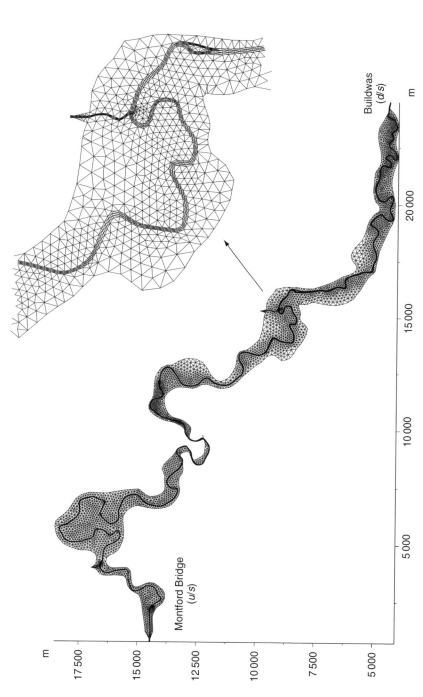

Figure 11.6 Two-dimensional unstructured mesh discretization constructed for a typical reach scale river–floodplain application. In this case the mesh was constructed by Bates et al. (1998) for use with the TELEMAC-2D finite element solution of the two-dimensional shallow water equations for a 60-km reach of the River Severn UK. (Published by Thomas Telford Publishing.)

Conservation of momentum

$$\frac{\partial Q}{\partial t} + \frac{\partial (Q^2/A)}{\partial x} + gA\left(\frac{\partial h}{\partial x} + S_\text{f} - S_\text{o}\right) = 0 \qquad (11.6)$$

where Q is the flow discharge [L^3T^{-1}]; A is the flow cross-section area [L^2], g is the gravitational acceleration [LT^{-2}], S_f is the friction slope [LL^{-1}] and S_o is the channel bed slope [LL^{-1}].

Conservation of mass

$$\frac{\partial Q}{\partial x} + \frac{\partial A}{\partial t} = q \qquad (11.7)$$

where q is the lateral inflow or outflow per unit length [L^2T^{-1}].

Equations (11.6) and (11.7) have no exact analytical solution, but with appropriate boundary and initial conditions they can be solved using numerical techniques (e.g. Preissmann, 1961; Abbott and Ionescu, 1967) to yield estimates of Q and h in both space and time. The river reach in 1D St Venant models is discretized as a series of irregularly spaced cross sections (Figure 11.7). Boundary conditions typically consist of the inflow hydrograph at the upstream cross section and, for sub-critical flow, the stage hydrograph at the downstream boundary. For super-critical flow where the Froude number, $Fr = \mathbf{u}/\sqrt{gh}$, exceeds 1, information from the downstream boundary cannot propagate upstream and no boundary condition needs therefore to be prescribed at the downstream boundary. These equations form the basis of most standard commercial hydraulic modelling software such as HEC-RAS, MIKE11, ISIS and SOBEK.

Horritt and Bates (2002) demonstrate for the simulation of a large flood event over the 60-km reach of the River Severn, UK, shown in Figure 11.6, that 1D, simplified 2D and full 2D models perform equally well in simulating flow routing and inundation extent given uncertainties over inflow, topography and validation data. This suggests that although gross assumptions are made regarding the flow physics incorporated in a 1D model applied to out-of-bank flows, the additional energy losses can be compensated for by using a calibrated effective friction coefficient.

One-dimensional approaches that attempt to correct some of these additional energy losses are reviewed by Knight and Shiono (1996, pp. 155–159) and are based on subdividing the channel in the cross-streamwise direction and then calculating the conveyance in each section using uniform flow formulae. The sub-area conveyances are then summed to give the total conveyance. Knight and Shiono (1996) identify three main variations on the channel division method which aim to simulate the channel–floodplain interaction more exactly. These are: (1) modification of the sub-area wetted perimeters (Wormleaton et al., 1982); (2) calculation of discharge adjustment factors for each sub-area based on a 'coherence' concept (Ackers, 1993); and (3) quantification of the apparent shear stresses on the sub-area division lines (Knight and Hamed, 1984). However, these methods have been developed to estimate the depth-discharge rating curve at particular cross sections and are largely yet to be incorporated in standard flood routing models. One exception here is the

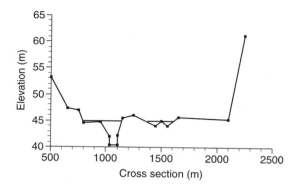
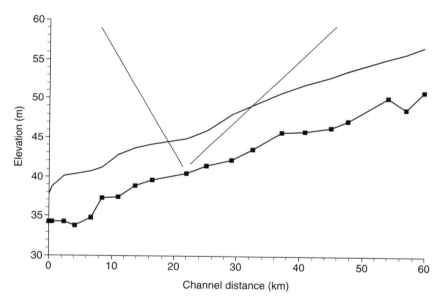

Figure 11.7 A typical discretization from a one-dimensional hydraulic model, in this case the HEC-RAS scheme applied to a 60-km reach of the River Severn in the UK by Horritt and Bates (2002). Shown are the bed elevation profile with model cross-section locations marked by squares and a typical simulated water free surface profile. A sample cross section is also shown. (Published by Elsevier.)

LISFLOOD-FF model of De Roo *et al.* (2001). Here, a correction of the Manning roughness value is applied to simulate the momentum exchange which occurs across the shear layer between main channel and floodplain flows.

11.3.4 Coupled 1D/2D models

Whilst 1D codes are computationally efficient, they do suffer from a number of drawbacks when applied to floodplain flows. These include the inability to simulate lateral spreading of the flood wave, the lack of a continuous treatment for topography and the subjectivity of cross-section location. Whilst all of these constraints can be overcome with higher-order codes, the computational cost of running a 2D or 3D simulation may be high. Consequently, recent research has begun to examine hybrid 1D/2D codes that seek to combine the best of each model class (for example Bechteler et al., 1994; Bladé et al., 1994; Estrela and Quintas, 1994; Romanowicz et al., 1996; Bates and De Roo, 2000; Dhondia and Stelling, 2002; Venere and Clausse, 2002).

Such models typically treat in-channel flow with some form of the 1D St Venant equations, but treat floodplain flows as two dimensional using a storage cell concept first described by Cunge et al. (1980). Here the floodplain is discretized as a series of regions, with flows between regions calculated using analytical uniform flow formulae such as the Manning equation. Initially, this approach was implemented in standard 1D river routing packages such as ISIS (Wicks et al., 2003) by defining the storage cells as large polygonal areas (surface areas of $\sim 10^0$ to 10^1 km^2) representing discrete flooding compartments (e.g. polders, storage basins, etc.) that are subjectively identified by the user. Recent developments in topographic data capture have, however, allowed high-resolution (cell size $\sim 10^0$ m) digital elevation models of floodplain areas to be produced. This has allowed storage cells to be discretized as a high-resolution grid (storage cells with surface area 10^{-5} to 10^{-4} km^2), for example the LISFLOOD-FP raster flood routing model of Bates and De Roo (2000). LISFLOOD-FP treats in-channel flow using either the simplified kinematic or diffusion wave forms of equations (11.6) and (11.7). Floodplain flows are similarly described in terms of continuity and mass flux equations, discretized over a grid of square cells, which allows the model to represent 2D dynamic flow fields on the floodplain. Flow between two cells is assumed simply to be a function of the free surface height difference between those cells (Estrela and Quintas, 1994):

$$Q_x^{i,j} = \frac{h_{\text{flow}}^{5/3}}{n} \left(\frac{h^{i-1,j} - h^{i,j}}{\Delta x}\right)^{1/2} \Delta y \qquad (11.8)$$

Water depths in each cell are then updated at each time step based on the sum of the fluxes over the four faces of the cell.

$$\frac{dh^{i,j}}{dt} = \frac{Q_x^{i-1,j} - Q_x^{i,j} + Q_y^{i,j-1} - Q_y^{i,j}}{\Delta x \Delta y} \qquad (11.9)$$

where $h^{i,j}$ is the free surface of water height at the node (i,j), Δx and Δy are the cell dimensions, n is the effective grid scale Manning's friction coefficient [$L^{1/3}T^{-1}$] for the floodplain, and Q_x and Q_y describe the volumetric flow rates between floodplain cells. Q_y is defined analogously to equation (11.8). The flow depth, h_{flow}, represents the depth through which water can flow between two cells, and is defined as the difference between the highest water free surface in the two cells and the highest bed elevation (this definition has been found to give sensible results both for wetting cells and for flows linking floodplain and channel cells). This approach is similar to diffusive wave propagation, but differs marginally due to the de-coupling of the x- and y-components of the flow. While this approach does not accurately represent diffusive wave propagation, it is computationally simple and has been shown to give very similar results to a more accurate finite difference discretization of the diffusive wave equation (Horritt and Bates, 2001a).

Equation (11.8) is also used to calculate flows between floodplain and channel cells, allowing floodplain cell depths to be updated using equation (11.12) in response to flow from the channel. These flows are also used as the source term, q, in the channel flow sub-model (equation 11.7), effecting the linkage of channel and floodplain flows. Thus only mass transfer between channel and floodplain is represented, and this is assumed to be dependent only on relative water surface elevations. While this neglects effects such as channel–floodplain momentum transfer and the effects of advection and secondary circulation on mass transfer, it is the simplest approach to the coupling problem and should reproduce the dominant behaviour of the real system.

The channel in the LISFLOOD-FP model occupies no floodplain pixels, but instead represents an extra flow path between pixels lying over the channel. Thus floodplain pixels lying over the channel have two water depths associated with them: one for the channel and one for the floodplain itself. This new scheme (referred to as the *Near Channel Floodplain Storage*, or NCFS, model by Horritt and Bates, 2001b) has proved more suitable for situations where large floodplain grid spacings are used in conjunction with a narrow (width $< \Delta x$) channel since channel width and raster grid size are de-coupled. Since the NCFS scheme will also calculate floodplain flows between cells occupied by the channel, extra flow routing in near channel regions will also be represented.

Such cellular approaches are capable of high-resolution application to relatively long river reaches up $\sim 10^2$ km in length and may provide an alternative to 1D codes. For example, the LISFLOOD-FP storage cell code has been successfully used to model flow routing along a 60-km reach of the River Severn in the UK (Horritt and Bates, 2002) and a 35-km reach of the River Meuse in The Netherlands (De Roo *et al.*, 2003; Figure 11.8).

11.4 Model parameterization, validation and uncertainty analysis

Section 11.3 has made clear the variety of models available for floodplain flow modelling. Choice of model has been shown to depend on the scale of the problem, the computational resources available and the needs of the user. However, any model is only as good as the data used to parameterize, calibrate and validate it, and in this section the data sources available for floodplain flow models are discussed along with the methods available to assimilate these data into hydraulic models. The data required by any hydraulic model are principally: (1) boundary condition data; (2) initial condition data; (3) topography data; (4) friction data; and (5) hydraulic data for use in model validation.

11.4.1 Boundary condition data

Boundary condition data consist of values of each model-independent variable at each boundary node at each time step for unsteady simulations. For 1D and 2D

286 *Computational Fluid Dynamics*

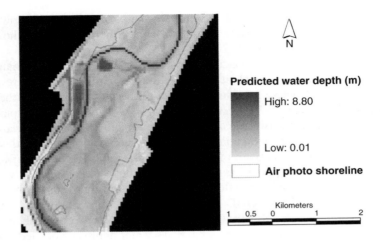

Figure 11.8 *Predicted water depths and inundation extent simulated using the LISFLOOD-FP two-dimensional storage cell model applied at 50-m resolution to a 35-km river–floodplain reach of the River Meuse, The Netherlands, over 20 days of a 1-in-63-year recurrence interval flood event that occurred in January 1995. Observed flow at the gauging station from 19th January onwards is used as a model boundary condition and predicted inundation is compared to that observed by aerial photography taken on 27th January. The model correctly classifies 85% of pixels in the above image as wet or dry.*

codes these are assigned from stage or discharge measured at river gauging stations, or measured in the field by the user. The precise data required depends on the model and the reach hydraulics, but for sub-critical flow will, as a minimum, consist of the flux rate into the model across each inflow boundary and the water surface elevation at each outflow boundary. These requirements reduce to merely the inflow flux rates for super-critical flow problems as, when the Froude number $Fr > 1$, information cannot propagate in an upstream direction. Rating curves may provide an alternative means of parameterizing the outflow water surface elevation in certain models. In addition, 3D codes require the specification of the velocity distribution at the inlet boundary and values for the turbulent kinetic energy. In most cases, hydrological fluxes outside the channel network, e.g. surface and subsurface flows from hillslopes adjacent to the floodplain and infiltration of flood waters into alluvial sediments, are ignored (for a more detailed discussion see Stewart *et al*., 1999).

11.4.2 Initial condition data

Initial conditions for a hydraulic model consist of values for each model-dependent variable at each computational node at time $t = 0$. In practice, these will be incompletely known, if at all, and some additional assumptions will therefore be necessary. For steady state (i.e. non-transient) simulations any reasonable guess at the initial conditions is usually sufficient, as the simulation can be run until the solution is in equilibrium with the boundary conditions and the initial conditions have ceased to have an influence. However, for dynamic simulations this will not be the case and whilst care can be taken to make the initial conditions as realistic as possible, a 'spin up' period during which model performance is impaired will always exist. For example, initial conditions for a flood simulation in a compound channel are often taken as the water depths and flow velocities predicted by a steady state simulation with inflow and outflow boundary conditions at the same value as those used to commence the dynamic run. Whilst most natural systems are rarely in steady state, careful selection of simulation periods to coincide their start with near steady state conditions can minimize the impact of this assumption.

11.4.3 Topography data

Topography is frequently considered the key data set for flow routing and inundation modelling. High-resolution, high-accuracy topographic data are essential to shallow water flooding simulations over low slope floodplains with complex microtopography, and such data sets are increasingly available from a variety of remotely mounted sensors. Traditionally, hydraulic models have been parameterized using ground survey of cross-sections perpendicular to the channel at spacings of between 100 and 1000 m. Such data are accurate to within a few millimetres in both the horizontal and the vertical and integrate well with 1D hydraulic models. However, ground survey data are expensive and time-consuming to collect and of relatively low spatial resolution. They hence require significant interpolation to enable their use in typical 2D and 3D models, whilst in 1D models the topography between cross-sections is effectively ignored and the results are sensitive to cross-section spacing (e.g. Samuels, 1990). Moreover, topographic data available on national survey maps tend to be of low accuracy with poor spatial resolution in floodplain areas. For example in the UK, nationally available contour data are recorded only at 5-m spacing to height accuracy of ± 1.25 m and for a hydraulic modelling study of a typical river reach Marks and Bates (2000) report finding only 3 contours and 40 unmaintained spot heights within a $\sim$6-km^2 floodplain area. When converted into a DEM such data lead to relatively low levels of inundation prediction accuracy in hydraulic models (see Wilson and Atkinson, in press).

Considerable potential therefore exists for more automated, broad-area mapping of topography from satellite and, more importantly, airborne platforms. Three techniques which currently show reasonable potential for flood modelling are aerial stereo-photogrammetry (Baltsavias, 1999; Lane, 2000; Westaway *et al.*, 2003), airborne laser altimetry or LiDAR (Krabill *et al.*, 1984; Gomes-Pereira and Wicherson, 1999) and airborne synthetic aperture radar interferometry (Hodgson *et al.*, 2003). Radar interferometry from sensors mounted on space-borne platforms, and in particular the Shuttle Radar Topography Mission (SRTM) data (Rabus *et al.*, 2003), may in the future also provide a viable topographic data source for hydraulic modelling in large, remote river basins where the flood amplitude is large compared to the topographic data error.

LiDAR in particular has recently attracted much attention in the hydraulic modelling literature (Marks and Bates, 2000; Bates *et al.*, 2003; Charlton *et al.*, 2003; French, 2003). Major LiDAR data collection programmes are underway in a number of countries, including The Netherlands and the UK; so far approximately

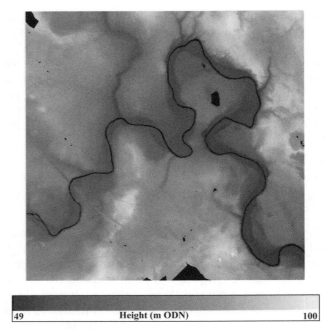

Figure 11.9 Digital elevation model at 10-m resolution derived from LiDAR data for a 6-km × 6-km region of the River Severn floodplain in the UK using the processing algorithm of Cobby et al. (2003). (Published by John Wiley & Sons Ltd.)

20% of the land surface area in England and Wales has been surveyed. In the UK, helicopter-based LiDAR survey is also beginning to be used to monitor in detail (~0.3 m spatial resolution) along critical topographic features such as flood defences, levees and embankments. LiDAR systems operate by emitting pulses of laser energy at very high frequency (~5–33 KHz) and measuring the time taken for these to be returned from the surface to the sensor. Global Positioning System data and an onboard Inertial Navigation System are used to determine the location of the plane in space and hence the surface elevation. As the laser pulse travels to the surface it spreads out to give a footprint of ~0.1 m^2 for a typical operating altitude of ~800 m. On striking a vegetated surface, part of the laser energy will be returned from the top of the canopy and part will penetrate to the ground. Hence, an energy source emitted as a pulse will be returned as a waveform, with the first point on the waveform representing the top of the canopy and the last point (hopefully) representing the ground surface. The last returns can then be used to generate a high-resolution 'bare earth' DEM (Figure 11.9).

11.4.4 Friction data

Friction is usually the only unconstrained parameter in a hydraulic model. Two- and three-dimensional codes which use a zero-equation turbulence closure may additionally require specification of an 'eddy viscosity' parameter which describes the transport of momentum within the flow by turbulent dispersion; however, this prerequisite disappears for most higher-order turbulence models of practical interest. Hydraulic resistance is a lumped term that represents the sum of a number of effects: skin friction, form drag and the impact of acceleration and deceleration of the flow. These combine to give an overall drag force C_d, which in hydraulics is usually expressed in terms of resistance coefficients such as Manning's n and Chezy's C and which are derived from uniform flow theory. This assumes that the rate of energy dissipation for non-uniform flows is the same as it would be for uniform flow at the same water surface (friction) slope. The precise effects represented by the friction coefficient for a particular model depend on the model's dimensionality, as the parameterization compensates for energy losses due to unrepresented processes, and the grid resolution. Thus, the extent to which form drag is represented depends on how the channel cross-sectional shape, meanders and long profile are incorporated into the model discretization. Similarly, a high-resolution discretization will explicitly represent a greater proportion of the form drag component than a low-resolution discretization using the same model (Chapter 10). Complex questions of scaling and dimensionality hence arise which may be somewhat difficult to disentangle. Certain components of the hydraulic resistance are, however, more tractable. In particular, skin friction for in-channel flows is a strong function of bed material grain size, and a number of relationships exist which express the resistance oefficient in terms of the bed material median grain size, D_{50} (e.g. Hey, 1979). Equally, on floodplain sections where conveyance rather than storage processes dominate, the drag due to vegetation is likely to form the bulk of the resistance term (Kouwen, 2000). Determining the drag coefficient of vegetation is, however, rather complex, as the frictional losses result from an interaction between plant biophysical properties and the flow. For example, at high flows the vegetation momentum-absorbing area will reduce due to plant bending and flattening (Chapter 15). To account for such effects Kouwen and Li (1980) and Kouwen (1988) calculated the Darcy-Weisbach friction factor f for short vegetation, such as floodplain grasses and crops, by treating such vegetation as flexible, and assuming that it may be submerged or non-submerged. The factor f is dependent on water depth and velocity, vegetation height and a product MEI, where M is the number of stems per unit area, E is the stem modulus of elasticity and I is the stem area's second moment of inertia. Whilst MEI often cannot be measured directly, it has been shown to correlate well with vegetation height (Temple, 1987).

Similar to topography, ground survey of grain size and vegetation parameters is extremely time-consuming and, recently, research has begun to consider the use of remotely sensed techniques to determine certain of the above data. For example, photogrammetric techniques to extract grain-size information from ground-based or airborne photography are currently under development (Butler et al., 2001). Similarly, recent research developments allow extraction of specific plant biophysical parameters from LiDAR data. For vegetation >4 m in height, this latter technique uses the timing difference between the first and last points on the returned LiDAR waveform to determine the height of the canopy. For vegetation <4 m the method uses the local standard deviation of the last return heights (Figure 11.10). Cobby et al. (2000, 2001) demonstrate that the height of short vegetation up to 1.4-m height can be estimated with such a technique to 0.14 m rmse. Taller vegetation (>10 m) is subject to greater height estimation error (~2–3 m) as canopies are typically denser

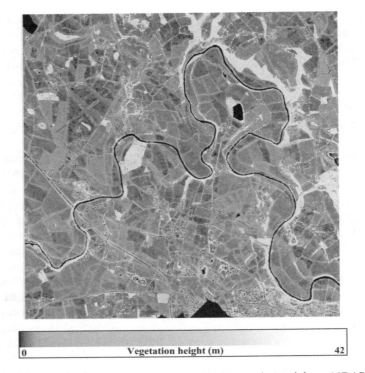

Figure 11.10 *Vegetation height map at 10-m resolution derived from LiDAR data for a 6-km × 6-km region of the River Severn floodplain in the UK using the processing algorithm of Cobby et al. (2003). (Published by John Wiley & Sons Ltd.)*

and it is less likely that the laser pulse will penetrate the full depth of the canopy; however, for the purpose of determining hydraulic resistance this is less of a problem. Given that other plant biophysical properties, e.g. *MEI*, correlate with plant height, Mason *et al.* (2003) have presented a methodology to calculate time- and space-distributed friction coefficients for flood inundation models directly from LiDAR data. Much further work is required in this area; however, such studies are beginning to provide methods to explicitly calculate important elements of frictional resistance for particular flow routing problems. This leads to the prospect of a much reduced need for calibration of hydraulic models and therefore a reduction in predictive uncertainty.

11.4.5 Model data assimilation

Increasing use of the remote sensing techniques described earlier has caused a rapid shift in floodplain hydraulic modelling from a data-poor to a data-rich and spatially complex modelling environment with attendant possibilities for model testing and development. Despite an increase in computational power, the relative resolution of model and topography data has now reversed for most codes typically used to simulate flood inundation at the reach scale (see Bates and De Roo, 2000 for a review).

A newly emergent research area is therefore how to integrate such massive data sets with lower-resolution numerical inundation models in an optimum manner that makes maximum use of the information content available. This is directly opposite to the problem that most environmental modellers have traditionally faced (see Grayson and Blöschl, 2001, for a general discussion).

For example, Marks and Bates (2000) describe the integration of a LiDAR data set with a 2D hydraulic model where the unstructured mesh discretization was derived independent of the topography. The topography was then assimilated into the model in an *a posteriori* step using weighted nearest neighbour interpolation to assign an elevation value to each computational node. This is typical of finite element mesh construction in many fields, but may not produce a mesh that captures those attributes of the original surface that are critical to the modelling problem in hand and may also lead to high data redundancy. To overcome these problems, Bates *et al.* (2003) describe a processing chain for high-resolution data assimilation into a lower-resolution unstructured model grid. This consists of: (1) variogram analysis to determine significant topographic length scales in the model; (2) identification of topographically significant points in this data set; (3) incorporation of these points into an unstructured model grid that provides a quality solution for the relevant numerical solver; and (4) use of the data left over from the mesh generation process to parameterize the Bates and Hervouet (1999) sub-grid-scale algorithm for dynamic wetting and drying. This method is demonstrated for the case of LiDAR topographic data but is general to any data type or model discretization. Cobby *et al.* (2003) take this process further and develop an automatic mesh generator that produces an unstructured grid refined according to vegetation features (hedges, stands of trees, etc.) on the floodplain (Figure 11.11) identified automatically from LiDAR. These methods show some promise; however, much further work is required in this area to more fully analyse the numerical quality of meshes generated by competing techniques.

Similarly for friction parameters, Mason *et al.* (2003) use an area-weighting method to calculate area-effective frictional resistance from the high-resolution height information contained in LiDAR data (Figure 11.12). This method aims to yield model parameters that are appropriate to the particular discretization used, rather than being scale- and discretization-independent, although this has yet to be fully tested.

11.4.6 Calibration, validation and uncertainty analysis

In all but the simplest cases (for example, the planar free surface 'lid' approach of Puech and Raclot, 2002), some form of calibration is required to successfully apply a floodplain flow model to a particular reach for a given flood event. Calibration is undertaken in order to identify appropriate values for parameters such that the model is able to reproduce observed data and, as previously mentioned, typically considers roughness coefficients assigned to the main channel and floodplain and values for turbulent eddy viscosity, ν_t, if a zero-equation turbulence closure is used. Though these values may sometimes be estimated in the field with a high degree of precision (Cunge, 2003), it has proven very difficult to demonstrate that such 'physically based' models are capable of providing accurate predictions

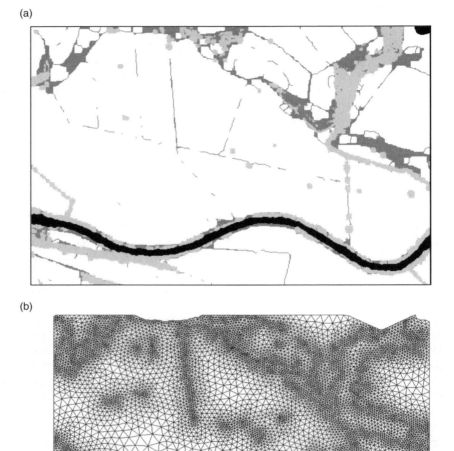

Figure 11.11 Results of the automatic unstructured two-dimensional mesh generator developed by Cobby et al. (2003). The software takes as input a LiDAR-derived vegetation height map (a) and produces an unstructured grid decomposed according to vegetation features (hedges, stands of trees, etc.) on the floodplain (b). (Published by John Wiley & Sons Ltd.)

from single realizations for reasons discussed in the critiques of Beven (1989, 1996, 2001) and Grayson *et al.* (1992). As such, values of parameters calculated by the calibration of models should be recognized as being effective values that may not have a physical interpretation outside the model structure within which they were calibrated. In addition, the process of estimating effective parameter values through calibration is further convoluted by a number of error sources inherent in the

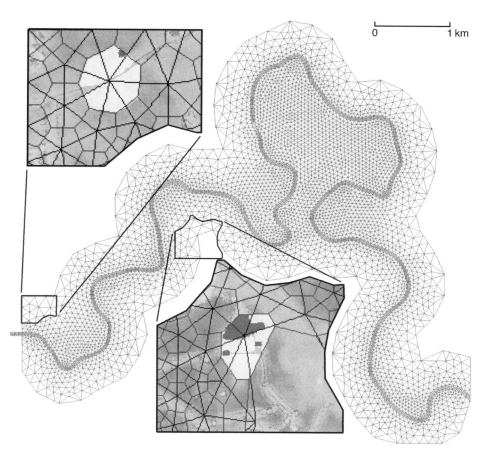

Figure 11.12 Assignment of an 'effective' friction value for each element of a finite element mesh using the vegetation height data shown in Figure 11.10 and the friction mapping algorithm of Mason et al. (2003). For each mesh node, an instantaneous friction factor is calculated at each time step, given the frictional material in the neighbourhood of the node and the current water depth and flow velocity there. A node's neighbourhood is defined as a polygon whose vertices are the centroids of the elements surrounding the node. Sub-regions of connected regions in the vegetation height map are formed by intersecting the node polygon map with the vegetation height map. Each sub-region in a polygon may contain either sediment or one of three vegetation height classes. If a sub-region contains vegetation, its region's average vegetation height is attributed to the sub-region (after Mason et al. 2003). (Published by John Wiley & Sons Ltd.)

inundation modelling process which cast some doubt on the certainty of calibrated parameters (Aronica *et al.*, 1998; Horritt, 2000; Bates *et al.*, 2004). Principally, these errors relate to the inadequacies of data used to represent heterogeneous river reaches (i.e. geometric integrity of floodplain topography and flow fluxes along the domain boundaries) but also extend to the observations with which the model is compared during calibration and the numerical approximations associated with the discrete solution of the controlling flow equations. The model will therefore require the estimation of effective parameter values that will, in part, compensate for these sources of error (Romanowicz and Beven, 2003).

Given that the number of degrees of freedom in even the simplest of numerical models is relatively large, it is no surprise that many different combinations of effective parameter values may fit sparse validation data equally well. Such equifinality in floodplain flow modelling has been well documented (e.g. Romanowicz et al., 1994, 1996; Aronica et al., 1998, 2002; Hankin et al., 2001; Romanowicz and Beven, 2003; Bates et al., 2004) and uncertainty analysis techniques, based on the Generalized Likelihood Uncertainty Estimation (GLUE) method of Beven and Binley (1992), have been developed and applied in response. Based on Monte Carlo simulation, GLUE explicitly recognizes this equifinality and seeks to make an assessment of the likelihood of a set of effective parameter values being an acceptable simulator of a system when model predictions are compared to observed field data (Chapter 17). In studies reported thus far, this level of 'acceptability' has typically been calculated by considering the deviation of simulated variables from one of the following classes of observed data (according to the terminology of Bates and Anderson (2000)):

1. External and internal bulk flow measures – for example, time series of stages and/or discharges recorded continuously at river gauging stations. Routinely available in national flow archives, these data have proven utility in testing the wave routing behaviour of flood models (e.g. Cunge et al., 1980; Klemeš, 1986) and have shown to be replicable by even the simplest of numerical schemes (Horritt and Bates, 2002). As Dietrich (2000) notes 'if the goal of a modelling exercise is to predict solely the time evolution of some bulk property of a complex system then spatial lumping (or spatial integration) can be invoked to filter out spatial variability'. As such, they are unlikely to provide a sufficiently rigorous test for competing model parameterizations (i.e. sets of effective parameter values) within spatially distributed schemes capable of directly simulating hydraulic patterns. Nevertheless, automatic collection of these data, often at hourly intervals or less, may provide records of sufficient temporal resolution to reasonably evaluate dynamic model performance and/or identify compensating errors operating within the model structure (e.g. Bates et al., 1998). However, given the considerable intermediate distances between gauging stations, internal observations are typically much less common than external ones. Clearly, bulk flow data, while being an important source of observational data, have only limited strength as a stand-alone piece of evidence for the evaluation of hydraulic models and are sometimes of questionable accuracy, particularly for large out-of-bank flows (Bates and Anderson, 2000). Recourse must therefore be made to additional sources of data, such as spatially distributed fields of model state variables (i.e. vector and point scale data) to further discriminate between competing models.

2. Vector data, for example, instantaneous observations of flood extent, processed either from remotely sensed data (Horritt, 1999) or from ground-survey methods (Simm, 1993), which may be considered an approximation to the zero water depth contour – observational data of this type has the advantage that for relatively flat floodplain topography, small changes in water surface elevation can result in

large changes in shoreline position. Hence, in order to replicate a flood shoreline adequately a model must generate, either directly or through secondary interpolation, an accurate distributed field of the flow depth. In studies thus far, inundation extent has proven a useful test of model performance as it is a relatively sensitive measure and its simulation requires a model capable of dealing with dynamic wetting and drying over complex topography (e.g. Bates et al., 1997; Bates and De Roo, 2000; Horritt, 2000). However, during events where the valley is entirely inundated, and the flood shoreline is constrained mostly by the slopes bounding the floodplain, changes in water levels may produce only small changes in flood extent (Horritt and Bates, 2001a,b). In these instances, internal validation based on hydrometric data may prove a more exacting test of model performance than extent data. Furthermore, inundation data have typically been limited to one synoptic view per event due to the limited temporal resolution of satellite-borne imaging radars and thus provide only single 'snapshots' of inundation extent which do not allow adequate checking of the inundation dynamics simulated by the model.

3. Point scale data – for example, water levels recorded directly during post-event surveys or velocities measured in the field. Maximum water levels can often be observed after the recession of a flood event, either as high-water marks on surviving structures or as trash or wracklines deposited at the limit of maximum inundation, and have been used in isolation as calibration data in a number of studies (e.g. Beven, 1989; Aronica et al., 1998). However, Lane et al. (1999) have drawn attention to the potential dangers of using these data for calibration purposes, observing that variables simulated by the model are integrated over both space and time and, as such, are rarely reconcilable with point scale observations collected in the field.

While it is clear from the preceding discussion that each class of observed data has some value in the calibration and uncertainty estimation processes, it is also apparent that the limitations of each data source are such that, when used in isolation, they may preclude any definitive judgements to be made about model performance. These limitations may be considered in direct relation to the dimensionality of the respective data sources and their failure to provide explicit definition in higher dimensions – for example, vector data which may be considered two dimensional in space but zero dimensional in time. However, by making use of more than one type of observational data in the calibration process, we should (hopefully) be able to overcome the inherent spatial or temporal limitations of any single source.

To date, only very limited attempts to calibrate distributed models against more than one particular data type have been made. Horritt and Bates (2002) tested the predictive performance of three industry-standard hydraulic codes on a 60-km reach of the River Severn, UK, using independent calibration data from hydrometric and satellite sources. They found that all models were capable of simulating inundation extent and floodwave travel times to similar levels of accuracy at optimum calibration, but that differences emerged according to the calibration data used when the models were used in predictive mode due to the different model responses to friction parameterizations. However, Horritt and Bates (2002) did not consider either the potential for combining both data sources in the calibration process or the uncertainties associated with the model predictions.

Multiple observation data sets for historical flood events are still exceedingly rare and, as such, the potential value of additional observations in the calibration process has yet to be explored in the case of floodplain flow models. Furthermore, the use of uncertainty estimation techniques during this conditioning process allows the relative value of individual (sets of) observations to be precisely quantified in terms of the reduction in uncertainty over effective parameter specification (cf. Beven and Binley, 1992; Hankin *et al.*, 2001). Interpretation of these uncertainty measures may also provide guidance over *how much* and *what type* of observed data would be required to achieve given levels of uncertainty reduction in simulated variables.

11.5 Future possibilities and research needs

Numerical modelling of floodplain flows has made major advances over the last decade as a result of increased computational power and the availability of rich and detailed remotely sensed data sources capable of accurate spatial parameterization of key input data. As a result, resources are now available for relatively inexperienced modellers to begin to run complex simulations on a desktop PC. Over the same period there has been relatively less progress on designing new models or solution algorithms, although laboratory experimental programmes continue to yield new insights into the physics of compound channel flow which serve to refine our perceptual model. These trends are likely to persist as the numerical solution of shallow water flow equations is now relatively well understood, whilst further advances in the field of floodplain flow modelling will continue to come from the development of new data sources. This in turn will lead to a better understanding of the uncertainties inherent in such codes when used in a calibration framework.

Examples of possible future research areas might include the use of increased computer power to simulate multi-dimensional flows over large domains with sophisticated turbulence closure. Application of such codes to flows in natural channels with relatively small grid sizes will also force researchers to consider how flow–vegetation interactions (Chapter 15) and flows in the near-wall region can be better represented. Increased computer power may also be used to extend the size of model domains, as well as enhance process representation. Here one could envisage basin-scale applications of simple 2D codes or the representation of significant portions of globally significant rivers to understand better the dynamics of the flood pulse in such environments (Coe *et al.*, 2002).

An improved understanding of the generation of friction and turbulence by vegetation combined with enhanced retrieval of plant biomechanical properties from remotely sensed data also has great potential. Mason *et al.* (2003) have already published an algorithm to determine first-order floodplain roughness coefficients from vegetation heights derived from airborne laser altimeter data, whilst Cobby *et al.* (2003) have used the same data to automatically generate unstructured grids for 2D floodplain models. With further research it is possible to imagine that we can move towards the automatic generation of uncalibrated hydraulic models that have reduced predictive uncertainty. Other types of remotely sensed data capture, for example surface flow velocity from along-track airborne radar interferometry (Srokosz, 1997) and synoptic sequences of inundation images through a flood, may also aid the model validation process and allow us to discriminate better between competing models, algorithms and data assimilation methods as suggested by Mason et al. (2003).

Lastly, whilst flow equations for floodplain flow modelling and their numerical solutions are in general well understood, one obvious exception is in the specification of appropriate schemes that are capable of handling dynamic wetting and drying effects over complex, low-lying topography (Chapter 6). Much research has already been undertaken in this area for moving and fixed grids (Lynch and Gray, 1980; Kawahara and Umetsu, 1986; Leclerc et al., 1990; Benkhaldoun and Monthe, 1994; Defina et al., 1994; Ip et al., 1998; Tchamen and Kawahita, 1998; Bates and Hervouet, 1999; Defina, 2000); however, a definitive solution has yet to be found. Such research is one area where further numerical analysis and new forms of the controlling equations are required, particularly in the case of 3D codes.

In conclusion, research into floodplain flow modelling has begun to enter a particularly dynamic phase as a result of new data and computing opportunities which will enable fundamental questions about distributed data assimilation, distributed calibration, the predictability of natural systems and the limits of physically based modelling to be critically assessed. Surface hydraulic models are a powerful tool with which to begin to answer these general environmental modelling questions as they are parametrically simple and we can actually make direct measurements of the process of interest. Floodplain flow modelling is also a key applied problem where distributed predictions are required and used to inform planning and insurance decisions that may have serious consequences for individuals but where, until recently, our ability to undertake such research was limited by the data available. Floodplain flow modelling is therefore an area where there is considerable potential for exciting and innovative science aimed at solving a major societal problem.

Acknowledgements

The research reported in this paper has been supported by a number of research grants including UK Natural Environment Research Council grant GR3 CO/030, UK Engineering and Physical Science Research Council grants and the EU Framework 5 project 'Development of a European Flood Forecasting System (EFFS)'. Matt Horritt is supported by a UK Natural Environment Research Council Fellowship.

References

Abbott, M. and Ionescu, F. 1967. On the numerical computation of nearly horizontal flows. *Journal of Hydraulic Research*, **5 (2)**, 97–117.
Ackers, P. 1993. Stage-discharge functions for 2-stage channels – the impact of new research. *Journal of the Institution of Water and Environmental Management*, **7 (1)**, 52–61.

Ali, S. and Goodwin, P. 2002. The predictive ability of 1D models to simulate floodplain processes. In R.A. Falconer, B. Lin, E.L. Harris and C.A.M.E. Wilson (eds), *Hydroinformatics 2002: Proceedings of the Fifth International Conference on Hydroinformatics. Volume One: Model Development and Data Management*, IWA Publishing, London, 247–252.

Aronica, G., Hankin, B.G. and Beven, K.J. 1998. Uncertainty and equifinality in calibrating distributed roughness coefficients in a flood propagation model with limited data. *Advances in Water Resources*, **22 (4)**, 349–365.

Aronica, G., Bates, P.D. and Horritt, M.S. 2002. Assessing the uncertainty in distributed model predictions using observed binary pattern information within GLUE. *Hydrological Processes*, **16**, 2001–2016.

Ashworth, P.J., Bennett, S.J., Best, J.L. and McLelland, S.J. (eds) 1996. *Coherent Flow Structures in Open Channels*, John Wiley & Sons, Chichester, UK, 733pp.

Baltsavias, E.P. 1999. A comparison between photogrammetry and laser scanning. *ISPRS Journal of Photogrammetry and Remote Sensing*, **54 (2–3)**, 83–94.

Bates, P.D. and Anderson, M.G. (eds) 2000. Validation of hydraulic models. *Model Validation, Perspectives in Hydrological Science*, John Wiley & Sons, Chichester, UK, 325–356.

Bates, P.D. and De Roo, A.P.J. 2000. A simple raster-based model for floodplain inundation. *Journal of Hydrology*, **236**, 54–77.

Bates, P.D. and Hervouet, J.-M. 1999. A new method for moving boundary hydrodynamic problems in shallow water. *Proceedings of the Royal Society of London, Series A*, **455**, 3107–3128.

Bates, P.D., Horritt, M.S., Smith, C. and Mason, D. 1997. Integrating remote sensing observations of flood hydrology and hydraulic modelling. *Hydrological Processes*, **11**, 1777–1795.

Bates, P.D., Stewart, M.D., Siggers, G.B., Smith, C.N., Hervouet, J.-M. and Sellin, R.H.J. 1998. Internal and external validation of a two-dimensional finite element model for river flood simulation. *Proceedings of the Institution of Civil Engineers, Water Maritime and Energy*, **130**, 127–141.

Bates, P.D., Stewart, M.D., Desitter, A., Anderson, M.G., Renaud, J.-P. and Smith, J.A. 2000. Numerical simulation of floodplain hydrology. *Water Resources Research*, **36**, 2517–2530.

Bates, P.D., Marks, K.J. and Horritt, M.S. 2003. Optimal use of high-resolution topographic data in flood inundation models. *Hydrological Processes*, **17**, 5237–5557.

Bates, P.D., Horritt, M.S., Aronica, G. and Beven, K.J. 2004. Bayesian updating of flood inundation likelihoods conditioned on flood extent data. *Hydrological Processes*, **18**, 3347–3370.

Bechteler, W., Hartmaan, S. and Otto, A.J. 1994. Coupling of 2D and 1D models and integration into Geographic Information Systems (GIS). In W.R. White and J. Watts, (eds), *Proceedings of the 2nd International Conference on River Flood Hydraulics*, John Wiley & Sons, Chichester, UK, 155–165.

Benkhaldoun, F. and Monthe, L. 1994. An adaptive nine-point finite volume Roe scheme for two-dimensional Saint Venant equations. In P. Molinaro and L. Natale (eds), *Modelling Flood Propagation Over Initially Dry Areas*, American Society of Civil Engineers, New York, 30–44.

Beven, K.J. 1989. Changing ideas in hydrology – The case of physically-based models. *Journal of Hydrology*, **105**, 157–172.

Beven, K.J. 1996. A discussion in distributed modelling. In J.-C. Refsgaard and M.B. Abbott (eds), *Distributed Hydrological Modelling*, Kluwer Academic Publishers, The Netherlands, 289–295.

Beven, K.J. 2001. *Rainfall-Runoff Modelling: The Primer*, John Wiley & Sons, Chichester, UK, 372pp.

Beven, K.J. and Binley, A. 1992. The future of distributed models: Model calibration and uncertainty prediction. *Hydrological Processes*, **6**, 279–298.

Bladé, E., Gómez, M. and Dolz, J. 1994. Quasi-two dimensional modelling of flood routing in rivers and flood plains by means of storage cells. In P. Molinaro and L. Natale (eds), *Modelling of Flood Propagation Over Initially Dry Areas*, American Society of Civil Engineers, New York, 156–170.

Bridge, J.S. and Gabel, S.L. 1992. Flow and sediment dynamics in a low sinuosity, braided river – Calamus River, Nebraska sandhills. *Sedimentology*, **39 (1)**, 125–142.

Burt, T.P., Bates, P.D., Stewart, M.D., Claxton, A.J., Anderson, M.G. and Price, D.A. 2002. Water table fluctuations within the floodplain of the River Severn, England. *Journal of Hydrology*, **262**, 1–20.

Butler, J.B., Lane, S.N. and Chandler, J.H. 2001. Automated extraction of grain-size data from gravel surfaces using digital image processing. *Journal of Hydraulic Research*, **39 (5)**, 519–529.

Castro, N.M. and Hornberger, G.M. 1991. Surface-subsurface water interactions in an alluviated mountain stream channel. *Water Resources Research*, **27**, 1613–1621.

Charlton, M.E., Large, A.R.G. and Fuller, I.C. 2003. Application of airborne LiDAR in river environments: The River Coquet, Northumberland, UK. *Earth Surface Processes and Landforms*, **28 (3)**, 299–306.

Cobby, D.M., Mason, D.C., Davenport, I.J. and Horritt, M.S. 2000. Obtaining accurate maps of topography and vegetation to improve 2D hydraulic flood models. *Proceedings of the EOS/SPIE Symposium on Remote Sensing for Agriculture, Ecosystems, and Hydrology II*, Barcelona, 25–29 September, 125–136.

Cobby, D.M., Mason, D.C. and Davenport, I.J. 2001. Image processing of airborne scanning laser altimetry for improved river flood modelling. *ISPRS Journal of Photogrammetry and Remote Sensing*, **56 (2)**, 121–138.

Cobby, D.M., Mason, D., Horritt, M.S. and Bates, P.D. 2003. Two-dimensional hydraulic flood modelling using a finite element mesh decomposed according to vegetation and topographic features derived from airborne scanning laser altimetry. *Hydrological Processes*, **17**, 1979–2000.

Coe, M.T., Costa, M.H., Botta, A. and Birkett, C. 2002. Long-term simulations of discharge and floods in the Amazon Basin. *Journal of Geophysical Research – Atmospheres*, **107 (D20)**, art. no. 8044.

Cunge, J.A. 2003. Of data and models. *Journal of Hydroinformatics*, **5 (2)**, 75–98.

Cunge, J.A., Holly, F.M. Jr and Verwey, A. 1980. *Practical Aspects of Computational River Hydraulics*, Pitman, London, 420pp.

De Roo, A.P.J, Odijk, M., Schmuck, G., Koster, E. and Lucieer, A. 2001. Assessing the effects of land use changes on floods in the Meuse and Oder catchment. *Physics and Chemistry of the Earth Part B – Hydrology, Oceans and Atmospheres*, **26 (7–8)**, 593–599.

De Roo, A.P.J., Bartholmes, J., Bates, P.D., Beven, K., Bongioannini-Cerlini, B., Gouweleeuw, B., Heise, E., Hils, M., Hollingsworth, M., Holst, B., Horritt, M., Hunter, N., Kwadijk, J., Pappenburger, F., Reggiani, P., Rivin, G., Sattler, K., Sprokkereef, E., Thielen, J., Todini, E. and Van Dijk, M. 2003. Development of a European flood forecasting system. *International Journal of River Basin Management*, **1 (1)**, 49–59.

Defina, A. 2000. Two-dimensional shallow flow equations for partially dry areas. *Water Resources Research*, **36 (11)**, 3251–3264.

Defina, A., D'Alpaos, L. and Matticchio, B. 1994. A new set of equations for very shallow water and partially dry areas suitable to 2D numerical models. In P. Molinaro and L. Natale (eds), *Modelling Flood Propagation Over Initially Dry Areas*, American Society of Civil Engineers, New York, 72–81.

Dhondia, J.F. and Stelling, G.S. 2002. Application of one-dimensional-two-dimensional integrated hydraulic model for flood simulation and damage assessment. In R.A. Falconer, B. Lin, E.L. Harris and C.A.M.E. Wilson (eds), *Hydroinformatics 2002: Proceedings of the Fifth International Conference on Hydroinformatics. Volume One: Model Development and Data Management*, IWA Publishing, London, 265–276.

Dietrich, C.R., 2000. On simulation, calibration and ill-conditioning with application to environmental system modelling. In M.G. Anderson and P.D. Bates (eds), *Model Validation, Perspectives in Hydrological Science*, John Wiley & Sons, Chichester, UK, 77–116.

Ervine, D.A. and Baird, J.I. 1982. Rating curves for rivers with overbank flow. *Proceedings of the Institution of Civil Engineers Part 2 – Research and Theory*, **73**, 465–472.

Estrela, T. and Quintas, L. 1994. Use of GIS in the modelling of flows on floodplains. In H.R. White and J. Watts (eds), *Proceedings of the 2nd International Conference on River Flood Hydraulics*, John Wiley & Sons, Chichester, UK, 177–189.

Feldhaus, R., Höttges, J., Brockhaus, T. and Rouvé, G. 1992. Finite element simulation of flow and pollution transport applied to a part of the River Rhine. In R.A. Falconer, K. Shiono and R.G.S. Matthews (eds), *Hydraulic and Environmental Modelling: Estuarine and River Waters*, Ashgate Publishing, Aldershot, 323–344.

Feng, Y.T. and Perić, D. 2003. A spatially adaptive linear space-time finite element solution procedure for incompressible flows with moving domains. *International Journal of Numerical Methods in Fluids*, **43**, 1099–1106.

French, J.R. 2003. Airborne LiDAR in support of geomorphological and hydraulic modelling. *Earth Surface Processes and Landforms*, **28 (3)**, 321–335.

Fukuoka, S. and Fujita, K. 1989. Prediction of flow resistance in compound channels and its application to the design of river courses. *Proceedings of the Japan Society of Civil Engineers*, **411**, 63–72.

Gee, D.M., Anderson, M.G. and Baird, L. 1990. Large scale floodplain modelling. *Earth Surface Processes and Landforms*, **15**, 512–523.

Ghisalberti, M. and Nepf, H.M. 2002. Mixing layers and coherent structures in vegetated aquatic flows. *Journal of Geophysical Research – Oceans*, **107 (C2)**, art. no. 3011.

Gomes-Pereira, L.M. and Wicherson, R.J. 1999. Suitability of laser data for deriving geographical data: A case study in the context of management of fluvial zones. *Photogrammetry and Remote Sensing*, **54**, 105–114.

Grayson, R. and Blöschl, G. 2001. *Spatial Patterns in Catchment Hydrology: Observations and Modelling*. Cambridge University Press, Cambridge, UK, 416pp.

Grayson, R.B., Moore, I.D. and McMahon, T.A. 1992. Physically-based hydrologic modelling: II. Is the concept realistic? *Water Resources Research*, **28 (10)**, 2659–2666.

Hankin, B.G., Hardy, R., Kettle, H. and Beven, K.J. 2001. Using CFD in a GLUE framework to model the flow and dispersion characteristics of a natural fluvial dead zone. *Earth Surface Processes and Landforms*, **26 (6)**, 667–687.

Haycock, N.E. and Burt, T.P. 1993. Role of floodplain sediments in reducing the nitrate concentration of subsurface run-off: A case study in the Cotswolds, UK. *Hydrological Processes*, **7**, 287–295.

Hervouet, J.-M. and Van Haren, L. 1996. Recent advances in numerical methods for fluid flows. In M.G. Anderson, D.E. Walling and P.D. Bates (eds), *Floodplain Processes*, John Wiley & Sons, Chichester, 183–214.

Hey, R.D. 1979. Flow resistance in gravel bed rivers. *Journal of Hydraulics Division*, American Society of Civil Engineers, **105 (4)**, 365–379.

Hodgson, M.E., Jensen, J.R., Schmidt, L., Schill, S. and Davis, B. 2003. An evaluation of LIDAR- and IFSAR-derived digital elevation models in leaf-on conditions with USGS Level 1 and Level 2 DEMs. *Remote Sensing of the Environment*, **84 (2)**, 295–308.

Horritt, M.S. 1999. A statistical active contour model for SAR image segmentation. *Image and Vision Computing*, **17**, 213–224.

Horritt, M.S. 2000. Calibration of a two-dimensional finite element flood flow model using satellite radar imagery. *Water Resources Research*, **36 (11)**, 3279–3291.

Horritt, M.S. and Bates, P.D. 2001a. Predicting floodplain inundation: Raster-based modelling versus the finite element approach. *Hydrological Processes*, **15**, 825–842.

Horritt, M.S. and Bates, P.D. 2001b. Effects of spatial resolution on a raster based model of flood flow. *Journal of Hydrology*, **253**, 239–249.

Horritt, M.S. and Bates, P.D. 2002. Evaluation of 1-D and 2-D numerical models for predicting river flood inundation. *Journal of Hydrology*, **268**, 87–99.

Ip, J.T.C., Lynch, D.R. and Friedrichs, C.T. 1998. Simulation of estuarine flooding and dewatering with application to Great Bay, New Hampshire. *Estuarine Coastal and Shelf Science*, **47 (2)**, 119–141.

Kawahara, M. and Umetsu, T. 1986. Finite element method for moving boundary problems in river flows. *International Journal of Numerical Methods in Fluids*, **6**, 365–386.

King, I.P. and Roig, L. 1988. Two-dimensional finite element models for floodplains and tidal flats. In K. Niki and M. Kawahara (eds), *Proceedings of an International Conference on Computational Methods in Flow Analysis*, Okayama, Japan, 711–718.

Klemeš, V. 1986. Operational testing of hydrologic simulation models. *Hydrological Sciences Journal*, **31**, 13–24.

Knight, D.W. and Hamed, M.E. 1984. Boundary shear in symmetrical compound channels. *Journal of Hydraulic Engineering*, American Society of Civil Engineers, **110 (10)**, 1412–1430.

Knight, D.W. and Shiono, K. 1996. River channel and floodplain hydraulics. In M.G. Anderson, D.E. Walling and P.D. Bates (eds), *Floodplain Processes*, John Wiley & Sons, Chichester, 139–182.

Kohane, R. and Welz, R. 1994. Combined use of FE models for prevention of ecological deterioration of areas next to a river hydropower complex. In A. Peter, G. Wittum, U. Meissner, C.A. Brebbia, W.G. Gray and G.F. Pinder (eds), *Computational Methods in Water Resources X*, Kluwer, The Netherlands, Volume **1**, 59–66.

Kouwen, N. 1988. Field estimation of the biomechanical properties of grass. *Journal of Hydraulic Research*, **26 (5)**, 559–568.

Kouwen, N. 2000. Closure of 'effect of riparian vegetation on flow resistance and flood potential'. *Journal of Hydraulic Engineering*, American Society of Civil Engineers, **126 (12)**, 954.

Kouwen, N. and Li, R.M. 1980. Biomechanics of vegetative channel linings. *Journal of Hydraulics Division*, American Society of Civil Engineers, **106 (6)**, 713–728.

Krabill, W.B., Collins, J.G., Link, L.E., Swift, R.N. and Butler, M.L. 1984. Airborne laser topographic mapping results. *Photogrammetric Engineering and Remote Sensing*, **50**, 685–694.

Lane, S.N. 2000. The measurement of river channel morphology using digital photogrammetry. *Photogrammetric Record*, **16 (96)**, 937–957.

Lane, S.N., Bradbrook, K.F., Richards, K.S., Biron, P.M. and Roy, A.G. 1999. The application of computational fluid dynamics to natural river channels: Three-dimensional versus two-dimensional approaches. *Geomorphology*, **29**, 1–20.

Leclerc, M., Bellemare, J.-F., Dumas, G. and Dhatt, G. 1990. A finite element model of estuarine and river flows with moving boundaries. *Advances in Water Resources*, **13**, 158–168.

Lopez, F. and Garcia, M.H. 2001. Mean flow and turbulence structure of open-channel flow through non-emergent vegetation. *Journal of Hydraulic Engineering*, American Society of Civil Engineers, **127 (5)**, 392–402.

Lynch, D.R. and Gray, W.G. 1980. Finite element simulation of flow deforming regions. *Journal of Computational Physics*, **36**, 135–153.

Ma, L., Ashworth, P.J., Best, J.L., Elliot, L., Ingham, D.B. and Whitcombe, L.J. 2002. Computational fluid dynamics and the physical modelling of an upland urban river. *Geomorphology*, **44 (3–4)**, 375–391.

Marks, K. and Bates, P.D. 2000. Integration of high resolution topographic data with floodplain flow models. *Hydrological Processes*, **14**, 2109–2122.

Mason, D., Cobby, D.M., Horritt, M.S. and Bates, P.D. 2003. Floodplain friction parameterization in two-dimensional river flood models using vegetation heights derived from airborne scanning laser altimetry. *Hydrological Processes*, **17**, 1711–1732.

McLelland, S.J., Ashworth, P.J., Best, J.L. and Livesey, J.R. 1999. Turbulence and secondary flow over sediment stripes in weakly bimodal bed material. *Journal of Hydraulic Engineering*, American Society of Civil Engineers, **125 (5)**, 463–473.

Mertes, L.A.K. 1997. Documentation and significance of the perirheic zone on inundated floodplains. *Water Resources Research*, **33**, 1749–1762.

Meselhe, E.A., Weber, L.J., Odgaard, A.J. and Johnson, T. 2000. Numerical modeling for fish diversion studies. *Journal of Hydraulic Engineering*, American Society of Civil Engineers, **126 (5)**, 365–374.

Molinaro, P., Di Filippo, A. and Ferrari, F. 1994. Modelling of flood wave propagation over flat dry areas of complex topography in presence of different infrastructures. In P. Molinaro and L. Natale (eds), *Modelling of Flood Propagation Over Initially Dry Areas*, American Society of Civil Engineers, New York, 209–225.

Nepf, H.M. 1999. Drag, turbulence, and diffusion in flow through emergent vegetation. *Water Resources Research*, **35 (2)**, 479–489.

NERC. 1975. *Flood Studies Report*. Natural Environment Research Council, London, 5 volumes.

Nezu, I., Tominaga, A. and Nakagawa, H. 1993. Field-measurements of secondary currents in straight rivers. *Journal of Hydraulic Engineering*, American Society of Civil Engineers, **119 (5)**, 598–614.

Nicholas, A.P. and Mitchell, C.A. 2003. Numerical simulation of overbank processes in topographically complex floodplain environments. *Hydrological Processes*, **17 (4)**, 727–746.

Nicholas, A.P. and Walling, D.E. 1997. Modelling flood hydraulics and overbank deposition on river floodplains. *Earth Surface Processes and Landforms*, **22**, 59–77.

Pinder, G.E. and Sauer, S.P. 1971. Numerical simulation of flood wave modification due to bank storage effects. *Water Resources Research*, **7**, 63–70.

Poole, G.C., Stanford, J.A., Frissell, C.A. and Running, S.W. 2002. Three-dimensional mapping of geomorphic controls on flood-plain hydrology and connectivity from aerial photos. *Geomorphology*, **48 (4)**, 329–347.

Preissmann, A. 1961. Propagation of translatory waves in channels and rivers. *Proceedings of the 1st Congress de l'Association Francaise de Calcul*, Grenoble, France, 433–442.

Puech, C. and Raclot, D. 2002. Using geographical information systems and aerial photographs to determine water levels during floods. *Hydrological Processes*, **16**, 1593–1602.

Rabus, B., Eineder, M., Roth, A. and Bamler, R. 2003. The shuttle radar topography mission – a new class of digital elevation models acquired by spaceborne radar. *ISPRS Journal of Photogrammetry and Remote Sensing*, **57 (4)**, 241–262.

Reynolds, O. 1895. On the dynamical theory of incompressible viscous fluids and the determination of the criterion. *Philosophical Transactions of the Royal Society*, **186A**, 123–164.

Rodi, W. 1993 *Turbulence Models and their Application in Hydraulics*. IAHR Special Publication, Delft, The Netherlands, 3rd edition, 104pp.

Rodi, W., Ferziger, J.H., Breur, M. and Pourquié, M. 1997. Status of Large Eddy Simulation: Results of a workshop. *Journal of Fluids Engineering*, Transactions of the American Society of Mechanical Engineers, **119**, 248–262.

Romanowicz, R. and Beven, K. 2003. Estimation of flood inundation probabilities as conditioned on event inundation maps. *Water Resources Research*, **39 (3)**, art. no. 1073.

Romanowicz, R.J., Beven, K.J. and Tawn, J.A. 1994. Evaluation of predictive uncertainty in nonlinear hydrological models using a Bayesian Approach. In V. Barnett and F. Turkman (eds), *Statistics for the Environment (2), Water Related Issues*, John Wiley & Sons, Chichester, UK, 297–318.

Romanowicz, R., Beven, K.J. and Tawn, J. 1996. Bayesian calibration of flood inundation models. In M.G. Anderson, D.E. Walling and P.D. Bates (eds), *Floodplain Processes*, John Wiley & Sons, Chichester, UK, 333–360.

Samuels, P.G. 1990. Cross section location in one-dimensional models. In W.R. White (ed.), *International Conference on River Flood Hydraulics*, John Wiley & Sons, Chichester, 339–350.

Sellin, R.H.J. 1964. A laboratory investigation into the interaction between the flow in the channel of a river and that over its floodplain. *La Houille Blanche*, **7**, 793–801.

Sellin, R.H.J. and Willets, B.B. 1996. Three-dimensional structures, memory and energy dissipation in meandering compound channel flow. In M.G. Anderson, D.E. Walling and P.D. Bates (eds), *Floodplain Processes*, John Wiley & Sons, Chichester, 255–298.

Sen, D. 2002. An algorithm for coupling 1D river flow and quasi-2D flood inundation flow. In R.A. Falconer, B. Lin, E.L. Harris and C.A.M.E. Wilson (eds), *Hydroinformatics 2002: Proceedings of the Fifth International Conference on Hydroinformatics. Volume One: Model Development and Data Management*, IWA Publishing, London, 103–108.

Shao, X.J., Wang, H. and Chen, Z. 2003. Numerical modelling of turbulent flow in curved channels of compound cross-section. *Advances in Water Resources*, **26 (5)**, 525–539.

Shiono, K. and Knight, D.W. 1991. Turbulent open-channel flows with variable depth across the channel. *Journal of Fluid Mechanics*, **222**, 617–646.

Shvidchenko, A.B. and Pender, G. 2001. Macroturbulent structure of open-channel flow over gravel beds. *Water Resources Research*, **37 (3)**, 709–719.

Simm, D.J. 1993. The deposition and storage of suspended sediment in contemporary floodplain systems: A case study of the River Culm, Devon. PhD Thesis, University of Exeter, 347pp.

Squillace, P.J. 1996. Observed and simulated movement of bank storage water. *Groundwater*, **34**, 121–134.

Srokosz, M. 1997. Ocean surface currents and waves and along-track interferometric SAR. *Proceedings of a Workshop on Single-pass Satellite Interferometry*, Imperial College, London, 22 July.

Stanford, J.A. and Ward, J.V. 1988. The hyporheic habitat of river ecosystems. *Nature*, **335**, 64–66.

Stewart, M.D., Bates, P.D., Anderson, M.G., Price, D.A. and Burt, T.P. 1999. Modelling floods in hydrologically complex lowland river reaches. *Journal of Hydrology*, **223**, 85–106.

Stoesser, T., Wilson, C.A.M.E., Bates, P.D. and Dittrich, A. 2003. Application of a 3D numerical model to a river with vegetated floodplains. *Journal of Hydroinformatics*, **5**, 99–112.

Tchamen, G.W. and Kawahita, R.A. 1998. Modelling wetting and drying effects over complex topography. *Hydrological Processes*, **12**, 1151–1183.

Temple, D.M. 1987. Closure of 'Velocity distribution coefficients for grass-lined channels'. *Journal of Hydraulic Engineering*, American Society of Civil Engineers, **113 (9)**, 1224–1226.

Thomas, T.G. and Williams, J.J.R. 1995. Large-Eddy Simulation of turbulent flow in an asymmetric compound open channel. *Journal of Hydraulic Research*, **33 (1)**, 27–41.

Venere, M. and Clausse, A. 2002. A computational environment for water flow along floodplains. *International Journal of Computational Fluid Dynamics*, **16 (4)**, 327–330.

Walling, D.E., Bradley, S.B. and Lambert, C.P. 1986. Conveyance loss of suspended sediment within a floodplain system. *IAHS Publication*, **159**, 119–132.

Westaway, R.M., Lane, S.N. and Hicks, D.M. 2003. Remote survey of large-scale braided, gravel-bed rivers using digital photogrammetry and image analysis. *International Journal of Remote Sensing*, **24 (4)**, 795–815.

Wicks, J., Mocke, R., Bates, P.D., Ramsbottom, D., Evans, E. and Green, C. 2003. Selection of appropriate models for flood modelling. *Proceedings of the 38th DEFRA Annual Flood and Coastal Management Conference*, 16–18 July, Department of Food and Rural Affairs, Keele University, London, UK.

Wilson, C.A.M.E. and Horritt, M.S. 2002. Measuring the flow resistance of submerged grass. *Hydrological Processes*, **16 (13)**, 2589–2598.

Wilson, C.A.M.E., Stoesser, T., Olsen, N.R.B. and Bates, P.D. 2003. Application and validation of numerical codes in the prediction of compound channel flows. *Proceedings of the Institution of Civil Engineers, Water Maritime and Energy*, **153**, 117–128.

Wilson, M. and Atkinson, P. (in press). The use of elevation data in flood inundation models: A comparison of ERS interferometric SAR and combined contour and GPS data. *IAHR International Journal of River Basin Management*.

Woessner, W.W. 2000. Stream and fluvial plain ground water interactions: Re-scaling hydrogeologic thought. *Groundwater*, **38**, 423–429.

Wormleaton, P.R., Allen, J. and Hadjipanos, P. 1982. Discharge assessment in compound channel flow. *Journal of Hydraulics Division*, American Society of Civil Engineers, **108 (HY9)**, 975–993.

Wroblicky, G.J., Campana, M.E., Valett, H.M. and Dahm, C.N. 1998. Seasonal variation in surface-subsurface water exchange and lateral hyporheic area of two stream-aquifer systems. *Water Resources Research*, **34**, 317–328.

12
Modelling water quality processes in estuaries

R.A. Falconer, B. Lin and S.M. Kashefipour

12.1 Introduction

Estuaries are complex ecosystems which can include wetlands and mangrove swamps along their inter-tidal shores. They are often dominated by complex channel networks, areas of salt marsh, fen or peatlands, or tidal floodplains. They serve as a meeting point between land and sea and occupy specific ecological niches of considerable importance to conservation groups, industry and local communities. They are often an important source of food for coastal aquaculture, whereby they act as natural filters for suspended material and pollutants, and offer effective flood protection for low-lying areas (Falconer and Lin, 1997).

An estuary is essentially a mixing zone through which a river discharges to the sea. The zone of mixing is not static, however, but moves in position according to factors such as tide, wind and freshwater flow. Consequently, water quality in estuaries usually exhibits marked spatial and temporal variations. For a well-mixed estuary, these variations produce a gradient in water quality along the estuary (HR Wallingford, 1996). The high suspended sediment load in most estuaries provides a large surface area for the attachment of chemical and biological species, including heavy metals arising from industrial waste discharges and bacterial micro-organisms arising from urban domestic wastewater discharges. With increasing awareness of all aspects of aqua-environmental and ecological pollution, in recent years there has been a marked increase in the development and application of numerical models to predict estuarine water quality characteristics. One of the main reasons for this increase is that numerical models offer a powerful tool for investigating the complex interactions between estuary morphology, sediment fluxes and water quality.

This chapter is devoted to the establishment of the governing differential equations and parameters describing the hydrodynamics, water quality and sediment transport processes occurring regularly in estuarine systems. Details are given of 1D, 2D and – to a lesser extent – 3D modelling approaches, with examples being given of the capabilities of such models as developed by the authors for predicting the distributions of sediment concentrations and faecal indicator levels in the Mersey and Ribble Estuaries, respectively, both located in North West England.

12.2 Governing equations

12.2.1 Hydrodynamic modelling

12.2.1.1 Three-dimensional flows

The numerical models used by water and environmental engineers and managers to predict the flow, water quality and contaminant and sediment transport processes in estuaries are based on first solving the governing hydrodynamic equations. For a Cartesian co-ordinate system, with the main body of the flow in the x-direction, the corresponding 3D Reynolds equations for mass and momentum conservation in the flow direction can be written in a general conservative form as (Falconer, 1993; Falconer et al., 2001):

$$\frac{\partial u}{\partial x} + \frac{\partial v}{\partial y} + \frac{\partial w}{\partial z} = 0 \qquad (12.1)$$

$$\underbrace{\frac{\partial u}{\partial t}}_{1} + \underbrace{\frac{\partial u^2}{\partial x} + \frac{\partial uv}{\partial y} + \frac{\partial uw}{\partial z}}_{2} = \underbrace{X}_{3} - \underbrace{\frac{1}{\rho}\frac{\partial P}{\partial x}}_{4} - \underbrace{\frac{\partial \overline{u'u'}}{\partial x} + \frac{\partial \overline{u'v'}}{\partial y} + \frac{\partial \overline{u'w'}}{\partial z}}_{5} \qquad (12.2)$$

where u, v, w = velocity components in x-, y-, z-co-ordinate directions, respectively, t = time, X = body force in x-direction, ρ = fluid density, P = fluid pressure, $\overline{u'u'}$, $\overline{u'v'}$, $\overline{u'w'}$ = Reynolds stresses in the x-direction on x, y, z planes, respectively.

Equations similar to (12.2) can be written to evaluate the velocity components v and w in the y- and z- directions respectively. The numbered terms in equation (12.2) refer to local acceleration (term 1), advective acceleration (2), body force (3), pressure gradient (4) and turbulent shear stresses (5).

In modelling estuarine flows in two and three dimensions the effects of the earth's rotation will need to be included, giving for the body force components:

$$X = 2v\varpi\sin\phi$$
$$Y = -2u\varpi\sin\phi \qquad (12.3)$$
$$Z = -g$$

where ϖ = speed of earth's rotation, ϕ = earth's latitude and g = gravitational acceleration. The main effects of the earth's rotation, giving rise to the Coriolis acceleration, are to set up transverse water surface slopes across the estuary and to enhance the effect of secondary currents and meandering.

For 3D flow predictions, either the full 3D governing equations are solved, which leads to a complex numerical formulation to evaluate the pressure P, or, more usually, a hydrostatic pressure distribution is assumed to occur in the vertical (z) direction, which leads to an expression for P of the following form:

$$P(z) = \rho g(\zeta - z) + P_a \qquad (12.4)$$

where ζ = water surface elevation above (positive) datum and P_a = atmospheric pressure. The corresponding derivative of equation (12.4), for inclusion in equation (12.2), gives:

$$\frac{\partial P}{\partial x} = \rho g \frac{\partial \zeta}{\partial x} + \frac{\partial P_a}{\partial x} \qquad (12.5)$$

A similar representation can be written for the pressure gradient in the y-direction. The effects of the atmospheric pressure gradient are generally small in estuarine flows and are neglected.

The only unknown terms remaining in equation (12.2) are the Reynolds stresses, which need to be related to the 3D velocity field before solving for the water levels and the 3D velocity components. In solving for these stresses, Boussinesq (Goldstein, 1938) proposed that they could be represented in a diffusive manner, giving:

$$-\overline{u'u'} = \nu_t \left[\frac{\partial u}{\partial x} + \frac{\partial u}{\partial x} \right]$$
$$-\overline{u'v'} = \nu_t \left[\frac{\partial u}{\partial y} + \frac{\partial v}{\partial x} \right] \qquad (12.6)$$
$$-\overline{u'w'} = \nu_t \left[\frac{\partial u}{\partial z} + \frac{\partial w}{\partial x} \right]$$

where ν_t = kinematic eddy viscosity.

The eddy-viscosity coefficient can be obtained in several ways (Falconer and Chen, 1996). The simplest approach is to assume a constant value based on field data. However, whilst this approach may be adequate for predicting velocity distributions in large water bodies, such as coastal basins, lakes or reservoirs, it is not particularly accurate for 3D model simulations in estuaries, where such models are generally used only for predicting complex velocity field distributions in the vicinity of structures (such as bridge piers) and short river reaches. Another approach is to apply a zero-equation turbulence model, similar to that prescribed by Prandtl's mixing-length hypothesis (Goldstein, 1938), wherein:

$$\nu_t = \ell^2 J \qquad (12.7)$$

where ℓ = a characteristic mixing length and J = magnitude of local velocity gradients in x-, y- and z-directions. Using this approach the mixing length can readily be determined for typical logarithmic-type velocity profiles, but for the more complex flow fields where 3D models are appropriate, the velocity distribution is unlikely to be primarily logarithmic in form, and strong secondary currents often lead to a more complex and less well defined mixing length.

Hence, for most practical problems where 3D models are appropriate for estuarine simulations, the turbulent stresses given in equation (12.2) need to be solved using either a two-equation turbulence model of the k-ε type, or a three-equation turbulence model of the algebraic stress type and wherein the Reynolds stress terms are solved directly. For the more usual approach, using either the linear or non-linear k-ε model, the eddy viscosity is defined as:

$$\nu_t = \frac{C_\mu k^2}{\varepsilon} \qquad (12.8)$$

where C_μ = turbulent model coefficient, k = turbulent kinetic energy and ε = dissipation rate of turbulent kinetic energy. Transport equations are derived for k and ε (Rodi, 1984), which in general include transport by advection, production and dissipation.

12.2.1.2 Two-dimensional flows

For many practical problems where there are significant variations across the streamwise flow direction, it is commonplace for the velocity field to be determined using a 2D depth-integrated numerical model. To determine the hydrodynamic velocity field using a 2D model, the governing 3D equations are integrated over the depth, giving for equations (12.1) and (12.2), respectively:

$$\frac{\partial \zeta}{\partial t} + \frac{\partial q_x}{\partial x} + \frac{\partial q_y}{\partial y} = 0 \qquad (12.9)$$

$$\frac{\partial q_x}{\partial t} + \beta \left[\frac{\partial u q_x}{\partial x} + \frac{\partial v q_x}{\partial y} \right] = f q_y - gH \frac{\partial \zeta}{\partial x} + \frac{\tau_{xw}}{\rho} - \frac{\tau_{xb}}{\rho} + 2\frac{\partial}{\partial x}\left[\bar{\nu}_t \frac{\partial q_x}{\partial x} \right] + \frac{\partial}{\partial y}\left[\bar{\nu}_t \left(\frac{\partial q_x}{\partial y} + \frac{\partial q_y}{\partial x} \right) \right] \qquad (12.10)$$

where q_x, q_y = discharge per unit width in x- and y-directions, β = momentum correction factor for non-uniform vertical velocity profile, τ_{xw}, τ_{xb} = surface and bed shear stress components respectively, in the x-direction, and $\bar{\nu}_t$ = depth-averaged eddy viscosity.

The momentum correction factor can be estimated either from field data, which is preferable, or alternatively by assuming a logarithmic velocity profile to give:

$$\beta = \left[1 + \frac{g}{C^2 \kappa^2}\right] \qquad (12.11)$$

where C = de Chezy bed roughness coefficient and κ = von Karman's constant (=0.41).

For the surface wind stress, a quadratic friction law is generally assumed (Wu, 1969), giving:

$$\tau_{xw} = C_s \rho_a W_x W_s \qquad (12.12)$$

where C_s = air–water resistance coefficient, ρ_a = air density, W_x = wind velocity component in the x-direction and W_s = wind speed ($=\sqrt{W_x^2 + W_y^2}$), where W_y = wind velocity component in the y-direction.

Bed friction is also generally represented in the form of a quadratic friction law, as given by:

$$\tau_{xb} = \rho g q_x \frac{V_s}{C^2 H} \qquad (12.13)$$

where V_s = depth-averaged fluid speed ($=\sqrt{U^2 + V^2}$), where U, V = depth-averaged velocities in x- and y-directions, respectively, and H = total depth of water.

To determine the de Chezy value for the bed roughness, most of the widely used models tend to use the Manning formula, which expresses C in terms of the local depth as follows:

$$C = \frac{H^{1/6}}{n} \qquad (12.14)$$

where n = Manning roughness coefficient, with typical values of n ranging from 0.012 for smooth-lined channelised rivers to 0.04 or more for meandering rivers with vegetation, etc. Although the Manning coefficient is primarily measured in practice for 1D river reaches, this parameter has been widely used in 2D flow fields with high levels of accuracy often being obtained for complex flow fields.

Although this approach is appropriate for most estuaries, the Manning representation assumes that the flow is rough turbulent flow and that the local head loss is dependent only on the size and characteristics of the bed roughness, i.e. form drag dominates totally. However, for low-velocity flows on shallow floodplains and wetlands, Reynolds number effects may be significant, reflecting the increased influence of skin friction. This complex hydrodynamic phenomenon can be represented using the comprehensive friction formulation given by Colebrook–White (Henderson, 1966) as:

$$C = -17.715 \log_{10} \left[\frac{k_s}{12H} + \frac{0.282 C}{R_e}\right] \qquad (12.15)$$

where k_s = Nikuradse equivalent sand grain roughness and R_e = Reynolds number for estuaries $= 4q_s/\nu$, where q_s = discharge per unit width along streamline and ν = kinematic laminar viscosity. The other advantage in using the Colebrook–White formulation to represent the bed roughness, rather than the Manning formulation, is that the physical roughness parameter k_s can be directly related to the height of bed features, such as ripples or dunes, rather than being based on a descriptive representation of the bed characteristics as for the Manning formulation.

Finally, the depth-averaged eddy viscosity, $\bar{v}_t$, can preferably be estimated from field data of the vertical velocity profile or, assuming bed-generated turbulence dominates over free shear layer turbulence, a logarithmic velocity profile can be assumed giving (Elder, 1959):

$$\bar{v}_t = 0.167 \kappa U_* H \tag{12.16}$$

where $U_* =$ shear velocity ($\sqrt{g}V_s/C$). However, field data by Fischer (1973) showed that the turbulent diffusion coefficient in straight, fairly uniform rivers is generally much higher and is more accurately represented by:

$$\bar{v}_t = 0.15 U_* H \tag{12.17}$$

For most estuaries, even this value is low compared to measured data recorded in rivers, with values for $\frac{\bar{v}_t}{U_* H}$ typically ranging from 0.42 to 1.61 (Fischer et al., 1979).

12.2.1.3 One-dimensional flows

In simulating flows within narrow estuaries, 1D models may be used where the longitudinal flow features dominate the system. The governing St Venant equations of motion are obtained by integrating the 3D equations of motion (i.e. equations 12.1 and 12.2) over the area of flow to give:

$$T\frac{\partial \zeta}{\partial t} + \frac{\partial Q}{\partial x} = q \tag{12.18}$$

$$\frac{\partial Q}{\partial t} + \beta \frac{\partial (Q^2/A)}{\partial x} = -gA\frac{\partial \zeta}{\partial x} - gAS_f \tag{12.19}$$

where $T =$ surface width of flow, $Q =$ discharge, $q =$ lateral inflow or outflow per unit length of river, $A =$ area of flow, and $S_f =$ friction slope, or slope of total energy line, which is given by the following representation:

$$S_f = \frac{Q|Q|}{C^2 A^2 R} \tag{12.20}$$

where $R =$ hydraulic radius for the conveyance segment and C (the de Chezy coefficient) may be estimated from either equation (12.14) or equation (12.15).

12.2.2 Water quality modelling

12.2.2.1 Three-dimensional flow fields

In modelling numerically the flux of water quality constituents, contaminants or sediments within an estuary, the conservation of mass equation can first be written in general terms for a 3D flow field as given by (Harleman, 1966):

$$\underbrace{\frac{\partial \varphi}{\partial t}}_{1} + \underbrace{\frac{\partial \varphi u}{\partial x} + \frac{\partial \varphi v}{\partial y} + \frac{\partial \varphi w}{\partial z}}_{2} + \underbrace{\frac{\partial}{\partial x}\overline{u'\varphi'} + \frac{\partial}{\partial y}\overline{v'\varphi'} + \frac{\partial}{\partial z}\overline{w'\varphi'}}_{3} = \underbrace{\varphi_s + \varphi_d + \varphi_k}_{4} \quad (12.21)$$

where φ = time-averaged solute concentration, φ_s = source or sink solute input (e.g. an outfall), φ_d = solute decay or growth term and φ_k = total kinetic transformation rate for solute. The individual terms in equation (12.21) – generally referred to as the advection–diffusion equation – refer to local effects (term 1), transport by advection (2), turbulence effects (3), and source (or sink), decay (or growth) and kinetic transformation effects (4).

The cross-produced terms $\overline{u'\varphi'}$, etc. represent the mass flux of solute due to the turbulent fluctuations and, by analogy with Fick's law of diffusion, it can be assumed that this flux is proportional to the mean concentration gradient and is in the direction of decreasing concentration (Chapter 3). Hence, the terms can be written as:

$$\left.\begin{aligned}\overline{u'\varphi'} &= -D_{tx}\frac{\partial \varphi}{\partial x}\\ \overline{v'\varphi'} &= -D_{ty}\frac{\partial \varphi}{\partial y}\\ \overline{w'\varphi'} &= -D_{tz}\frac{\partial \varphi}{\partial z}\end{aligned}\right\} \quad (12.22)$$

where D_{tx}, D_{ty}, D_{tz} = turbulent diffusion coefficients in x-, y-, z-directions. These coefficients are often associated with the eddy viscosity ν_t by a Schmidt number, whose value is found to vary between 0.5 and 1.0 (Lin and Shiono, 1995). For estuarine flows it is common to assume isotropic turbulence and to approximate the horizontal diffusion terms to the depth-mean coefficients as given by Fischer (1973), whereby in the absence of field data, these terms are often equated to:

$$D_{tx} = D_{ty} = 0.15 U_* H \quad (12.23)$$

Likewise, for the vertical diffusion coefficient, in the absence of stratification and field data it is common to assume a linear shear stress distribution and a logarithmic velocity profile giving (Vieira, 1993):

$$D_{tz} = U_* \kappa z \left(1 - \frac{z}{H}\right) \quad (12.24)$$

However, as indicated for hydrodynamic modelling, 3D models tend to be used only for simulating water quality processes in which the vertical variations of a water quality indicator, e.g. density, are significant.

12.2.2.2 Two-dimensional flow fields

For more practical problems where there are significant variations in the flow field across the estuary, such as over floodplains or mangrove forests, a 2D numerical model solution is more commonly used, together with the depth integrated advection–diffusion equation (Falconer, 1991):

$$\frac{\partial \phi H}{\partial t} + \frac{\partial \phi q_x}{\partial x} + \frac{\partial \phi q_y}{\partial y} - \frac{\partial}{\partial x}\left[HD_{xx}\frac{\partial \phi}{\partial x} + HD_{xy}\frac{\partial \phi}{\partial y}\right] - \frac{\partial}{\partial y}\left[HD_{yx}\frac{\partial \phi}{\partial x} + HD_{yy}\frac{\partial \phi}{\partial y}\right]$$
$$= H[\phi_s + \phi_d + \phi_k] \tag{12.25}$$

where ϕ = depth-average solute concentration, k = rate constant, etc., ϕ_s, ϕ_d, ϕ_k = depth-average concentrates corresponding to φ_s, φ_d, φ_k, in equation (12.21) and D_{xx}, D_{xy}, D_{yx}, D_{yy} = depth-average longitudinal dispersion and turbulent diffusion coefficients in x-, y-directions.

The dispersion–diffusion terms, these coefficients can be shown to be of the following form (Preston, 1985):

$$\left.\begin{array}{l} D_{xx} = \dfrac{(D_\ell U^2 + D_t V^2)H\sqrt{g}}{V_s C} + D_w \\[2mm] D_{yy} = \dfrac{(D_\ell V^2 + D_t U^2)H\sqrt{g}}{V_s C} + D_w \\[2mm] D_{xy} = D_{yx} = \dfrac{(D_\ell - D_t)UVH\sqrt{g}}{V_s C} + D_w \end{array}\right\} \tag{12.26}$$

where D_ℓ = depth average longitudinal dispersion constant, D_t = depth-average turbulent diffusion constant, and D_w = wind-induced dispersion coefficient. The values of D_ℓ and D_t, which are dimensionless constants, can be obtained preferably from field data, or minimum values can be obtained by assuming a logarithmic velocity profile, wherein $D_\ell = 5.93$ (Elder, 1959) and $D_t = 0.15$ (Fischer, 1973). However, in practical studies these values tend to be rather low (Fischer et al., 1979), with measured values of D_ℓ and D_t ranging from 8.6 to 7500 and 0.42 to 1.61, respectively. In the absence of field data, undertaken in the form of extensive dye dispersion studies, the authors have found that the most accurate results have generally been obtained using their DIVAST (Depth Integrated Velocities And Solute Transport, Falconer, 1991) model with typical values of $D_\ell = 13.0$ and $D_t = 1.2$.

12.2.2.3 One-dimensional flow fields

As for hydrodynamic studies, water quality studies of narrow well-mixed estuaries are often based on a 1D numerical model. For this purpose the 3D advection–diffusion equation (12.21) is integrated over an arbitrary cross-sectional area A to give:

$$\frac{\partial \phi A}{\partial t} + \frac{\partial \phi Q}{\partial x} - \frac{\partial}{\partial x}\left[AD_1\frac{\partial \phi}{\partial x}\right] = S_s + S_d + qS_L \tag{12.27}$$

where ϕ = area-averaged solute concentration, S_s = area-averaged source term, S_d = area-averaged decay term and S_L = solute concentration of lateral input (or output).

12.3 Water quality processes

In modelling water quality processes in estuaries a range of water quality parameters are often modelled, including physical, chemical and biological indicators (Falconer *et al.*, 2001). In this chapter, details of the procedures used for modelling Faecal Coliform (FC) and suspended sediments are given.

12.3.1 Faecal coliform modelling

Total coliform (TC) has been used for many years as the main indicator in evaluating bathing water quality with respect to domestic waste. However, because of the difficulties associated with the occurrence of non-faecal bacteria in its test, the use of the TC test is being gradually replaced by using FC and faecal streptococci as the main bacteriological indicators (Thomann and Muller, 1987). The FC bacteria group is indicative of organisms from the intestinal tract of humans and other animals. In recent years FC has been used by several investigators to assess the quality of bathing water and urban streams for outfall and/or non-outfall sources (Wyer *et al.*, 1997; Thackston and Murr, 1999; Young and Thackston, 1999).

In modelling FC the decay terms in equations (12.25) and (12.27) are generally expressed as a first-order decay function and are included in the 1D FASTER (Flow And Solute Transport for Estuaries and Rivers, Kashefipour *et al.*, 2002) and 2D DIVAST models, respectively, via the following equations:

$$S_d = -K_B \phi A \quad \text{and} \quad \phi_d = -K_B \phi \qquad (12.28)$$

where K_B = coliform decay rate (day^{-1}). The response time, i.e. the time required for the water body (i.e. river, estuary, etc.) to complete a fixed percentage of its recovery, is also commonly used to represent the growth/decay of bacteria. This parameter can be easily derived from a first-order decay formulation of the form $\phi = \phi_0 e^{-kt}$, where ϕ_0 = initial concentration at $t = 0$. In practice, T_{90} is the commonly used parameter for this purpose and is defined as the time during which the original organism population would reduce by 90%. It can readily be shown that T_{90} can be calculated from (Chapra, 1997):

$$T_{90} = \frac{2.303}{K_B} \qquad (12.29)$$

Several factors may influence the population of the organisms in a water body and thus in reporting the decay rate, sampling conditions are usually specified. These factors are mainly: sunlight intensity and duration, temperature and salinity levels, suspended particulate matter, and concentrations of toxic substances. A large range of decay rates for TC, FC and faecal streptococci have been reported in the literature. For example, for TC and FC, decay rate values have been reported from 0 to 2.4 day^{-1} in 2–18% salinity and dark samples, and from 2.5 to 6.1 day^{-1} in 15% salinity and sunlight samples. A range of 37–110 day^{-1} has been also reported for FC decay rates in seawater under good sunlight conditions (Thomann and Muller, 1987, Table 5.9).

12.3.2 Sediment transport

Sediment characteristics are morphologically and biologically important for estuaries. The movement of sediments can cause siltation in waterways and harbours, or erosion of estuarine and river banks. To deal with these problems, expensive dredging operations or bank protection are generally needed. The heavily contaminated sediments, resulting from industrial and municipal effluents, or accidental oil spills, could also release heavy metals, mineral oils and other toxic contaminants, which may then be re-adsorbed onto sediments or be available for resuspension by strong tidal currents, short wave action and dredging operations. This desorption of contaminants from their particulate phase can have a severe impact on the ecological balance of estuarine and coastal waters. Therefore, understanding and determining sediment dynamics in estuarine and coastal waters is of vital importance for the environmental management of estuarine and coastal waters.

Based on the general advection–diffusion equation (12.21), the governing equation for sediment transport processes can be written as:

$$\frac{\partial S}{\partial t} + \frac{\partial}{\partial x}(uS) + \frac{\partial}{\partial y}(vS) + \frac{\partial}{\partial z}[(w-w_s)S] - \frac{\partial}{\partial x}\left(D_{tx}\frac{\partial S}{\partial x}\right)$$
$$- \frac{\partial}{\partial y}\left(D_{ty}\frac{\partial S}{\partial y}\right) - \frac{\partial}{\partial z}\left(D_{tz}\frac{\partial S}{\partial z}\right) = S_T \tag{12.30}$$

where S = suspended sediment concentration and S_T = source or sink term. It should be noted that the source or sink term is introduced through the bed conditions. For cohesive sediments the following bed conditions are used (Wu *et al.*, 1998):

$$-w_s S - D_{tz}\frac{\partial S}{\partial z} = q_{\text{dep}} \quad \text{when } \tau_b \leq \tau_d \quad \text{(deposition)} \tag{12.31a}$$

$$-w_s S - D_{tz}\frac{\partial S}{\partial z} = q_{\text{ero}} \quad \text{when } \tau_b \geq \tau_e \quad \text{(erosion)} \tag{12.31b}$$

$$-w_s S - D_{tz}\frac{\partial S}{\partial z} = 0 \quad \text{when } \tau_d < \tau_b < \tau_e \quad \text{(equilibrium)} \tag{12.31c}$$

where τ_b = bed shear stress; τ_d = critical shear stress beyond which no further deposition occurs; τ_e = critical shear stress for erosion; and q_{dep}, q_{ero} = deposition and erosion rates, respectively, at the bed.

In recent years considerable effort has been made to investigate the mechanisms of cohesive sediment transport, and many experimental and field studies have been carried out to investigate deposition and erosion rates of such sediments (e.g. Krone, 1962; Mehta, 1973; Thorn and Parsons, 1977; Thorn, 1981; Parchure and Mehta, 1985; Mehta, 1988; Lick *et al.*, 1995). A widely used deposition rate first proposed by Krone (1962) is given as:

$$q_{\text{dep}} = \begin{cases} -w_s S_b \left(1 - \frac{\tau_b}{\tau_d}\right), & \tau_b < \tau_d \\ 0, & \tau_b \geq \tau_d \end{cases} \quad (12.32)$$

where S_b = near-bed cohesive sediment concentration.

Likewise, a widely used erosion rate given by Partheniades (1965) is:

$$q_{\text{ero}} = \begin{cases} E\left(\frac{\tau_b}{\tau_e} - 1\right), & \tau_b > \tau_e \\ 0, & \tau_b \leq \tau_e \end{cases} \quad (12.33)$$

where E = erosion constant (kg m^{-2} s^{-1}).

For non-cohesive sediments, the boundary of the sediment bed is usually given at a small reference level above the bed, and the reference concentration or gradient of the sediment concentration is prescribed by its equilibrium value at this reference level. A variety of relationships exist in the literature for predicting the near-bed reference concentration of suspended sediment, from which the entrainment rate of bed sediment flux into suspension can be obtained. The equilibrium reference concentration used in the studies presented in this chapter was proposed by van Rijn (1984), given as:

$$S_{\text{ae}} = 0.015 \frac{D_{50} T^{1.5}}{a D_*^{0.3}} \quad (12.34)$$

where S_{ae} = equilibrium reference sediment concentration; D_{50} = sediment diameter than which 50% of the bed material is finer; T = transport stage parameter; a = reference level of the sediment concentration profile, normally defined as the upper edge of the bedload layer and D_* = particle parameter. The definitions of the transport stage T and particle parameter D_* can be found in van Rijn (1984).

The upward diffusive sediment flux is given by:

$$E_{\text{ae}} = w_s S_{\text{ae}} \quad (12.35)$$

At the free surface, zero flux is normally used as the boundary condition.

12.4 Modelling applications

12.4.1 Modelling bathing water quality in Ribble Estuary

The Ribble Estuary is located along the north-west coast of England, in the county of Lancashire. At the mouth of the estuary there are two well-known seaside resorts, namely Lytham St Annes and Southport, with both being designated EU (European Union) bathing waters. The Fylde Coast, which is bounded between Fleetwood in the north and the Ribble Estuary in the south, includes one of the most famous beaches for tourism in England, namely Blackpool, with an average of more than 17 million visitors per annum. The area has four main centres of population, namely Blackpool and Lytham St Annes to the north of the Ribble Estuary, Southport to the south of the estuary and the town of Preston, which is inland and straddles the river Ribble at the tidal limit (Figure 12.1).

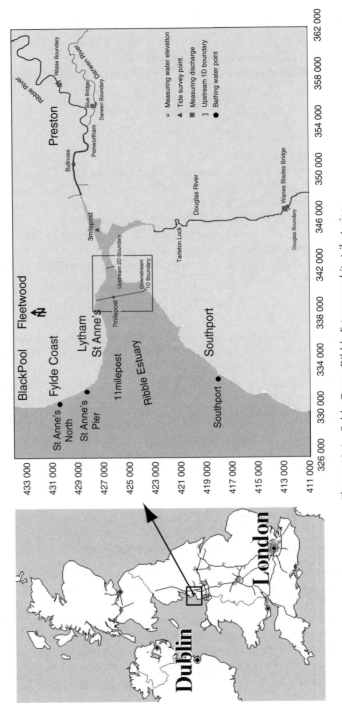

Figure 12.1 Fylde Coast, Ribble Estuary and its tributaries.

In order to enhance the bathing water quality along the Fylde coast, a major civil engineering investment programme has been undertaken in recent years to reduce the bacterial input to the estuary. About £600 million has been invested over the past 10 years in new sewerage works and treatment plants along the Fylde Coast and Ribble Estuary. Examples include: upgrading the wastewater treatment works at Clifton Marsh from primary treatment to include UV disinfection; and reducing storm water discharges from the wastewater network by constructing 260 000 m^3 of additional storage. Although the reduction in input bacterial loads has resulted in a marked decrease in the concentration of bacterial indicators in the coastal receiving waters, occasional elevated FC counts have still been measured. As a result, the bathing waters still occasionally fail to comply with the EU mandatory water quality standards (see Council of the European Communities, 1976).

A numerical modelling study was undertaken to establish the water quality of the EU-designated bathing waters located at the mouth of the Ribble Estuary. In order to reduce the possible inaccuracies caused by setting up the boundary conditions required by the numerical models, the upstream boundaries were set up at the tidal limits of the rivers Ribble, Darwen and Douglas (Figure 12.1) and the downstream boundary was located around the 25-m depth contour in the Irish Sea. The length of the seaward boundary was 41.2 km, with the width of the river boundaries being generally less than 10 m. Such a great variation in the modelling dimensions made it impractical to use either a 1D or 2D model alone. In applying either of these two models individually or linking the models statically would have introduced inaccuracies in the model prediction results and would also have required a considerable amount of effort in exchanging the data. Thus, in this study a 1D and 2D model (namely FASTER and DIVAST) were linked dynamically to create a single model, in which the mathematical computations of the hydrodynamic variables and water quality indicators were undertaken simultaneously across the entire modelling region. More details regarding this linked model may be found in Kashefipour *et al.* (2002).

An extensive programme of data collection was also undertaken by the UK Environment Agency to provide data for model calibration and verification. Six sets of hydrodynamic and water quality data were collected during the winter period of 1998 and the summer period of 1999, to include a combination of different weather and tide conditions. During each survey, measurements were taken at all discharge locations and upstream river boundaries (i.e. tidal limits) for two consecutive days. Comprehensive data sets were collected, including: water depths, current speed and directions, salinity levels and concentrations of suspended solids, faecal and total coliforms, and faecal streptococci. In total there were 34 input sources identified that contributed to the pollution loads of the estuary. These included: direct discharges of treated wastewater from treatment plants, inputs of the three major rivers from the upstream boundary and inputs from several smaller rivers and combined sewer overflows. Four calibration points were chosen along the main channel of the Ribble, with these sites being referred to as 11milepost, 7milepost, 3milepost and Bullnose (Figure 12.1). Measurements at these locations were taken only for the second day of a survey.

The 2D area was represented horizontally using a mesh of 618 × 454 uniform grid squares, each with a length of 66.7 m. A total of 1075 cross sections was designed for

318   Computational Fluid Dynamics

the 1D area with the minimum and maximum distances between two consecutive cross sections being 10 and 50 m, respectively.

Measured discharges at the upstream tidal limits of the rivers Ribble, Darwen and Douglas were used as the upstream open boundary conditions. The water levels for the seaward boundary conditions were acquired from the Proudman Oceanographic Laboratory (POL), predicted using their Irish Sea model.

The main parameter used for calibration of the hydrodynamic model was the bed roughness. By changing the values of the bed roughness and comparing the model predictions with measured data, the model was calibrated by choosing the best fit between the predicted results and measured data. Different roughness values were used along different reaches of the model domain to reflect the local conditions. For the open coastal waters it was found that the most appropriate value for the Nikuradse equivalent sand grain roughness was 20 mm. For the 1D model region the Manning roughness coefficient was used, with the optimum values ranging from 0.021 for the lower part of the river to 0.028 for the upper part of the river.

As an example, the results obtained for the survey on 3 June 1999 are discussed herein. This survey was carried out during a wet period with a mean tidal range. Figure 12.2 shows comparisons at 7milepost between the predicted water elevations and current speeds and directions, respectively. Figure 12.2a shows that good agreement has been achieved between the predicted and measured water levels at

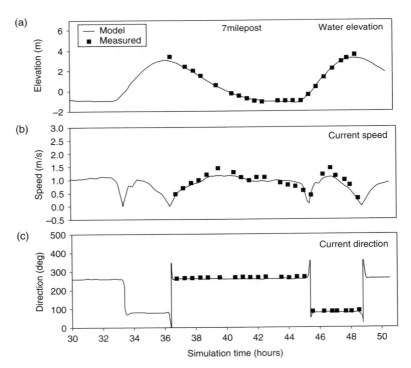

Figure 12.2   *Comparison of predicted and measured: (a) water levels; (b) current speeds; and (c) current directions at 7milepost, on 3 June 1999.*

this site, with an average error of 0.11 m, i.e. 2.1% error, when compared with the measured tidal range. Figures 12.2b and 12.2c illustrate relatively good agreement between the predicted current speeds and directions and the corresponding measured data, with an average error of 0.13 m s^{-1} (i.e. 8.8% error relative to the maximum measured speed) and 8.1°, respectively.

In this study, FC was used as the main water quality indicator. Since the fate of FC is influenced by many factors, calibration of the water quality model was generally more difficult than that for the hydrodynamic model. The accuracy of the model predictions depends upon an accurate representation of the advection and dispersion processes, as well as on the decay rate. To reflect all of the environmental conditions affecting the fate of micro-organisms, different decay rates were used for: (i) day and night times, (ii) dry and wet weather conditions, and (iii) sea and river waters. For the survey of 3 June 1999, it was found that for sea waters the optimum value of T_{90} was 44.0 and 64.8 hours for day- and night-time, respectively. The values of 51.9 and 86.7 hours were obtained for the day- and night-time T_{90} values for freshwaters.

Faecal coliform inputs during the 3 June 1999 survey along the rivers and estuary were categorised into four groups, including: open boundaries, Wastewater Treatment Works (WwTW), Combined Sewer Overflows (CSOs) and the other inputs. Figure 12.3 shows a pie chart representing the percentage contribution of each input. As can be seen, the most significant FC load came from the river inputs. Statistical analysis showed similar results for the other surveys, both wet and dry events.

Predicted and measured FC concentrations at 7milepost for the survey on 3 June 1999 are compared in Figure 12.4. Due to the large variation in the values of faecal coliform concentrations, it is plotted using a logarithmic scale. Relatively good agreement between both sets of data can be seen from this Figure, with an average error of 35 000 cfu 100 ml^{-1}, i.e. 30.5%.

Figure 12.5 shows another example for the linked model application to the Ribble Estuary and compares the predicted and measured FC concentrations at 3milepost for the 19 May 1999 survey, which was carried out for a dry event and a spring tidal range. As can be seen from this Figure both sets of data agreed well, with an average

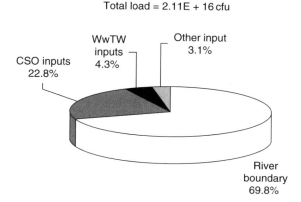

Figure 12.3 *Faecal coliform inputs to Ribble Estuary on 3 June 1999.*

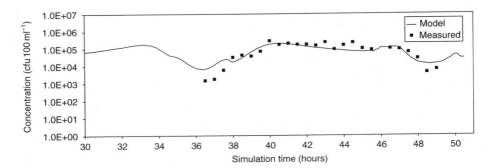

Figure 12.4 Comparison of predicted and measured FC concentrations at 7milepost, on 3 June 1999.

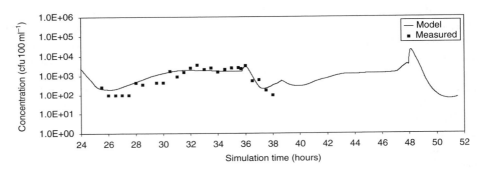

Figure 12.5 Comparison of predicted and measured FC concentrations at 3milepost, on 19 May 1999.

error of 477 cfu 100 ml^{-1}, i.e. 27.6%. For this dry event the T_{90} values for the 2D area were found to be 14.7 and 20.2 hours for day- and night-time periods, respectively. The corresponding values for the 1D area were obtained and found to be 21.4 and 26.9 hours, respectively. For this event an average T_{90} value was considered to be 20.8 hours. All of these values were for a water temperature of 20 °C.

12.4.2 Modelling salinity and sediment transport in Mersey Estuary

The Mersey Estuary is one of the largest estuaries in the UK, with a catchment area of some 5000 km^2 of north-western England, and includes the major conurbations of Liverpool and Manchester. The Upper Estuary, between Warrington and Runcorn Gap, is a narrow meandering channel of about 15 km in length. Below the Gap, it opens into a large shallow basin to form the Inner Estuary of about 20 km in length, with extensive inter-tidal banks and salt marshes on its southern margin. Further downstream of the Inner Estuary, the estuary converges to form the Narrows, a straight narrow channel up to 30-m depth even at low water. Seaward of the Narrows, the channel widens to form the Outer Estuary, consisting of a large area of inter-tidal sand and mud banks (NRA, 1995). The Mersey Estuary is a macrotidal estuary, with tidal ranges at Liverpool varying from 10.5 m (extreme spring) to 3.5 m (extreme neap) over a typical spring–neap cycle. Freshwater flows from the river into the Mersey Estuary vary from 10 to 600 m^3 s^{-1} at the extremes. These extreme values occur infrequently and more typical flows lie in the range 20–40 m^3 s^{-1}.

The present models have been set up and applied to simulate the tidal flow, and salt and sediment fluxes in the Mersey Estuary from Gladstone to Howley Weir (Figure 12.6). The data from the latest Mersey Estuary bathymetry, surveyed by HR Wallingford Ltd and ABP in 1997, have been used to generate the data files for the present numerical model. The model area was represented horizontally in two grid systems, i.e. the 2D and 1D parts. The 2D part covered the region from New Brighton to Hale, and was represented horizontally using a mesh of 216 × 116 uniform grid squares, each with a length of 100 m. The 1D part covered the region from Hale to Howley Weir and was represented using 80 segments.

Six sets of hydrodynamic data, provided by the Environment Agency, were used to calibrate the hydrodynamic model. Three sets were collected in spring, two in autumn, and one in summer. Four of these data sets were collected during spring tides and two were collected during neap tides. The freshwater input from the River

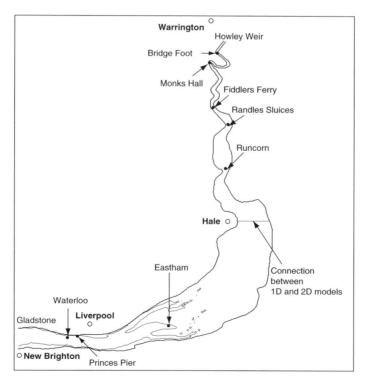

Figure 12.6 Map of the Mersey Estuary and location of sampling sites.

Mersey for these sets of data covered both wet season and dry season conditions. The present hydrodynamic model was calibrated against observation data recorded along the estuary at Waterloo, Eastham, Runcorn and Fiddlers Ferry (Figure 12.6).

The water elevation data collected at Gladstone tide gauge were used as the seaward boundary condition. All runs were started at high water associated with the tidal curve used to drive the model. The velocities in the estuary were initially set to zero. The 10-day average freshwater input from the River Mersey was provided by the Environment Agency and used to provide the upstream boundary condition.

Comparisons of model predictions and field-measured water levels on 18 September 1989 are shown in Figure 12.7. It can be seen that good agreement was obtained between the model predicted water elevations and field data at Waterloo. The predicted high water level at Waterloo was 10.19 m, which was 0.19 m higher than the observed value. The predicted low water level at Waterloo was 0.6 m, which was 0.1 m lower than the observed value. The model predicted water levels agreed reasonably well with the observation levels, except that the flow was blocked slightly at Runcorn during ebb tide in the numerical model, which gave slightly higher water levels during ebb tide. This was caused possibly by using the bathymetry surveyed during 1997 to model an event in 1989, as the bathymetry of the Mersey Estuary undergoes frequent changes, especially around Runcorn Gap. The predicted high water level at Runcorn was 11.07 m, which was 0.29 m higher than the observed value of 10.78 m. The predicted low water level was 5.42 m, which was 0.13 m lower than the observed value of 5.55 m. The model predicted tidal range at Runcorn was 5.65 m, which was 8% larger than that from the field observations.

The predicted and measured salinity levels at Runcorn from 6:30 to 19:30 hours on 18 September 1989 are shown in Figure 12.8. Predicted salinity levels agreed well with the observed values, except that the model gave slightly higher salinity values during the ebb tide. The predicted maximum salinity level at Runcorn was 25.50 ppt, which was 0.17 ppt higher than the observed value of 25.33 ppt. The predicted minimum salinity level at Runcorn was 4.09 ppt, which was 1.66 ppt higher than the observed value of 2.43 ppt.

Figure 12.9 shows comparisons of the maximum, tidally averaged and minimum salinity levels between the model predictions and observed values along the estuary, from 6:30 to 19:30 hours on 21 March 1990. It can be seen that the model predictions agreed well with the field data along the estuary, except near Eastham (about 32.5 km downstream of Howley Weir), where the model over-predicted the tidally averaged and minimum salinity levels. This was thought to be caused by the neglection of freshwater inputs around Eastham in the model, due to a lack of accurate data.

To simulate sediment fluxes in the Mersey Estuary, both cohesive and non-cohesive sediments needed to be considered. In the Upper Estuary, between Warrington and Runcorn Gap, the bed sediment was considered to be cohesive, whilst along the rest of the estuary the bed sediment was considered as non-cohesive. The grain sizes of the sediments in the Inner Estuary were provided by the Environment Agency. Based on these data, values of $D_{16} = 12$ μm, $D_{50} = 75$ μm, $D_{84} = 195$ μm and $D_{90} = 225$ μm were used in the model. A cohesive sediment (floc) size of 20 μm was assumed. Three sets of suspended sediment data provided by the Environment Agency were used to calibrate the sediment transport module. The model was calibrated by trial and error, based on the suspended sediment concentrations at

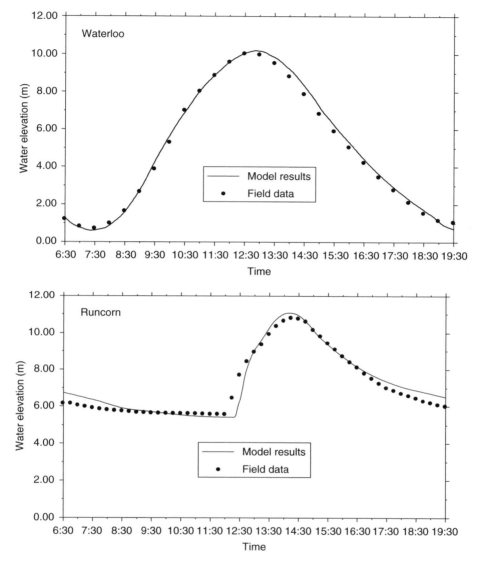

Figure 12.7 Comparison of predicted and field-measured water elevations at Waterloo and Runcorn on 18 September 1989.

Princes Pier, Eastham, Runcorn, Randles Sluices, Fiddlers Ferry, Monks Hall and Bridge Foot. The calibrated critical shear stress for erosion τ_{ero}, the critical shear stress for deposition τ_{dep} and the erosion constant E were $1.0\,N\,m^{-2}$, $0.25\,N\,m^{-2}$ and $0.00004\,kg\,m^{-2}\,s^{-1}$, respectively.

A comparison between the predicted and observed suspended sediment concentrations at Eastham, from 6:30 to 19:30 hours on 18 September 1989, is shown in Figure 12.10. The predicted suspended sediment concentrations agreed well with the

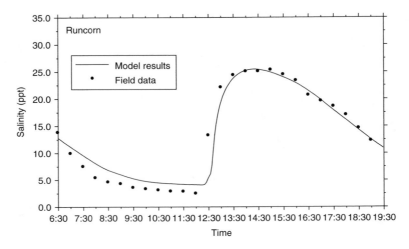

Figure 12.8 Comparison of predicted and measured salinity distributions at Runcorn on 18 September 1989.

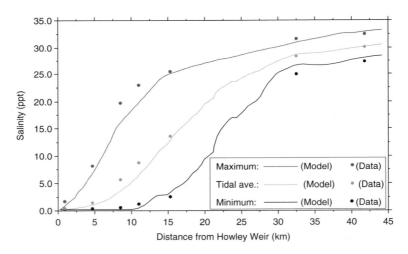

Figure 12.9 Comparison of predicted and measured salinity distributions along the Mersey Estuary on 18 September 1989.

field data, with the maximum concentration of $0.42\,\mathrm{kg\,m^{-3}}$ being only $0.01\,\mathrm{kg\,m^{-3}}$ lower than the measured value.

Likewise, comparisons of predicted and measured suspended sediment concentrations at Runcorn from 6:30 to 19:30 hours on 18 September 1989 are shown in Figure 12.11. The predicted suspended sediment concentrations again agreed well with the observed values, but the model gave a sharper gradient at about 13:30 hours, when the flooding velocity decreased. The predicted maximum suspended sediment concentration at Runcorn was $2.72\,\mathrm{kg\,m^{-3}}$, which was $0.32\,\mathrm{kg\,m^{-3}}$ higher than the observed value.

Modelling water quality processes in estuaries 325

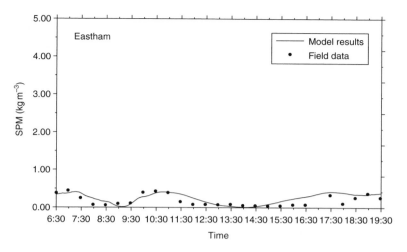

Figure 12.10 *Comparison of predicted and measured sediment concentrations at Eastham on 18 September 1989.*

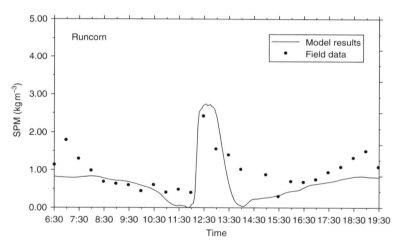

Figure 12.11 *Comparison of predicted and measured sediment concentrations at Runcorn on 18 September 1989.*

12.5 Conclusions

Numerical models have been increasingly used to predict flow, water quality, and sediment and contaminant transport processes in estuarine waters. Details are given in this chapter of the governing hydrodynamic equations for 3D, 2D and 1D flow field predictions and the corresponding advection–diffusion equation for the transport of water quality indicators, suspended sediments and contaminant solutes within the water column. In applying the general advection–transport equation to solute transport processes, details are given of the kinetic reactions and decay of FC. For sediment transport simulations, details are given of procedures of modelling erosion and deposition for both cohesive and non-cohesive sediments.

Models developed by the authors for predicting estuarine flow and solute concentration distributions are summarised, together with details of the treatment of the boundary conditions within such models. These models have been applied to many estuaries, with examples given herein of model applications to two sites, i.e., the Ribble and Mersey estuaries, both located in the north-west of England. In modelling the FC concentration distributions along the Ribble Estuary, it has been shown that having accurate data to specify the discharge loads from boundary and outfall inputs, the decay rate is the key factor that controls the accuracy of the model results. In this study a variable decay rate for day- and night-time and freshwater and seawater has been shown to be critical with regard to the accuracy of the model predictions. In modelling the sediment concentration distributions along the Mersey Estuary, the specification of sediment erosion criteria was believed to be critical with regard to the accuracy of the model results.

Acknowledgement

The numerical model studies reported herein were funded through a number of grants including those from: North West Water Ltd, the Environment Agency (North West Region), the Engineering and Physical Sciences Research Council WITE programme (grant GR/M99774) and the Natural Environmental Research Council Environmental Diagnostics programme (grant GST/04/1567). The authors would also like to acknowledge the support and assistance given by Professor D. Kay, University of Wales, Aberystwyth, Professor G.E. Millward, Plymouth University and Dr Y. Wu, Hyder Consulting Ltd.

References

Chapra, S.C. 1997. *Surface Water Quality Modelling*. McGraw-Hill Companies Inc., USA, 844pp.

Council of the European Communities. 1976. Council directive of 8th December 1975 concerning the quality of bathing waters (76/160/EEC). *Official Journal of the European Communities*, L31, 1–7.

Elder, J.W. 1959. The dispersion of marked fluid in a turbulent shear flow. *Journal of Fluid Mechanics*, Vol. 5, No. 4, pp. 544–560.

Falconer, R.A. 1991. Review of modelling flow and pollutant transport processes in hydraulic basins. *Proceedings of First International Conference on Water Pollution: Modelling, Measuring and Prediction*, Computational Mechanics Publications, Southampton, UK, pp. 3–23.

Falconer, R.A. 1993. An introduction to nearly horizontal flows. In: *Coastal, Estuarial and Harbour Engineers' Reference Book*, M.B. Abbott and W.A. Price (eds), E.&F.N. Spon Ltd, London, Chapter 2, pp. 27–36.

Falconer, R.A. and Chen, Y. 1996. Modelling sediment transport and water quality processes on tidal floodplains. In: *Floodplain Processes*, M.G. Anderson, D.E. Walling and P.D. Bates (eds), John Wiley & Sons Ltd, Chichester, Chapter 11, pp. 361–398.

Falconer, R.A. and Lin, B. 1997. Three-dimensional modelling of water quality in the Humber Estuary. *Water Research*, IAW, Vol. 31, No. 5, pp. 1092–1102.

Falconer, R.A., Lin, B. and Keshefipour, S.M. 2001. Modelling water quality processes in riverine systems. In: *Model Validation: Perspectives in Hydrological Science*, M.G. Anderson and P.D. Bates (eds), John Wiley & Sons, Chichester, Chapter 14, pp. 358–387.

Fischer, H.B. 1973. Longitudinal dispersion and turbulent mixing in open channel flow. *Annual Review of Fluid Mechanics*, Vol. 5, pp. 59–78.

Fischer, H.B., List, E.J., Koh, R.C.J., Imberger, J. and Brooks, N.H. 1979. *Mixing in Inland and Coastal Waters*, Academic Press Inc., San Diego, 483pp.

Goldstein, S. 1938. *Modern Development in Fluid Dynamics*, Oxford University Press, Oxford, Vol. 1.

Harleman, D.R.F. 1966. Diffusion processes in stratified flow. In: *Estuary and Coastline Hydrodynamics*, A.T. Ippen (ed.), McGraw-Hill Book Co. Inc., New York, Chapter 12, pp. 575–597.

Henderson, F.M. 1966. *Open Channel Flow*. Collier-Macmillan Publishers, London, 522pp.

HR Wallingford, 1996. *Estuary Processes and Morphology Scoping Study*. Report SR 446, 104pp.

Kashefipour, S.M., Lin, B., Harris, E.L. and Falconer, R.A. 2002. Hydro-environmental modelling for bathing water compliance of an estuarine basin. *Water Research*, IWA, Vol. 36, pp. 1854–1868.

Krone, R.B. 1962. *Flume Studies of the Transport of Sediment in Estuarial Processes*. Final Report, Hydraulic Engineering Laboratory and Sanitary Engineering Research Laboratory, University of California, Berkeley.

Lick, W., Xu, Y.J. and Mcneill, J. 1995. Resuspension properties of sediments from the Fox, Sagina and Buffalo rivers. *Journal of Great Lakes Research*, Vol. 21, pp. 257–274.

Lin, B. and Shiono, K. 1995. Numerical modelling of solute transport in compound channel flows. *Journal of Hydraulic Research*, IAHR, Vol. 33, pp. 773–788.

Mehta, A.J. 1973. *Deposition Behaviour of Cohesive Sediment*, PhD Thesis, University of Florida, Gainesville.

Mehta, A.J. 1988. Laboratory studies on cohesive sediment deposition and erosion. In: *Physical Processes in Estuaries*, J. Dronkers and W. van Leussen (eds), Springer, New York, pp. 427–455.

NRA. 1995. *The Mersey Estuary: A Report on Environmental Quality*. Water Quality Series No. 23, HMSO, London, p. 44.

Parchure, T.M. and Mehta, A.J. 1985. Erosion of soft cohesive sediment deposits. *Journal of Hydraulic Engineering*, ASCE, Vol. 111, pp. 1308–1326.

Partheniades, E. 1965. Erosion and deposition of cohesive soils. *Journal of the Hydraulics Division*, ASCE, Vol. 91(HY1), pp. 105–137.

Preston, R.W. 1985. *The Representation of Dispersion in Two-Dimensional Shallow Water Flow*. Central Electricity Research Laboratories, Report No. TPRD/U278333/N84, 13pp.

Rodi, W. 1984. *Turbulence Models and Their Application in Hydraulics*. Second edition, International Association for Hydraulics Research, Delft, Netherlands, 104pp.

Thackston, E.L. and Murr, A. 1999. CSO control model modifications based on water quality studies. *Journal of Environmental Engineering*, ASCE, Vol. 125, pp. 979–987.

Thomann, R.V. and Muller, J.A. 1987. *Principles of Surface Water Quality Modelling Control*. Harper-Collins Publisher Inc., New York, 644pp.

Thorn, M.F.C. 1981. Physical processes of siltation in channels. *Proceedings of the Conference on Hydraulic Modelling Applied to Maritime Engineering Problems*, Institution of Civil Engineers, London, pp. 47–55.

Thorn, M.F.C. and Parsons, J.G. 1977. *Properties of Grangemouth Mud. Hydraulic Research Station*. Report No. EX 781, Wallingford, UK.

van Rijn, L.C. 1984. Sediment transport part 2: Suspended load transport. *Journal of Hydraulic Engineering*, ASCE, Vol. 10, pp. 1613–1641.

Vieira, J.K. 1993. Dispersive processes in two-dimensional models. In: *Coastal, Estuarial and Harbour Engineers' Reference Book*, M.B. Abbott and W.A. Price (eds), E.&F.N. Spon Ltd, London, Chapter 14, pp. 179–190.

Wu, J. 1969. Wind-stress and surface roughness at air-sea interface. *Journal of Geophysical Research*, Vol. 74, pp. 444–455.

Wu, Y., Falconer, R.A. and Uncles, R.J. 1998. Modelling of water flows and cohesive sediment fluxes in the Humber Estuary, UK. *Marine Pollution Bulletin*, Vol. 37, pp. 182–189.

Wyer, M.D., Oneill, G., Kay, D., Crowther, J., Jackson, G. and Fewtrell, L. 1997. Non-outfall sources of faecal indicator organisms affecting the compliance of coastal waters with directive 76/160/EEC. *Water Science and Technology*, Vol. 35, pp. 151–156.

Young, K.D. and Thackston, E.L. 1999. Housing density and bacterial loading in urban streams. *Journal of Environmental Engineering*, ASCE, Vol. 125, pp. 1177–1180.

13

Roughness parameterization in CFD modelling of gravel-bed rivers

A.P. Nicholas

13.1 Introduction

Computational fluid dynamics models that solve the Reynolds-averaged Navier–Stokes equations provide a physically based framework for investigating 3D flows in channels with a range of geometrical configurations (e.g. engineered, laboratory and natural). That such models are not entirely physically based in the strictest sense reflects the fact that they incorporate a number of semi-empirical components, most notably in their treatment of turbulence and boundary roughness. Applications involving relatively simple channel geometries (e.g. backward-facing steps, ducts, confluences and compound channels) have been tested rigorously using data obtained under controlled laboratory conditions (Nakagawa and Nezu, 1987; Younis, 1996; Sofialidis and Prinos, 1998; Bradbrook *et al.*, 1998, 2001). It is for these flows that the majority of turbulence and roughness schemes have been developed and validated (Launder *et al.*, 1975; Rodi, 1980; Patel *et al.*, 1984; Launder, 1989). However, even in situations where boundary conditions are well defined and relatively uncomplicated, existing parameterization schemes have known limitations (Launder, 1989; Gad-el-Hak and Bandyopadhyay, 1995; Bradshaw *et al.*, 1996; Patel, 1998).

Over the past 5–10 years a growing body of research has demonstrated the potential for applying 3D CFD models in natural fluvial systems. Flows within a variety of settings have been simulated, including single-thread rivers (Hodskinson,

1996; Hodskinson and Ferguson, 1998; Booker et al., 2001), braided channels (Sinha et al., 1998; Lane et al., 1999; Nicholas and Sambrook Smith, 1999) and river floodplains (Nicholas and McLelland, 1999; Nicholas, 2002). Despite this progress, there remains considerable uncertainty surrounding the appropriate representation of turbulence and boundary roughness in such environments. With a few exceptions (e.g. Olsen and Stoksteth, 1995; Nicholas, 2001) field-based CFD applications have used conventional parameterization schemes developed for laboratory flows. However, the extent to which such approaches are adequate has yet to be established, largely because of the difficulty of obtaining high-resolution spatially distributed datasets of bed topography and mean and turbulent flow characteristics in natural settings. Recent theoretical and field-based research has led to greater awareness of the complexity of roughness conditions within natural rivers, and has both highlighted the existence of multiple roughness length scales (Robert, 1988; Clifford et al., 1992; Nikora et al., 1998; Butler et al., 2001) and demonstrated the relationship between small-scale bed topography and near-bed flow structures (Kirkbride, 1993; Roy et al., 1999; Buffin-Bélanger et al., 2000; Lawless and Robert, 2001). This work has assisted the interpretation of CFD model output and has helped to identify some potential limitations of current roughness parameterization schemes. This chapter examines these issues with particular reference to the representation of roughness in gravel-bed rivers.

13.2 Basic flow equations and numerical methods

The results presented here were obtained using CFD models that solve the equation of mass conservation and the Reynolds-averaged Navier–Stokes momentum equation in two or three dimensions. These equations may be written in Cartesian form as

$$\frac{\partial}{\partial x_i}(u_i) = 0 \tag{13.1}$$

$$\frac{\partial}{\partial x_j}(u_i u_j) = \frac{\mu}{\rho}\frac{\partial}{\partial x_j}\left(\frac{\partial u_i}{\partial x_j} + \frac{\partial u_j}{\partial x_i}\right) + \frac{F_i}{\rho} - \frac{1}{\rho}\frac{\partial p}{\partial x_i} + \frac{\partial}{\partial x_j}(-\overline{u_i' u_j'}) \tag{13.2}$$

where p is pressure, μ and ρ are the dynamic viscosity and density of water, u_i is the velocity in the x_i-direction, F_i represents external forces, and the final term in equation (13.2) represents the Reynolds stresses resulting from the decomposition of instantaneous velocities into their mean and fluctuating components. The latter are modelled using the Boussinesq approximation

$$-\overline{u_i' u_j'} = \nu_t\left(\frac{\partial u_i}{\partial x_j} + \frac{\partial u_j}{\partial x_i}\right) - \frac{2}{3}k\delta_{ij} \tag{13.3}$$

where k is the turbulent kinetic energy, δ_{ij} is the Kronecker delta and ν_t is the eddy viscosity, which is a function of k and the energy dissipation rate (ε)

$$\nu_t = C_\mu \frac{k^2}{\varepsilon} \tag{13.4}$$

A two-equation Renormalisation Group (RNG) turbulence closure is used here which solves the following transport equations for k and ε

$$\frac{\partial}{\partial x_j}(u_j k) = \frac{\partial}{\partial x_j}\left(\frac{\nu_t}{\sigma_k}\frac{\partial k}{\partial x_j}\right) + G_k - \varepsilon \tag{13.5}$$

$$\frac{\partial}{\partial x_j}(u_j \varepsilon) = \frac{\partial}{\partial x_j}\left(\frac{\nu_t}{\sigma_\varepsilon}\frac{\partial \varepsilon}{\partial x_j}\right) + C_{\varepsilon 1}\frac{\varepsilon}{k}G_k - C_{\varepsilon 2}\frac{\varepsilon^2}{k} - R \tag{13.6}$$

where the R term reduces the destruction of ε in areas of high strain and streamline curvature. The production term (G_k) is given by

$$G_k = \nu_t S^2 \tag{13.7}$$

where S is the modulus of the mean rate-of-strain tensor and is defined as

$$S = \sqrt{2 S_{ij} S_{ij}} \tag{13.8}$$

$$S_{ij} = \frac{1}{2}\left(\frac{\partial u_i}{\partial x_j} + \frac{\partial u_j}{\partial x_i}\right) \tag{13.9}$$

The parameters C_μ, $C_{\varepsilon 1}$, $C_{\varepsilon 2}$, σ_k and σ_ε are constants that are assigned the standard values for the RNG k-ε model (cf. Yakhot and Orszag, 1986). In the applications reported here equations (13.1)–(13.9) are solved using a control volume technique employing second-order upwind discretization. Pressure–velocity coupling is carried out using the SIMPLEC algorithm of Vandoormaal and Raithby (1984).

13.3 Wall functions

Most CFD applications involving natural channels have represented boundary roughness using standard wall functions which take a variety of similar forms in different codes. In the current application the following expressions are applied within the near-bed cell of the model mesh

$$\frac{u}{u_*} = \frac{1}{\kappa}\ln\left(\frac{\rho u_* E z}{\mu}\right) - \Delta B \tag{13.10}$$

$$\frac{\partial k}{\partial n} = 0 \tag{13.11}$$

$$\varepsilon = \frac{u_*^3}{\kappa z} \tag{13.12}$$

where u is the velocity at height z above the bed, u_* is the shear velocity, n is the local coordinate normal to the wall, κ is the von Karman constant, E is a roughness constant that takes a value of 9.8 and ΔB is a roughness function that is defined using the equations of Cebeci and Bradshaw (1977)

$$k_s^+ < 2.25 \quad \Delta B = 0 \qquad (13.13)$$

$$2.25 < k_s^+ < 90 \quad \Delta B = \frac{1}{\kappa} \ln\left(\frac{k_s^+ - 2.25}{87.75} + C_{ks} k_s^+\right) \qquad (13.14)$$
$$\sin[0.4258(\ln k_s^+ - 0.811)]$$

$$k_s^+ > 90 \quad \Delta B = \frac{1}{\kappa} \ln(1 + C_{ks} k_s^+) \qquad (13.15)$$

where C_{ks} is a constant that takes a value in the range 0.3–1, $k_s^+ = k_s \rho u_*/\mu$, and k_s is a roughness length scale. Equations (13.13)–(13.15) define the roughness function for conditions that are smooth, transitional and rough, respectively. In gravel-bed rivers flows are generally within the rough regime represented by equation (13.15). Where $C_{ks} = 0.33$, this is equivalent to Nikuradse's expression for the law of the wall.

One of the main difficulties in applying equations (13.10)–(13.15) within natural rivers involves the determination of an appropriate value for the roughness length scale (k_s). Several previous studies have used $k_s = 3.5 D_{84}$ (e.g. Hodskinson, 1996; Hodskinson and Ferguson, 1998; Booker et al., 2001), where D_{84} is the 84th percentile of the bed sediment diameter frequency distribution. However, this relationship is based on the evidence of field investigations (e.g. Hey, 1979; Bray, 1982) and represents a measure of the total roughness contributed by sources operating over a range of spatial scales (Wiberg and Smith, 1987; Clifford et al., 1992). In CFD applications it is appropriate to separate these contributions into sub-grid-scale and supra-grid-scale components, where the former are represented by the wall function and the latter by the topography of the model mesh. As a result, one might expect k_s to take a lower value, and indeed several studies have found that $k_s = D_{50}$ yields good results (e.g. Nicholas and McLelland, 1999; Nicholas and Sambrook Smith, 1999). However, the clear implication of roughness partitioning between sub-grid-scale and supra-grid-scale components is that k_s is also a function of model mesh resolution. In addition, since the topographic content of the model mesh is generally also a function of the mesh resolution, a three-way interaction exists between mesh resolution, topographic representation and the appropriate value of k_s. Although these possibilities have been recognized (Lane and Richards, 1998; Nicholas and Sambrook Smith, 1999; Nicholas, 2001; Lane et al., 2002; Lane, pers. comm.), they have yet to be explored fully. In addition, several studies have suggested that it is not possible to characterize bed roughness using a single length scale (e.g. Nikora et al., 1998; Butler et al., 2001). For example, effective roughness may be controlled by both the size and the spacing of obstacle clasts (Nowell and Church, 1979; Gomez, 1993). Consequently, there remains considerable uncertainty surrounding roughness representation in natural channels, and it may be necessary to accept that, even where wall functions are applicable, k_s is not a simple function of bed sedimentology and must be determined by calibration. Such an approach may be possible where high-quality mean velocity, turbulence and water surface elevation data are available. However, calibration procedures are complicated greatly by the spatial heterogeneity of roughness in natural rivers, and by the high degree of local spatial variability that is a characteristic of hydraulic variables in field environments, especially within the near-bed region.

Beyond the problem of identifying suitable value(s) for the roughness length scale, the wall function approach is associated with a number of additional limitations when applied to natural channels. Strictly speaking it is not valid in the presence of strongly 3D flows or where separation occurs (Patel *et al.*, 1991; Patel, 1998). Furthermore, field evidence suggests that it may be unsuitable for situations involving large roughness elements (Wiberg and Smith, 1987), sharp transitions in bed sedimentology (Robert *et al.*, 1992), intense bedload transport (Pitlick, 1992), and vegetative roughness (Ikeda and Kanazawa, 1996). Field and laboratory measurements of velocity profiles over gravel beds also demonstrate that the wall function approach may not provide a realistic representation of turbulent flow characteristics in the near-bed region of gravel-bed rivers. For example, in the absence of other sources of turbulent kinetic energy production, CFD calculations based on the wall function approach predict turbulence intensities to be highest at the bed and to decline towards the water surface. In contrast, several studies have demonstrated that peak turbulence intensities occur above the bed and typically at the top of roughness elements (Wang *et al.*, 1993; Lawless and Robert, 2001). Below this level, turbulence intensities decline and flow separation commonly occurs in the lee of obstacle clasts (Kirkbride, 1993). In such situations, near-bed roughness elements introduce mass blockage effects that may be represented poorly using the wall function approach. Indeed, realistic simulation of 3D grain-scale flow structures may be possible only by incorporating these blockage effects explicitly (Lane *et al.*, 2002).

In addition to issues concerning the ability of wall functions to provide a realistic representation of flow characteristics in some situations, a number of technical problems may be experienced with this approach. For example, where shallow flows occur over coarse-grained riffles (Hodskinson, 1996) or braid bar margins (Nicholas and Sambrook Smith, 1999) the ratio z/k_s becomes small. This may lead to problems in terms of either numerical stability or solution accuracy for flows characterized by high relative roughness, particularly where structured model grids are used. It may also result in an upper limit on the maximum shear velocity that can be modelled. Hodskinson (1996) addressed this problem by increasing the height of the near-bed cell. However, the centroid of this cell should lie within the bottom 20% of the flow and above $z^+ = 30$–60 for the log law to be valid (where $z^+ = \rho u_* z / \mu$). In addition, such a coarse near-bed mesh restricts the potential for resolving complex flow fields. These observations suggest that where flow depths are shallow and roughness elements occupy a substantial proportion of the profile, parameterization schemes are required that use a fine near-bed mesh and account for roughness effects in regions other than just the wall-adjacent cell. Clearly, the physical realism and applicability of the wall function approach is questionable in these situations.

13.4 Alternative roughness models

Two modified or alternative approaches to representing boundary roughness in gravel-bed rivers have received some attention in the literature and appear to be worth pursuing further. The first reflects the partitioning of roughness between sub-grid-scale and supra-grid-scale components, and involves adding random elevation perturbations to the topographic mesh to increase the contribution of supra-grid-scale roughness. The second involves the modification of equation (13.2), and possibly also equations (13.1), (13.5) and (13.6), to represent drag on roughness elements and account for the fact that model cells may contain part fluid and part bed material.

Nicholas (2001) conducted CFD simulations to investigate the effects of random variability in bed elevation for 2D flows under approximately longitudinally uniform conditions. Simulations using the standard wall function approach in a channel with a uniform slope were compared with simulations in which random perturbations with a Gaussian distribution ($\sigma = k_s$) were added to the bed. This relationship between σ and k_s represents a simple first approximation that is based on data presented by Nikora *et al.* (1998) supporting $\sigma = 0.3$–$0.4 D_{50}$, and the grain-scale roughness length expression ($k_g = 0.4 D_{50}$) derived by Clifford *et al.* (1992). Figures 13.1 and 13.2 show modelled velocity and turbulent kinetic energy profiles

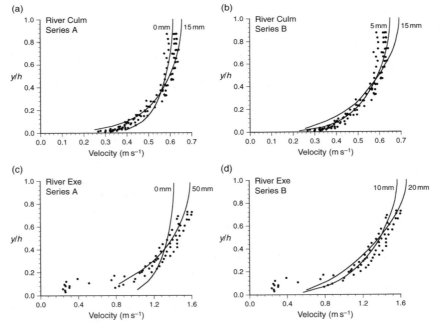

Figure 13.1 *Simulated streamwise-averaged velocity profiles for the River Culm (a and b) and River Exe (c and d). Labels indicate roughness length scales in model runs using the wall function approach alone (a and c) and random bed elevation model (b and d). Closed circles indicate ADV data.*

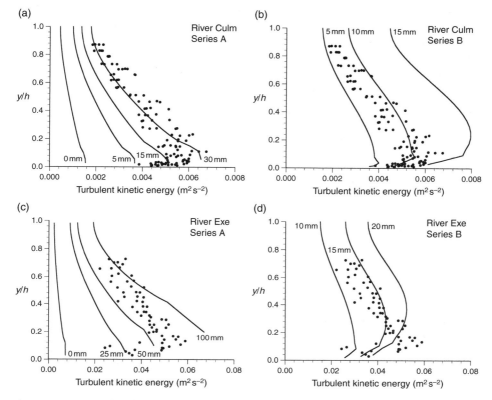

Figure 13.2 Simulated turbulent kinetic energy profiles for the River Culm (a and b) and River Exe (c and d). Labels indicate roughness length scales in model runs using the wall function approach alone (a and c) and random bed elevation model (b and d). Closed circles indicate ADV data.

derived using these two approaches for two sites with contrasting bed characteristics and flow conditions (Table 13.1). Model calculations are compared with field measurements obtained using an array of Acoustic Doppler Velocimeters (ADVs). These data illustrate that optimum model results are obtained using a lower value of k_s where random perturbations in bed elevation are introduced. This has positive

Table 13.1 Study site characteristics

	River Culm	River Exe
Mean width	7.5 m	21 m
Profile flow depth	0.525 m	0.5 m
Depth-mean velocity	0.56 m s^{-1}	1.30 m s^{-1}
Water surface slope	2.4×10^{-4}	2.2×10^{-3}
Bed material D_{50}	15 mm	32 mm
Froude number	0.25	0.59
Reynolds number	2.94×10^5	6.5×10^5

implications for vertical cell dimensions at the bed and, consequently, the ability to resolve hydraulic gradients within the near-bed region. In addition, the random bed elevation model is able to simulate the displacement of peak turbulent kinetic energy values above the bed resulting from velocity shear in the lee of topographic highs.

Despite these potential benefits, the introduction of random bed elevation perturbations may have several negative consequences. For example, where k_s is large, the peak in the turbulent kinetic energy profile may be displaced too far above the bed unless a high-resolution mesh is used, at least in the near-bed region. In addition, this approach generates local spatial variations in hydraulic variables with associated implications for model validation. This local variability results from topographic highs and lows on the bed. However, this does not imply that the model is able to resolve the flow structures associated with these features in a physically realistic way. Furthermore, both the increase in the effective roughness of the bed and the degree of local variability in simulated flow conditions are sensitive to the resolution of the model mesh. These limitations may mean that the random bed elevation model is of limited utility in 3D field applications. However, this approach does provide a framework for investigating the relative contribution of sub-grid-scale and supra-grid-scale roughness components. In addition, it highlights the effect on simulated hydraulics of high-frequency variability in mesh topography, which has important implications for natural channel applications since this variability is an inherent characteristic of field-derived topographic datasets used in mesh construction.

The second alternative method of representing surface roughness noted above involves what is termed a 'discrete element model' (cf. Taylor *et al.*, 1985). In this approach the net effect on the flow of a collection of individual roughness elements is represented by the inclusion of a form drag term in the momentum equation. In addition, mass and momentum equations may be modified to include blockage terms that account for the proportion of model cells occupied by the roughness elements. Finally, the turbulence model may be adjusted to reflect resulting changes in the production and dissipation of turbulent kinetic energy, although this modification is commonly neglected.

A number of discrete element models are reported in the literature for regular shaped roughness elements such as cylinders, cones and hemispheres (e.g. Finson, 1982; Christoph and Pletcher, 1983; Taylor *et al.*, 1985). This approach has also been adopted to represent the effects of vegetative roughness both in laboratory channels (e.g. Tsujimoto *et al.*, 1996; Fischer-Antze *et al.*, 2001; López and García, 2001) and on natural floodplains (Nicholas, 2002). In contrast, discrete element models have been used less widely to represent gravel-bed roughness, although Wiberg and Smith (1987) present an analytical model based on this approach. Olsen and Stoksteth (1995) describe a porosity model, developed by Engelund (1953), that involves the addition of a sink term to the momentum equation, and is used to represent the effects of large roughness elements. More recently, Lane *et al.* (2002) used a porosity-based approach to represent individual roughness elements when modelling flow structures over gravel beds in a laboratory flume. These applications illustrate the benefits of the discrete element model and demonstrate that it may have considerable potential for use in gravel-bed applications. For example, this approach is able to represent the effects of roughness elements that extend into regions of the flow located above the bed-adjacent cell and has been shown to replicate the displacement of peak turbulence intensity values above the roughness layer (in the case of vegetated surfaces). In addition, it appears that this method overcomes some of the numerical limitations of the wall function approach that may be experienced in shallow flows.

13.5 Development of a discrete element model for gravel-bed rivers

Despite the potential benefits of the discrete element approach in gravel-bed river applications, its use has been restricted mainly to situations involving large roughness elements in steep mountain rivers (e.g. Wiberg and Smith, 1987; Olsen and Stoksteth, 1995). Lane *et al.* (2002) recently extended this approach to represent roughness and bed topography in a high-resolution simulation of flow over individual gravel particles. However, a model that is applicable to gravel-bed rivers in general has yet to be presented. A first approximation of such a model can be derived by combining existing discrete element approaches with information quantifying the small-scale topographic characteristics of gravel beds. The main component of this model is a sink term that is added to the momentum equation to represent the form drag generated by roughness elements

$$F_i = -\frac{\rho C_D}{2L} u_i |u| \qquad (13.16)$$

where F_i is the drag force per unit volume, L is the reciprocal of the area of roughness elements projected into the flow per unit volume (equivalent to the mean element spacing), C_D is a drag coefficient, u_i is the velocity in the x_i-direction and $|u|$ is the resultant velocity magnitude. In order to implement this approach, expressions are required for the drag coefficient and roughness element spacing. These are derived here using a simple stochastic model to generate artificial gravel-bed profiles.

The stochastic model simulates the change in bed height (Δz) between two locations separated by a horizontal step length (Δx). Values of Δz are drawn at random from a frequency distribution of height changes. The direction of the height change (i.e. up or down) is determined by the current elevation within the bed layer, so that the probability (P) that the elevation increases over the next downstream step is given by

$$P = 1 - \frac{z_i}{H} \qquad (13.17)$$

where z_i is the current height above the lowest point on the bed and H is the total thickness of the bed profile (i.e. the difference between the maximum and minimum elevations). Values for H, Δz and Δx can be estimated from the results of field and laboratory investigations of gravel-bed topography. For example, Gomez (1993) describes bed profiles for which $\Delta x \approx D_{50}$, while data obtained by Kirchner *et al.* (1990) and Nikora *et al.* (1998) indicate that bed elevation distributions can be described by a Gaussian model with a standard deviation $\sigma_z = 0.3$–$0.5 D_{50}$. Defining the bed profile thickness (H) to be equivalent to the region that contains 99% of bed elevation points suggests that $H \approx 2D_{50}$ is a reasonable approximation, and this is consistent with the bed profiles presented by Kirchner *et al.* (1990) and Gomez (1993). These profiles are characterized by regions of relatively uniform elevation separated by steep transitions. Consequently, although elevation distributions can be approximated by a Gaussian model, distributions of local slope or height change tend to be strongly leptokurtic. With this in mind, frequency distributions of Δz were described by

$$N_i < 0.5 \quad C_i = N_i(1 - |N_i - 0.5|)^\alpha \qquad (13.18)$$
$$N_i \geq 0.5 \quad C_i = 1 - (1 - N_i)(1 - |N_i - 0.5|)^\alpha \qquad (13.19)$$

where C_i is the cumulative frequency of bed height changes less than Δz_i, which is related to the normal cumulative distribution function (N_i) with mean $<\Delta z> = AD_{50}$ and standard deviation $\sigma_{\Delta z} = BD_{50}$ by a shape parameter (α) which determines the kurtosis of the height change frequency distribution. Gomez (1993) presents measurements of roughness height (which is approximately equivalent to Δz), for gravel-bed profiles with median diameters in the range 10–20 mm. His data for $<\Delta z>$ and Δz_{95} suggest values of 0.3 for A and 2.65 for the ratio $\Delta z_{95}/<\Delta z>$. This stochastic model might be improved by incorporating an explicit treatment of the scale dependence of roughness (cf. Nikora *et al.*, 1998; Butler *et al.*, 2001). However, it provides a first approximation of gravel-bed topography that can be used to parameterize the discrete element model of bed roughness developed here.

The model was applied using a range of values for B and α to generate artificial bed profiles. Values of $B = 1.6$ and $\alpha = 1$ were found to yield leptokurtic distributions of Δz characterized by the correct value of the ratio $\Delta z_{95}/<\Delta z>$. Figure 13.3 shows a typical gravel-bed profile derived using this approach. Bed profiles can be used to determine values for the mean obstacle spacing (L) at any elevation above the bed. However, specification of the drag coefficient (C_D) is more problematic. Fischer-Antze *et al.* (2001) treat C_D as a constant with a value of one, while others

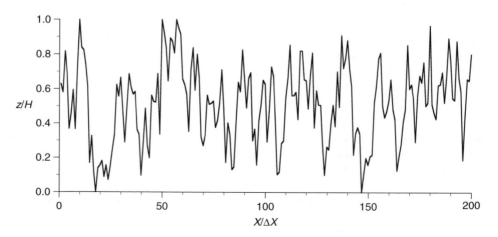

Figure 13.3 Part of a bed elevation profile simulated using the stochastic model.

have related this parameter to the Reynolds number (e.g. Taylor et al., 1985; Naot et al., 1996). In addition, several studies have recognized the importance of wake effects which promote reduced drag forces at high roughness densities where obstacle spacing is small (Nowell and Church, 1979; Liu et al., 1995). Nepf (1999) has quantified the associated reduction in C_D for flows through aquatic vegetation. However, the resulting relationship is unlikely to be directly transferable to flows over gravel beds. A relatively simple approach is adopted here in which equation (13.16) is rewritten in the form

$$F_i = -\frac{\rho}{2D_{50}} N \hat{C}_D S_F u_i |u| \tag{13.20}$$

where N is the number of obstacles per unit length along the random bed profile (measured in dimensionless horizontal steps), $\hat{C}_D$ is initially assumed to take a value close to one (although a range of drag coefficients are examined below) and S_F is a shading factor which accounts for reduced drag due to wake effects. The latter is given by an equation, equivalent to that presented by Naot et al. (1996), that is based on the work of Schlichting (1962) for randomly distributed rods:

$$S_F = 1 - \lambda N (1 - 0.5\sqrt{\lambda N}) \tag{13.21}$$

where λ is the mean obstacle size (measured in dimensionless horizontal steps). Bed profiles generated by the stochastic model can be used, in conjunction with equation (13.21), to calculate the function $(N\hat{C}_D S_F)$ for any height above the bed. In doing this the shading factor can be considered to account for the increasing significance of wake effects and flow separation lower in the profile, although it must be recognized that this is a crude approximation since equation (13.21) is based on the bulk properties of the flow.

Figure 13.4 shows the relationship between this drag function and the dimensionless height above the bed (non-dimensionalized by the effective height of the bed layer (H_e)). A fourth-order polynomial was fitted to this relationship to allow the drag term in the momentum equation to be calculated as

$$F_i = -\frac{\rho}{2D_{50}} u_i |u| \frac{1}{(z_{*_{j+1}} - z_{*_j})} \int_{z_{*_j}}^{z_{*_{j+1}}} N\hat{C}_D S_F(z_*) \, dz_* \tag{13.22}$$

where F_i is the force per unit flow volume in the x_i-direction that acts in the cell extending between dimensionless elevations z_{*_j} and $z_{*_{j+1}}$, and $z_* = (z - z_{bed})/H_e$. When applying this approach the bed elevation in the model mesh (z_{bed}) is taken to represent the mean bed level. The implication of this is that although the actual bed levels within a mesh cell cover a range of elevations (H) extending both above and below the mean elevation, for the purpose of model calculations the volume fraction of fluid should be treated as unity in all cells. Consequently, where the model mesh is coarser than the horizontal scale of the roughness elements, there is no need

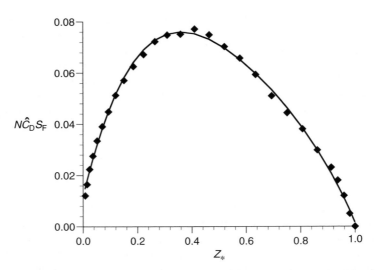

Figure 13.4 Relationship between dimensionless height above the bed and values of the drag function ($N\hat{C}_D S_F$) calculated for the bed elevation profile shown in Figure 13.3.

to introduce extra terms into mass and momentum conservation equations representing horizontal blocking effects. In addition, the effective height of the bed layer (defined in the context of the model domain as the height above the bed over which the drag force is non-zero) becomes $H_e = D_{50}$. The function ($N\hat{C}_D S_F$) shown in Figure 13.4 and integrated in equation (13.22) is expressed in terms of this effective height, which is related to the bed elevation profiles generated using the stochastic model by a simple transformation. No modifications to turbulence model equations (13.5) and (13.6) are implemented in the discrete element model at this stage since preliminary tests indicated that such changes did not improve model performance significantly. In the applications reported here the boundary conditions at bed cells are satisfied using equations (13.10)–(13.12) with $\Delta B = 0$.

As an initial test of the discrete element model it was used in a series of 2D calculations to simulate the flow profiles monitored on the Rivers Culm and Exe (Figure 13.1). Simulations were performed using high-resolution meshes, containing 10 cells within the bed layer and 40 cells above, with the intention of assessing the ability of the discrete element model to resolve the shape of velocity and turbulence profiles in the near-bed region. In the form outlined above, two parameters are required to implement the model (D_{50} and $\hat{C}_D$). Estimates of the D_{50}, based on bed sediment samples obtained using a size-by-number approach, are provided in Table 13.1. Simulations were carried out using values of $\hat{C}_D$ in the range 0.5–1.5. Modelled velocity profiles (Figure 13.5) appear similar to those obtained using other methods (e.g. in Figure 13.1). However, they are characterized by a marked break in slope at the top of the roughness layer, above which they fit a semi-logarithmic relationship with height above the bed. These results illustrate that the discrete element model is able to represent deviations from the log law, which are known to occur both over artificial beds (Nowell and Church, 1979) and in natural gravel-bed channels (Wiberg and Smith, 1987), although it does not reproduce the very low

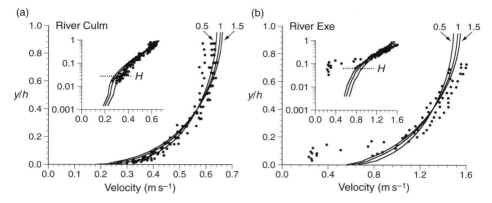

Figure 13.5 Simulated streamwise-averaged velocity profiles for the River Culm (a) and River Exe (b). Labels indicate value of drag coefficient in model runs using the discrete element model. Closed circles indicate ADV data.

near-bed velocities measured in the River Exe (Figure 13.5b). Simulated turbulent kinetic energy profiles (Figure 13.6) exhibit a marked peak at the top of the bed layer (i.e. at relative depths $y/h \approx 0.1$) and show excellent agreement with field data. Optimum values of the drag coefficient are approximately $\hat{C}_D = 1$ for the River Culm and $\hat{C}_D = 1.25$ for the River Exe. Overall these results are encouraging and suggest that the discrete element approach is able to replicate important characteristics of velocity and turbulence profiles in gravel-bed rivers that are not resolved using the wall function approach.

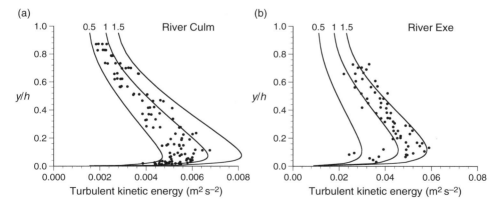

Figure 13.6 Simulated turbulent kinetic energy profiles for the River Culm (a) and River Exe (b). Labels indicate value of drag coefficient in model runs using the discrete element model. Closed circles indicate ADV data.

13.6 Simulation of 3D flow in a braided channel

As a further test of the discrete element model it was used to parameterize bed roughness in a simulation of 3D flow in the braided Ryton River, New Zealand. Bed topography and water surface elevation were surveyed in detail to allow the construction of a model mesh composed of 356 000 hexahedral cells with 10 cells in the vertical and a horizontal resolution of 0.3 m. The mesh is characterized by a regular grid structure, with all internal vertical cell faces aligned parallel to the x- and y-axes of the model domain. At the boundary of inundation body-fitted coordinates are used to represent external cell faces. Given the lack of spatially distributed grain-size data for the reach, a constant value of the effective height of the bed layer (H_e) was used within the model domain. This was set equal to the median grain size for the reach of 30 mm, estimated on the basis of sediment samples obtained from bar surfaces and areas of shallow flow. Figure 13.7 shows the spatial pattern of flow depth within the reach. Figure 13.8a shows the proportion of the model domain containing different numbers of cells within the roughness layer (i.e. within the layer $z < z_{bed} + H_e$). In 75% of locations, the roughness layer extends above the near-bed cell highlighting the limited applicability of the wall function approach.

Initial model runs were conducted using a range of spatially uniform values of $\hat{C}_D$. Optimization of this parameter can be achieved by comparing the surveyed drop in water surface elevation along the reach (2.62 m) with the quantity $\Delta P/\rho g$, where ΔP is

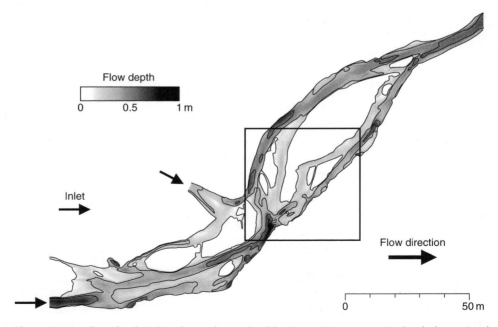

Figure 13.7 *Flow depth (m) in the study reach of the Ryton River, New Zealand, determined by field survey of river bed topography and water surface elevation. The box indicates the region for which results are shown in Figures 13.10–13.14.*

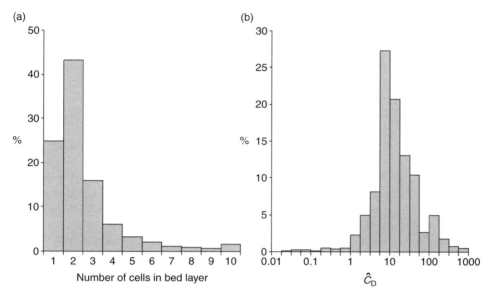

Figure 13.8 Frequency distributions (in terms of the proportion of the planform area of the model domain) for: (a) the number of cells lying within the bed layer and (b) the drag coefficient $\hat{C}_D$, for the simulation involving spatially variable bed roughness.

the modelled drop in pressure at the water surface along the reach. Table 13.2 shows the value of this quantity for a range of values of $\hat{C}_D$. The sensitivity of $\Delta P/\rho g$ to $\hat{C}_D$ clearly declines as the drag parameter increases. This, in part, reflects the fact that increases in $\hat{C}_D$ are compensated by reductions in near-bed velocity. On the basis of these results values of $\hat{C}_D$ in the range 6–6.5 appear to represent the optimum parameterization. These values are higher than those identified for the River Exe and River Culm datasets, and are also higher than experimental drag constants for large particles and objects such as cylinders (for which values are typically 0.5–1.5). The high optimum value of $\hat{C}_D$ for the Ryton River might reflect underestimation of the average D_{50} within the reach, underestimation of the drag contribution of areas of coarser bed sediment (i.e. the assumption of a uniform value of the D_{50}) or underestimation of the ratio H/D_{50} in the stochastic model used to generate the artificial bed elevation profiles. Alternatively, the high value of $\hat{C}_D$ may be a product of the under-representation of macro-scale form roughness within the model domain, which in turn reflects the resolution of the topographic data used to construct the model mesh.

Table 13.2 Calculated drop in $P/\rho g$ at the water surface along the study reach for simulations conducted using a range of spatially uniform values of the drag constant ($\hat{C}_D$)

$\hat{C}_D$	0.5	2	6	6.5	7
$\Delta P/\rho g$ (m)	0.91	1.59	2.57	2.67	2.76

Spatially distributed values of the pressure at the water surface can also be used to calculate values of the local water surface elevation residual (E) defined as:

$$E = z_{\text{ref}} + \frac{P_i - P_{\text{ref}}}{\rho g} - z_i \qquad (13.23)$$

where z_i and z_{ref} are the water surface elevations at cell i and at a reference cell (at the downstream reach boundary) and P_i and P_{ref} are the pressures at the water surface at these locations. Olsen and Kjellesvig (1998) present a procedure, based on an equation of this form, for adjusting the position of the water surface to minimize E. However, this procedure would be difficult to implement in the current application where a change in water surface elevation might lead to a substantial change in inundation of bar surfaces. Furthermore, since the water surface within the model mesh is based on detailed field survey information (>500 data points) and the residual term (E) is a product of a number of factors other than the specification of the water surface position, it is not clear that it would be appropriate to adopt this approach here. Instead it can be argued that non-zero values of E reflect the existence of spatial heterogeneity in bed roughness that is not incorporated within the model parameterization, rather than errors in the water surface location. If one assumes that this is the case it is possible to use distributed values of E to map spatial patterns of bed roughness within the reach.

Spatial variations in bed roughness were modelled by implementing a three-stage procedure to determine local values of the drag coefficient $\hat{C}_D$. This procedure is based on the principle that for a given flow depth and unit discharge, bed roughness and water surface slope are positively correlated. In stage one the reach was divided into 16 regions, largely on the basis of the position of mid-channel bars, confluences and diffluences. The value of $\hat{C}_D$ within each region was optimized by comparing the surveyed drop in water surface elevation along the channel thalweg with the equivalent quantity ($\Delta P/\rho g$) simulated by the model. Values of $\hat{C}_D$ were then adjusted (i.e. where the mesh water surface slope along the thalweg exceeds $\Delta P/\rho g$ the value of $\hat{C}_D$ is increased) and a new model solution was obtained. This manual calibration procedure was repeated approximately 20 times until the optimum values of $\hat{C}_D$ within each region were obtained (i.e. until the mean water surface height drop within each region was equal to $1/\rho g$ times the drop in pressure). Following this, in a second stage of the procedure, local values of the drag constant at individual cells were determined as:

$$\hat{C}_{Di} = \hat{C}_{Dj} \left[\alpha \left(\frac{S_{Mi}}{S_{Pi}} \right) + (1 - \alpha) \right] \qquad (13.24)$$

where $\hat{C}_{Di}$ is the value of the drag coefficient at cell i, $\hat{C}_{Dj}$ is the value of the drag coefficient within region j (determined by the first stage of the procedure), S_{Mi} is the local slope of the water surface in the model mesh (in the direction of the depth-mean resultant flow vector), S_{Pi} is the local slope of the quantity ($P/\rho g$) in the same direction, and α is a parameter set equal to 0.5 to minimize the mean absolute value of E within the reach. Equation (13.24) is used to incorporate spatial variations in bed roughness at scales smaller than the 16 regions used in stage one. In the final stage of the procedure sharp transitions in the value of $\hat{C}_D$ at the boundaries between these regions are removed by applying a smoothing function that recalculates values of $\hat{C}_D$ as the weighted sum of values at local and neighbouring cells.

Figure 13.8b shows the frequency distribution of drag coefficient values within the reach determined using this procedure. Figure 13.9 shows the frequency distribution of water surface elevation residuals within the reach for simulations using constant ($\hat{C}_D = 6.5$) and spatially variable values of the drag coefficient. Water surface elevation residual summary statistics are presented in Table 13.3 along with details of best-fit lines between model predictions of resultant velocity and turbulent kinetic energy and measurements obtained using an array of ADVs. Use of spatially distributed values of $\hat{C}_D$ removes approximately 60% of the residual in water surface elevation and is particularly effective at eliminating the tail of the residual distribution associated with constant bed roughness. Despite these improvements the quality of fit between field measurements and model predictions of velocity and turbulence kinetic energy is not significantly different for the two simulations. This reflects the high degree of spatial variability in flow conditions within the field (as is also reflected by the relatively low values of the coefficient of determination for turbulent kinetic energy) and highlights the difficulty of using such hydraulic data to discriminate between different model parameterizations. Levels of unexplained variance for velocity are similar to those reported in previous CFD applications involving braided gravel-bed rivers (e.g. Lane *et al.*, 1999; Nicholas and Sambrook Smith, 1999). Furthermore, one might expect coefficients of determination to be lower for the current application since relative roughness is high and ADV data were obtained close to the bed. Improved representation of such near-bed flow structures would

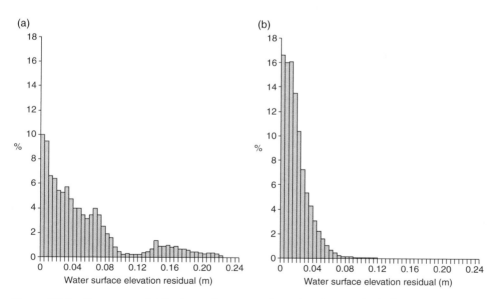

Figure 13.9 Frequency distributions of water surface elevation residuals for: (a) calculations using a constant value of $\hat{C}_D = 6.5$ and (b) calculations using a spatially variable value of $\hat{C}_D$.

Table 13.3 *Summary statistics for simulations conducted using constant and spatially variable values of $\hat{C}_D$. Rows 1–3 contain mean, median and 95th percentile of water surface elevation residuals (E). Rows 4–9 contain the gradient (B), intercept (C) and coefficient of determination (r^2) of relationships between measured and predicted flow variables determined by structural analysis (based on a sample of 97 ADV data points)*

	Constant $\hat{C}_D = 6.5$	Spatially variable $\hat{C}_D$
$\bar{E}$	52 mm	19 mm
E_{50}	37 mm	16 mm
E_{95}	169 mm	50 mm
B (resultant velocity)	1.04	1.02
C (resultant velocity)	0.07	0.09
r^2 (resultant velocity)	0.71	0.70
B (turbulent kinetic energy)	1.74	1.55
C (turbulent kinetic energy)	−0.011	−0.009
r^2 (turbulent kinetic energy)	0.40	0.37

necessitate the use of a higher-resolution approach, the success of which is likely to be dependent upon the availability of topographic data at a similar resolution.

Although 90% of drag coefficient values lie within the range 1–100, values of $\hat{C}_D$ up to a maximum of 2000 occur in limited areas of the reach. Figure 13.10 shows the spatial pattern of $\hat{C}_D$ for the central portion of the study reach highlighted in Figure 13.7. Flow is separated between two main branches, each of which is subdivided further by a number of mid-channel bars. The majority of flow is routed down the far left-hand channel (Figure 13.11) and it is in this area that roughness values are highest. Although independent quantitative data with which to validate the spatial patterns of roughness mapped here are not available, these patterns are consistent with qualitative visual observations that the left-hand channel branch was characterized by a steeper slope, a more broken water surface and a larger proportion of coarse bed sediment.

Results for the two simulations discussed here also afford the potential to evaluate the sensitivity of model predictions to the parameterization of bed roughness. Previous CFD applications in gravel-bed rivers have often used a single bed roughness parameter value over a whole reach (e.g. Hodskinson, 1996; Lane et al., 1999; Nicholas and Sambrook Smith, 1999) despite the fact that spatial variations in roughness are a characteristic of natural channels. The extent to which hydraulic predictions are sensitive to the use of spatially distributed versus uniform bed roughness has yet to be considered fully, although this issue has important implications for the prediction of sediment transport and channel evolution. Figure 13.12 shows modelled patterns of bed shear stress within the central part of the study reach (determined by integrating the force per unit bed area throughout the roughness layer) within a part of the study reach for simulations conducted using uniform and spatially variable values of the drag coefficient. Broad spatial patterns remain unchanged as one might expect, but significant differences between simulations are evident in some areas. For example, in the upstream half of the left-hand channel branch shear stresses are up to 50% lower for the simulation conducted using

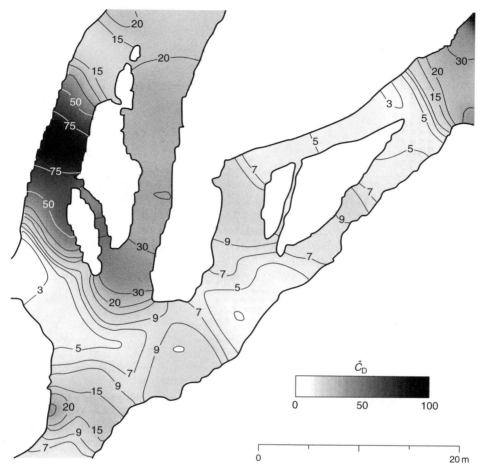

Figure 13.10 Spatial pattern of $\hat{C}_D$ for the part of the study reach highlighted in Figure 13.7.

spatially variable roughness, while in the downstream half of this channel they are up to three times greater than those simulated using a uniform value of $\hat{C}_D$. Differences in shear stress between simulations reflect the spatial pattern of bed roughness (Figure 13.10), the influence of this on velocity profiles and on the distribution of flow between individual channels. The order of magnitude increase in bed roughness within a part of the left-hand channel is compensated by a marked reduction in near-bed velocity (Figure 13.13) which moderates the increase in bed shear stress. Increased bed roughness in this area also promotes greater velocities at the water surface and hence a significant increase in velocity profile gradients. This is also evident in the adjacent channel to the right, but both surface and near-bed velocities are lower in this channel reflecting redistribution of flow to other areas as a result of increased roughness. Differences in patterns of simulated turbulent kinetic energy between the two simulations are also evident (Figure 13.14), particularly near the bed

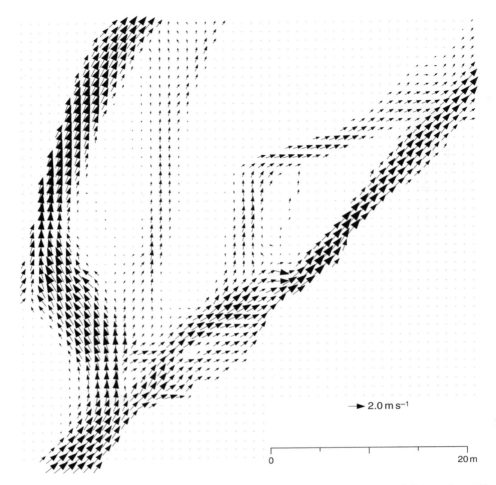

Figure 13.11 *Modelled flow vectors at the water surface for the part of the study reach highlighted in Figure 13.7. In the interest of clarity only every third vector in the model domain is shown.*

and in the left-hand channel due to the increased near-bed velocity gradients in this area. The effect of this increase in near-bed turbulence production is also apparent at the water surface further downstream. Overall, the relative sensitivity of model results to the use of spatially distributed roughness appears greater for turbulence intensity than for velocity. In most areas of the model domain resultant velocities change between the two simulations by <10% at the water surface and <20% near the bed. Relative changes in turbulent kinetic energy are typically greater than this and are similar at the bed and water surface. For example, 20% of model cells experience >50% change in turbulent kinetic energy.

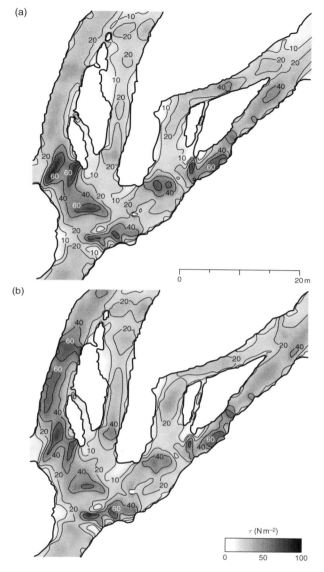

Figure 13.12 Modelled pattern of bed shear stress for the part of the study reach highlighted in Figure 13.7 for: (a) calculations using a constant value of $\hat{C}_D = 6.5$ and (b) calculations using a spatially variable value of $\hat{C}_D$.

13.7 Discussion

The procedure outlined above provides a method of reducing the deviations between surveyed water surface topography and elevations determined from the simulated pressure field at the water surface. In so doing it is also possible to derive distributed information quantifying spatial patterns of bed roughness within the model domain.

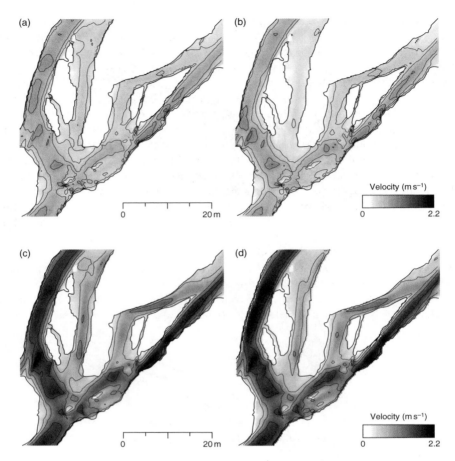

Figure 13.13 Modelled pattern of resultant velocity for the part of the study reach highlighted in Figure 13.7: (a) in the bottom 20% of the flow for calculations using a constant value of $\hat{C}_D = 6.5$; (b) in the bottom 20% of the flow for calculations using a spatially variable value of $\hat{C}_D$; (c) at the water surface for calculations using a constant value of $\hat{C}_D = 6.5$ and (d) at the water surface for calculations using a spatially variable value of $\hat{C}_D$. Contour intervals are 0.5, 1.0, 1.5 and 2.0 m s^{-1} (a and b) and 0.25, 0.5, 0.75 and 1.0 m s^{-1} (c and d).

Water surface elevation residuals, as defined here, represent the sum of all sources contributing to the pressure field, in addition to errors in the specification of the mesh water surface topography. Consequently, although these residuals may be expressed in units of length, this does not imply that these data represent true inaccuracies in the position of the mesh water surface. Furthermore, where direct comparisons are to be made between simulated and measured flow velocities or turbulence characteristics, and where data are available with which to specify the water surface position, it may be inappropriate to adjust the water surface position or implement a procedure intended to account for a change in its position. Apparent errors in the water surface position may reflect topographic errors in the model mesh, or may be largely a product of the inadequate representation of boundary

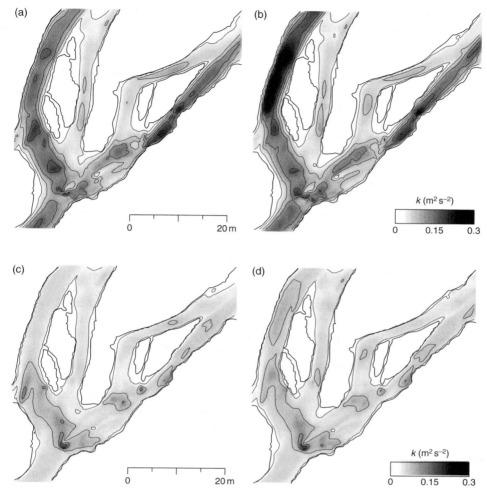

Figure 13.14 Modelled pattern of turbulent kinetic energy for the part of the study reach highlighted in Figure 13.7; (a) in the bottom 20% of the flow for calculations using a constant value of $\hat{C}_D = 6.5$; (b) in the bottom 20% of the flow for calculations using a spatially variable value of $\hat{C}_D$; (c) at the water surface for calculations using a constant value of $\hat{C}_D = 6.5$ and (d) at the water surface for calculations using a spatially variable value of $\hat{C}_D$. Contour intervals are 0.01, 0.05, 0.1, 0.15, 0.2 and 0.25 $m^2\,s^{-2}$.

roughness, where the latter is assumed to be uniform due to a lack of spatially distributed roughness information. Consequently, while the application of a water surface correction procedure may yield a physically realistic solution, this may be achieved by introducing errors into the representation of the water surface topography to compensate for errors in the parameterization of bed roughness.

In the case of simulations conducted for the Rivers Culm and Exe, the optimum drag coefficient value is close to 1, whereas for the braided channel simulation using spatially uniform roughness the optimum value of this parameter is approximately 6.5. This reflects the fact that this parameter is sensitive to both hydraulic conditions and grid resolution. The difference between the optimum values of $\hat{C}_D$ for the simulations conducted here may also be due to the greater contribution of form roughness at intermediate spatial scales in the braided channel. Such form roughness is not represented fully in either the channel bed topographic survey or the bed profile stochastic model used in the discrete element approach. Field measurements on the Rivers Culm and Exe were obtained from straight sections with near plane-bed conditions to minimize the contribution of form roughness. In the interest of simplicity, and because of the lack of spatially distributed grain-size information, spatially non-uniform bed roughness has been represented in the braided channel simulation by using distributed values of the drag coefficient ($\hat{C}_D$) in equation (13.22), rather than by adjusting the depth of the bed roughness layer. As a result one might expect values of this parameter to vary over a wider range than would otherwise be the case, and indeed values of up to 2000 occur, although most values lie in the range 1–100. Extreme values of $\hat{C}_D$ also reflect the fact that changes in the magnitude of the momentum sink term resulting from changes in $\hat{C}_D$ are moderated by adjustments in the flow field. Consequently, a large change in $\hat{C}_D$ is required to generate a smaller change in bed shear stress.

The three-stage procedure implemented here to derive distributed values of the drag coefficient removed approximately 60% of the residual in the water surface elevation field. However, given the nature of this procedure it is undoubtedly possible to derive other, probably similar, roughness parameter maps that perform equally well. Distinguishing between the quality of simulations conducted using alternative spatially distributed parameterizations appears to be highly problematic, particularly in shallow, braided gravel-bed channels. Indeed it is unlikely that this can be accomplished conclusively even using a reasonably large dataset of mean and turbulent flow measurements. The high level of unexplained variance in turbulent kinetic energy, and to some extent velocity, reflects a difference between the scale of bed topography represented by the model mesh and that which controls near-bed flow structures in the field. High-resolution approaches may address this situation (e.g. Lane *et al.*, 2002), but are unlikely to be applicable to large river reaches in the near future. In addition to these problems it is apparent, from results presented here and in many other studies, that roughness parameter values are application-dependent and are a function of model grid resolution, the availability of topographic data and, in the case of the discrete element model, flow characteristics. These issues have yet to be resolved fully; however, they highlight the fact that while CFD models are sophisticated (in geomorphological terms) they require calibration since their physical basis is clearly scale dependent. Consequently, attempts to use the results of such models to predict patterns of sediment transport and channel change should proceed with caution, particularly where bed roughness is assumed to be spatially homogeneous.

Acknowledgements

This work is a product of a number of separate projects funded by NERC, the Royal Society and the Nuffield Foundation. I am grateful to Tim Quine, Rhian Thomas and Miles Marshall who provided field assistance in New Zealand, and to Stuart Lane and Rob Ferguson for their comments on the manuscript.

References

Booker, D.J., Sear, D.A. and Payne, A.J. 2001. Modelling three-dimensional flow structures and patterns of boundary shear stress in a natural pool-riffle sequence. *Earth Surface Processes and Landforms*, **26**, 553–576.

Bradbrook, K.F., Biron, P.M., Lane, S.N., Richards, K.S. and Roy, A.G. 1998. Investigation of controls on secondary circulation in a simple confluence geometry using a three-dimensional numerical model. *Hydrological Processes*, **12**, 1371–1396.

Bradbrook, K.F., Lane, S.N., Richards, K.S., Biron, P.M. and Roy, A.G. 2001. Flow structures and mixing at an asymmetrical open-channel confluence: A numerical study. *Journal of Hydraulic Engineering*, ASCE, **127**, 351–368.

Bradshaw, P., Launder, B.E. and Lumley, J.L. 1996. Collaborative testing of turbulence models. *Journal of Fluids Engineering*, **118**, 243–247.

Bray, D.I. 1982. Flow-resistance in gravel-bed rivers. In: *Gravel-bed Rivers*, Hey, R.D., Bathurst, J.C. and Thorne, C.R. (eds), Wiley, Chichester, 109–133.

Buffin-Bélanger, T., Roy, A.G. and Kirkbride, A.D. 2000. On large-scale flow structures in gravel-bed rivers. *Geomorphology*, **32**, 417–435.

Butler, J.B., Lane, S.N. and Chandler, J.H. 2001. Characterisation of the structure of river-bed gravels using two-dimensional fractal analysis. *Mathematical Geology*, **33**, 301–330.

Cebeci, T. and Bradshaw, P. 1977. *Momentum Transfer in Boundary Layers*, Hemisphere Publishing Corporation, New York.

Christoph, G.H. and Pletcher, R.H. 1983. Prediction of rough-wall skin friction and heat transfer. *American Institute of Aeronautics and Astronautics Journal*, **21**, 509–515.

Clifford, N.J., Robert, A. and Richards, K.S. 1992. Estimation of flow resistance in gravel-bedded rivers: A physical explanation of the multiplier of roughness length. *Earth Surface Processes and Landforms*, **17**, 111–126.

Engelund, F. 1953. On the laminar and turbulent flows of groundwater through homogeneous sand. *Transactions of the Danish Academy of Technical Sciences*, No. 3.

Finson, M.L. 1982. A model for rough wall turbulent heating and skin friction. *American Institute of Aeronautics and Astronautics Paper 82-0199*.

Fischer-Antze, T., Stoesser, T., Bates, P. and Olsen, N.R.B. 2001. 3D numerical modelling of open-channel flow with submerged vegetation. *Journal of Hydraulic Research*, **39**, 303–310.

Gad-el-Hak, M. and Bandyopadhyay, P.R. 1995. Field versus laboratory turbulent boundary layers. *American Institute of Aeronautics and Astronautics Journal*, **33**, 361–364.

Gomez, B. 1993. Roughness of stable, armored gravel beds. *Water Resources Research*, **29**, 3631–3642.

Hey, R.D. 1979. Flow resistance in gravel-bed rivers. *Journal of the Hydraulics Division*, ASCE, **105**, 365–379.

Hodskinson, A. 1996. Computational fluid dynamics as a tool for investigating separated flow in river bends. *Earth Surface Processes and Landforms*, **21**, 993–1000.

Hodskinson, A. and Ferguson, R.I. 1998. Numerical modelling of separated flow in river bends: Model testing and experimental investigation of geometric controls on the extent of flow separation at the concave bank. *Hydrological Processes*, **12**, 1323–1338.

Ikeda, S. and Kanazawa, M. 1996. Three-dimensional organized vortices above flexible water plants. *Journal of Hydraulic Engineering*, **122**, 634–640.

Kirchner, J.W., Dietrich, W.E., Iseya, F. and Ikeda, H. 1990. The variability of critical shear stress, friction angle, and grain protrusion in water-worked sediments. *Sedimentology*, **37**, 647–672.

Kirkbride, A. 1993. Observations of the influence of bed roughness on turbulence structure in depth limited flows over gravel beds. In: *Turbulence: Perspectives on Flow and Sediment Transport*, Clifford, N.J., French, J.R. and Hardisty, J. (eds), Wiley, Chichester, 185–196.

Lane, S.N. and Richards, K.S. 1998. Two-dimensional modelling of flow processes in a multi-thread channel. *Hydrological Processes*, **12**, 1279–1298.

Lane, S.N., Bradbrook, K.F., Richards, K.S., Biron, P.A. and Roy, A.G. 1999. The application of computational fluid dynamics to natural river channels: three-dimensional versus two-dimensional approaches. *Geomorphology*, **29**, 1–20.

Lane, S.N., Hardy, R.J., Elliott, L. and Ingham, D.B. 2002. High resolution numerical modelling of three-dimensional flows over complex river bed topography. *Hydrological Processes*, **16**(11), 2261–2272.

Lawless, M. and Robert, A. 2001. Three-dimensional flow structure around small-scale bedforms in a simulated gravel-bed environment. *Earth Surface Processes and Landforms*, **26**, 507–522.

Launder, B.E. 1989. Second-moment closure: present and future? *International Journal of Heat and Fluid Flow*, **10**, 282–300.

Launder, B.E., Reece, G.J. and Rodi, W. 1975. Progress in the development of a Reynolds-stress turbulence closure. *Journal of Fluid Mechanics*, **68**, 537–566.

Liu, J., Tominaga, A. and Nagao, M. 1995. Numerical study of turbulent structure over strip roughness in open channel flow. *Hydra 2000*, Thomas Telford, London, 165–170.

López, F. and García, M.H. 2001. Mean flow and turbulence structure of open-channel flow through non-emergent vegetation. *Journal of Hydraulic Engineering*, **127**, 392–402.

Nakagawa, H. and Nezu, I. 1987. Experimental investigation on turbulent structure of backward-facing step flow in an open channel. *Journal of Hydraulic Research*, **25**, 67–88.

Naot, D., Nezu, I. and Nakagawa, H. 1996. Hydrodynamic behavior of partly vegetated open channels. *Journal of Hydraulic Engineering*, **122**, 625–633.

Nepf, H.M. 1999. Drag, turbulence, and diffusion in flow through emergent vegetation. *Water Resources Research*, **35**, 479–489.

Nicholas, A.P. 2001. Computational fluid dynamics modelling of boundary roughness in gravel-bed rivers: An investigation of the effects of random variability in bed elevation. *Earth Surface Processes and Landforms*, **26**, 345–362.

Nicholas, A.P. 2002. Modelling flood hydraulics in topographically-complex lowland floodplain environments. Proceedings of the IAHR Conference, *River Flow 2002*, Balkema.

Nicholas, A.P. and McLelland, S.J. 1999. Hydrodynamics of a floodplain recirculation zone investigated by field monitoring and numerical simulation. In: *Floodplains: Interdisciplinary Approaches*, Marriott, S., Alexander, J. and Hey, R. (eds), Geological Society, London, Special Publications, **163**, 15–26.

Nicholas, A.P. and Sambrook Smith, G.H. 1999. Numerical simulation of three-dimensional flow hydraulics in a braided channel. *Hydrological Processes*, **13**, 913–929.

Nikora, V.I., Goring, D.G. and Biggs, B.J.F. 1998. On gravel-bed roughness characterisation. *Water Resources Research*, **34**, 517–527.

Nowell, A.R.M. and Church, M. 1979. Turbulent flow in a depth-limited boundary layer. *Journal of Geophysical Research*, **84**, 4816–4824.

Olsen, N.R.B. and Kjellesvig, H.M. 1998. Three-dimensional numerical flow modelling for estimation of maximum local scour depth. *Journal of Hydraulic Research*, **36**, 775–784.

Olsen, N.R.B. and Stoksteth, S. 1995. Three-dimensional numerical modelling of water flow in a river with large-bed roughness. *Journal of Hydraulic Research*, **33**, 571–581.

Patel, V.C. 1998. Perspective: Flow at high Reynolds number and over rough surfaces – Achilles heel of CFD. *Journal of Fluids Engineering*, **120**, 435–444.

Patel, V.C., Rodi, W. and Scheuerer, G. 1984. Turbulence models for near-wall and low Reynolds number flows: A review. *American Institute of Aeronautics and Astronautics Journal*, **23**, 1308–1319.

Patel, V.C., Tyndall Chon, J. and Yoon, J.Y. 1991. Turbulent flow in a channel with a wavy wall. *Journal of Fluids Engineering*, **113**, 579–586.

Pitlick, J. 1992. Flow resistance under conditions of intense gravel transport. *Water Resources Research*, **28**, 891–903.

Robert, A. 1988. Statistical properties of sediment bed profiles in alluvial channels. *Mathematical Geology*, **20**, 205–225.

Robert, A., Roy, A.G. and De Serres, B. 1992. Changes in velocity profiles at roughness transitions in coarse grained channels. *Sedimentology*, **39**, 725–735.

Rodi, W. 1980. *Turbulence Models and Their Application in Hydraulics – A State of the Art Review*, IAHR, Delft, 104pp.

Roy, A.G., Biron, P.M., Buffin-Bélanger, T. and Levasseur, M. 1999. Combined visual and quantitative techniques in the study of natural turbulent flows. *Water Resources Research*, **35**, 871–877.

Schlichting, H. 1962. *Boundary Layer Theory*. Fourth Edition, McGraw-Hill, New York.

Sinha, S.K., Sotiropoulos, F. and Odgaard, A.J. 1998. Three-dimensional numerical model for flow through natural rivers. *Journal of Hydraulic Engineering*, ASCE, **124**, 13–24.

Sofialidis, D. and Prinos, P. 1998. Compound open-channel flow modelling with non-linear low-Reynolds k-ε models. *Journal of Hydraulic Engineering*, **124**, 253–262.

Taylor, R.P., Coleman, H.W. and Hodge, B.K. 1985. Prediction of turbulent rough-wall skin friction using a discrete element approach. *Journal of Fluids Engineering*, **107**, 251–257.

Tsujimoto, T., Kitamura, T., Fujii, Y. and Nakagawa, H. 1996. Hydraulic resistance of flow with flexible vegetation in open channel. *Journal of Hydroscience and Hydraulic Engineering*, **14**, 47–56.

Vandoormaal, J.P. and Raithby, G.D. 1984. Enhancements of the SIMPLE method for predicting incompressible fluid flows. *Numerical Heat Transfer*, **7**, 147–163.

Wang, J., Chen, C., Dong, Z. and Xia, Z. 1993. The effects of bed roughness on the distribution of turbulent intensities in open-channel flow. *Journal of Hydraulic Research*, **31**, 89–98.

Wiberg, P.L. and Smith, J.D. 1987. Initial motion of coarse sediment in streams of high gradient. *Erosion and Sedimentation in the Pacific Rim*, International Association of Hydrological Sciences Publication, **165**, 299–308.

Yakhot, V. and Orszag, S.A. 1986. Renormalisation group analysis of turbulence: I. Basic theory. *Journal of Scientific Computing*, **1**, 1–51.

Younis, B.A. 1996. Progress in turbulence modelling for open-channel flows. In: *Floodplain Processes*, Anderson, M.G., Walling, D.E. and Bates, P.D. (eds), Wiley, Chichester, 299–332.

14

Modelling of sand deposition in archaeologically significant reaches of the Colorado River in Grand Canyon, USA

S. Wiele and M. Torizzo

14.1 Introduction

The construction and application of computational models provide a rigorous framework for the study of river mechanics as well as for quantitative predictions that can serve as the basis for informed management decisions regarding environmental issues. Computational models of fluvial processes are typically tailored for particular applications. A balance between data availability, computational efficiency and sufficient accuracy to meet application ends is typically required, especially for models intended to predict the evolution of some fluvial process over time. In modelling the Colorado River in the Grand Canyon, this balance must account for data limitations resulting from the remoteness of the study sites and the computational demands stemming from the multi-dimensional modelling of flow fields with recirculation zones, suspended sand transport and time stepping of changes in sand deposits. The purpose of the Grand Canyon model is to predict deposition rates, volumes and location over a range of discharges and sand supplies to anticipate the consequences of the operation of Glen Canyon Dam and tributary events.

The Colorado River corridor in Grand Canyon National Park has been severely altered from its free-flowing state by the construction and closure of Glen Canyon Dam. The once turbid, sediment-laden flow is now clear except during tributary flows. The current average annual sand volume in transport below the first major sediment-contributing tributary below the dam, the Paria River, is about 1.5 million metric tons, or about 6% of the pre-dam sediment load (Topping et al., 2000a). Pre-dam average annual flood peaks of $2420\,m^3/s$ have been typically reduced to the powerplant capacity of $930\,m^3/s$ except on several occasions during exceptional hydrologic conditions or, as of this writing, a single experimental release (see Webb et al., 1999). Winter base flows have been increased from less than about $80\,m^3/s$ prior to the dam construction to about $140\,m^3/s$ under current dam operations. These changes to the flow and sediment characteristics of the river have had direct impacts on dependent resources, such as the preservation of archaeological artefacts, riparian vegetation and endangered species such as the humpback chub and the Kanab Amber snail. The retention of the main-stem sand supply behind Glen Canyon Dam also has led to the erosion of downstream sand deposits that form the substrate for riparian flora and fauna, are used as campsites by riverside visitors to the national park and are an important aesthetic attribute of the natural, formerly unregulated river in the park.

The only tool currently under consideration for mitigating the effects of Glen Canyon Dam on the downstream river corridor is the dam itself. An Environmental Impact Study (EIS; US Department of Interior, 1995) proposed releasing water at discharges in excess of powerplant capacity to entrain and suspend sediment stored at lower elevations within the channel for redeposition at higher elevations along the channel sides. An experimental release from Glen Canyon Dam in 1996 demonstrated that the dam can be operated to restore deposits along the channel sides if sufficient sand is available. The experimental release consisted of a steady discharge of $225\,m^3/s$ for 4 days rising to a steady $1270\,m^3/s$ for 7 days, followed by another steady $225\,m^3/s$ for about 3 1/2 days. As summarized by Schmidt (1999), many of the objectives for a high release contained in an EIS (US Department of Interior, 1995) were achieved by the 1996 release. River guides reported improved camping conditions after the 1996 test flow (Thompson et al., 1997). Kearsley et al. (1999) reported an increase in usable campsites from 218 to 299. Studies of aerial photos (Schmidt et al., 1999) and bathymetric measurements before, during and after the test flow (Andrews et al., 1999) documented significant deposition at many sites, along with considerable spatial and temporal variability. Although many sandbars were replenished, sandbars closest to the dam, and therefore with the smallest available sand supply, were more likely to erode (Schmidt, 1999). More recent studies, moreover, have concluded that the channel sand-storage capacity is much less than was posited in the EIS (Topping et al., 2000b) and have suggested that the timing of high flows should be adjusted to follow shortly after tributary inputs (Lucchitta and Leopold, 1999; Rubin et al., 2002) to maximize the use of the available sediment.

Studies of the geomorphology of near-river environments where archaeological sites are located (Hereford et al., 1991, 1993) have concluded that the lower peak discharges and lower sand concentrations since the closure of Glen Canyon Dam have directly or indirectly damaged some of those sites. Hereford et al. (1991, 1993) concluded that the high discharges released from the dam during 1983 directly eroded archaeological sites. They also describe a more pervasive process related to the lowering of base level for side channels. One type of stream channel they describe, their Type 1, consists of short (300–400-m long), steep, ephemeral streams that drain small near main-stem catchments during rainstorms. They concluded that the erosion of sand deposits at the base of these streams since dam closure has lowered the base level of the streams and led to the upstream migration of nickpoints. As side streams have deepened and widened, they have encroached upon archaeological sites, and in some cases, the erosion has exposed and damaged the sites. Hereford *et al.* (1991, 1993) proposed that the periodic restoration of sand deposits near river level would raise the base level for these side channels, promoting refilling of the channels that would in turn help preserve the archaeological sites. Thompson and Potochnik (2000), in their extensive study of reaches with abundant artefacts and active gullies, concluded that renewed sand deposition could help preserve some, but not all, sites. Observations soon after the 1996 test flow showed that under the conditions at that time, the terraces containing resources gained sand in some cases, and no harm to these sensitive terrace deposits was reported (Yeatts, 1996).

Hereford *et al.* (1991, 1993) focused on water transport of sediment in gullies in their suggestion that erosion of sand bars near the main-stem shorelines is linked to erosion of archaeological sites. Evidence of local reworking of sand by wind, however, has been documented by several workers and suggests an alternative link between the erosion of sand bar deposits along the channel sides and the formation of artefact-damaging gullies. Aeolian deposits are widespread in many areas within Grand Canyon (Schmidt and Graf, 1990; Hereford *et al.*, 1991, 1993, 1998; Schmidt and Leschin, 1995; Hereford, 1996). Thompson and Potochnik (2000) found that half of the 199 catchments they studied in Marble Canyon, Furnace Flats, the Aisles and Western Grand Canyon had some kind of aeolian deposition, and 42% had active aeolian deposition. The possible significance of aeolian processes in the erosion of streamside sand deposits has been noted in studies of the deposition and longevity of streamside deposits (Howard and Dolan, 1981; Beus *et al.*, 1985; Hereford *et al.*, 1993, 1996; US Department of Interior, 1995; Yeatts, 1998; Schmidt, 1999). Thompson and Potochnik (2000) suggested that wind may be a major mechanism in restoring sand to gullies owing to the evidence of aeolian deposition at many of their sites. They further suggested that aeolian reworking of newly deposited sand onto higher terraces would be significant as long as the supply of sand deposited by the river is available and is not cut off from upper slopes by vegetation. Powell (1897), although concerned more about survival than sediment transport on his pioneering journeys in Grand Canyon, noted in his diaries that fierce flames erupted from campfires as a result of high winds near river level. An implication of the observations of these studies is that streamside sand deposits are an important source for windblown sand. Although some of this windblown sand would be immediately lost to the river, some would be redistributed over the nearby slopes.

The effects of windblown sand would have been made more significant in the pre-dam era as a result of the lower winter river stages, which would expose larger subaerial portions of deposits, coincide with high winter winds and follow the prime season for tributary contributions of bar-forming sand (Cluer, 1995; David J. Topping, US Geological Survey, oral communication, 1999). The infrequent, local, intense rainstorms associated with the initiation and development of gullying

(Hereford et al., 1991, 1993; Thompson and Potochnik, 2000) may have been offset in the pre-dam era by the infilling and healing of incipient gullies by windblown sand.

The mechanism linking channel-side sand deposits with the erosion of upslope artefacts is uncertain and likely to be the subject of ongoing investigation. Nevertheless, the mechanisms described above provide motivation for evaluating the volume and locations of sand deposition in response to sand supplies and dam releases beyond the simple sand-storage concerns that have dominated past sand studies in Grand Canyon. Short of experimenting with a variety of high flows, which is not feasible and would in any case run the risk of permanently damaging cultural resources, the best option is the application of a model that has been demonstrated to represent well the complex flow, sand transport, and erosion and deposition patterns in previous studies.

14.2 Flow and sediment transport model

14.2.1 Model

The numerical methods used in the model are described by Wiele et al. (1996) and are based on Patankar's finite volume method (1980), which features a staggered grid and upwind differencing to solve the three equations of motion for the vertically averaged, 2D flow field. A key attribute of the finite volume method is that it conserves mass, a crucial requirement for sediment transport applications. The flow algorithms must contend with complex flow fields, typically including recirculation zones, which generate strong velocity gradients. A diffusion–advection equation is used to calculate the 3D sand concentration field from which the local suspended sand transport can be determined. A turbulence closure is applied to recover the vertical structure of the turbulent mixing and the velocity profile. Local bedload also is calculated on the basis of local shear stress at the bed, critical shear stress and local bed slope. The change in bed configuration over a small time step is calculated from the divergence of the total sand transport rate.

The flow field is calculated by numerically solving the momentum equations in the downstream direction,

$$\frac{\partial u}{\partial t} + u\frac{\partial u}{\partial x} + v\frac{\partial u}{\partial y} - \frac{\partial}{\partial x}\varepsilon\frac{\partial u}{\partial x} - \frac{\partial}{\partial y}\varepsilon\frac{\partial u}{\partial y} + g\left(\frac{\partial h + \eta}{\partial x} - S\right) + \frac{\tau_x}{\rho h} = 0 \quad (14.1)$$

and cross-stream direction,

$$\frac{\partial v}{\partial t} + v\frac{\partial v}{\partial y} + u\frac{\partial v}{\partial x} - \frac{\partial}{\partial y}\varepsilon\frac{\partial v}{\partial y} - \frac{\partial}{\partial x}\varepsilon\frac{\partial u}{\partial x} + g\frac{\partial h + \eta}{\partial y} + \frac{\tau_y}{\rho h} = 0 \quad (14.2)$$

and with the continuity equation

$$\frac{\partial h}{\partial t} + \frac{\partial uh}{\partial x} + \frac{\partial vh}{\partial y} = 0 \quad (14.3)$$

where x is the direction normal to the upstream boundary, y is the direction normal to x, u is the vertically averaged velocity in the x-direction, v is the vertically averaged velocity in the y-direction, h is the flow depth, η is the bed surface elevation, S is the average reach slope, ε is the eddy viscosity, g is gravity, ρ is density of water, and τ_x and τ_y are the shear stresses in the x- and y-directions, respectively. Equations (14.1)–(14.3) are in Cartesian coordinates. For applications to reaches with significant curvature, the equations were modified with the metric of Smith and McLean (1984) for calculations based on an orthogonal curvilinear coordinate system with a variable radius of curvature.

A friction coefficient, c_f, is used to relate the resolved shear stress, τ, to the resolved velocity,

$$\tau = \rho c_f |U| U \tag{14.4}$$

where $U = (u^2 + v^2)^{1/2}$ is the magnitude of the resolved velocity. The x and y components are determined from the relations:

$$\tau_x = \rho c_f u U \tag{14.5}$$

and

$$\tau_y = \rho c_f v U \tag{14.6}$$

The friction coefficient is defined by

$$c_f = \left(\frac{\kappa}{\ln \frac{h}{z_0} - 1} \right)^2 \tag{14.7}$$

where κ is von Karman's constant and z_0 is the roughness parameter. Equation (14.7) is derived by vertically averaging the logarithmic velocity profile (Keulegan, 1938). The value of z_0 at each node depends in part on the thickness of the sand cover. A value for z_0 was initially computed based on bathymetric measurements with $z_0 = 0.1 b_{84}$, where b_{84} is the 84th percentile of the deviations of the local bathymetric measurement from a straight line drawn between two adjacent nodes. The coefficient 0.1 is typically used to relate z_0 to a distribution of gravel sizes. Where the local sand thickness exceeds the z_0 computed from bathymetric records, then the local z_0 was computed based on estimated bedform dimensions, as will be discussed in more detail later.

The eddy viscosity, ε, is defined by

$$\varepsilon(z) = u_* \kappa z (1 - z/h) \tag{14.8}$$

where u_* is the shear velocity $(\tau/\rho)^{1/2}$, and z is the distance above the bed. Equation (14.8) is vertically averaged for use in equations (14.1) and (14.2):

$$\varepsilon = \frac{u_* \kappa h}{6} \tag{14.9}$$

Total load sediment transport equations were found to be far too sensitive to the flow field and yielded physically unreasonable results. Instead, a 3D suspended sand field is calculated using a near-bed boundary condition that is a function of local boundary shear stress. This combination is computationally robust and yields predictions that agree well with measurements. The transport of suspended sand is governed by an advection–diffusion equation:

$$\frac{\partial c}{\partial t} + \frac{\partial cu}{\partial x} + \frac{\partial cv}{\partial y} + \frac{\partial cw}{\partial z} - \frac{\partial}{\partial x}\varepsilon\frac{\partial c}{\partial x} - \frac{\partial}{\partial y}\varepsilon\frac{\partial c}{\partial y} - \frac{\partial}{\partial z}\varepsilon\frac{\partial c}{\partial z} + w_s\frac{\partial c}{\partial z} = 0 \qquad (14.10)$$

where c is sand concentration and w_s is sediment settling velocity. The sediment eddy viscosity, ε, in equation (14.10) is assumed to be equal to the momentum eddy viscosity represented by equation (14.8). As with equations (14.1)–(14.3), equation (14.10) was modified for applications to the Lower Tanner and Upper Unkar reaches, which have significant curvature, for application to an orthogonal curvilinear coordinate system using the metric of Smith and McLean (1984).

Equation (14.10) is solved for a given flow field with 11 points in the vertical that are concentrated near the water surface and near the bed to resolve the gradients near the boundaries. The sand transport is represented by the median grain size, d_{50}. The eddy viscosity as a function of z is calculated with equation (14.8). The velocity as a function of z is computed from the logarithmic velocity profile (Keulegan, 1938) from which equation (14.7) is derived. The numerical method used to solve equation (14.10) is similar to the one used for the flow equations (Patankar, 1980) extended to three dimensions.

The lower sediment-concentration boundary condition used in the solution of equation (14.10) is calculated by first determining a reference concentration, c_a, at the top of the bedload layer, where $z = z_a$. The reference concentration, c_a, is determined from the relations of Smith and McLean (1977):

$$c_a = \frac{c_b \gamma s}{1 + \gamma s} \qquad (14.11)$$

where c_b is the bed concentration and s is the normalized excess shear stress:

$$s = \frac{\tau_{sf} - \tau_c}{\tau_c} \qquad (14.12)$$

where the subscript sf indicates skin friction shear stress and τ_c is critical shear stress for the initiation of significant particle motion (Shields, 1936). The value of the constant γ has been updated to 0.004 by Wiberg (reported by McLean, 1992). The distance above the bed corresponding to c_a, namely z_a, is determined from the expression presented by Dietrich (1982) with coefficients a_1 and a_2 as updated by Wiberg and Rubin (1985)

$$z_a = d\frac{a_1 + T_*}{1 + a_2 T_*} \qquad (14.13)$$

where d is grain diameter, T_* is the ratio of the skin friction shear stress to the critical shear stress (τ_{sf}/τ_c), and a_1 is a constant ($a_1 = 0.68$).

The coefficient a_2 is a function of the grain size in centimetres:

$$a_2 = 0.02035(\ln(d))^2 + 0.02203 \ln(d) + 0.07090 \tag{14.14}$$

The boundary condition at the water surface is $c=0$, which is consistent with equation (14.8).

The evolution of the bed over time is calculated from the sediment continuity equation:

$$\frac{\partial \eta}{\partial t} = -\frac{1}{c_b}\left(\frac{\partial q_s}{\partial x} + \frac{\partial q_s}{\partial y}\right) \tag{14.15}$$

where η is the bed elevation. The sediment discharge, q_s, is the sum of the sand transported by bedload and in suspension. The suspended sand discharge is determined by vertically integrating the product of the flow velocity and the sand concentration.

The bedload is determined by applying the Meyer-Peter and Müller (1948) formula modified with the critical shear stress of the given grain size in place of their constant of 0.047:

$$\phi = 8(\tau_* - \tau_{*c})^{3/2} \tag{14.16}$$

where ϕ is the nondimensional bedload transport ($\phi = q/[(\rho_s/\rho - 1)gd^3]^{0.5}$), τ_* is the nondimensional boundary shear stress ($\tau_* = \tau/[(\rho_s - \rho)gd]$), and τ_{*c} is the nondimensional critical shear stress ($\tau_{*c} = \tau_c/[[(\rho_s - \rho)gd]$). The grain diameter is represented by d; d_{50}, the median grain diameter, is used in the model results presented later. The density of the sand is represented by ρ_s.

The boundary shear stress used in equation (14.16) is the magnitude of the vector sum of the shear stress calculated from the flow equations and an apparent stress due to gravity. The apparent stress due to gravity is calculated with a method proposed by Nelson and Smith (1989a) in which

$$\tau_g = \tau_c \frac{\sin \xi \nabla \eta}{\sin \phi |\nabla \eta|} \tag{14.17}$$

where τ_g is the apparent gravitational stress, ξ is the local maximum bed slope, and ϕ is the grain angle of repose. The x and y components of the bedload are determined from the respective components of the flow velocity and components of the local bed slope. The magnitude of τ_g is zero where the bed is horizontal, and approaches τ_c where deposition increases the bed slope to the grain angle of repose.

A large fraction of the total shear stress at the bed is exerted as form drag on large roughness elements, such as the extreme irregularity of the bedrock channel, bedforms, and boulder-size talus and bed material. This form drag must be deducted from the total shear stress to arrive at the skin friction portion of the total shear stress that transports sediment. Wiele et al. (1996) calculated the fraction of the total shear stress active in transporting sediment in reaches of the Colorado River in the vicinity of the reaches in this study. They made their calculation by determining the skin friction required to match measured transport rates during periods in which the sand supply was stable. This calculation yielded a skin friction that was 15% of the total shear stress. This low value is consistent with the extremely large channel roughness and associated form drag and was used in the calculations in this study in areas of the reach where the sand depth is less than the bed roughness.

The procedure used in this study for determining the skin friction in portions of the channel where the sand is sufficiently thick to cover the bed roughness is different from the one used by Wiele *et al.* (1996). Sand thickness tends to be greatest in recirculation zones, which are isolated from the tumultuous flow in the main channel, and where the flow and sand transport more closely resemble that of alluvial streams. In this region, the local channel resistance and skin friction were calculated as functions of local flow, depth and sand size. This procedure used the methods described by Bennett (1995) to estimate bedform dimensions and form drag. Bennett drew on the work of van Rijn (1984) who used u_* to distinguish between ripples, dunes and upper plane bed and to estimate the dimensions of the bedforms, if present. Given bedform height and wavelength, the local friction and skin friction are determined in Bennett's algorithm using the relations of Smith and McLean (1977) and Nelson and Smith (1989b). Relating local flow resistance and skin friction to bedforms is an improvement over the use of values derived only from the local hydraulics, but errors may be induced by uncertainties in the relations used and in the assumption of equilibrium between the local flow and the bedforms.

The model has been used to examine depositional processes and rates in the Colorado River main stem during a flood on the Little Colorado River (Wiele *et al.*, 1996), to examine the effect of sand supply on depositional patterns and magnitudes, and to compare the effects of natural tributary flooding with flooding caused by increased dam releases (Wiele, 1997; Wiele *et al.*, 1999). The model has shown good agreement with cross-section measurements from before and after the Little Colorado River flood in 1993 (Wiele *et al.*, 1996) and has replicated depositional patterns during the 1996 test flow (Wiele *et al.*, 1999). In one of the study reaches during the Little Colorado River flood, high sand concentrations led to massive deposition in the main channel (up to 12 m) and formed a large bar along the left side of the main channel in the recirculation zone. In contrast, with lower concentrations and higher water discharge during the test flow, the main channel scoured, and the focus of the deposition was near the recirculation zone reattachment point. The model replicated these differences in depositional pattern with no calibration. The accuracy of the model without calibration in these studies supports the use of the model to predict results for the hypothetical cases in this study.

14.3 Study site selection and morphology

Four modelling sites (Figure 14.1) were selected, which are within the study area of Hereford *et al.* (1991, 1993). Each of these reaches also contains gullies studied by Thompson and Potochnik (2000). The modelling site farthest downstream, the Upper Unkar reach, contains an especially sensitive archaeological site with abundant artefacts concentrated in a small area.

Debris flows and floods from streams in side canyons (Howard and Dolan, 1981; Schmidt, 1987; Webb *et al.*, 1989; Schmidt and Graf, 1990; Melis *et al.*, 1994; Schmidt and Rubin, 1995) form debris fans that partially constrict the channel, and recirculation zones are generated in the lee of the channel constriction. The spacings between debris fans are controlled to a large degree by bedrock structure (Dolan *et al.*, 1978). The bed of the Colorado River in Grand Canyon National Park is about 60% bedrock, talus blocks or boulders. Sand supplied to the mainstem river by tributary floods is stored primarily along the channel sides and in recirculation zones. Sand is stored temporarily on the channel bottom (Howard and Dolan, 1981; Wilson, 1986; Schmidt and Graf, 1990).

The four study reaches discussed in this report are located between 112 and 140 km below the dam, starting about 6 km downstream from the confluence with the Little Colorado River (Figure 14.1), one of the two main sand-supplying tributaries. The reaches modelled in previous studies tended to have narrow constrictions with large, abrupt expansions that produced large, well-developed recirculation zones. These large zones can be effective at storing large volumes of sand. Reaches modelled in this study have channel constrictions and recirculation zones, but the expansions are narrower or more constrained and the resulting recirculation zones are smaller than the previously modelled reaches. In the two downstream reaches in this study, the recirculation-zone sand deposits are less significant than the channel-margin deposits. To clarify the discussion and to avoid repeated, lengthy descriptions of morphological features in the reaches, these features are designated with letters in the accompanying figures.

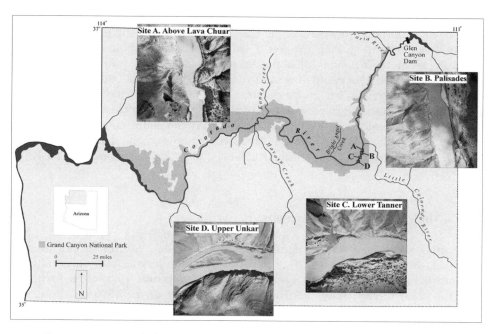

Figure 14.1 Aerial photos of study sites and map showing study site locations.

14.4 Model grids

Channel bathymetry was measured in the study reaches by the Grand Canyon Monitoring and Research Center (GCMRC). A fathometer mounted on a manoeuvrable boat recorded local water depth. The location of the fathometer, including elevation, was tracked from a shore station with a theodolite manually trained on a target mounted directly above the fathometer. The boat followed streamwise and cross-stream lines spaced about 10 m apart (see Andrews *et al.*, 1999, for a more detailed description of surveying methods). In addition, the shoreline was surveyed to outline channel shape; additional measurements were made around and over sand deposits (Figure 14.2).

The water discharge was about $425\,\text{m}^3/\text{s}$ during the measurements of channel shape. In order to form topographic maps that extend to elevations in excess of the river stage at $2830\,\text{m}^3/\text{s}$, the field-surveyed bathymetric and shoreline data were combined with GCMRC photogrammetrically generated contour data (Figure 14.2). Where the data sets overlapped, which typically occurred near the channel margin, the field-surveyed data were used to generate the topographic surface because these data were considered to be more accurate. A Triangulated Irregular Network (TIN) surface model was created using the Delauney method of triangulation in which topographic features are developed into a series of connected triangles where the nodes of the triangles correspond to measured locations and the facets of the triangles correspond to changes in slope. Contours were generated from the TIN surface and corresponded well with the photogrammetric contours in the areas of overlap. The TIN surface was then interpolated using a bivariate quintic interpolation scheme, implemented by the ARC/INFO Geographical Information System software (Environmental Systems Research Institute, Inc., 1991), in order to generate 10-m resolution grids used as the basis for model calculations (Figure 14.3).

The choice of grid spacing requires a balance between computational efficiency, especially given the demands of a time-stepping model, and sufficient detail. The 10-m grid is sufficient for computing sand volumes and locations, but does not capture metre-scale detail such as backwater channels. Comparisons with model results using a 5-m grid showed smoother representation of channel shape, but provided no significant improvement on calculated volumes. Tighter grids are more compatible with the calculation of flow fields only, without time-stepping erosion and deposition, and can be helpful for some applications such as habitat studies (e.g. Korman *et al.*, 2004).

14.5 Model application and results

In addition to channel shape, application of the model requires specification of discharge, downstream water surface elevation at that discharge, sand flux and grain size into the reach, and the initial thickness of sand deposits. Two water discharges were modelled in this study: $1270\,\text{m}^3/\text{s}$, corresponding to the 1996 test flow, and $2830\,\text{m}^3/\text{s}$, which is close to the pre-dam average annual flood and the highest discharge (in 1983) since the closure of Glen Canyon Dam in 1963. Three sand conditions (Figure 14.4) were modelled in each reach. These sand conditions are

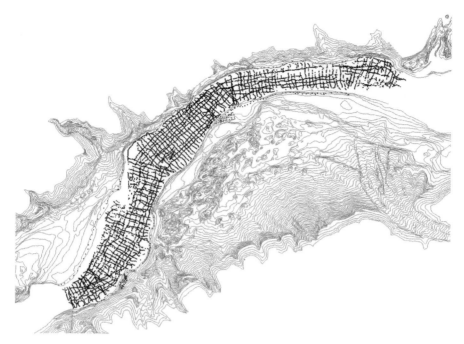

Figure 14.2 Illustration of the Lower Tanner reach showing the locations of the field-surveyed bathymetric and shoreline surveys (dots in the main channel and along the channel sides) combined with GCMRC photogrammetrically generated contour data.

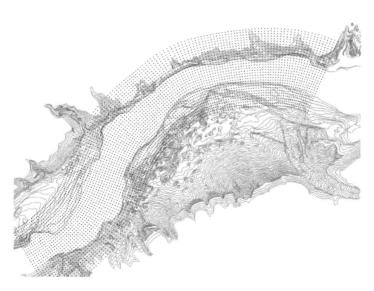

Figure 14.3 Computational grid (dots) of the Lower Tanner reach superimposed on the topography. Elevations at the grid points are interpolated from the TIN surface.

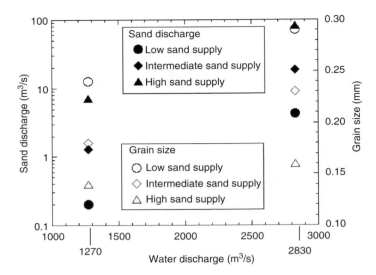

Figure 14.4 Sand discharge and grain size as functions of water discharge for three sand conditions used as boundary conditions for the model runs: high sand supply during 1956, intermediate sand supply represented by the conditions during the 1996 test flow and low sand supply during 1983. These values for sand discharge and grain size were contributed by Topping (USGS, written communication, 1998).

representative of historical high sand supplies, historical low sand supplies and an intermediate condition corresponding to that which occurred during the 1996 test flow. The historical high sand supply corresponds to conditions represented by measurements in 1956, prior to the dam construction. The historical low was taken from sand conditions in 1983, after the dam had been eliminating main channel inputs for nearly 20 years and during flows approaching $2830 \, m^3/s$. These two conditions, high discharge and low sediment supply, combined to winnow the bed to a coarser state and thereby reduce suspended sand transport (Topping *et al.*, 2000a,b). The 1996 sand conditions are representative of typical contemporary conditions. The upstream sand boundary conditions corresponding to the three sand supplies were taken from sand rating curves supplied by David J. Topping (US Geological Survey, written communication, 1998). Topping (1997) related sand concentrations and sand sizes to antecedent conditions and the evolving status of sand on the channel bed, which forms the source from which sand in suspension is derived.

One of the initial conditions that must be specified is the sand coverage on the bed at the start of the simulation. Previous model applications have been in reaches in which bathymetric measurements have been repeated, allowing for the estimation of a minimum bed elevation by combining the lowest elevations from different surveys. Another approach in modelling events in which the change in sand deposits is dominantly depositional, and where there is insufficient information to synthesize a minimum bed elevation, is to neglect erosion and make the starting bed shape the minimum bed elevation. In this study, the local initial sand conditions on the bed were estimated by running the model in each reach at a discharge characteristic of dam operation ($481 \, m^3/s$ was chosen for these simulations) with a sand boundary condition at the inlet appropriate for the discharge and sand supply. The model was run until the sand coverage reached near-equilibrium with the sand influx. For the cases with low and intermediate sand supply, corresponding sand boundary conditions were used to model the initial sand coverage. The highest sand supply, however, is produced by relatively brief events (David J. Topping, US Geological Survey, oral communication, 1999), such as local tributary flooding. As a result, a bed in equilibrium with the intermediate sand supply was judged a more reasonable antecedent condition for the high sand supply than a bed configuration in equilibrium with the high sand supply.

The effects of the different combinations of water and sand discharge on sand deposition are considered in the following sections. The final results are taken from the end of the simulation. The length of the simulation varies as a result of differences in the real time required for the model run. The length of the model runs depends on the time step, which is determined within the model on the basis of the rate at which deposition or erosion occurs. As a result, events with rapid changes in bed morphology, such as would occur with the high sand supply, generally progress in the model much more slowly in real time than events in which there is little change in bed morphology. Nevertheless, direct comparisons between model results under different conditions can be made at different times because most of the changes within a reach occur rapidly. The events with the shortest elapsed simulation time (as a result of small time steps causing slow rates of advance in time) tend to be those with the most rapid changes and, therefore, reach near-equilibrium in the least amount of simulated time.

Although maximum sand deposition is generally favourable for the restoration and preservation of the riparian environment, deposition within the main channel is of little long-term value because the sand storage is short-lived. Reduction in sand supply, by the cessation of tributary flooding (which is the only mechanism currently capable of producing high sand concentrations), will lead to the rapid erosion of main channel deposits. For the purposes of this study, we emphasize near-shore environments and deposits above the 708 m^3/s stage. In the two upstream reaches, most of the deposition that would be likely to fill gully mouths occurs in recirculation zones, and so these environments are considered in detail. In the two downstream reaches, in which deposition is not dominated by recirculation zones, deposits along the channel sides above the 708 m^3/s stage are considered. The depositional patterns and volumes with the intermediate sand supply are generally bracketed by the results with the high and low sand supplies. Consequently, only the model predictions with the high and low sand supplies are shown in the figures.

14.6 The four modelling sites

14.6.1 Above Lava Chuar reach

14.6.1.1 *Initial conditions*

The Above Lava Chuar reach contains an expansion along river left (i.e. left side of the river looking downstream) in the upper part of the reach (A) that shows no recirculation at either 1270 (Figure 14.5a) or 2830 m³/s (Figure 14.6a). The debris fan at the downstream end of this expansion (B), however, generates a small recirculation zone in the embayment (C) formed between it and another debris fan (D) immediately downstream. The downstream debris fan (D) also partly forms the Lava Chuar rapid. The bank along river right consists of a steep cliff along most of its length; a side channel forms an indentation (E) about 500 m from the reach inlet and a camping beach (F) is at the downstream end of the bank. Gullies mapped by Hereford *et al.* (1991, 1993) appear in two locations in the Above Lava Chuar reach, in the vicinity of B and D and downstream from F. The greatest concentration is in C, and two gullies reach the water level in F.

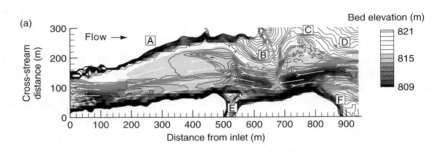

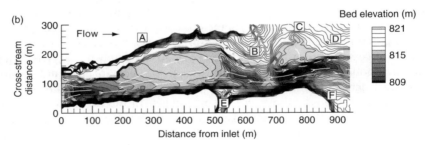

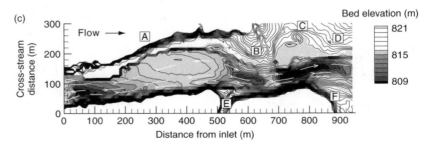

Figure 14.5 *Contour maps of the Above Lava Chuar reach at 1270 m³/s showing (a) the initial bed morphology (at time = 0), (b) the model prediction after 72 hours with the low (1983) sand supply and $d_{50} = 0.24$ mm, and (c) the model prediction after 68 hours with the high (1956) sand supply and $d_{50} = 0.16$ mm.*

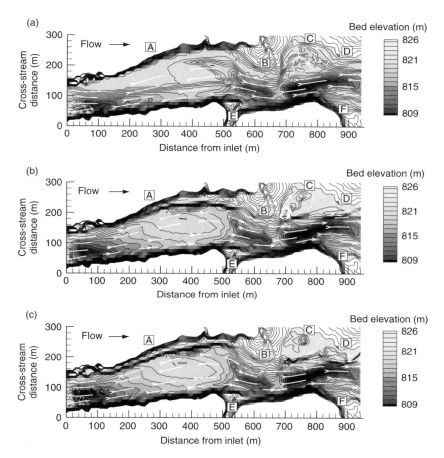

Figure 14.6 Contour maps of the Above Lava Chuar reach at 2830 m³/s showing (a) the initial bed morphology (at time = 0), (b) the model prediction after 72 hours with the low (1983) sand supply and $d_{50} = 0.29$ mm, and (c) the model prediction after 37 hours with the high (1956) sand supply and $d_{50} = 0.14$ mm.

14.6.1.2 Model predictions

The total change in sand deposit volume in the Above Lava Chuar reach was negative for all cases (Figures 14.5, 14.6 and 14.7) largely as a result of erosion of the sand deposit in the channel expansion marked A under all combinations of discharge and sand supply (Figure 14.8). The modelled erosion of the sand bar in A may, in part, be a consequence of the model's functioning with just one grain size. If the initial sand deposit within the expansion were of significantly coarser material than the modelled grain size, it would be more resistant to erosion. The combination of high discharge and finest grain-size resulted in the most rapid erosion (Figure 14.8). During the course of this erosion, bars that deflect flow towards the bank formed within the expansion, resulting in erosion of sand deposits along the bank. After the

372   *Computational Fluid Dynamics*

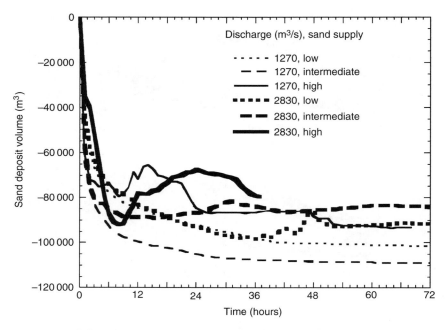

Figure 14.7 Modelled change in sand volume as a function of time in the Above Lava Chuar reach.

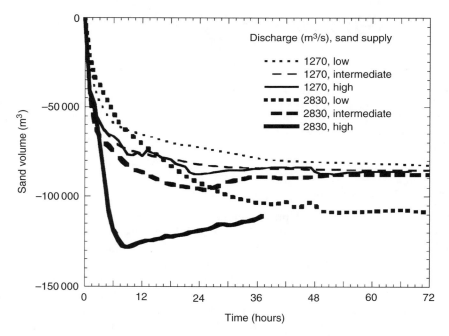

Figure 14.8 Modelled change in sand volume as a function of time at location A in the Above Lava Chuar reach.

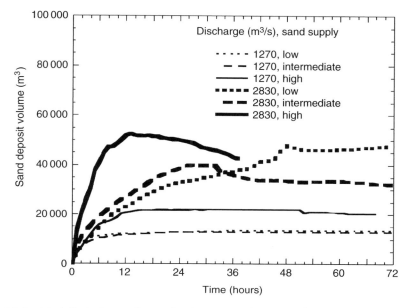

Figure 14.9 Modelled change in sand volume as a function of time at location C in the Above Lava Chuar reach.

bars migrated downstream, the sand deposits in A were restored slightly as sand was redeposited along the bank.

Deposition occurred in all cases within C (Figures 14.5, 14.6 and 14.9), the most critical area for archaeological resources. The sand volume within C increased with higher sand concentration in the flow at 2830 m³/s during the first day (Figure 14.9). There was a rapid initial increase in sand volume, followed by a steady decline after about 12–14 hours. This pattern is related to the increase in sand flux that resulted from the erosion of the sand deposit in A, followed by partial erosion of the deposit in C after most of the sand in A had been evacuated. A similar pattern, but with a smaller amplitude, occurs for the intermediate sand supply. At 1270 m³/s, model results show volumes similar to the low and intermediate sand supplies, and the largest sand volume coincided with the high sand supply. At 2830 m³/s, more sand was deposited within the embayment and to a higher elevation, although backwaters remained in the embayment.

14.6.2 Palisades reach

14.6.2.1 *Initial condition*

The Palisades reach has a simpler form than the Above Lava Chuar reach, with a recirculation zone (G) on river left downstream from the rapid that forms the reach inlet (Figures 14.10a and 14.11a). The channel expands (H) downstream from the recirculation zone. A cobble bar is located midway in the channel adjacent to H

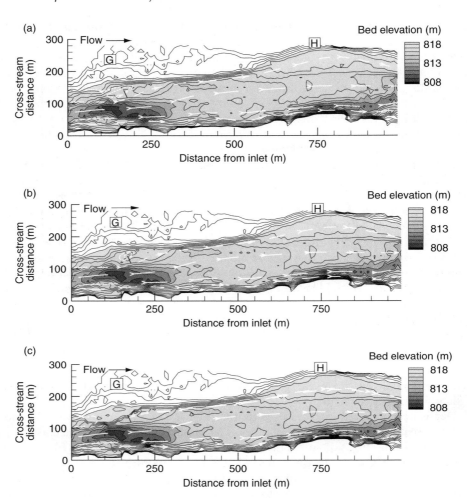

Figure 14.10 Contour maps of the Palisades reach at 1270 m³/s showing (a) the initial bed morphology (at time = 0), (b) the model prediction after 72 hours with the low (1983) sand supply and $d_{50} = 0.24$ mm, and (c) the model prediction after 72 hours with the high (1956) sand supply and $d_{50} = 0.16$ mm.

where the thalweg is close to the right bank. The deepest part of the channel is downstream from the inlet near the centre of the channel. The right bank consists of a cliff along most of its length.

Gullies mapped by Hereford *et al.* (1991, 1993) in this reach are near G. The gently sloping surface above G contains many artefacts (Yeatts, 1996; K. Thompson and A. Potochnik, SWCA, oral communication, 1999). The gullies within G in the Palisades reach have a lower gradient and appear less deeply incised than gullies in the Above Lava Chuar reach. If further incision of gullies is to be arrested or past incision healed by deposition along the water's edge, these sites may be the best candidates for benefits from high releases.

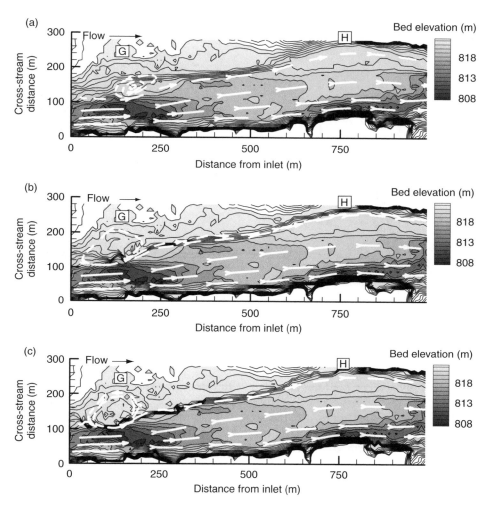

Figure 14.11 Contour maps of the Palisades reach at 2830 m³/s showing (a) the initial bed morphology (at time = 0), (b) the model prediction after 72 hours with the low (1983) sand supply and $d_{50} = 0.29$ mm, and (c) the model prediction after 31 hours with the high (1956) sand supply and $d_{50} = 0.14$ mm.

14.6.2.2 Model predictions

The most important depositional site in the Palisades reach is G, the recirculation zone on river left near the reach inlet. Hereford *et al.* (1991, 1993) mapped gullies extending to the 142 m³/s water surface in this area. Modelling results (Figures 14.10, 14.11 and 14.12) show that some deposition will occur in G under all combinations of water discharge and sand conditions. The initial deposit volume increased with the amount of sand available for transport at 2830 m³/s, and the volumes were larger for all cases at 2830 m³/s than at 1270 m³/s (Figure 14.13). At 1270 m³/s with the low sand supply, minimal deposition occurred in G, and most of that extended upstream from the

376   *Computational Fluid Dynamics*

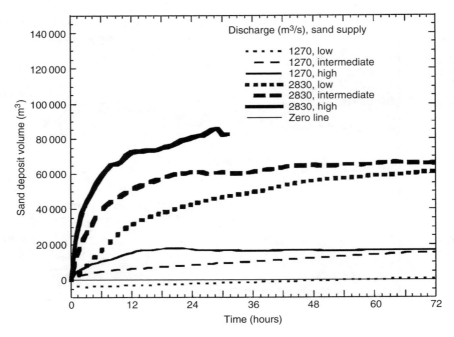

Figure 14.12   Modelled change in sand volume as a function of time in the Palisades reach.

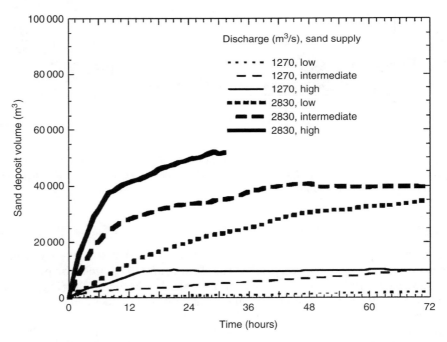

Figure 14.13   Modelled change in sand volume as a function of time at location G in the Palisades reach.

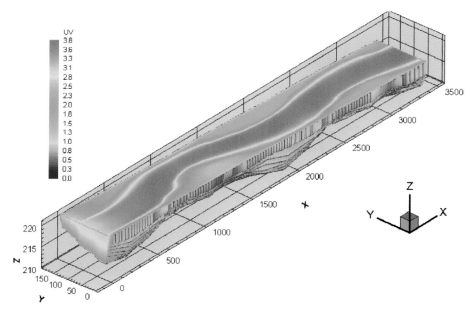

Plate 1 (Figure 11.5) Scalar product of u and v velocities in ms^{-1} predicted by Stoesser *et al.* (2003) for the 3.5 km reach of the River Rhine shown in Figure 11.4. (Published by IWA Publishing.)

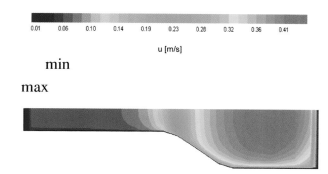

Plate 2 (Figure 15.15) See main text for caption.

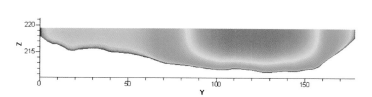

Plate 3 (Figure 15.20) See main text for caption.

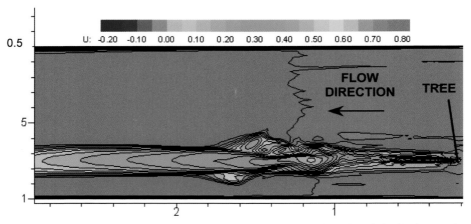

Plate 4 (Figure 15.16) Instantaneous velocity distribution, length unit for both axes in metres.

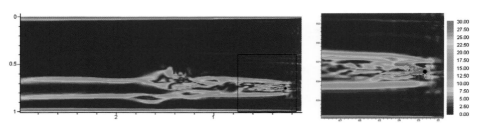

Plate 5 (Figure 15.17) Stress tensor distribution (in 1/s), length unit in metres.

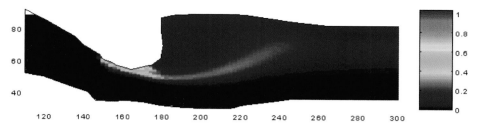

Plate 6 (Figure 17.6) Map of prediction bounds. The normalised weighted deviation from the mean concentration field as an indicative measure of risk that model predictions are more uncertain. A measure of 0 (dark blue) represents locations where there is little deviation between all the different model predictions, and a measure of 1 means there is a large deviation, which gives rise to wider prediction limits.

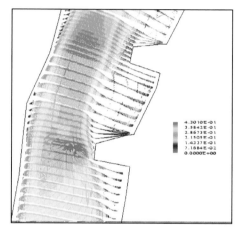

Plate 7 (Figure 18.11) Velocities 20% below surface at deflector 3f.

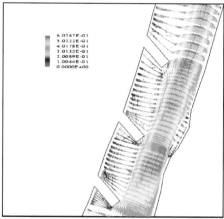

Plate 8 (Figure 18.12) Velocities 20% below surface at deflector 3c.

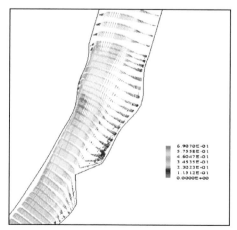

Plate 9 (Figure 18.13) Velocities 20% below surface at deflector 6a.

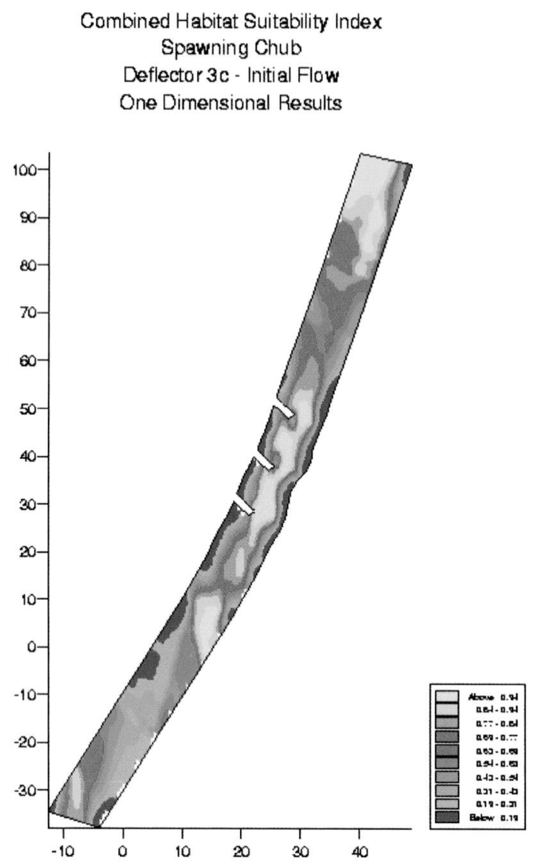

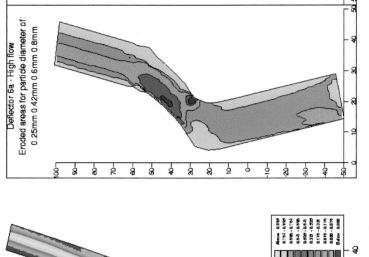

Plate 10 (Figure 18.14) Spatial Plots of WUA for spanning chub at deflector 3c for 1D model.

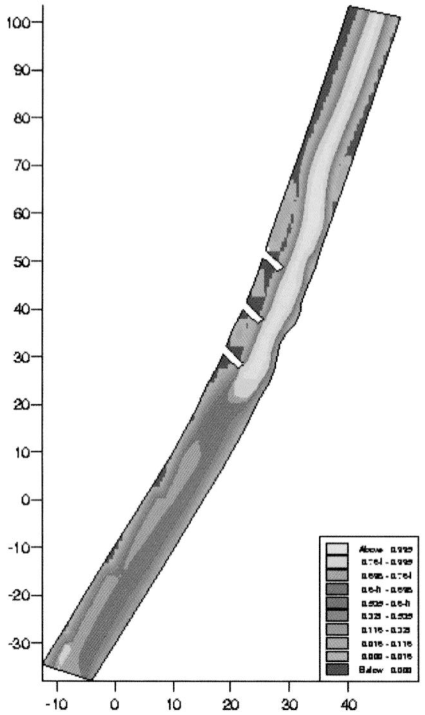

Plate 11 (Figure 18.15) Spatial Plot of WUA for spanning chub at deflector 3c for 3D model.

Plate 12 (Figure 18.18) Erosion patterns for different grain sizes.

reattachment point; deposition in the region of the gully mouths was less than in all other cases. The intermediate and high sand supplies at 1270 m³/s produced deposits better suited to archaeological site preservation because deposition was greatest close to the water's edge. With the intermediate sand supply, sand accumulated at a steady rate over the course of the simulation and the volume matched the sand volume obtained with the high sand supply by the end of the simulation. With the high sand supply, however, most of the accumulation occurred within the first 14 hours. All three cases at 2830 m³/s deposited more sand than the 1270 m³/s cases, and the deposit volume increased with increased sand supply. Most of the deposition was near the reattachment point and along the cobble bench that forms the left bank at the 2830 m³/s stage.

14.6.3 Lower Tanner reach

14.6.3.1 Initial conditions

The Lower Tanner reach (Figures 14.14a and 14.15a) is unlike the Above Lava Chuar and Palisades reaches in several respects. The Lower Tanner reach is longer (about 1200 m), has significant curvature and does not contain a large recirculation zone that dominates the sand storage in the reach. The outside of the bend, along river right, is bordered by bedrock cliffs along the upper half of the reach. The area near the outlet on river right consists of a low gravelly bench. The interior of the bend, along river left, is bounded by fans. Coppice dunes were mapped by Hereford

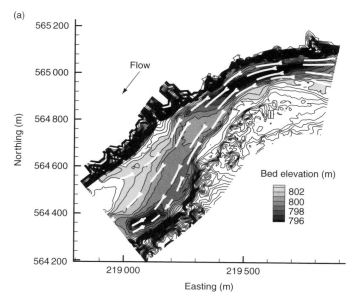

Figure 14.14 Contour maps of the Lower Tanner reach at 1270 m³/s showing (a) the initial bed morphology (at time = 0), (b) the model prediction after 72 hours with the low (1983) sand supply and $d_{50} = 0.24$ mm, and (c) the model prediction after 72 hours with the high (1956) sand supply and $d_{50} = 0.16$ mm.

378 Computational Fluid Dynamics

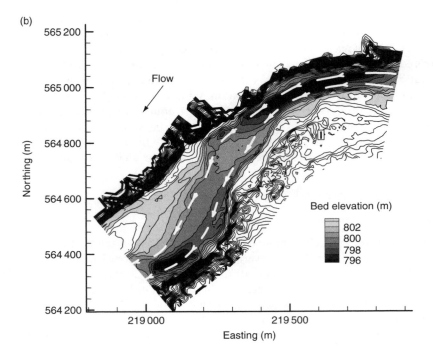

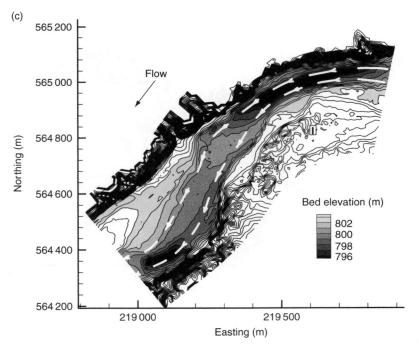

Figure 14.14 (Continued).

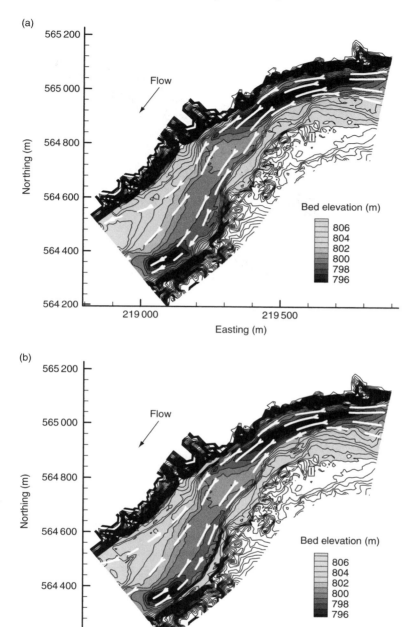

Figure 14.15 Contour maps of the Lower Tanner reach at 2830 m³/s showing (a) the initial bed morphology (at time = 0), (b) the model prediction after 72 hours with the low (1983) sand supply and $d_{50} = 0.29$ mm, and (c) the model prediction after 35 hours with the high (1956) sand supply and $d_{50} = 0.14$ mm.

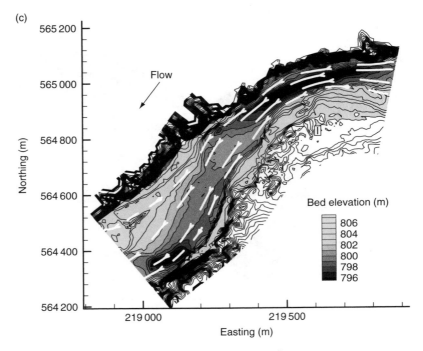

Figure 14.15 (Continued).

et al. (1991, 1993) in the area marked as "I" on Figures 14.14 and 14.15, and most of the gullies are contained in this area.

Most of the gullies mapped by Hereford *et al.* (1991, 1993) in this reach are on river left near the middle and downstream parts of the reach (Figures 14.14 and 14.15). Deposition predicted by the model was primarily along the channel sides, especially in the lee of bank irregularities that were too small to form well-developed recirculation zones. Deposition in the lee of bank irregularities may correspond to the deposition in microeddies proposed by John C. Schmidt (Utah State University, oral communication, 1999) as a significant depositional process, but eddies on the scale of the computational grid are represented in the model by the channel roughness.

14.6.3.2 Model predictions

Model predictions of deposition rates for the entire channel show a straightforward relation between sand supply and discharge and the consequent deposition (Figure 14.16). At both discharges, deposit volume increased with increased sand supply, and after 12 hours, the higher discharge produced larger deposits for all sand supplies.

At $1270 \, m^3/s$, deposition rates for sand deposits above the $708 \, m^3/s$ stage follow similar patterns, asymptotically approaching a maximum value (Figure 14.17). The magnitude of the deposit volume is proportional to the sand supply at $1270 \, m^3/s$. With the low and intermediate sand supplies, the maximum deposit volume is reached after about 48 hours, and only slight increases occur subsequently. With

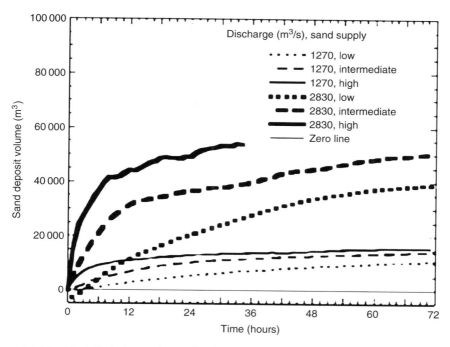

Figure 14.16 Modelled change in sand volume as a function of time in the Lower Tanner reach.

the high sand supply, the maximum value was reached after about 1 day. In contrast to the smoothly increasing deposit volume at 1270 m³/s, sand deposits above the 708 m³/s stage at 2830 m³/s accumulated in irregular temporal patterns. The initial deposition rate was highest with the high sand supply, but dropped off after about 6 hours to match the sand volume accumulated with the intermediate sand supply. With the low sand supply, an initial scouring was followed by a rapid increase in volume until about 12 hours, followed by a steady increase in sand volume over the rest of the simulation. A sand deposit on river left near the reach inlet was inundated at 2830 m³/s and eroded at that discharge with all sand supplies. With the low sand supply, deposition elsewhere was initially insufficient to offset erosion at this location. Sand volume at 2830 m³/s with the low sand supply exceeded all the sand deposited at 1270 m³/s after about 36 hours.

The volume of deposits above the 708 m³/s stage, which have the most potential for preservation, is only about a third of the sand volume stored in the recirculation zone in the Above Lava Chuar reach (zone C) and the recirculation zone in the Palisades reach (zone G). This relatively small volume reflects the absence of significant recirculation zones in this reach. The volume of sand with preservation potential in this reach is small compared to that in the two upstream reaches, but there is some compensation in that these deposits are above typical dam-release stages. The recirculation deposits in the upstream reaches, although removed from the main channel, are subject to flow-induced erosion, albeit at a rate much lower than that for main channel deposits.

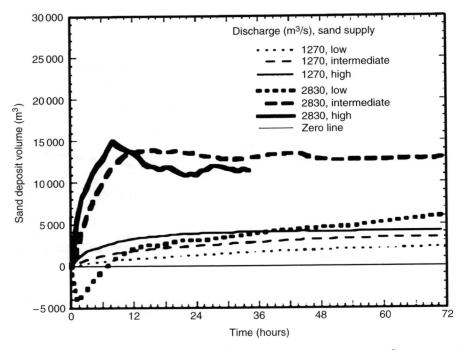

Figure 14.17 Modelled change in sand deposit volume above the 708 m³/s stage in the Lower Tanner reach.

14.6.4 Upper Unkar reach

14.6.4.1 Initial conditions

The Upper Unkar reach (Figures 14.18a and 14.19a) is about 1.6-km long down the channel centreline. A small recirculation zone is on river left near the reach inlet, and a deep hole in the main channel typically associated with recirculation zones in the upstream reaches is present here as well. The reach bends to the left and maintains a more consistent channel width at higher flow than the other study reaches. The most striking feature of this reach is the mid-channel bar that is inundated at higher flows, but forms an island at most typical dam releases (below about 500 m³/s). This bar was formed by a debris flow that spilled gravels into the reach that have been reworked into the present configuration. At low flow, when the bar is an island, most of the discharge occupies the left channel.

A "No Visitation" label in river guides of Grand Canyon (e.g. Stevens, 1990) dramatically marks this reach as one of the most sensitive in Grand Canyon. The part of the reach labelled "J" on Figures 14.18 and 14.19 has been declared off limits to the public by the National Park Service because of the abundance of artefacts at that location. The gullies mapped by Hereford *et al.* (1991, 1993) also are located in this area (Figures 14.18 and 14.19).

Modelling of sand deposition in Colorado River 383

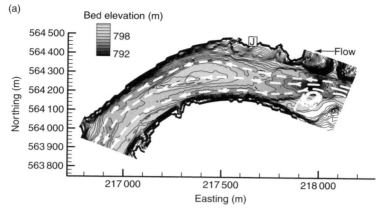

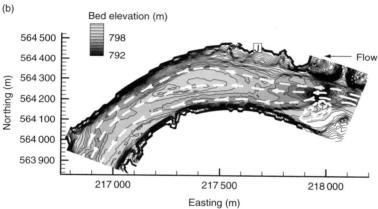

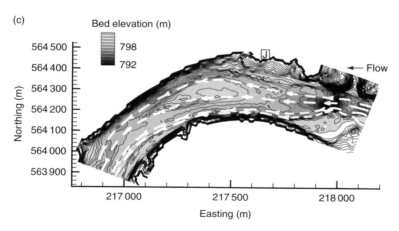

Figure 14.18 Contour maps of the Upper Unkar reach at $1270\,m^3/s$ showing (a) the initial bed morphology (at time = 0), (b) the model prediction after 72 hours with the low (1983) sand supply and $d_{50} = 0.24\,mm$, and (c) the model prediction after 72 hours with the high (1956) sand supply and $d_{50} = 0.16\,mm$.

14.6.4.2 Model predictions

The Upper Unkar reach contains a recirculation zone on river left near the reach inlet, but the area of interest with respect to gullies and vulnerable archaeological sites is on river right near the upper part of the mid-channel bar. The total sand deposition showed a clear separation between the deposition volume at 2830 m³/s with the high sand supply and the other conditions (Figure 14.18, 14.19 and 14.20). At 2830 m³/s with the high sand supply, substantial bars formed along the channel

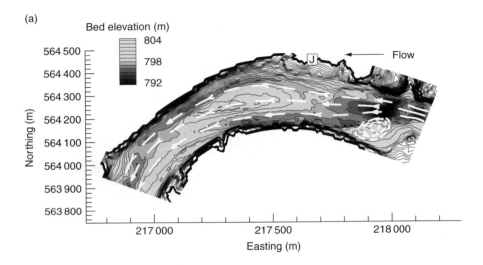

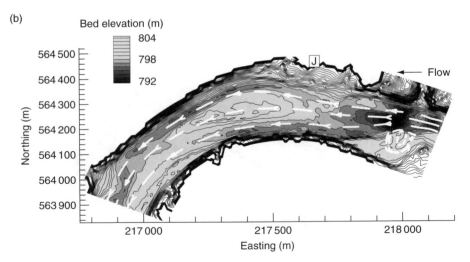

Figure 14.19 Contour maps of the Upper Unkar reach at 2830 m³/s showing (a) the initial bed morphology (at time = 0), (b) the model prediction after 72 hours with the low (1983) sand supply and $d_{50} = 0.29$ mm, and (c) the model prediction after 34 hours with the high (1956) sand supply and $d_{50} = 0.14$ mm.

Modelling of sand deposition in Colorado River 385

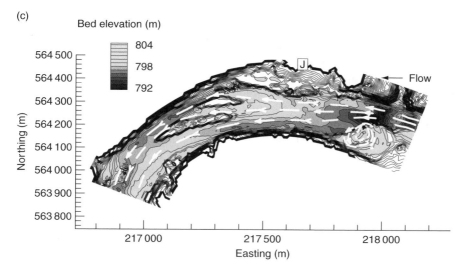

Figure 14.19 (Continued).

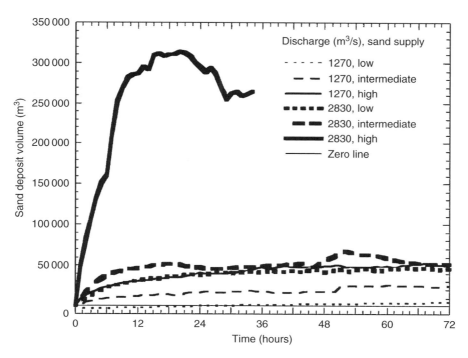

Figure 14.20 Modelled change in sand volume as a function of time in the Upper Unkar reach.

sides and around the tail end of the mid-channel gravel bar. At 2830 m³/s with the low and intermediate sand supplies, the results were similar, with the deposit reaching capacity more rapidly with the intermediate sand supply. At 1270 m³/s, the sand volume with the high sand supply matched the 2830 m³/s results with the low and intermediate sand supplies. Total deposition at 1270 m³/s with the low and intermediate sand supplies was relatively small.

Considering only the sand deposited above the 708 m³/s stage, the differences between the volumes deposited with the three sand supplies at 1270 m³/s were relatively small (Figure 14.21). At 1270 m³/s, all three sand conditions converged to a similar sand volume; the deposit reached capacity within about 6–10 hours with the high and intermediate sand supplies, whereas the deposit steadily grew over most of the simulation with the low sand supply. The significance of accommodation space created by high stage was especially apparent in the results at 2830 m³/s. At 2830 m³/s, the sand deposits with the low and intermediate sand supplies were close to capacity after about 12 hours and reached a volume about seven times larger than with the lower discharge. With the high sand supply at 2830 m³/s, sand deposit volume was nearly double the volume obtained with the two other sand conditions at 2830 m³/s (Figure 14.21).

A more narrow focus on the deposits in region J, the critical region near the archaeological sites (Figure 14.22), showed a proportionately larger gap between the deposit volume at 2830 m³/s with the high sand supply and the other cases than was evident for total depositional volumes for the entire reach (Figure 14.20). Total sand deposition near J for all cases except for the high sand supply at 2830 m³/s and the low sand supply at 1270 m³/s showed similar results at the end of the simulation.

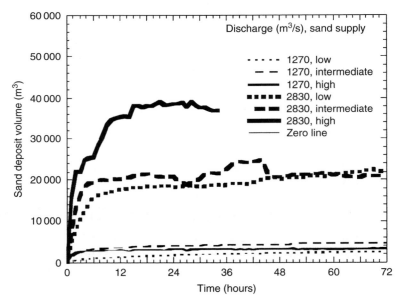

Figure 14.21 Modelled change in sand deposit volume above the 708 m³/s stage in the Upper Unkar reach.

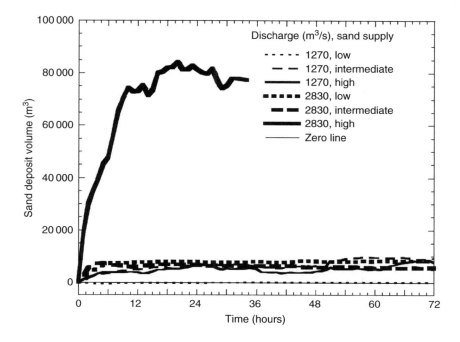
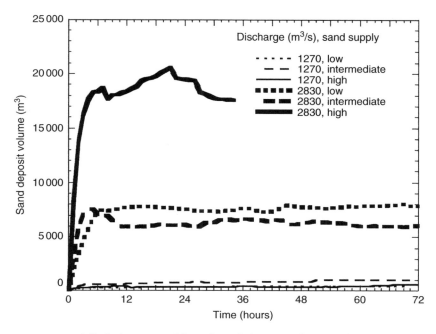

Figure 14.22 Modelled changes in (a) total sand deposit volume and (b) the sand deposit volume above the 708 m³/s stage as functions of time at location J in the Upper Unkar reach.

Sand deposits most likely to be of greatest benefit to artefact preservation were the deposits near J above the 708 m³/s stage (Figure 14.22). Sand deposit volumes near J above 708 m³/s were small at 1270 m³/s for all sand supplies compared with the results at 2830 m³/s. The increased accommodation space at 2830 m³/s allowed for much larger deposits with the low and intermediate sand supplies than those formed at 1270 m³/s. With the high sand supply, sand deposit volume near J above the 708 m³/s stage was three times the sand deposit volume at 2830 m³/s with the two other lower sand supplies. About half of the increase in total deposit volume for the reach shown in Figure 14.21 at 2830 m³/s with the high sand supply was a result of increased deposition near J. In all cases, some deposition occurred in that region along the bank, aided in part by the presence of debris that was deposited upstream near the right bank during the same event that formed the mid-channel bar. With the highest sand supply, there was sufficient deposition to initiate substantial bars in the lee of that debris; thus large deposits were formed at the most advantageous site for the preservation of archaological artefacts.

14.7 Discussion

Modelling results show significant variability in deposition volumes relating to channel shape as well as to discharge and sand supply (Figure 14.23). In reaches

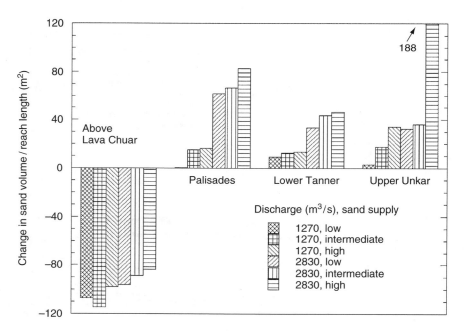

Figure 14.23 *Change in sand volume divided by reach length at the end of the modelling period. The change in sand volume is divided by reach length to facilitate the comparison of the effect of channel morphology on sand deposition and erosion.*

without significant recirculation zones, sand was stored in narrow bands along the channel sides under most conditions, primarily in the lee of channel irregularities. Sand volumes deposited above the 708 m^3/s stage in the Lower Tanner reach and in the critical right bank region (J) in the Upper Unkar reach were, at best, half the deposit volumes in the Above Lava Chuar and Palisades recirculation zones. Recirculation zones appear more consistent in response to increased discharge and variations in sand supply, and store more sediment for a given reach length. Trends in overall sand storage in reaches dominated by recirculation zones can be generalized, but in reaches where sand storage is dependent on finer-scale morphology of the channel sides, the trends are more variable.

High discharge releases. Higher flows, such as the 2830 m^3/s flow, tend to deposit sand in sheltered areas even with low sand supply. This is especially evident in the two upstream reaches in this study. Higher water discharges are significantly more effective in depositing sand in critical areas than lower discharges. This is a result of the greater sand transport rate for given sand conditions and, perhaps more significantly, the larger accommodation spaces created by higher stages.

High-flow duration. Deposition rate depends strongly on the volume of sand already present at a depositional site. Deposition rate falls rapidly as the site fills. As a result, high flows are most effective within the first day or two in filling depositional sites. Longer-duration high releases distribute sand more thoroughly within a depositional site, as pointed out by Anima *et al.* (1998), but are less efficient at utilizing sand in transport.

Sand supply. High sand concentrations, such as would occur during significant tributary flow, would be most effective in forming significant deposits in critical areas. The possibility and potential advantages of dam releases timed to coincide or shortly follow tributary flows were considered by the authors of the EIS (US Department of Interior, 1995), and were reiterated by Lucchitta and Leopold (1999) with particular emphasis on flows in the Little Colorado River. Careful analyses of sand deposits, sand transport processes and suspended sand measurements (Rubin *et al.*, 1998; Topping *et al.*, 2000a,b) have provided a process-based explanation of the importance of tributary inputs to replenishing sand resources, especially those in Marble Canyon that rely primarily on the sand inputs from the Paria River, and a quantification of the potential benefits of high releases associated with tributary inputs. Releasing high flows during or shortly following significant tributary flows would also increase the supply of fine-grained sediment, which would increase the stability of the deposits (as pointed out by Richard Hereford, USGS, oral communication, 1998). High discharges increase the total volume of the deposit and place the sand higher up the channel bank and, therefore, would place a larger volume within the gully mouths.

It is important to note that these study sites are in a relatively sand-rich reach compared with the channel above the confluence with the Little Colorado River. High discharges may deposit sand at many sites in this region, even with a low sand supply; but upstream of the confluence with the Little Colorado River, the lower sand supply in the absence of significant flow in the Paria River would make that reach more vulnerable to erosion (Schmidt, 1999; Topping *et al.*, 2000a,b).

Influence of channel shape on deposition volume. Results in region A in the Above Lava Chuar reach are consistent with the conceptual model of Melis (1997) for the relation between shoreline types and fan–eddy attributes and the potential for sand storage. Melis (1997) suggested that the greatest potential for sand storage would occur in reaches with the greatest density of debris fans and downstream from the tightest channel constrictions, and that the least potential for sand storage would occur in reaches with steep shorelines and few debris fans. The model results for these four reaches support the conclusion that recirculation zones are the most effective sites for storing large volumes of sand that are likely to endure on a time scale of months to years. The parts of these deposits that are beneath the water surface during normal dam operation, however, are still subject to erosion at lower discharges by scour, although at a much slower rate than occurs in the main channel. These inundated deposits may also erode rapidly during routine dam operation (Cluer, 1995). Deposits along the channel sides that are perched above the $708\,\mathrm{m^3/s}$ stage, albeit smaller than the recirculation deposits, may have greater potential for preservation. Little is currently known about the rates at which erosive processes, such as aeolian transport, operate.

Potential negative impacts of high flows. The cases studied so far have led to the generalizations listed above. Particular sites may respond differently. Schmidt *et al.* (1999) and Hazel *et al.* (1997, 1999) documented considerable variation in response to the 1996 test flow within reaches with similar morphology. Even at a given site, periodic mass failure of rapidly accumulated sand deposits can lead to a temporal variability in sand deposit volume (Andrews *et al.*, 1999). Widening of the main channel flow at higher stages may diminish deposits formed at lower discharges in some reaches, as shown in A in the Above Lava Chuar reach and as was suggested by Melis (1997). This is especially likely if high releases are necessary to lower Lake Powell, as happened during 1983–1984 when sustained high flows peaking at $2720\,\mathrm{m^3/s}$ caused erosion at some archaeological sites (US Department of Interior, 1995). Although results so far have shown deposition under most conditions in sheltered areas, such as recirculation zones, under some conditions erosion can occur within recirculation zones (Joseph Hazel, Northern Arizona University, oral communication, 1999). Erosion in recirculation zones is most likely to occur with deeper flows, caused by some combination of increased stage and initial low elevation of the sand deposit, if the channel morphology is conducive to high flow velocities and is combined with low sand supply in the main stem. Erosion and deposition in reaches with gradually varying channel width are more likely to show greater sensitivity to variations in water and sand discharge.

14.8 Conclusion

The model used in this study provides a physically based, predictive method for examining the effects of dam releases and sand supplies on sand deposits. Predictions of changes in sand bars can be made at specific sites and general trends can be inferred from modelling results over a range of conditions and at multiple sites. The model was constructed with sufficient complexity to represent processes with an accuracy commensurate with the purpose of the study using available computational power.

Acknowledgements

The authors thank Ruth Lambert, Cultural Resources Program Manager at the GCMRC, for funding and for her support of this project. Billy Elwanger, Mark Gonzalez, Steve Lamphear, Mimi Murov, Nels Niemi, Barbara Ralston and Don Scarbrough made indispensable contributions to the field work. Ted Melis, David Topping, Kate Thompson and Andre Potochnik contributed with helpful scientific discussions and insights. Reviews by Ned Andrews, Paul Bates, Rob Ferguson, Julie Graf, Steve Longsworth, Ted Melis, Melissa Ray and Tracey Suzuki improved the manuscript.

References

Andrews, E.D., C.E. Johnston, J.C. Schmidt and M. Gonzales, 1999, Topographic evolution of sand bars, in Webb, R.H., Schmidt, J.C., Marzolf, G.R. and Valdez, R.A. (eds), *The Controlled Flood in Grand Canyon: Washington, D.C.*, American Geophysical Union, Geophysical Monograph 110, pp. 117–130.

Anima, R.J., M.S. Marlow, D.M. Rubin and D.J. Hogg, 1998, Comparison of Sand Distribution between April 1994 and June 1996 along Six Reaches of the Colorado River in Grand Canyon, Arizona, *U.S. Geological Survey Open-File Report 98-141*, 33pp.

Bennett, J.P., 1995, Algorithm for resistance to flow and transport in sand-bed channels, *Journal of Hydraulic Engineering*, ASCE, 121(8), pp. 578–590.

Beus, S.S., S.W. Carothers and C.C. Avery, 1985, Topographic changes in fluvial terrace deposits used as campsite beaches along the Colorado River in Grand Canyon, *Journal of the Arizona–Nevada Academy of Science*, 20, pp. 111–120.

Cluer, B.L., 1995, Cyclic fluvial processes and bias in environmental monitoring, Colorado River in Grand Canyon, *Journal of Geology*, 95(103), pp. 411–421.

Dietrich, W.E., 1982, Flow, boundary shear stress, and sediment transport in a river meander, Unpublished PhD dissertation, University of Washington, Seattle, Washington.

Dolan, R., A.D. Howard and D. Trimble, 1978, Structural control of rapids and pools of the Colorado River in the Grand Canyon, *Science*, 202, pp. 629–631.

Environmental Systems Research Institute, Inc., 1991, ARC/INFO User's Guide – Surface Modeling with TIN^{TM}, Environ. Syst. Res. Inst., Inc., Redlands, Calif.

Hazel, J.E., M. Kaplinski, M.F. Manone, R.A. Parnell, A.R. Dale, J. Ellsworth and L. Dexter, 1997, The effects of the 1996 Glen Canyon Dam beach/habitat-building test flow on Colorado River sand bars in Grand Canyon, *Final Report to Glen Canyon Environmental Studies*.

Hazel, J.E., M. Kaplinski, R.A. Parnell, M.F. Manone and A.R. Dale, 1999, Topographic and bathymetric changes at thirty-three long-term study sites, in Webb, R.H., Schmidt, J.C., Marzolf, G.R. and Valdez, R.A. (eds), *The Controlled Flood in Grand Canyon: Washington, D.C.*, American Geophysical Union, Geophysical Monograph, 110, pp. 161–183.

Hereford, R., 1996, Map showing surficial geology and geomorphology of the Palisades Creek area, Grand Canyon, National Park, Arizona, Miscellaneous Investigations Series, Map I-2449, *U.S. Geological Survey*, 1 sheet, scale 1:2000.

Hereford, R., H.C. Fairley, K.S. Thompson and J.R. Balsom, 1991, The effect of regulated flows on erosion of archeologic sites at four areas on eastern Grand Canyon National Park, Arizona: A preliminary analysis, *U.S. Geological Survey Administrative Report*, 36pp.

Hereford, R., H.C. Fairley, K.S. Thompson, J.R. Balsom, 1993, Surficial geology, geomorphology, and erosion of archaeological sites along the Colorado River, eastern Grand

Canyon, Grand Canyon National Park, Arizona, *U.S. Geological Survey Open-File Report 93-517*, 46pp. 2 sheets.

Hereford, R., K.S. Thompson, K.J. Burke and H.C. Fairley, 1996, Tributary debris fans and the late Holocene alluvial chronology of the Colorado River, eastern Grand Canyon, Arizona, *Geological Society of America Bulletin*, 108(1), 3–19.

Hereford, R., K.J. Burke and K.S. Thompson, 1998, Quaternary geology and geomorphology of the Nankoweap Rapids area, marble Canyon, Arizona, Geologic Investigations Series, Map I-2608, *U.S. Geological Survey*, 18pp. 1 sheet, scale 1:2000.

Howard, A.D. and R. Dolan, 1981, Geomorphology of the Colorado River in the Grand Canyon, *Journal of Geology*, 89(3), pp. 269–298.

Kearsley, L.H., R.D. Quartaroli and M.C. Kearsley, 1999, Changes in the number and size of campsites as determined by inventories and measurement, in Webb, R.H., Schmidt, J.C., Marzolf, G.R. and Valdez, R.A. (eds), *The Controlled Flood in Grand Canyon: Washington, D.C.*, American Geophysical Union, Geophysical Monograph 110, pp. 147–159.

Keulegan, G.H., 1938, Laws of turbulent flows in open channels, Res. Paper RP1151, National Bureau of Standards, 21, pp. 707–741.

Korman, J., S.M. Wiele and M. Torizzo, 2004, Modeling the Effect of Discharge on Habitat Quality and Dispersal of Juvenile Humpback Chub (*Gila cypha*) in the Colorado River in Grand Canyon, Regulated Rivers, 20, pp. 379–400.

Lucchitta, I. and L.B. Leopold, 1999, Floods and sandbars in the Grand Canyon, *GSA Today*, 9(4), pp. 1–7.

McLean, S.R., 1992, On the calculation of suspended sediment load for noncohesive sediments, *Journal of Geophysical Research*, 97(C4), pp. 5759–5770.

Melis, T.S., 1997, Geomorphology of debris flows and alluvial fans in Grand Canyon National Park and their influence on the Colorado River below Glen Canyon Dam, Arizona, PhD dissertation, University of Arizona, Tucson, 490pp.

Melis, T.S., R.H. Webb, P.G. Griffiths and T.J. Wise, 1994, Magnitude and frequency data for historic debris flows in Grand Canyon National Park and vicinity, Arizona, *U.S. Geological Survey Water Resources Investigation Report 94-4214*, 285pp.

Meyer-Peter, E. and R. Müller, 1948, *Formulas for Bed-Load Transport*, International Association for Hydraulic Research, Second Meeting, Stockholm.

Nelson, J.M. and J.D. Smith, 1989a, Evolution and stability of erodible channel beds, in Ikeda, S. and Parker, G. (eds), *River Meandering*, pp. 321–377.

Nelson, J.M. and J.D. Smith, 1989b, Mechanics of flow over ripples and dunes, *Journal of Geophysical Research*, 89(C6), pp. 8146–8162.

Patankar, S.V., 1980, *Numerical Heat Transfer and Fluid Flow*, Hemisphere, Washington, DC, 197pp.

Powell, J.W., 1897, *The Exploration of the Colorado River and Its Canyons*, Penguin Books, New York, 291pp.

Rouse, H., 1937, Modern conceptions of the mechanics of turbulence, *Transactions of the American Society of Civil Engineers*, 102, pp. 463–543.

Rubin, D.M., J.M. Nelson and D.J. Topping, 1998, Relation of inversely graded deposits to suspended-sediment grain-size evolution during the 1996 flood experiment in Grand Canyon, *Geology*, 26(2), pp. 99–102.

Rubin, D.M., D.J. Topping, J.C. Schmidt, T. Melis, J. Hazel and M. Kaplinski, 2002, Results of the 1996 experimental flow in Grand Canyon, EOS.

Schmidt, J.C., 1987, Geomorphology of alluvial-sand deposits, Colorado River, Grand Canyon National Park, Arizona, Unpublished PhD dissertation, Johns Hopkins University, Baltimore, Maryland.

Schmidt, J.C., 1999, Summary and synthesis of geomorphic studies conducted during the 1996 controlled flood in Grand Canyon, in Webb, R.H., Schmidt, J.C., Marzolf, G.R. and

Valdez, R.A. (eds), *The Controlled Flood in Grand Canyon*: Washington, D.C., American Geophysical Union, Geophysical Monograph 110, pp. 329–341.

Schmidt, J.C. and J.B. Graf, 1990, Aggradation and degradation of alluvial sand deposits, 1965 to 1986, Colorado River, Grand Canyon National Park, Arizona, *U.S. Geological Survey Professional Paper 1493*, 74pp.

Schmidt, J.C. and D.M. Rubin, 1995, Regulated streamflow, fine-grained deposits, and effective discharge in canyon with abundant debris fans, in J.E. Costa *et al.* (eds), *Natural and Anthropogenic Influences in Fluvial Geomorphology: The Wolman Volume*, Geophysical Monograph Series, American Geophysical Union, Washington, DC, 89, pp. 177–195.

Schmidt, J.C. and M.F. Leschin, 1995, Geomorphology of post-Glen Canyon Dam fine-grained alluvial deposits of the Colorado River in the Point Hansbrough and Little Colorado River confluence study reaches in Grand Canyon National Park, Arizona, *Report to Glen Canyon Environmental Studies*.

Schmidt, J.C., P.E. Grams and M.F. Leschin, 1999, Variation in the magnitude and style of deposition and erosion in three long (8–12 km) reaches as determined by photographic analysis, in Webb, R.H., Schmidt, J.C., Marzolf, G.R. and Valdez, R.A. (eds), *The Controlled Flood in Grand Canyon*: Washington, D.C., American Geophysical Union, Geophysical Monograph 110, pp. 185–283.

Shields, A., 1936, Application of similarity principles and turbulence research to bed-load movement (in German), Mitteil. Preuss. Versuchanst. Wasser, Erd, Schiffsbau, Berlin, 26.

Smith, J.D. and S.R. McLean, 1977, Spatially averaged flow over a wavy surface, *Journal of Geophysical Research*, 82(12), pp. 1735–1746.

Smith, J.D. and S.R. McLean, 1984, A model for flow in meandering streams, *Water Resources Research*, 20(9), pp. 1301–1315.

Stevens, L.E., 1990, *The Colorado River in Grand Canyon, A Guide*, Red Lake Books, Flagstaff, Arizona, 115pp.

Thompson, K. and A. Potochnik, 1997, Proposal to test and apply a geomorphic model related to erosion of pre-dam river terraces in the Colorado River ecosystem containing cultural materials, proposal to the Cultural Resources Program, Grand Canyon Monitoring and Research Center.

Thompson, K. and A. Potochnik, 2000, Development of a geomorphic model to predict erosion of pre-dam Colorado River terraces containing archaeological resources, prepared for Grand Canyon Monitoring and Research Center, Flagstaff, Arizona.

Thompson, K., K. Burke and A. Potochnik, 1997, Effects of the beach-habitat building flow and subsequent interim flows from Glen Canyon Dam on Grand Canyon camping beaches, 1996: A repeat photography study by Grand Canyon river guides (adopt-a-beach program), *Administrative Report to the Grand Canyon Monitoring and Research Center*, Flagstaff, Arizona.

Topping, D.J., 1997, Flow, sediment transport, and channel geomorphic adjustment in the Grand Canyon, Arizona gage reach of the Colorado River during the Grand Canyon flood experiment: Abstracts and executive summaries, Symposium on the 1996 Glen Canyon Dam Beach/Habitat-Building Flow, Flagstaff, Arizona.

Topping, D.J., D.M. Rubin and L.E. Vierra Jr, 2000a, Colorado river sediment transport, 1, Natural sediment supply limitation and the influence of Glen Canyon Dam, *Water Resources Research*, 36(2), pp. 515–542.

Topping, D.J., D.M. Rubin, J.M. Nelson, P.J. Kinzel III and I.C. Corson, 2000b, Colorado river sediment transport, 2, Systematic bed-elevation and grain-size effects of sand supply limitation, *Water Resources Research*, 36(2), pp. 543–570.

US Department of Interior, 1995, Operation of Glen Canyon Dam: Final Environmental Impact Statement, *Colorado River Storage Project*, Coconino County, Arizona, 337pp. and appendices, Salt Lake City, Utah.

van Rijn, L.C., 1984, Sediment transport. Part III: suspended load transport, *Journal of Hydrological Engineering*, ASCE, 110(11), pp. 1613–1641.

Webb, R.H., P.T. Pringle and G.R. Rink, 1989, Debris flows from tributaries of the Colorado River, Grand Canyon National Park, Arizona, *U.S. Geological Professional Paper 1492*, 39pp.

Webb, R.H., D.L. Wegner, E.D. Andrews, R.A. Valdez and D.T. Patten, 1999, Downstream effects of Glen Canyon Dam on the Colorado River in Grand Canyon: A review, in Webb, R.H., Schmidt, J.C., Marzolf, G.R. and Valdez, R.A. (eds), *The Controlled Flood in Grand Canyon: Washington, D.C.*, American Geophysical Union, Geophysical Monograph 110, pp. 131–145.

Wiberg, P.L. and D.M. Rubin, 1985, Bed roughness produced by saltating sediment, *Journal of Geophysical Research*, 94(C4), pp. 5011–5016.

Wiele, S.M., 1997, Modeling of flood-deposited sand distributions for a reach of the Colorado River below the Little Colorado River, Grand Canyon, Arizona, *U.S. Geological Survey Water-Resources Investigation Report 97-4168*, 15pp.

Wiele, S.M. and J.D. Smith, 1996, A reach-averaged model of diurnal discharge wave propagation down the Colorado River through the Grand Canyon, *Water Resources Research*, 32(5), pp. 1375–1386.

Wiele, S.M., J.B. Graf and J.D. Smith, 1996, Sand deposition in the Colorado River in the Grand Canyon from flooding of the Little Colorado River, *Water Resources Research*, 32(12), pp. 3579–3596.

Wiele, S.M., E.D. Andrews and E.R. Griffin, 1999, The effect of sand concentration on depositional rate, magnitude, and location, in Webb, R.H., Schmidt, J.C., Marzolf, G.R. and Valdez, R.A. (eds), *The Controlled Flood in Grand Canyon: Washington, D.C.*, American Geophysical Union, Geophysical Monograph 110, pp. 131–145.

Wilson, R.T., 1986, *Sonar patterns of Colorado River bed, Grand Canyon*, Proceedings of the Fourth Federal Interagency Sedimentation Conference, Las Vegas, Nevada, 2, pp. 5-133–5-142.

Yeatts, M., 1996, High elevation sand deposition and retention from the 1996 spike flow: An assessment for cultural resources stabilization, *Final Report to Glen Canyon Environmental Studies*.

Yeatts, M., 1998, 1997, Data recovery at five sites in the Grand Canyon, *Final Report to the Bureau of Reclamation*, Salt Lake City, Utah and Grand Canyon Monitoring and Research Center, Flagstaff, Arizona.

15

Modelling of open channel flow through vegetation

C.A.M.E. Wilson, T. Stoesser and P.D. Bates

15.1 Introduction

Both aquatic and riparian vegetation have become central to river and coastal restoration schemes, the creation of flood retention space and coastal protection projects. Aquatic and riparian plants obstruct the flow and reduce the mean flow velocities relative to non-vegetated regions. The additional drag exerted by plants strongly influences the transport processes, and hence in restoration and managed retreat schemes the system's morphology is critically determined by the colonisation and establishment of vegetation. Yet, currently there are few models within CFD codes that conceptually or physically represent the hydrodynamic impact of a vegetation canopy on the velocity field and the resulting flood storage, conveyance characteristics and water levels within an estuarine or riverine system. Within such applications, hydraulic models are typically calibrated by adjusting a bulk energy loss coefficient (e.g. Manning–Strickler's n or Chezy's C) to achieve an acceptable match between sparse observed data and model predictions. Yet, as the possible range of friction parameters for any given environment is wide, and the appropriate distributions are model- and grid-dependent, the hydraulic impact of vegetation is largely uncharacterised in current approaches which use a roughness coefficient.

Few recommendations are available which directly characterise vegetation–flow interaction to assess the impact of vegetation reinstatement on the velocity distribution, flood storage and conveyance characteristics, the geomorphological processes, the water quality and habitat diversity of a riverine system. The rapid erosion of salt marsh and mud flat fronting defences has led to the development of sustainable defence strategies (DEFRA & EA, 2001), but currently we do not have the knowledge or tools within standard CFD codes to execute such tasks.

We should explore both conceptually and physically based representations of vegetation–flow interactions and aim to reduce calibration uncertainty rather than add further intangible parameters to our modelling framework. The development of such CFD codes will potentially have enormous consequences for environmental modellers, not least as it will remove or reduce the ability to subsume under-represented processes and data uncertainties within the calibration process.

Whether the vegetation–flow interaction model is conceptually or physically based it is important that an appropriate level of detail is chosen, which satisfies the level of approximations made by the governing equations and numerical schemes on which the CFD code is based. Furthermore, the model complexity should be appropriate for the processes and application being examined. A 3D CFD code may be more appropriate in applications where there are considerable velocity gradients (both laterally and vertically) and strong secondary currents. However, an emergent homogenous plant canopy may require a simpler vegetation model than a submerged heterogeneous plant canopy where there may be considerable variation in velocity over the flow depth resulting in shear layer formation between the canopy and surface-flow zones. As scientists and engineers, ultimately we need to examine and explore many ways to explain and describe flow through natural vegetation and evaluate a suitable level of representation necessary in order to deliver accurate predictions.

This chapter is in three parts. Firstly, we review approaches to characterise the impedance of flow due to vegetation and this will involve both relationships derived from experimental investigation and their numerical representation within models. Secondly, we focus on two case studies which examine the hydraulic resistance characteristics of both simulated and real vegetation in the laboratory. Finally, physically based numerical concepts for the representation of vegetation are illustrated within a 3D Finite Volume (FV) model framework, although all these numerical concepts could be implemented within a 2D modelling framework of differing numerical discretisation.

15.2 Previous work

Most research activity into vegetation in the benthic and riparian environment has been devoted to laboratory flumes of simple cross section, although there has been a considerable contribution from agricultural engineers and meteorologists working on terrestrial systems (e.g. Ree, 1958; Plate and Quraishi, 1965). This section aims to cover the various approaches taken by researchers to characterise the hydraulic resistance of vegetation within the benthic and riparian environment.

15.2.1 Divided channel method (DCM)

Although the DCM is an empirically based hand-calculation method it is worth mentioning here as it is a method which acknowledges the lateral variation in boundary roughness, flow depth and hence velocity across a river/floodplain section. The channel is divided into sub-regions of different roughness properties or flow behaviour, and the total discharge is equated to the sum of discharges for each sub-region. Overbank areas are subdivided into different regions that have roughness properties relatively different to that of the main channel. Many variants of the DCM, which was first proposed by Lotter (1933), have been suggested (e.g. Masterman and Thorne, 1992; Greenhill and Sellin, 1993). Some approaches also account for the turbulent interaction between the faster velocities in the main channel and the slower velocities on the vegetated floodplain (Wormleaton *et al.*, 1982; Ervine and Ellis, 1987). The DCM has generally been applied to compound channel sections; however, the German research community (e.g. Kaiser, 1984; Pasche, 1984; Bertram, 1985; Mertens, 1989; Nuding, 1991) has refined the DCM for use in partially vegetated channels to account for both the turbulent momentum exchange between zones and the production of macro-scale turbulence caused by vegetal obstructions to flow (Figure 15.1). In this approach, the four zones are classified according to their hydrodynamics (Figure 15.2). *Zone I* is a emergent vegetated zone with no interactive energy loss, *zone II* a submerged vegetated zone which is influenced by flow interaction with the main channel zone, *zone III* a main channel zone influenced by the flow interaction with the vegetation zone and *zone IV* a main channel zone which remains unaffected by the flow interaction.

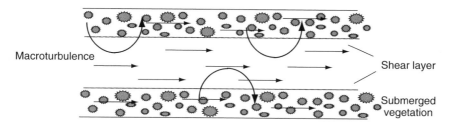

Figure 15.1 *The mechanisms for energy loss within a vegetated compound channel flow system where channel is partially vegetated and vegetation is submerged.*

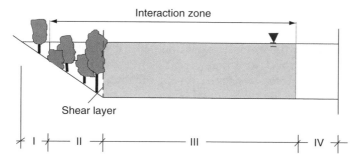

Figure 15.2 *Characteristic zones in an open-channel with vegetation. (After Bertram, 1985.)*

15.2.2 Empirical n-UR method

Some of the earliest experimental work conducted by the US Soil Conservation Service on vegetated channels focused on a design approach called the n-UR method (e.g. USDA, 1947). This method was based on the premise that for a particular vegetation type there is a unique relationship between the Manning–Strickler roughness coefficient

$$n = \frac{1}{U} R^{2/3} S^{1/2} \qquad (15.1)$$

and the product of the mean area velocity (U) and hydraulic radius (R), regardless of the relative values of U, R and the gradient of the energy line (S), where $U=$ mean area velocity, $R=$ hydraulic radius, $S=$ slope of the energy line. No scientific justification is given for the n-UR approach and it has been shown by Kouwen *et al.* (1969) that for the same vegetation type and height, plots of Manning's n versus UR do not fall on a single line and hence this empirical approach may be seriously flawed. Ree (1958) has also shown that the n-UR correlation does not hold for emergent plant canopies. Further, Kouwen and Li (1980) claims that this method is invalid when the vegetation is short and stiff or when the slope is smaller than 0.05, conditions encountered on many lowland rivers prone to flood inundation.

15.2.3 Roughness coefficient as a function of flow depth

For inbank flows the Manning–Strickler value (n) often decreases with increasing stage (e.g. Sargent, 1979; Sellin *et al.*, 2001). When the flow depth is shallow within a river the irregularities of the channel bottom are exposed and their effects become pronounced. For overbank flow conditions the value of n varies with the level of plant submergence, and hence the degree of plant submergence or emergence is an important parameter in defining the relationship between roughness coefficient and flow depth. Wu *et al.* (1999) examined this relationship for both emergent and submerged vegetation conditions within a laboratory flume. For emergent conditions n decreased but as the plant became completely submerged the coefficient significantly decreased tending to an asymptotic constant. For a deeply submerged canopy, Temple (1987) and Wilson and Horritt (2002) also observed the roughness coefficient tending towards a constant. Wu *et al.* (1999) have further shown that the asymptotic constant is a function of the vegetation height.

15.2.4 Force-equilibrium approach and drag force approach

Ree (1958) and Petryk and Bosmajian (1975) addressed the problem that Manning's n is a function of flow depth and vegetation density. The latter researchers used a force-equilibrium approach for the emergent plant condition where the gravitational force can be equated to two resistance forces, the drag force of the vegetal obstruction and the shear force on the channel bottom and sides:

$$LBy\rho s = \frac{1}{2}\rho C_D \lambda U^2 + F_{bed\ shear} \qquad (15.2)$$

Where L is the length of control volume, B width of channel, y is the flow depth, ρ is the fluid density, S is the longitudinal gradient, C_D is the drag coefficient, U is the area mean velocity, $F_\text{bed shear}$ is the shear force on the channel bed and sides and λ is the vegetation coefficient which is equal to the area of the plants projected normal to the streamwise flow direction. The drag coefficient can be determined from experiments or evaluated for simple geometries from the plants' base shape and the Reynolds number. The effect of plant bending with flow is ignored. For conditions where the energy losses due to vegetation dominate, the drag force increases as the square of the area mean velocity and Manning's n is proportional to the square root of the plant density. However, Fathi-Maghadam and Kouwen (1997), through their experimental work on non-submerged pine and cedar saplings and branches have directly measured the drag force for a variety of flow conditions. They have found that the relationship between drag force and flow velocity is probably linear. This is similar to the findings from Wilson and Horritt (2002) for a flexible grass canopy where the drag force per unit area was calculated from a force-equilibrium approach. Dunn *et al.* (1996) simulated a rigid emergent plant canopy using vertical cylinders set in a staggered arrangement with variable density. The drag coefficient calculated from a force-equilibrium approach was found to be a function of the vertical distance from the bed reaching a maximum value at one-third of the flow depth and a mean C_D value of 1.13 ± 0.15.

The German Association for Hydraulic Engineering (DVWK, 1991) recommends a hydraulic resistance method directly based on the drag force approach calculated per unit area to be used for tree- and shrub-vegetated floodplains. Drag force is related to the vegetation geometric properties and their spacing. Plant density is quantified by the vegetation parameters a_x (average spacing between obstructions in the streamwise direction), a_y (average spacing between obstructions in the cross-streamwise direction) and d_p (average plant stem or tree trunk diameter) (Figure 15.3) such that λ in equation (15.2) becomes:

$$\lambda = \frac{d_p}{a_x \cdot a_y} \quad (15.3)$$

15.2.5 Velocity profiles and relative roughness height approach

Researchers (e.g. Kouwen *et al.*, 1969; Tsujimoto *et al.*, 1992; Nepf and Vivoni, 1999) have observed that the mean streamwise velocity profile within a submerged vegetated layer (irrespective of whether the vegetation is rigid or flexible) no longer follows the universal logarithmic law. In the surface flow layer above the vegetated layer, researchers (for example Kouwen *et al.*, 1969) have generally assumed the logarithmic law prevails. Kouwen *et al.* (1969) empirically fitted the logarithmic law by adjusting the origin intercept and values of the roughness parameter. They advocate the use of the log law using an undeflected height of the roughness elements and concluded that the origin intercept and slope were functions of the plant density and flexibility. The deflected roughness height is related to the undeflected height of the grass, local boundary shear, flow depth and the *MEI* (the product of the stem density M, stem modulus of elasticity E and the stem area's second moment of inertia I). This was defined from laboratory experiments using flexible plastic strips of vegetation where *MEI* was a measurable parameter. In natural vegetation, the parameter *MEI* is more difficult to define and quantify. For example, Temple (1987)

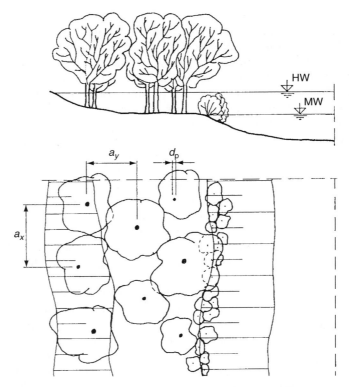

Figure 15.3 Characterisation of vegetation parameters a_x, a_y and d_p. (Published by DVWK, 1991.)

undertook laboratory experiments on grasses correlating MEI to undeflected vegetation height and found a strong relationship for growing grasses but no significant correlation for dormant grasses.

15.2.6 Plant stiffness and flexibility classification

Freeman *et al.* (1998) quantified plant stiffness in terms of the horizontal force necessary to bend the plant to a 45° inclination from a vertical emergent position. They evaluated the horizontal force for five species (including willow varieties) and related it to the modulus of elasticity (E). Separate empirical relationships were developed for submerged and emergent conditions relating Manning's n to a number of plant properties (e.g. E, momentum-absorbing area, deflected and undeflected plant height and density) where Manning's n is proportional to the area mean velocity squared. The application of these formulae outside their empirical domain is questionable and remains untested; however, Freeman *et al.*'s contribution lies in the documenting of stem stiffness values in a tractable framework. Kutija and Hong (1996) introduce a flexible vegetation approach that models the process of bending analogous to a cantilever beam. The model takes into account the effects of shear stresses at the bed and the additional forces induced by the flow.

15.2.7 More detailed investigations of turbulence structure

More recent work has focused on detailed examination of the flow field and turbulence structure within a plant canopy. These studies have involved modelling vegetation as an array of rigid cylinders of the same height and diameter at regular spacing (Pasche, 1984; Tsujimoto et al., 1992; Dunn et al., 1996; Meijer and Van Velzen, 1999; Nepf, 1999). Nepf (1999), for example, has tested a physically based model which links vegetation form drag, turbulence intensity and turbulent diffusion and uses Raupach and Shaw's (1982) spatial and temporal averaging approach to account for the heterogeneous nature of the flow field and to present the problem in a 1D context. For both rigid and flexible vegetation types in emergent conditions, the Reynolds stresses within the flow are negligible and the streamwise turbulence fluctuations are found to be small. However, when the vegetation becomes submerged, a horizontal shear layer forms in the surface flow region above the canopy and also penetrates to some depth within the plants. The Reynolds stress profile reaches a peak at the interface and decays within and above the canopy. It seems that this characterises the flow irrespective of the vegetation being rigid or flexible (Tsujimoto et al., 1992; Dunn et al., 1996; Nepf and Vivoni, 1999). Through experiments on flexible simulated vegetation Nepf and Vivoni (1999) observed that the turbulent production within the canopy was made up of two components: that generated from the horizontal shear layer at the canopy top and that generated within the stem wakes. The relative dominance of these sources across the transition from emergent to submerged conditions was also addressed.

15.2.8 Higher-dimensional modelling approaches

More complex methods for multi-dimensional flow problems have been developed by Burke and Stolzenbach (1983), Nakagawa et al. (1992), Shimizu and Tsujimoto (1994) and Lopez and Garcia (1997) using a two-equation turbulence closure approach. This was first introduced by Wilson and Shaw (1977) to model atmospheric flows over plant canopies. A modified k-ε turbulence closure model is used, introducing drag-related sink terms into the momentum as well as into the turbulent transport equations. Laboratory experiments by Tsujimoto et al. (1992) have been used to validate the model. Shimizu and Tsujimoto (1994) calibrated their model by adjusting two additional weightings, that appear by introducing the vegetative sink into the turbulence equations, to reproduce observed mean velocities and Reynolds stresses. Lopez and Garcia (1997) numerically modelled similar test cases (those experimentally investigated by Dunn et al., 1996) and modified the same weighting factors of the k-ε model but reported calibrated values which differed by about 500% from those of Shimizu et al. More recently Lopez and Garcia (2001) tested the k-ε model and another two-equation closure known as the k-ω formulation (Wilcox, 1988). A constant C_D value of 1.13 (Dunn et al., 1996) was used in the drag force sink term in the momentum equations and no modifications were made to the five standard constants of the k-ε model. However, values for the weighting coefficients for the drag-related source terms for the k- and ε-equations ($C_{fk} = 1.0$ and $C_{f\varepsilon} = 1.33$) were again found to be inconsistent with those previously reported. These researchers claim that the inconsistencies in weighting coefficient values are due to the spatial and temporal averaging technique used within this study to account for the heterogeneous nature of the velocity and turbulence intensity fields. No significant difference was found between the numerical performances of either turbulence model. To eliminate the calibration from the validation procedure, Fischer-Antze *et al.* (2001) developed a similar drag force approach but only introduced sink terms into the Navier–Stokes equations and hence made no modification to the k-ε model. The turbulence closure thus used the standard values for the weighting coefficients (after Rodi, 1980) and a C_D equal to unity. The model was used to simulate rigid and emergent vegetation in simple-section and compound-section channel arrangements (experimentally conducted by Tsujimoto *et al.* [1992] and Pasche [1984], respectively) and good agreement was found between the computed and the observed streamwise velocity distributions. Naot *et al.* (1996) were the first who have used a higher-order anisotropic closure, the Reynolds-stress model (RSM), to simulate the flow through rigid submerged vegetation elements. They have also used the drag force approach but have additionally accounted for the effect of shading, i.e. the effect of upstream rods resulting in weaker ambient flow and reducing the drag force on the vegetational elements. Furthermore they have related the two vegetative weighting coefficients to a characteristic length scale that can be estimated from the configuration and geometry of the plants. However, according to Naot *et al.* the sensitivity of this characteristic length and the shading factors indicated the importance of these terms. Choi and Kang (2001) also used a higher-order anisotropic closure (RSM) and compared these results against the k-ε model for the data set collected by Dunn *et al.* (1996). Both models employed a C_D value of 1.13 (Dunn *et al.*, 1996). Within the vegetated layer there was no significant difference between the performances in the prediction of the mean velocity profile for the one condition examined. However, within the surface layer region there was a significant improvement in the computed mean velocity, turbulent intensity and Reynolds stress profiles for the RSM relative to the k-ε model.

15.3 Relationships derived from experimental investigations

In order to build conceptual and process-based vegetation–flow interaction models we need to understand how types of vegetation which differ in terms of biomechanical properties, physical scale, plant form, natural density and level of submergence impede the flow and the resulting change in velocity and momentum exchange processes.

The majority of the experimental work has looked at the hydraulics of rigid emergent vegetation, where the vegetation has been simulated by a group of cylinders of the same height and diameter at regular spacing (Pasche, 1984; Tsujimoto *et al.*, 1992; Nepf, 1999). The few experimental studies that have been conducted on flexible vegetation have generally not documented the bending stiffness or flexural rigidity of the simulated plants and how this can be scaled from real vegetation. Further, the research on both rigid and flexible vegetation has focused on only one type of plant form per study.

In this section two case studies which fill in some of these research gaps are presented. The first will examine the hydraulic resistance of a flexible grass canopy and examines and compares the characteristics of real grass versus simulated grass. The second study looks at the effect of different forms of flexible submerged vegetation on the velocity field and turbulence characteristics. The biomechanical properties of the simulated vegetation are presented and briefly compared to their natural prototypes.

15.3.1 Case study 1 – Hydraulic resistance properties of a flexible grass canopy

Often simplified and generalised plant forms are used to simulate vegetation within the laboratory. While these models provide well-defined geometries, and ensure experimental repeatability, it is questionable whether simulated plants behave in a manner analogous to real vegetation. The biomechanical properties of simulated plants need to be measured and quantified and then related back to their prototype. This is further compounded by the fact that little biomechanical information is available for real plants. Using real plants eliminates these latter issues. This section will outline the findings from experiments conducted with both common garden grasses and two types of artificial grasses (display grass and astroturf). The real grasses examined have average blade heights (h) of 20 and 70 mm, respectively. The results for the longer grass height are from a study conducted by Wilson and Horritt (2002). Preliminary blade counts and photographs suggest that the real grasses have a very similar density and a relatively smaller blade density compared to the artificial grasses used in this study. The display grass and astroturf have blade heights of 6 and 16 mm, and blade densities of 1.28×10^6 and $1.83 \times 10^6 \, \text{m}^{-2}$ respectively.

15.3.1.1 *Experimental description*

Experiments were conducted in a tilting flume 300 mm in width and 10 m in length with a bed slope of 1 in 2000. The water surface profile was controlled by the downstream tailgate weir which can be raised and lowered by a gear system allowing its height to be set with a high degree of accuracy. To establish uniform conditions, flow depths were measured along the flume for either four or six different tailgate settings. An approach based on Horton (1933) was used to eliminate the vertical shear exerted by the flume's sidewalls. The streamwise velocity component was measured in the flume study with a downward-looking acoustic Doppler velocimeter at a frequency of 25 Hz for a 180-s sampling period. For each grass lining, three vertical profiles to account for the heterogeneity in flow field were measured at differing locations along the centreline of the flume.

15.3.1.2 *Background*

The total drag force generated by submerged vegetation is composed of two components: the drag force due to its form and that created from the form's boundary roughness. If the total boundary shear force (per unit area) is caused by the total drag force generated by the plants, then the average boundary shear stress, τ_0, is equal to the average drag force of the plants per unit plan area:

$$\tau_0 = \frac{F_D}{(bl)} \qquad (15.4)$$

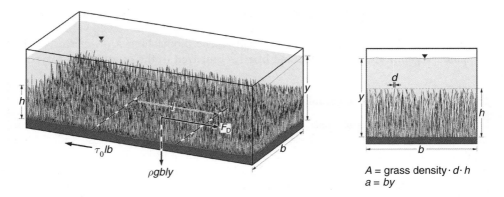

Figure 15.4 Definition of notation used by Wilson and Horritt (2002).

Using the general formula for fluid flowing past an immersed body, the drag force may be described by

$$F_D = \frac{1}{2} C_D \rho A U^2 \qquad (15.5)$$

in which C_D = drag coefficient of the vegetation; U = average longitudinal velocity; A = momentum-absorbing area of the vegetation; and ρ = density of water. The momentum-absorbing area, A, and the ratio A/a (where a is the cross sectional flow area) were phrases first coined by Fathi-Maghadam and Kouwen (1997). For submerged grass the projected area or momentum-absorbing area is defined in Figure 15.4. The shear stress τ_0 and the drag force F_D can easily be converted to resistance coefficients. From the definition of the Darcy–Weisbach friction factor, f,

$$f = 8 \frac{\tau_0}{\rho U^2} \qquad (15.6)$$

equation (15.6) becomes:

$$f = 4 C_D \left(\frac{A}{bl} \right) \qquad (15.7)$$

or in terms of the Manning–Strickler friction coefficient:

$$n = \left(\frac{1}{2g} R^{1/3} C_D \left(\frac{A}{bl} \right) \right)^{1/2} \qquad (15.8)$$

15.3.1.3 Results and discussion

For all grass heights and types, the hydraulic resistance significantly decreases with increasing flow depth, tending towards an asymptotic constant which is reached at

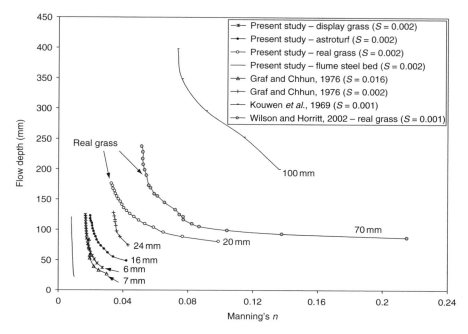

Figure 15.5 Variation of Manning's n with flow depth for various grass linings. Undeflected grass height is annotated for each curve.

different levels of submergence depending on the grass properties (Figure 15.5). This can be explained by the fact that the exposure of the grass or the momentum-absorbing area is more pronounced as the depth decreases. The ratio A/a decreases with increasing flow depth and therefore as the cross sectional flow area increases Manning's n will decrease. Graf and Chhun (1976) and Kouwen *et al.* (1969) conducted similar experiments using artificial grass and arrays of flexible elements, respectively, and they observed a similar relationship between n and flow depth (see Figure 15.5).

For the shorter grass height (7 mm) examined in the Graf and Chhun study the asymptotic Manning's n value tends towards the same value as observed in this study ($n = 0.017-0.019$). However, as the flow depth decreases, the Manning's n curves corresponding to heights 6 and 7 mm deviate. This is probably related to the thickness of the interaction zone directly above the grass canopy which is dependent on grass height and grass properties. For taller grass heights ($h > 7$ mm) there is a significant increase in Manning's n with decreasing flow depth and for the real grasses the constant Manning's n value is dependent on the grass height whereby for undeflected grass heights 20 and 70 mm the value varies from 0.033 to 0.052, respectively. The actual value of n for the taller grass is 40% greater than that usually given for short grassy floodplains in well-established texts (Chow, 1959), but relatively lower than values predicted using the USDA handbook method and the method of Green and Garton (1983). These methods over-predict the Manning–Strickler coefficient by 140 and 167% respectively (Figure 15.6). This value also

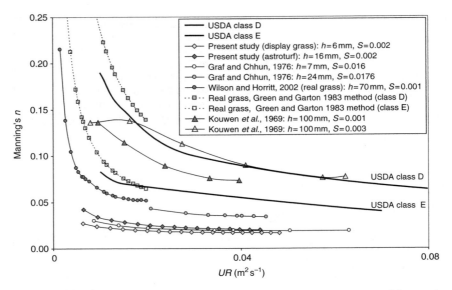

Figure 15.6 Relationship of Manning's n roughness coefficient as a function of the product of mean area velocity and hydraulic radius.

corresponds well with the values found from calibration studies of reach length (10–60 km) 2D hydraulic models of compound channel flow (Horritt and Bates, 2001).

For all grass types examined, the vertical profile of time-averaged streamwise velocity is similar (Figure 15.7). Profiles for all canopies show significant variation in the mean velocity and indicate the generation of a horizontal shear layer. Within the canopy the mean flow is much retarded and immediately above the grass layer there is a significant increase in mean velocity with increasing flow depth. The profiles corresponding to the display grass and astroturf tend towards convergence

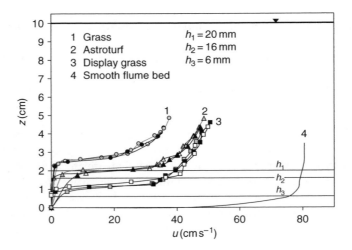

Figure 15.7 Time-averaged streamwise velocities for different grass linings.

at higher flow depths as the distance from the momentum exchange interface increases. The real grass is marginally longer but less dense (in terms of grass blades per square metre) than the astroturf; however, the mean velocity is significantly smaller (25%) at half the flow depth compared to that observed for the astroturf. Although more tests are needed, the decrease in projected plant area due to pronation of the real grass may be a factor in distinguishing the differences in hydraulic behaviour between real and artificial grasses.

15.3.2 Case study 2 – Open channel flow through different forms of flexible submerged vegetation

Research on both rigid and flexible vegetation has mainly focused on only one type of plant form per study. Also the few experimental studies that have been conducted on flexible vegetation (Dunn *et al.*, 1996; Tsujimoto and Kitamura, 1998) have generally not documented the bending stiffness or flexural rigidity of the simulated plants and how this can be scaled from real vegetation. This case study addresses the issue of scaling biomechanical properties from real vegetation and outlines methods that can be used to quantify these parameters. It also explores the effect of two forms of flexible vegetation on the turbulence structure within the submerged canopy and in the surface flow region.

The turbulence characteristics of uniform flow through flexible vegetation were investigated under two simulated plant forms: flexible rods (stipes) of constant height, and density, and the same rods with a frond foliage attached (Figure 15.8). The rods were configured in a staggered arrangement at a stipe density (A or λ_s) of $1.67 \, \mathrm{m}^{-1}$ (given by equation 15.3). The simulated plants were 1/10 scale replicas of a species of kelp (*Laminaria hyperborea*) and all parameters, both geometric and kinematic, were scaled using the Froude law (see Tables 15.1 and 15.2 for scalar relationships and physical properties). They were manufactured from a liquid plastic of the necessary density to cast the 1/10 scale plants with the appropriate stiffness. This work was carried out at the Norwegian University of Science and Technology (by A. Torum and his team). Whilst we here use a marine species as a model plant, the simulated vegetation does bear a morphologic and biomechanical resemblance to commonly encountered riverine plants (Larsen *et al.*, 1990). Without foliage, the rod-like vegetation could be considered equivalent to long grasses or reeds, whilst the plants with foliage bear a resemblance to a number of species of aquatic macrophyte.

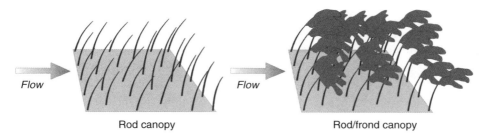

Figure 15.8 *Plant forms investigated in experiments described in section 15.3.2.*

Table 15.1 Scalar relationships for Froudian law of scaling

Quantity	Dimensions	Froude Law model scale 1:x
Geometric		
Length	L	x
Area	L^2	x^2
Kinematic		
Velocity	LT^{-1}	$x^{1/2}$
Dynamic		
Force	MLT^{-2}	x^3
Bending stiffness K	MT^{-2}	x^2
Flexural rigidity EI	$ML^{-3}T^{-2}$	x^5

Table 15.2 Summary of physical and biomechanical parameters

Stipe/rod parameters		
Stipe length	h_{stipe} or h	85 mm
Stipe base diameter	ϕ_b	4 mm
Stipe top diameter	ϕ_t	2 mm
Mean bending stiffness	K	11.0 N m^{-1} (in linear range)
		45.1 N m^{-1} (based on all measurements in both linear and non-linear ranges)
Flexural rigidity	J	$1 \times 10^{-5} - 2 \times 10^{-5}$ N m^2
Frond parameters		
Frond surface area	A_{frond}	0.006 m^2
Approximate height of frond when stretched	a_z	70 mm
Approximate width of frond when stretched	a_y	100 mm
Rod array parameters		
Array height	h_{stipe} or h	85 mm
Stipe/rod density	λ_s	1.67 m^{-1}
Rod/frond canopy parameters		
Thickness of pronated fronds	h_e	20 mm
Canopy height in pronation	$h_{stipe} + h_e$	105 mm
Plant density	λ_p	22.4 m^{-1}
Hydraulic parameters		
Flow depth	H	0.128–0.290 m
Depth ratio	H/h	1.5–3.4
Area mean Reynolds number	$Re_a = ((Q/A)H)/\nu$	6000–20 000

15.3.2.1 *Biomechanical properties*

The bending stiffness of the prototype stipes was measured both in the field and in the laboratory by Dubi (1995). In both cases the stipes were tested in cantilever bending. The larger diameter end was fixed horizontally and weights were attached to the smaller diameter end.

Seven kelp samples were tested in the field and a total of 58 force–deflection measurements were taken. Of these 84% fell within the linear range giving a mean bending stiffness of 16.7 N m^{-1} with a standard deviation of 9.2 N m^{-1}. Three manufactured stipes were tested in the laboratory and a total of 22 measurements conducted. The manufactured stipes exhibited a relatively non-linear relationship compared to the natural kelps, with 27% of measurements being within the linear range. Within this range the mean bending stiffness was 11.0 N m^{-1} (when scaled up to prototype) with a standard deviation of 3.6 N m^{-1}; when measurements in the non-linear and linear ranges are included the mean bending stiffness becomes 45.1 N m^{-1} with a relatively larger spread of values (standard deviation was 30.3 N m^{-1}). This highlights one of the difficulties in comparing properties of different material types.

For the flow conditions examined in the experimental investigation, the flexural rigidity (J) of the model stipes, which is defined as the product of the modulus of elasticity (E) and the second moment of area (I), was evaluated as approximately in the range of 6.8–11.3 × 10^{-5} N m^2, which corresponds to a full-scale flexural rigidity in the range of 6.8–11.3 N m^2. For the natural stipes the flexural rigidity was evaluated as being in the range of 1.0–6.1 N m^2. This was computed using the relationship between the force (F), deflection (w), beam length (L), modulus of elasticity (E) and the second moment of area (also referred to as the moment of inertia) (I):

$$EI = \frac{F}{w} \frac{L^3}{3} \qquad (15.9)$$

where the bending stiffness (F/w) or gradient of the force–deflection curves is defined from the linear range of the curves and not using the maximum force and corresponding deflection measured.

The drag force–velocity relationship of the model frond (without the stipe attached) was measured in the laboratory in a similar manner to their prototypes in the field (see Figure 15.9, results are scaled up to full scale). The drag forces on the

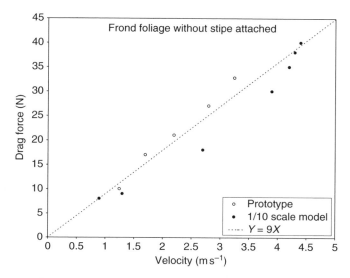

Figure 15.9 Drag force–velocity relationship for the 1/10 scale model frond and one prototype frond, shown at prototype scale according to the Froudian law of scaling.

artificial and prototype fronds were found to be linearly proportional to the velocity in the practical range of 0.75–3 m s^{-1}, and both can be represented by a linear function $Y = CX$ where $C = 9$ N m^{-1} s (where Y = drag force and X = velocity); see Dubi (1995) for further details.

15.3.2.2 Experimental description

The experiments were conducted in a flume 0.5-m wide and 10-m long, with longitudinal bed slope set at 1/1000. Uniform flow conditions were established in a similar manner to that outlined in section 15.3.1.1. The flow depth was varied to produce depth ratios (H/h where h = plant stem height, H = water depth) from 1.5 to 3.4. A 3D sideways-looking acoustic Doppler velocimeter, measuring at a frequency of 25 Hz, was then used to measure the velocity and turbulence statistics over a 240-s sampling period.

15.3.2.3 Results and discussion

The velocity profiles (Figure 15.10) show that the mean flow in the plant layer is much retarded and no longer follows the logarithmic law profile. This is in agreement with findings for both rigid and flexible vegetation (Tsujimoto *et al.* [1991] and Nepf and Vivoni [1999], respectively). Profiles for both the rod and the frond canopies show significant variation in mean velocity and indicate the generation of a horizontal shear layer. Within the plant layer the magnitude of the mean velocity for the frond canopy is less than half that observed for the simple rod canopy. The additional superficial area of the fronds alters the momentum transfer between the canopy and the surface flow region. The fronds shift the peak Reynolds stress to a higher level in the flow, above the canopy (Figure 15.11). The turbulent shear layer also penetrates to a relatively larger proportion of the rod canopy than the frond canopy and does so at a higher level of relative flow depth ($z/h > 1.9$). The distribution of turbulence in the upper, surface flow region is unaffected by the plant form. This suggests that the fronds at the top of the canopy inhibit the momentum exchange between the interior of the canopy and the surface flow layer by confining the lower canopy layer. The drag of the frond surfaces induces additional turbulence which shifts the maximum turbulent stresses to a level above the top of the canopy. For flow conditions where the depth ratio (H/h) is greater than or equal to 2.4, the turbulent stress profiles converge at a similar depth in the flow, at between 60 and 70% of the total flow depth, and the turbulence structure is unaffected by additional drag imposed by the frond foliage. The higher level of turbulence stress generated by the rod array over its submerged depth implies that the shear interaction and turbulent mixing between the plant canopy and surface flow region is greater for the rods alone than for the rod/foliage combination. So although the fronds induce larger drag forces, shear generated turbulence is reduced, possibly due to inhibition of vertical momentum exchange by the greater surface area of the fronds.

The thickness of the active momentum exchange layer, h_p, can be defined as the distance from the top of the canopy to the point within the plant canopy or array by which the turbulent stress has decayed to 10% of its maximum value (Nepf and

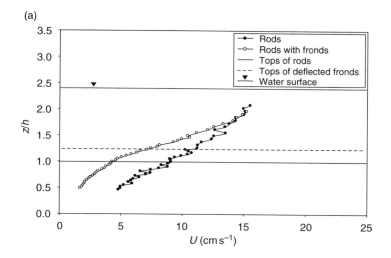

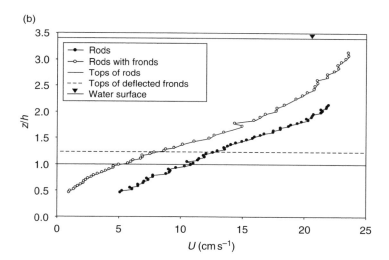

Figure 15.10 *Mean velocity profile for depth ratios (H/h): (a) 2.4 and (b) 3.4.*

Vivoni, 1999). The plants used by Nepf and Vivoni (1999) are of a different form (six blades attached to a short basal stem) and the flexural rigidity of the blades ($J = 1.0 \times 10^{-5}\,\mathrm{N\,m^2}$) is lower than that of the model stipes used in our experiments, but similar to natural prototype stipes. Dunn *et al.* (1996) used commercial drinking straws of constant diameter to simulate a flexible stipe array.

The results from all these studies are compared in Figure 15.12, where h_c is the height of the canopy top (h_c is greater for the frond canopy compared to the stipe-alone canopy). One assumption made in these plots is that relatively small turbulent stresses are produced in the emergent condition. For the same stipe density but differing magnitudes of momentum-absorbing area due to plant form, the penetration

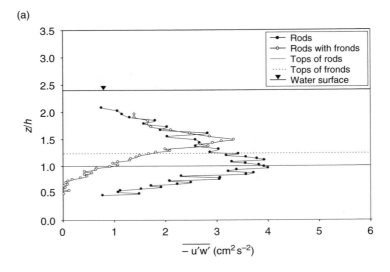

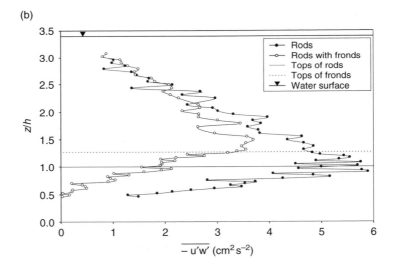

Figure 15.11 *Representative profiles of Reynolds stress for the simple rod array and rod/frond canopy for depth ratios (H/h): (a) 2.4 and (b) 3.4.*

ratio (h_p normalised by h_c) decreases with the addition of foliage and the increase in the momentum-absorbing area. This has a similar effect to increasing stipe density and hence increasing the momentum-absorbing area over the full height of the plants per unit volume: a decrease in the thickness of the active momentum exchange layer within the canopy (see Figure 15.12).

The addition of the plant foliage at the top of the stems inhibits the turbulent mixing between the two flow regions (the canopy layer and the surface flow layer)

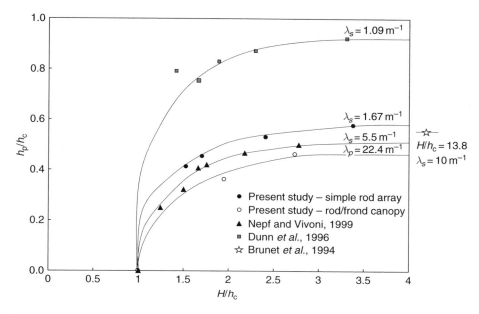

Figure 15.12 Transition from emergent (H/h$_c$ = 1) to deeply submerged conditions (H/h$_c$ > 1). Penetration thickness, h$_p$, was based on the Reynolds stress profiles. And h$_c$ is plant canopy thickness.

and shifts the turbulent stress peak to a level above the canopy top surface. Whilst it is known that the additional drag exerted by plants reduces the mean flow velocity within the vegetated regions relative to unvegetated ones, this research indicates that the greater momentum-absorbing area provided by some plant forms can have a significant effect on the mean flow field. This has implications for restoration schemes. The establishment of 'with foliage' species relative to non-foliage species will lead to larger contaminant retention times within the plant canopy, and greater drag forces in near-bed regions will lead to a greater likelihood of sediment deposition. Hence plant species with fronds may have potential in protecting vulnerable areas prone to scour (for example, surrounding bridge piers and bank-protection works) and in retaining agricultural runoff within buffer zones for longer treatment times relative to their non-frond equivalents.

15.4 Numerical representation – Some examples

As introduced in section 15.2.6, a number of mathematical approaches exist, which can quantify the effects of vegetation on the flow resistance and hence the hydrodynamics of the flow. From a practical engineering perspective, a 1D description of the flow is often sufficient for many channel design purposes and for flood protection measures. However, a 1D mathematical formulation is not necessarily an accurate prediction, since it is based on a number of assumptions and empirical parameters which lump together the effects of a number of distinct energy loss

414 Computational Fluid Dynamics

mechanisms. Hence, more reliable prediction tools are recommended in order to comprehend the full three dimensionality of the flow through and above vegetation. Such a 3D representation of the flow physics is fairly new, since the mathematical formulation of flow resistance due to vegetation and the accompanying effects of flow hampering, turbulence production, energy dissipation, bed shear stress redistribution and relocation of turbulence intensities is still a topic of ongoing research. Moreover, high-quality data sets including detailed time-dependent velocity and turbulence measurements are needed in order to clarify resistance laws and turbulence structures as well as to provide a database for numerical model validation (Nezu and Onitsuka, 2001). Preliminary model development has concentrated on a drag-force approach (after Wilson and Shaw, 1977). Further research into basic mechanisms of the flow interaction between vegetation and the accompanying correlations between plant biomechanical properties, plant bending and plant movement due to the hydrodynamic forces will allow more sophisticated mathematical formulations for roughness closure to be developed. Prior to this, the drag-force approach provides a useful and practical tool with which to begin exploring the simulation of flow over and through vegetation canopies.

15.5 Numerical representation

In this section the implementation of two variants of the drag-force approach in two different 3D Finite Volume (FV) numerical models, SSIIM (Olsen, 2002) and HYDRO3D (Stoesser, 2002), will be described. These models are then applied to a number of laboratory flow conditions which involve both submerged and emergent vegetation and the results are compared.

a. Modification of Navier–Stokes equations
As suggested by Wilson and Shaw (1977), a drag-related force term can be included into the Navier–Stokes equation to account for the flow resistance of vegetation. A subgrid drag force F_D per fluid mass unit in a FV cell is calculated with the definition of plant density in Figure 15.13.

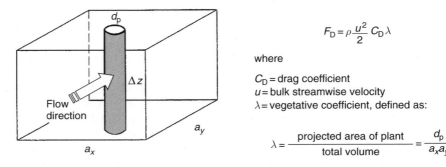

$$F_D = \rho \frac{u^2}{2} C_D \lambda$$

where

C_D = drag coefficient
u = bulk streamwise velocity
λ = vegetative coefficient, defined as:

$$\lambda = \frac{\text{projected area of plant}}{\text{total volume}} = \frac{d_p}{a_x a_y}$$

Figure 15.13 *Definition of drag force in an FV cell. Arrow indicates mean flow direction.*

The implementation of the drag force into the FV discretised form of the Navier–Stokes equations is accomplished using the iterative nature of the code and similar to the suggestion of Patankar (1980) as:

$$\left(A_P + \rho \frac{u_P}{2} C_D \lambda \Delta z\right) \cdot u_P = \sum_{m=1}^{6} A_m \cdot u_m + QP_u, \quad m = E, W, N, S, T, B \qquad (15.10)$$

Here the drag force is added to the matrix coefficient A_P of the numerical point P, which has to equal the surrounding matrix coefficients A_m at locations East (E), West (W), North (N), South (S), Top (T) and Bottom (B), plus an external source QP_u.

b. Modification of k-ε turbulence model
Based on the assumption that the drag force produces additional turbulent energy in the flow and increases the dissipation rate, the turbulent production term P_k is modified as (Wilson and Shaw, 1977):

$$P_k = P_k + C_{fk}(F_{Di} \cdot u_i) \qquad (15.11)$$

for the *k*-equation, and as:

$$P_k = P_k + C_{f\varepsilon}(F_{Di} \cdot u_i) \qquad (15.12)$$

for the ε-equation, where subscript k refers to the turbulence energy.

Since the weighting coefficients C_{fk} and $C_{f\varepsilon}$ have to be calibrated (Tsujimoto *et al.*, 1992; Lopez and Garcia, 1997), which is undesirable for practical applications, Fischer-Antze *et al.* (2001) suggested that one should omit the additional production terms P_k in both the *k*- and the ε-equations. This allowed Fischer-Antze *et al.* (2001) to obtain satisfactory results for simulations of flow through both submerged and emergent vegetation without a recalibration of the weighting coefficients. Another possibility for modelling vegetational flow resistance is by explicit representation of the plants analogous to the flow around a circular cylinder (e.g. Breuer, 1998) or over a matrix of cubes (Mathey *et al.*, 1998). This can be achieved by relating the plan-view mesh to plant geometry and out-blocking particular cells to represent the presence of vegetal obstructions. However, this is restricted to well-defined singular elements and not applicable to a tree, submerged to the level of its branches. Nevertheless, the rigid part of a submerged plant (i.e. the stem) may be distinct in terms of its location and geometry; hence this part can be modelled as a discrete element and the branches and leaves are treated by the subgrid drag force approach. This combined approach is implemented in HYDRO3D (developed by Stoesser, 2002) in addition to the standard drag force approach and will be referred to hereafter as the hybrid approach.

The HYDRO3D and SSIIM codes have a similar numerical basis (3D FV solution of the Navier–Stokes equations with *k*-ε turbulence closure) and hence their comparison to experimental data allows the testing of the validity of the vegetational roughness closures. In the following section, the results of several calculations performed with both HYDRO3D and SSIIM are compared with laboratory

Table 15.3 Experimental and numerical boundary conditions for channel flows through vegetation

	Tsujimoto et al.	Lopez and Garcia	Pasche	Schnauder
Experiment ref.	A31	Exp9	P224	E1A
Geometry				
Type	Rectangular flume		Compound channel flume	
Depth H (m)	0.0936	0.214	0.2 (main)	0.21
Slope I (10^{-3})	2.60	3.60	0.50	1.0
Vegetation				
Type	Fully vegetated, submerged		Emergent cylinders	Emergent artificial tree
Plant height h (m)	0.046	0.12	0.076	0.21
Plant diameter D (mm)	1.5	6.4	12	≈ 80
Plant density λ (m^{-1})	3.75	2.46	2.69	–
Simulation details				
No. of cells	15 000	12 000	38 304	239 760
Turbulence closure	k-ε model (no modification)		k-ε model (no modification)	LES (no modification)
Vegetative resistance	Subgrid drag force		Subgrid drag force	Hybrid
Time marching	Steady state		Steady state	Crank–Nicholson

measurements of submerged and emergent vegetation in laboratory channels. Simulation and laboratory data set details are given in Table 15.3. Later in this chapter the drag-force approach is up-scaled for a practical approach used in the field.

15.5.1 Results – Rigid submerged vegetation in a straight flume

Numerical simulations of the flume experiments of Tsujimoto *et al.* (1992) and Lopez and Garcia (1997) were carried out to simulate submerged vegetation. The drag force coefficient C_D was evaluated for all experiments to be equal to 1.0 for the given Reynolds numbers. Calculated velocity distributions were compared to observed data. Good agreement between measured and calculated velocities for both models could be achieved for Tsujimoto *et al.*'s experiment A31 and for Lopez *et al.*'s experiment no. 9 (Figure 15.14). As is evident, a better match to the observed data in the vegetation layer could be achieved by HYDRO3D although there is slight over-prediction of the velocity near the surface compared to the SSIIM model. However, no information on bed roughness of the flume was available and a Manning's n value of $n = 0.0143$ was chosen for the HYDRO3D study. For the Tsujimoto *et al.* (1992) test case, HYDRO3D computed the velocity at the first grid point at the boundary with an appreciably better accuracy than that of Fischer-Antze *et al.* (they selected a Manning's n value of 0.02). Further, HYDRO3D gives a better prediction in the vegetation layer. For both simulations the weighing factors C_{fk} and $C_{f\varepsilon}$ were chosen to be zero; however, it was found that the difference in predictions with and without these additional terms was negligible.

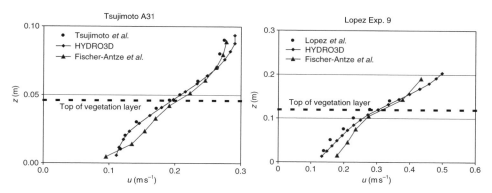

Figure 15.14 Calculated and observed vertical velocity distribution of Tsujimoto et al.'s A31 experiment and Lopez et al.'s Exp. 9.

15.5.2 Results – Rigid emergent vegetation on the floodplain of a compound channel

The experiments undertaken by Pasche (1984) examined the influence of a vegetated floodplain on the horizontal velocity distribution in a compound channel with emergent rigid cylinders. Figure 15.15 shows a comparison of depth-averaged velocities predicted by HYDRO3D and the simulations conducted by Fischer-Antze *et al.* (2001) using SSIIM against observed data. It can be seen that both models give reasonably good predictions and that the reduction of flow velocities through the vegetational elements can be simulated fairly well. The effect of vegetation on the floodplain is evident: additional flow resistance on the floodplain causes a further reduction of discharge capacity, which is accompanied by a steeper velocity gradient in the transition region, higher velocities in the main channel and a shift of the location of highest velocities towards the right bank of the main channel. However, no turbulence-driven secondary flows were simulated and this is a likely attribute of the linear k-ε model. Nevertheless, the major effect of emergent vegetation on floodplains on the lateral velocity distribution could be reproduced with a reasonable accuracy. This is especially important for practical hydraulic problems where the overall hydrodynamic behaviour and discharge capacities are of interest.

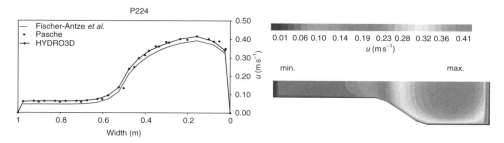

Figure 15.15 Calculated and observed depth-averaged main flow velocities of the compound channel flow with vegetated floodplain at $\lambda_P = 2.69$. (See Plate 2.)

15.5.3 Discussion and conclusion

In all three test cases of flow through vegetation that were simulated using a drag force approach it was apparent that the channel bed roughness seemed to be an important issue from a mathematical point of view (not from a physical one as stated by Nepf, 1999). The reason for this lies in the mathematical formulation and the iterative solution scheme employed. Since the calculation of the velocity field is dependent on the calculation of the drag force, which in turn depends on the calculated oncoming velocity, the drag force is only correctly evaluated at the end of the iterative procedure. Hence, if for instance external forces like the bed shear stress are falsely predicted (due to an over-estimation or underestimation of bed roughness), the velocity at the first grid point is erroneous and hence the evaluation of the drag force is erroneous as well, leading to an incorrect flow field. This was apparent in all simulations of flow through vegetation performed in the course of a validation series (Stoesser, 2002). Furthermore, it is important to note that the inclusion of drag-related sink terms in the momentum equations is physically incorrect, since it only accounts for the advective fluxes, ignoring diffusive fluxes and local pressure gradients. Hence a scheme that includes the effects of mean velocity gradients and pressure differences generated by the vegetational obstacles in the flow would be ideal. An example of how to remedy the above-mentioned shortfall is given below.

15.6 Emergent artificial tree on the floodplain of a compound channel

The effect of emergent vegetation of natural shape on the flow was studied using a single artificial tree placed in the middle of the floodplain. Since no velocities and turbulent quantities were measured, a video animation of tracer spreading due to advection and turbulent mixing was used for a visual comparison. This was employed in order to qualitatively test a hybrid discrete element/drag-force approach together with a Large Eddy Simulation (LES), in order to overcome the limitations of the subgrid drag force as discussed earlier in section 15.5.3. The beneficial use of dye tracing for testing turbulence model performance has previously been demonstrated by Forkel (1995). The primary mechanism for dispersion in shear flows is the variation in convective velocity, which in turn is dependent on the correct treatment of turbulent quantities and hence an accurate description of turbulent Reynolds stresses. The transport of a neutral quantity is described with the advection–diffusion equation (see Chapter 3).

Following the suggestions of Fathi-Maghadam and Kouwen (1997), who have conducted flume experiments on pine and cedar tree saplings and branches for emergent conditions, the momentum-absorbing area is based on total foliage area (in the flow direction) per unit volume. This definition is, however, extended here for use with submerged plant forms of stem/foliage structure where the cross sectional area varies as a function of the plant height. The total surface area of the foliage was thus used instead of the projected frontal area, since a considerable area of foliage or frond is hidden behind the frontal area and this absorbs momentum in addition to the projected area. The hybrid discrete element/drag-force approach is implemented such that the stem of the plant is explicitly modelled by means of the discrete element technique (i.e. blocking pressure cells) and the foliage is represented using the subgrid momentum sink via the drag force analogous to the treatment in RANS models. Figure 15.16 shows the near surface instantaneous velocity distribution. The effect of the modelled discrete stem of the single vegetational element is clearly visible. The results are similar to the flow around a circular cylinder, with a stagnation point in front of the cylinder, a recirculation zone behind the cylinder, and periodic vortices shed downstream of the element and act to decelerate the flow in the wake. Moreover, introducing vegetation into a turbulent channel flow results in additional production of turbulence energy and energy transfer to smaller scales until dissipation into heat occurs. This is represented in Figure 15.17 showing the instantaneous stress tensor S due to the variation of mean velocities. The formation of two horizontal shear layers due to the presence of the vegetational element can be observed. One forms along the rigid stem of the element and another forms on the verge of the branches. The stresses reach a peak at the interfaces and decay in the free flow behind the element.

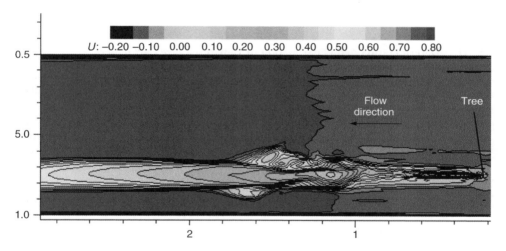

Figure 15.16 Instantaneous velocity distribution, length unit for both axes in metres. (See Plate 4.)

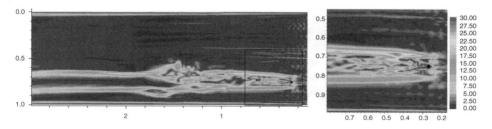

Figure 15.17 Stress tensor distribution (in 1/s), length unit in metres. (See Plate 5.)

15.7 Up-scaling to the field – A practical approach

As pointed out earlier, central to any river design project is the accurate prediction of the flow field in and around riparian vegetation. However, the approaches introduced in the previous sections to model the flow resistance due to vegetation need to be simplified to be used at field scale in order to provide a practical but physically based roughness closure approach for vegetative resistance which involves a minimum of parameter calibration. The German Association for Hydraulic Engineering (DVWK, 1991) recommends a method for defining the roughness parameter λ at large scale. This relates the vegetation geometric properties, including plant diameter and spacing, to the drag force. To implement the method, geometric vegetation parameters of an existing established riparian forest were quantified and used as input to the drag force roughness closure. For the field scale HYDRO3D was used with the classical drag force approach since the hybrid approach would require both considerable mesh resolution and an exact definition of the locations of every single tree along the river reach. This approach was then validated in two stages. First, computed water levels were compared against measured water levels. Second, the floodplain flow velocity for a 1-in-100-year event was measured by dilution gauging to allow the computed floodplain velocities to be verified.

15.7.1 Determination and classification of flow resistance parameters

An established groyne field was used as a reference site in order to quantify the vegetation parameters of the existing plant communities. It was composed of a well-developed riparian forest with a rich mixture of rigid trees and flexible bushes. The vegetation was arranged in strips of different thickness, interspersed with open areas of grass, reeds and nettles. The vegetated strip adjacent to the main channel consisted of a dense group of willows. Other species found were cottonwood trees, elm trees, ash trees, privet, elder and blackberry bushes. An area of 77 m × 45 m was surveyed in terms of vegetation type and its relative location. The vegetation varied over the entire monitoring patch and was classified into five characteristic strips, for which average values of vegetation parameters were determined (Figure 15.18). The patch surveyed was composed of 33 plants and the average vegetative flow resistance parameters per strip are given in Table 15.4.

15.7.2 Floodplain velocity measurements

Velocity measurements are complicated in the case of vegetated floodplains as the plants act as vertical obstructions to the flow. This results in the production of vortices immediately downstream of the vegetational elements and generates a strongly heterogeneous flow field. Hence, dilution gauging was performed to determine the mean velocity u_m as a bulk parameter in submerged or emergent vegetation, as this appears to be a good alternative to conventional point velocity measurement methods (e.g. Nepf, 1999). The gulp injection or integration method was employed and the tracer was released from a boat. Three locations in the groyne field were selected for the release station and sampling stations 1 and 2. The injection station and sampling station 1 (SaSt1) were 54 m apart, whilst the injection station and the sampling station 2 (SaSt2) were 99.5 m apart. The concentration was sampled at

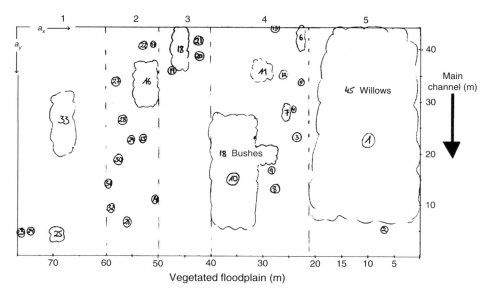

Figure 15.18 Quantification of riparian forest vegetation at the test site (Dittrich and Hartmann, 1998). Strip numbering is given on top of the diagram.

Table 15.4 Summary of vegetation parameters for each monitored strip (see Figure 15.18 for strip numbering)

Strip number	1	2	3	4	5	Average
a_x (m)	18.0	4.5	1.1	5.0	20.0	9.7
a_y (m)	3.5	2.0	1.0	3.5	10.0	4.0
d_P (m)	1.9	1.4	1.5	3.7	13.0	4.3

intervals of 10 s for a 10-minute period. In order to acquire an accurate measurement of the mean flow velocities, the tracer injection and concentration measurement was repeated four times. The centroid of the time–concentration curve was evaluated in order to estimate the travel times from the injection point to the sampling stations. The average value of mean flow velocity was evaluated as $u_m \cong 1.1 \text{ m s}^{-1}$ with a standard deviation of 0.07 m s^{-1}.

15.7.3 Numerical simulations

The test reach of the Rhine between 190.0 and 193.46 km was discretised according to 26 cross sections, which were surveyed in 1986. The grain roughness k_s was calculated from the average gradation curve of the bed material determined through particle sieve analysis. To provide increased confidence in the hydraulic resistance values, the bed roughness of the main channel and the form roughness of the vegetated floodplains were independently verified using two different flow conditions. These correspond to flow situations where both the water level observations and the discharge measurements are available:

High flow event on 2 Nov 1998 where Q = 700 m³ s⁻¹
Flood event on 29 May 1994 where Q = 3040 m³ s⁻¹

During the 1998 event most of the vegetated floodplains were not inundated, hence this flow condition was used to verify the value of main channel bed roughness. For this flow the bed friction parameter was adjusted, and an optimum value of $n = 0.032$ was found. The difference between calibrated ($n = 0.032$) and uncalibrated ($n = 0.033$) values was negligible and good correspondence could be obtained between the measured and computed water surface profiles for this event (Figure 15.19) for both

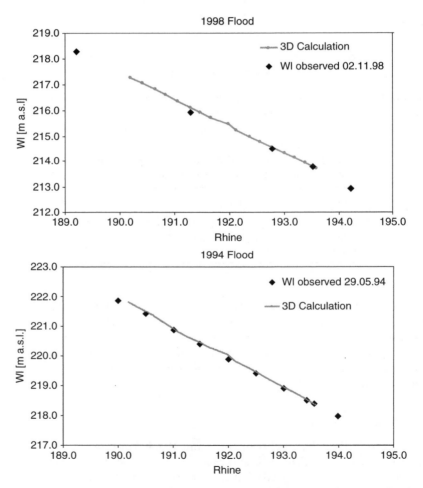

Figure 15.19 Comparison of observed and computed water surface elevation after the main channel bed roughness and the floodplain vegetative parameters were calibrated for the in-bank (1998) and over-bank (1994) flood events.

parameterisations. The vegetative roughness coefficient λ was verified using the larger flow event (3040 m³ s⁻¹) where all floodplains were inundated. The parameter λ was adjusted from 0.11 to 0.082 m⁻¹ in order to achieve a good match between the observed and computed water levels (Figure 15.19). If we assume that the distribution of vegetation is the same as in the reference site (i.e. that a_x and a_y are held constant), this value corresponds to a reduction in average plant diameter from 4.3 to 3.2 m. If we assume that this need for recalibration is predominantly due to errors in determining vegetation geometric properties, this indicates that within a framework where the drag coefficient is assumed to be unity, this is the level of precision necessary to achieve these measurements. These calibrated values for bed roughness and form roughness were used to simulate the 1999 flood. As Figure 15.20 shows, the agreement between observed and calculated water levels for the 1999 event is likewise satisfying and provides a simple split-sample test of model prediction skill.

Figure 15.20 shows the distribution of flow velocities in the cross section of the Rhine corresponding to the test site where the dilution gauging was undertaken. There is a marked reduction in velocity at the main channel/riparian floodplain

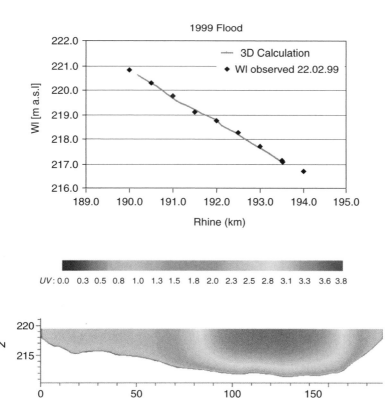

Figure 15.20 *Independent verification of the computed water surface elevation for the 1999 event and the resultant velocity in a Rhine cross section. Velocity magnitude parameter UV is given in m² s⁻². (See Plate 3.)*

interface, where floodplain flow velocities are approximately between 1.0 and 1.3 m s^{-1}. In contrast to the main channel velocity distribution with depth, the floodplain velocity follows a relatively uniform vertical velocity profile. This is in keeping with findings from experimental investigations whereby the velocity profile within both rigid and flexible vegetation layers no longer follows the logarithmic law (Tsujimoto et al., 1992; Nepf and Vivoni, 1999). The uniformly distributed velocity (over floodplain width and depth) as computed, corresponds well with the observed mean velocity from the tracer field tests. Whilst the roughness closure approach presented herein does not eliminate entirely the process of calibration, we have presented and applied a method which is based on the physics of flow resistance and which significantly reduces the modelling uncertainties associated with traditional methods of calibration.

15.8 Summary and conclusions

This chapter has demonstrated how recent research studies are at last beginning to generate an understanding of the interaction of flow with natural vegetation and how this is beginning to inform new conceptual and mathematical representations of these effects that can be incorporated in numerical models. These initial studies confirm the complexity of vegetation–flow interaction and make clear why this particular area of hydraulics has for so long defied effective treatment. Vegetation–flow interactions are, clearly, an area where our hydraulic understanding is weak, yet they are central to many problems of practical interest to hydraulic engineers including flood risk studies, sediment transport studies and the analysis of the hydraulic performance of river restoration schemes. However, the increasing availability of sophisticated velocity measurement equipment (such as acoustic Doppler velocimeters) as standard tools in hydraulic laboratories, and the hosting of CFD code on desktop computers rather than mainframes, has in recent years has allowed considerable progress to be made.

Through analysis of vegetation–flow interaction, and in particular the energy losses generated, we are now beginning to unpack the various components of hydraulic resistance in natural channels. In doing so the primary aim has been to move away from lumped, and often unknowable, friction parameterisations to physically based laws describing each component contributing to the energy loss source term in the Navier–Stokes equations. This process marks the beginning of the development of what could be best described as 'roughness closure' schemes and these are deserving of the same attention in hydraulics as turbulence closure schemes have had over the last 20 years. Given the availability now of a number of adequate, if not perfect, methods of turbulence closure such as the k-ε scheme and LES methods, roughness closure is arguably a more important and less well understood problem. The comparison with turbulence modelling is instructive, as both are in effect subgrid-scale closures with strong scale and model dimension dependency.

Whilst the role of vegetation in generating energy losses is key, this chapter has also shown that vegetation–flow interactions play a critical role in turbulence generation and, hence, sediment transport. It is therefore clear that we need to consider the generation of resistance and turbulence as linked processes in numerical models.

Indeed, this is just one of many potential research areas that require investigation and, although progress is being made, in general our understanding of vegetation–flow interaction is still limited. We need to elucidate many basic hydraulic mechanisms and show how these depend on specific plant biophysical properties. In terms of numerical modelling and the incorporation of these processes into CFD codes, we require better model validation data sets for both flume and field situations. Moreover, in beginning to develop roughness closure methods driven by physically measurable plant parameters we need to develop methods to generate the necessary input data, perhaps through remote sensing. For example, Mason *et al.* (2003) have used airborne laser altimetry data to determine vegetation height at a horizontal resolution of ~5 m and then used this information to directly calculate the Manning–Strickler coefficient at this scale using a number of literature-based techniques. Finally, although progress has been made in understanding the hydraulics of flow–vegetation interaction, almost no research has yet been conducted into the influence of vegetation on sediment transport and this is likely to be a key area for studies of the hydraulic behaviour and morphologic development of river restoration schemes.

In summary, the hydraulic impact of vegetation is a new and exciting area of hydraulics research where there is considerable scope for the development of better empirical understanding and theory. These tasks need to be pursued in conjunction with a programme of numerical model development both as a further facet of scientific enquiry and in order to translate the knowledge gained into practical tools for hydraulic engineers.

Acknowledgements

We are grateful to Professor Dittrich of University of Braunschweig (Germany) and to Othmar Hupmann of the Gewässerdirektion Südl for providing data and valuable support for the Rhine case study. We are indebted to Professor Torum of Trondheim University for his interest and invaluable help in the experiments conducted with 1:10 scale replicas of a species of kelp. We would like to thank both SINTEF for use of their laboratory facilities for these experiments at the Norwegian Hydrotechnical Laboratory and Dr N.R.B. Olsen (University of Trondheim) who coordinated this visit. We acknowledge the assistance of G. Challinor and G. Westmore of Cardiff University in the experimental study conducted with different real and simulated grasses. Finally we would like to thank C. Wigley for his proof reading.

References

Bertram, H.-U. 1985. Über den Abfluß in Trapezgerinnen mit extremer Böschungsrauheit. Mitteilungen aus dem Leichtweiß-Institut für Wasserbau, TU Braunschweig, Heft 86.

Breuer, M. 1998. Numerical and modelling influences on large eddy simulations for the flow past a circular cylinder. *International Journal of Heat and Fluid Flow*, 19: 512–521.

Brunet, Y., Finnigan, J. and Raupach, M. 1994. A wind tunnel study of air-flow in waving wheat: Single-point velocity statistics. *Boundary Layer Meterology*, 70: 95–132.

Burke, R.W. and Stolzenbach, K.D. 1983. Free surface flow through salt marsh grass. *MIT-Sea Grant Report MITSG 83-16*, Cambridge, MA. USA. 252ff.

Choi, S.-U. and Kang, H. 2001. Reynolds Stress Modeling of Vegetated Open Channel Flows. XXIX IAHR Congress Conference Proceedings, Beijing, China. Theme D. Vol. 1, 264–269.

Chow, V.T. 1959. *Open Channel Flow*. McGraw-Hill Book Company, Singapore.

DEFRA & EA. 2001. Fluvial Estuarine and Coastal Processes Theme – Sediments and Habitats Concerted Action. *Concerted Action Report*. August.

Dittrich, A. and Hartmann, G. 1998. *Untersuchungen zum Vorlandabtrag zwischen Maerkt und Karpfenhod*. Schlussbericht. Institut für Wasserwirtschaft und Kulturtechnik. Universitaet Karlsruhe.

Dubi, A.M. 1995. Damping of water waves by submerged vegetation: A case study on laminaria hyperborea, PhD Thesis, University of Trondheim, Norway.

Dunn, C., Lopez, F. and Garcia, M. 1996. Mean flow and turbulence structure induced by vegetation: Hydraulic Engineering Series No. 51, UILU-ENG 96-2009, Department of Civil Engineering, University of Illinois at Urbana-Champaign, Urbana, Illinois, USA.

DVWK. 1991. Hydraulische Berechnung von Fließgewaessern. Merkblaetter. 220/1991. Verlag Paul Parey. Hamburg und Berlin.

Ervine, D.A. and Ellis, J. 1987. Experimental and computational aspects of overbank floodplain flow. *Transactions of the Royal Society of Edinburgh: Earth Sciences*, 78: 315–325.

Fathi-Maghadam, M. and Kouwen, N. 1997. Non-rigid, non-submerged, vegetative roughness on floodplains. *Journal of Hydraulic Engineering*, ASCE, 123: 51–57.

Fischer Antze, T., Stoesser, T., Bates, P.B. and Olsen, N.R. 2001. 3D Numerical modelling of open-channel flow with submerged vegetation. *IAHR Journal of Hydraulic Research*, 39: 303–310.

Forkel, C. 1995. Die Grobstruktursimulation abgeloester turbulenter Stroemungs,- und Stoffausbreitungsprozesse in komplexen Geometrien. Mitteilungen Nr. 102. Institut fuer Wasserbau und Wasserwirtschaft. RWTH Aachen.

Freeman, G.E., Copeland, R.R., Rahmeyer, W. and Derrick, D.L. 1998. Field Determination of Manning's *n* Value for Shrubs and Woody Vegetation. ASCE Wetlands Engineering and River Restoration Conference 1998, Engineering Approaches to Ecosystem Restoration. CD-ROM.

Graf, W.H. and Chhun, V.H. 1976. Manning's roughness for artificial grasses. *Journal of Irrigation and Drainage Division*, ASCE, 102: 413–423.

Green, J.E.P. and Garton, J.E. 1983. Vegetation lined channel design procedures. *Transactions of ASAE*, 26: 437–439.

Greenhill, R.K. and Sellin, R.H.J. 1993. *Development of a Simple Method to Predict Discharges in Compound Meandering Channels*, Proceedings of the Institution of Civil Engineers – Water Maritime and Energy, 101: 37–44.

Horritt, M.S. and Bates, P.D. 2001. Predicting floodplain inundation: Raster-based modelling verses the finite element approach. *Hydrological Processes*, 15: 825–842.

Horton, R. 1933. Separate Roughness coefficients for channel bottom and sides. *Engineering News Record*, p. 652, November.

Kaiser, W. 1984. Fließwiderstandsverhalten in Gerinnen mit durchströmten Ufergehölzen. Wasserbaumitteilungen der TH Darmstadt, Heft 23.

Kouwen, N. and Li, R.M. 1980. Biomechanics of vegetated channel linings. *Journal of the Hydraulics Division*, ASCE, 106: 1085–1203.

Kouwen, N., Unny, T.E. and Hill, H.M. 1969. Flow retardance in vegetated channels. *Journal of Irrigation and Drainage Division*, ASCE, 95: 329–342.

Kutija, V. and Hong, H. 1996. A numerical model for addressing the additional resistance to flow introduced by flexible vegetation. *Journal of Hydraulic Research*, 34: 99–114.

Larsen, T., Frier, J.-O. and Vestergaard, K. 1990. Discharge–stage relations in vegetated Danish streams. In White, W.R. (ed.), *International Conference on River Flood Hydraulics*, John Wiley & Sons, Chichester, 187–195.

Lopez, F. and Garcia, M. 1997. *Open Channel Flow Through Simulated Vegetation: Turbulence Modelling and Sediment Transport*. Hydrosystems Laboratory, Department of Civil Engineering, University of Illinois.

Lopez, F. and Garcia, M. 2001. Mean flow and turbulence structure of open-channel flow through non-emergent vegetation. *Journal of Hydraulic Engineering*, ASCE, 127: 392–402.

Lotter, G.K. 1933. Considerations on hydraulic design of channels with different roughnesses of walls. *Transactions All Union Scientific Research*, Institute for Hydraulic Engineering, Leningrad, No. 9: 238–241.

Mason, D., Cobby, D.M., Horritt, M.S. and Bates, P.D. 2003. Floodplain friction parameterization in two-dimensional river flood models using vegetation heights derived from airborne scanning laser altimetry. *Hydrological Processes*, 17: 1711–1732.

Masterman, R. and Thorne, C.R. 1992. Predicting influence of bank vegetation on channel capacity. *Journal of Hydraulic Engineering*, ASCE, 118: 1052–1058.

Mathey, F., Fröhlich, J. and Rodi, W. 1998. *Large Eddy Simulation Over a Matrix of Surface-mounted Cubes*. Proceedings of Workshop on Industrial and Environmental Applications of Direct and Large Eddy Simulation, Istanbul, Turkey.

Meijer, D. and Van Velzen, E.H. 1999. *Prototype-scale Flume Experiments on Hydraulic Roughness of Submerged Vegetation*, XXVIII IAHR Congress Conference Proceedings, Graz, Austria. CD-Rom Theme D1. Abstract Volume p. 221.

Mertens, W. 1989. Zur Frage hydraulischer Berechnungen naturnaher Fließgewässer. Wasserwirtschaft 79, Heft 4: 170–179.

Nakagawa, H., Tsujimoto, T. and Shimizu, Y. 1992. *Sediment Transport in Vegetated Bed Channel*. 5th International Symposium on River Sedimentation. Karlsruhe.

Naot, D., Nezu, I. and Nakagawa, H. 1996. Hydrodynamic behaviour of partly vegetated open-channels. *Journal of Hydraulic Engineering*, 122: 625–633.

Nepf, H.M. 1999. Drag, turbulence, and diffusion in flow through emergent vegetation. *Water Resources Research*, 35: 479–489.

Nepf, H. and Vivoni, E.R. 1999. *Turbulence Structure in Depth-Limited Vegetated Flow: Transition between Emergent and Submerged Regimes*, XXVIII IAHR Congress Conference Proceedings, Graz, Austria. CD-Rom Theme D1. Abstract Volume p. 217.

Nezu, I. and Onitsuka, K. 2001. Turbulent structures in partly vegetated open-channel flows with LDA and PIV measurements. *Journal of Hydraulic Research*, 39(6): 629–642.

Nuding, A. 1991. Fließverhalten in Gerinnen mit Ufergebüsch. Wasserbaumitteilungen der TH Darmstadt, Heft 35.

Olsen, N.R. 2002. *SSIIM Users Manual*. Norwegian University of Science and Technology. <www.bygg.ntnu.no/~nilsol/ssiimwin>.

Pasche, E. 1984. Turbulenzmechanismen in natürlichen Fliessgewässern und die Möglichkeit Ihrer mathematischen Erfassung. Mitteilungen Institut für Wasserbau and Wasserwirtschaft, No. 52. RWTH Aachen.

Patankar, S.V. 1980. *Numerical Heat Transfer and Fluid Flow*. Hemisphere Publishing Corporation. McGraw-Hill, New York.

Petryk, S. and Bosmajian, G. 1975. Analysis of flow through vegetation. *Journal of Hydraulic Engineering*, 101(7): 871–884.

Plate, E.J. and Quraishi, A.A. 1965. Modeling of velocity distributions inside and above tall crops. *Journal of Applied Meteorology*, 4: 400–408.

Raupach, M.R. and Shaw, R.H. 1982. Averaging procedures for flow within vegetation canopies. *Boundary-layer Meteorology*, 22: 79–90; Conditional Statistics of Reynolds stress in rough-wall and smooth-wall turbulent boundary layers. *Journal of Fluid Mechanics*, 108: 363–382.

Ree, W.O. 1958. Retardation coefficients for row crops in division terraces. *Transactions of ASAE*, 1(1): 78–80.

Rodi, W. 1980. *Turbulence Models and their Application in Hydraulics: A State of the Art Review*. IAHR Monograph. AA Balkema, Rotterdam.

Sargent, R.J. 1979. Variation of Manning's n roughness coefficient with flow in open river channels. *Journal of Institution of Water Engineers and Scientists*, 33: 290–294.

Sellin, R.H.J., Wilson, C.A.M.E. and Naish, C. 2001. Model and prototype results for a sinuous two-stage river channel design, *Journal of Chartered Institution of Water and Environmental Management*, 13: 207–216.

Shimizu, Y. and Tsujimoto, T. 1994. Numerical analysis of turbulent open channel flow over a vegetation layer using a k-ε turbulence model. *Journal of Hydroscience and Hydraulic Engineering*, 11(2): 57–67.

Stoesser, T. 2002. Development and validation of a CFD code for open-channel flows, PhD Thesis, Department of Civil Engineering. University of Bristol.

Temple, D.M. 1987. Closure of velocity distribution coefficients for grass lined channels. *Journal of the Hydraulics Division*, ASCE, 113: 1224–1226.

Tsujimoto, T. and Kitamura, T. 1998. *Interaction Between River Bed Degradation and Growth of Vegetation in Gravel Bed River*. ASCE Water Resources Engineering Conference, August, 2: 580–585.

Tsujimoto, T., Shimizu, T. and Okada, T. 1991. Turbulent structure of flow over rigid vegetation-covered bed in open channels. *KHL Progressive Report 1*. Hydraulic Laboratory, Kanazawa University, Japan.

Tsujimoto, T., Shimizu, Y., Kitamura, T. and Okada, T. 1992. Turbulent open-channel flow over bed covered by rigid vegetation. *Journal of Hydroscience and Hydraulic Engineering*, 10: 13–25.

USDA. 1947. *Handbook of Channel Design for Soil and Water Conservation*. United States Department of Agriculture – Soil Conservation Service SCS-TP-61.

Wilcox, D.C. 1988. Reassessment of the scale determining equation for advanced turbulence models. *American Institute of Aeronautics and Astronautics Journal*, 26: 1299–1310.

Wilson, C.A.M.E. and Horritt, M. 2002. Measuring the flow resistance of submerged grass. *Hydrological Processes*, 16: 2589–2598.

Wilson, N.R. and Shaw, R.H. 1977. A higher order closure model for canopy flow. *Journal of Applied Meteorology*, 16: 1198.

Wormleaton, P.R., Allen, J. and Hajipanos, P. 1982. Discharge assessment in compound channel flow. *Journal of Hydraulic Division*, ASCE, 108: 975–994.

Wu, F., Shen, H. and Chou, Y. 1999. Variation of roughness coefficients for unsubmerged and submerged vegetation. *Journal of Hydraulic Engineering*, ASCE, 125: 934–942.

16
Ecohydraulics: A new interdisciplinary frontier for CFD

M. Leclerc

16.1 The river system in perspective

The *river continuum concept* states that the evolution of abiotic variables along a spatial gradient (upstream/downstream) is the main feature defining fish assemblages and distribution in streams (Vannote *et al.*, 1980). The notion of riverine habitat belongs to this conceptual framework. The biological functions of aquatic species and their communities are accomplished locally in a stream reach or river facies, but may vary in location in the hydrographic network depending on the spatial distribution of limiting abiotic factors and their relationship with hydrology and water use. The habitat requirements can include open sea or coastal zones for particular phases of the species life cycle. For most fish species, one can distinguish the following life phases and the corresponding habitats: *reproduction* and *egg incubation* in the spawning grounds, *fry emergence* and *feeding of the juveniles* (rearing habitats), *upstream and downstream migrations* often defined by the swimming capability of fish to pass through obstacles such as steep rapids, falls or civil works, *rest*, *predator avoidance* and *overwintering* which require some shelter habitat.

From a numerical hydraulics standpoint, the simulation of abiotic factors defining riverine habitats represents a fairly new challenge (Leclerc, 2002) because of the increased level of accuracy required to provide results at a local microhabitat scale, corresponding to the instream flow needs of aquatic species or in studies of fish behaviour. The main variables traditionally considered to represent potential aquatic habitats are physical (velocity, stage, depth) and morphological (including riverbed composition). But distance to cover, hydraulic gradients, possibility of stranding, territoriality and competition are all factors that influence physical habitat quality and which can be investigated using CFD models. Physico-chemical attributes of water can contribute to actualize or inhibit this potential. Again, study of such variables can benefit from CFD through transport–diffusion–kinetics models.

In this chapter, we will adopt a general perspective with regard to the ecological aspects of habitat modelling. We will try to demonstrate the conceptual framework and mathematical background defining the notion of habitat and focus on the new challenges offered to CFD practitioners, specifically those involved in environmental hydraulics. Even though the mathematical formulation of abiotic preferences of aquatic species and of habitat availability is fairly simple compared to the sophisticated algebraic systems defining the dynamics of the flow, several underlying hypotheses limit drastically the interpretation of results. The collaboration of knowledgeable ecologists with numerical modellers is absolutely essential to avoid misunderstanding of model outcomes.

Some of the proposed methods or tools used in microhabitat modelling methodology are fairly new, even in the academic context, and they still need some proofing in operational situations (e.g. development of habitat time series). For this reason, among others, this applied science has a very promising future for the next generation of hydraulicians. Public policies will contribute significantly to the emergence and development of this methodology. For example, the European Water Framework Directive (WFD) requires catchment-wide habitat assessment. In the UK, habitat assessments are required for designing abstraction management plans. In the US, Environmental Impact Assessment is the driver, especially in relation to the relicencing of hydroelectric power plants.

16.2 Conceptual modelling of habitats

Among existing types of models representing individuals, populations, communities, ecosystems and/or their habitats, one can identify microhabitat modelling, population dynamics, bioenergetics, stock recruitment, nutrient dynamics, individual-based ecological modelling, etc. The mathematical formulation of these models varies drastically from one option to another but some of them do not need a local representation of the flow domain (e.g. population dynamics) and consequently their mathematical framework does not rely on Partial Differential Equations (PDEs) to take into account the distributed aspect of the flow field. However, some avenues relate to distributed variables, such as the Instream Flow Incremental Methodology (IFIM) with its Physical Habitat Simulation Module (PHABSIM) proposed by Bovee (1978) and Bovee and Milhous (1978). However, this approach is based on a simplistic definition of the flow based on traditional 1D hydraulic tools originally developed for engineering purposes and mostly used to predict water levels and rough estimates of mean velocities. More recently, Leclerc *et al.* (1994, 1995, 1996) and several other contributors have started to make use of modern CFD and discretization methods to provide more detailed flow fields to river ecologists. This improvement has gained in popularity in the last decade and it has become a strong component of the state-of-the-art in this domain (Leclerc *et al.*, 2003). Several chapters of this book present in detail all aspects related to

CFD. In this chapter, we will focus on physical habitat modelling and aggregate formulations which suppose that fields of abiotic variables are available through field measurements and/or numerical simulation, and that these variables are discretized and can be numerically processed in terms of habitat values for each target species.

Habitat modelling attempts to predict the quality, usability or suitability of physical conditions but does not aim at predicting explicitly population dynamics or the presence/absence of species. Usually, habitat modelling does not explicitly take into account the ecological processes underlying the preference formulae defining the calculated values. Mostly based on the observation of presence or absence of the species in specific abiotic conditions, habitat models look more like "black boxes" than sophisticated conceptual tools (Leclerc, 2002). Thus, such models usually remain relatively simplistic and sometimes frustrating for river ecologists who often perceive them as highly reductionist engineering tools. In fact, several underlying hypotheses, unfortunately not always recognized and formulated, and rarely verified with proper validation protocols, can limit the predictive success of this methodology. Nevertheless, scientists who have used habitat models to improve their knowledge of rivers and related ecology would admit that such techniques also helped to introduce ecological considerations into the negotiation process of water resources allocation in a quantitative way.

16.2.1 The paradigm of numerical habitat modelling

The three main components of aquatic habitat models are (Figure 16.1):

1. One or more *living species* (fish, invertebrate, plant) with their specific abiotic preferences for each flow-dependent phase of their life cycle.
2. A proper description of *environmental hydraulics* and of other abiotic factors contributing to the habitat occurrence.
3. A *drainage basin including water uses* and *allocation schemes* contributing not only a hydrological regime, natural or influenced, but also physico-chemical water attributes such as water quality, substrate grain size, temperature, nutrient loads and contaminants.

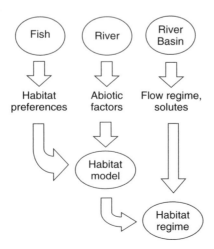

Figure 16.1 *The paradigm of habitat modelling (after Leclerc, 2002).*

Distributed habitat models are the result of the post-treatment of the flow field (item 2) by making use of specific transfer functions defining the habitat *preferendum* (item 1). Taking into account the hydrological regime (item 3), whether influenced or natural, provides an essential relationship between habitat availability at the reach scale and catchment influences.

16.2.2 Modelling the aquatic species

16.2.2.1 Ecological role and selection of abiotic factors for habitat models

Abiotic factors can play several ecological roles during a life cycle and thus contribute to define the habitat value. For example, velocity exerts a very important influence at different phases of the life cycle of salmonids, and for several other fish species. For example, the summer feeding behaviour of juvenile salmon (0^+- to 2^+-years old) relies very much on the speed of the current carrying the food drift (invertebrates). The fish catches its prey by swimming against the current, often by climbing from its stand-still watching position into the recirculation zone located behind its "home rock", through the turbulent boundary layer often up to the water surface. To achieve a net gain for its own growth, the total energy expenditure (including the basal metabolism) has to remain below the amount of energy contained in the prey.

Some debate can take place about which velocity to consider as a priority in habitat models: the *nose (or snout) velocity* or the *mean value over the water column*. This choice is highly relevant to the selection of a hydrodynamic modelling option, especially its discretization scheme (2D versus 3D). Some modellers argue that only 3D models can describe the vertical profile of the flow which would allow one to analyse accurately the fish behaviour, or specifically for understanding the fish bioenergetics (feeding and resting). Supporters of 2D models argue that the mean velocity is more integrated with respect to the feeding ecology of fish even though no direct or indirect description of the vertical velocity profile is provided by 2D tools. They also argue that even 3D models fail to take into account the random distribution of larger substrate elements and their transient effects (turbulence) on the flow patterns close to the bottom.[1] In fact, the nose velocity is a fish property rather than a flow characteristic. Moreover, this specific range of velocity occurs somewhere at the bottom of the turbulent boundary layer, and in practical decision-making contexts, one does not really need to know the exact position where the fish can rest between two prey catches. However, if one seeks to develop detailed bioenergetic versions of habitat model (e.g. individual-based ecological modelling) such a detailed 3D representation might become necessary.

Similar debates could also be mentioned for substrate grain size and composition which are also very important for feeding because mid-size rocks and stones provide very local flow features (microhabitats) which allow the fish to rest between its attacks on prey. Moreover, spawning activity, egg incubation, predator avoidance and fish over-wintering (notably, in cold climates) are partly governed by the grain size and composition of substrate elements which provide shelter between interstices

---

[1] Section 16.2.3.5 proposes an interpretation of the inherent flow complexities that can contribute to habitat occurrence in rivers and the abilities of hydrodynamic models to represent them.

and thus, oxygenation of interstitial waters. Granulometric distribution of substrates is also important in order to avoid the presence of fine materials which result in the filling up of interstices and, occasionally, the deleterious paving of the riverbed. This is harmful not only for reproduction activity, but also for feeding of fish because the clogging of substrates restricts the production of benthic invertebrates that represent the main food source for many fish species (Milhous, 1996).

Traditionally, microhabitats of fish or other species living in river are mostly modelled by three key abiotic variables: mean velocity of the water column, depth and riverbed substrate (mean diameter) (Bovee, 1978). These generic variables can be considered as the basic template of aquatic habitats: velocity mainly determines the bioenergetics of individuals through feeding and swimming; substrate influences the shelter opportunities, the production of invertebrates and the quality of spawning grounds; and depth relates to light penetration and may influence the predation process.

Other important physical and/or physico-chemical attributes, such as water temperature, availability of light in the water column or close to the riverbed, turbidity can also play an important role and must often be considered (Bechara *et al.*, 2003). The slope of the riverbed and the presence of tree cover are other examples of variables that may be measured or modelled for a more thorough assessment (Hardy and Addley, 2003; Morin *et al.*, 2003). In the case of larger fluvial systems with long wind fetches, wave climate can also become an important habitat variable, as can the organic content of sediments (Morin *et al.*, 2003).

Lamouroux and Souchon (2002) have proposed aggregated abiotic variables as the main explanatory factors of habitat distribution over the river network (mesohabitat). For a given segment, the governing processes belong to the flow regime defined by the mean Froude number of the reach, the average rate of flow per unit of width (specific discharge) and the median grain size. The Froude number reflects the lentic (slow currents)/lotic (fast currents) character of the flow regime which obviously connects strongly with distinct ecological strategies of fish species. As to the specific discharge, this variable aggregates in one single term the combined effects of velocity and depth, both dominant factors of the standard microhabitat. Mean substrate size obviously reflects the role of this variable at the local microhabitat scale (shelter, home rock, feeding position).

16.2.2.2 Defining the preference range for abiotic factors among species

The classic approach (Figure 16.2) to quantifying habitat value consists of estimating local habitat indices (the so-called preference curves) based on a knowledge of the optimum ranges of abiotic conditions (the so-called *preferendum*) for the targeted species during a specific life stage (Bovee, 1982). In a river, the life stages considered often relate to those potentially influenced by variations of the flow regime (Bovee, 1978). A Habitat Suitability Index (HSI or Global index) is thus calculated by assigning a weight (in the form of a power factor), between 0 and 1, to each abiotic variable defining habitat, and combining these values in a global index, usually by way of a simple geometric or weighted geometric mean (equation (16.1)). The geometric mean is often used because it implies that each independent value can play the role of a limiting factor (i.e. if the value of one independent variable is 0,

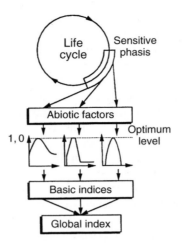

Figure 16.2 Building a classical fish preferendum model.

HSI = 0). If the local index for a given variable is one, then the requirement is completely met for the target species. Due to its great simplicity, such an algebraic formulation must be interpreted with caution. Further explanations with regard to this will be provided later (section 16.2.2.3).

$$\mathrm{HSI}(x, y) = I_V^a \cdot I_H^b \cdot I_S^c \quad \text{with} \quad a + b + c = 1.0 \tag{16.1}$$

where I_V, I_H and I_S represent basic indices for velocity (V), depth (H) and substrate (S), respectively and a, b, c are weights powering the basic indices and influencing the relative ecological importance given to each basic index in the model.

Most often, the weights are assumed to have an equal influence upon each of the factors considered ($a = b = c = 1/3$). In some cases, statistical techniques such as principal component methods or regression can be employed to discriminate the relative explanatory importance of each abiotic factor upon the fish distribution. The percent of residual explanation of each variable can then be used as the weighting factor (Leclerc et al., 1995). However, as the mentioned variables are obviously autocorrelated through the logarithmic representation of the velocity profile forming the turbulent boundary layer, one cannot be sure of the adequacy of this approach and this criticism has often been mentioned as a drawback of this modelling approach (Shirvell, 1986).

Each basic index forming the HSI formulation is represented by preference curves. They can be built for the range of abiotic conditions encountered, based on field observation data of the local fish presence or absence in the stream, and on the value of controlling abiotic factors measured at the corresponding location. Fish observations are realised by snorkeling or electro-fishing means (Schneider and Jorde, 2003). In some situations where observation is restricted by turbidity, sediment load or water colour, expert advice can replace measurements. A mixed measurement–expert approach can also be mobilized for adjusting empirical curves, especially in the extreme ranges of physical variables, and used to obtain more meaningful results with respect to effective habitat value.

When building the preference curves (or basic indices), one must take into account the availability of the entire range of flow conditions within the flow domain. More frequent conditions might artificially appear preferable if very abundant. If the statistical method used to determine the *preferendum* takes into account only the conditions where the presence of fish was observed, one obtains a "utilization curve" which is still a useful piece of ecological information but does not strictly represent the *preferendum*.

With field data available, the most common methodology (Figure 16.3) to establish the *preferendum* consists firstly of building histograms of the presence of fish as a function of the abiotic variables observed at the fish location. A second histogram representing the total number of presences and absences of fish with respect to local abiotic conditions is also prepared. The latter histogram is considered to represent the availability of conditions. The *preferendum* is obtained by dividing the first histogram by the second (classwise) and normalizing the result by setting the maximum value to 1.0.

In the hypothetical situation proposed in Figure 16.3, one can observe that the maximum number of fish presences occurs in class 3 of the independent variable while the maximum availability of abiotic conditions occurs in class 4. It is also interesting to observe that class 2 offers half of the amount of river space available in class 4 but the number of fish presences in class 2 is superior to class 4 which means a greater fish density, and consequently more favourable habitats. As already stated, in Figure 16.3, a normalized graph of the fish presence histogram would be interpreted as a "utilization curve", which, when employed instead of the preference curve, could lead to inaccurate results.

One of the main criticisms of this classic HSI approach is that it does not take into account the fact that the so-called independent variables are not truly independent (e.g. flow and depth). The information content provided by some of the abiotic variables may be redundant.

In spite of its drawbacks, the HSI approach is still commonly used and can often be validated by the measurement of the density distribution of the target species in the field (e.g. Guay *et al.*, 2000). However, it remains difficult to transfer this local modelling parameterization to other rivers.

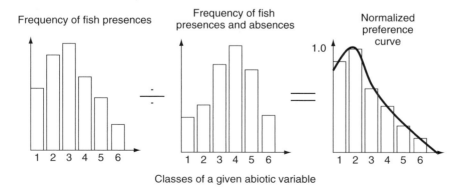

Figure 16.3 *The general methodology for establishing fish preference curves.*

16.2.2.3 The interpretation of preference curves

Let us look more carefully at the interpretation of the algebraic formulation of HSI, specifically when considering the combination of a geometric mean and basic indices defined in the range of 0–1. As already mentioned, a zero value for a given basic index produces a zero HSI value as a result, which means that this index plays the ecological role of a limiting factor. A value of 1.0 for a basic index does not influence the HSI value, which means that local conditions comply with the fish abiotic preferences. These statements can play an important role when building preference curves. If no or a very small number of fish are observed in a given class of abiotic variable, does this mean that this range is restrictive for the presence of fish? Possibly not. For example, the river reach might not be occupied optimally by the fish population considered. In turn, this could be related to the presence of predators, a low population with respect to the current habitat availability during the field characterization, or to adverse factors such as the presence of contaminants inhibiting biological functions. Moreover, the habitat availability for this particular phase of the fish life cycle may not be the limiting factor for the population size.

Consequently, it is recommended that one adapts the preference curves obtained from the standard method in the extreme ranges of the variables by taking into account the existing ecological knowledge of the species under study. Expert advice is recommended in this case. Ending the curve at an intermediate value (e.g. 0.3–0.5) could be acceptable in order to keep these areas under consideration, especially if all other basic indices behave satisfactorily.

16.2.2.4 Defining habitat value with a probabilistic approach

Probabilistic multivariate approaches such as logistic regression have recently been described in the literature and offer better possibilities for model validation (see section 16.3) and could even allow for better model transferability from one river to another (Guay *et al.*, 2000, 2003; Boisclair, 2003; Parasiewicz, 2003). Logistic regression can be used to model dichotomous variables (presence–absence) as a function of independent variables. If independent variables are standardized, logistic regression coefficients can be compared to identify which variables cause the greatest increase or decrease in the odds ratio. The protocol for physical habitat characterization is essentially the same in this case as for HSI and preference curves. A Probabilistic Habitat Index (PHI), which refers to the probability of presence of individuals of the target species, can thus be calculated. These results can then be used to assess microhabitat or mesohabitat (reach value, see Parasiewicz, 2003).

$$\text{PHI}(x, y, Q) = \frac{1}{1 + e^{-P(V,H,S)}} \qquad (16.2)$$

where P is a polynomial composed of linear and/or quadratic terms that are functions of controlling abiotic variables V, H, S, etc. where V is the velocity, H the depth and S the substrate grain size (Figure 16.4).

If the application of logistic regression appears to give promising results (Guay *et al.*, 2003), it must, however, be kept in mind that the interpretation of the regression

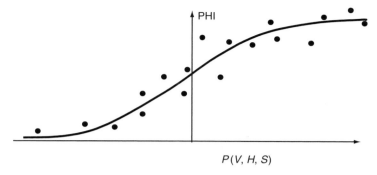

Figure 16.4 The logistic regression approach to the aquatic species preferendum.

coefficients is different than that of linear Ordinary Least Squares (OLS) models. Moreover, it must be noted that the coefficients of determination used to assess the goodness of fit of logistic regression models are also different from those of OLS models. The so-called pseudo-R^2 metrics are based on log-likelihood ratios (e.g. McFadden R^2) and are not the equivalent of the OLS R^2 used in linear regression models (Scheiner and Gurevitch, 2001).

The order (linear and/or quadratic) of the algebraic terms forming the polynomial P can have some importance for the predictive power of the model. The linear term often establishes a certain proportionality between the abiotic variable considered (independent variable) and the probability of species presence, especially in the lower range of the variable. The addition of quadratic terms as a function of the same variables transforms the polynomial P into a parabola which will damp the ever-increasing or ever-decreasing effects of the linear terms. This statement has a corollary regarding the range of value of explanatory variables considered by the measurement protocol. If this range is limited to a narrow domain of representativeness, the linear component might be sufficient to reproduce accurately the fish distribution in this limited region but with low predictability outside this specific range.

16.2.2.5 *Fuzzy logic for modelling habitat* preferendum

Alternate approaches have also been proposed to model fish preference, based on fuzzy or imprecise information available on the selective behaviour of target species (Jorde *et al.*, 2001; Schneider and Jorde, 2003). Unlike previous methods that do not really allow one to consider the existing knowledge being offered by fisheries biologists, the fuzzy-rule-based method operates with combinations of qualitative and semi-quantitative criteria to which a suitability level is attributed according to specialists who try to achieve a consensual interpretation. The procedure starts by setting up check lists with possible combinations of relevant physical criteria and lets the specialists define in natural language whether habitat quality is low, medium or good for each of the proposed combinations. The basic information used is often classified by using common language (low, medium, fast velocity) (Figure 16.5). The

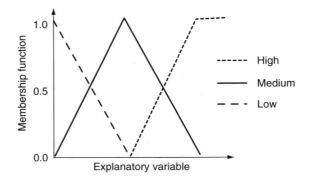

Figure 16.5 *Illustration of fuzzy delineation of habitat variable classes (adapted from Schneider and Jorde, 2003).*

number of combinations depends upon the number of variables considered, J, and the number of linguistic classes, k_j, defining each of them.

$$N = \prod_{j=1}^{J} k_j \qquad (16.3)$$

In order to take into account the imprecise definition of the intervals of classification, fuzzy logic uses overlapping "membership functions" allowing one to classify a given variable value in a combination of two or more intervals of definition. The numerical habitat value being associated with a combination of basic explanatory variables takes into account the value of the membership functions which serve further as weighting factors. Thus, in practice, a given value of velocity can be partly low (e.g. 0.25) but mostly medium (0.75).

For further explanations about the numerical implementation of this method, the reader is advised to consult Schneider and Jorde (2003). This method has proved to perform better at the validation step than the standard *preferendum* method. It is fully implemented in a software tool called CASIMIR (Computer Assisted Simulation Model for Instream Flow Requirements) which forms an integrated toolbox for habitat simulation in rivers.

16.2.2.6 *Guilds: Aggregating fish* preferendum *models for similar species*

Most habitat preference models have been developed for single species and these models are considered specific to a particular river with low transferability to other, even similar rivers. By combining data sets collected on different, but similar, river systems, it is possible to build more generic models offering a better representation of the entire range of preference of the target species. It is even possible to combine data sets formed by different species sharing similar ecological behaviour and niches. For example, Lamouroux and Souchon (2002) have suggested the use of *guilds*, a well-known ecological concept that allows for the combination of a number of species and even of various life stages that have similar habitat requirements. A multi-specific habitat model can thus be built. Grouping criteria are generally based on the assumption that those species share common feeding or reproduction strategies. Guild *preferenda* can be associated with more general reach characteristics such as riffle, runs, flats and pools (Lamouroux and Souchon, 2002). This approach may also allow for inter-basin transfer of *preferendum* parameters. This gain is counterbalanced by a loss of precision for a given species within a guild. The main advantage of this transferability is that general guild preference curves can be used in sites with scarce local data.

16.2.3 Characterizing and/or simulating abiotic factors controlling habitats

Data acquisition for some of the abiotic factors describing habitats is possible via field surveys, but in many cases it is not feasible to do so and simulation model outputs are used instead (e.g. detailed spatial velocity fields, temperature distribution, wave climate in larger systems). In fact, a combination of field data and simulation results is mostly employed. For example, a hydrodynamic simulation requires field measurements such as the substrate composition (roughness), discharge values, stage–discharge relationships at open boundaries, and topography to feed into the model.

16.2.3.1 The classical 1D approaches

One-dimensional or semi two-dimensional models of the PHABSIM type (Bovee, 1978, 1982; Bovee and Milhous, 1978; Bovee *et al.*, 1998; EVHA simulator by Ginot *et al.*, 1998) have been and are still used by specialists. This type of model, which constitutes the most popular module of the IFIM (proposed by the US Geological Survey) modelling system, is mainly characterized by field measurements at a limited number of sampling points spatially organized on transects (Figure 16.6). Each sampling point is centred on a rectangular cell and this discretization scheme forms a structured grid covering the entire flow domain. This approach aims to represent the global variability of the morphology and the flow characteristics (depth, velocity) over a flow domain. Typically, a minimum number of transects (4–6) is necessary to capture a minimum number of different river *facies* within a given reach (succession of pools and riffles, meanders). The position of sampling points requires exact location in order to ensure some positional consistency between flow events. Whilst this logistic requirement does not represent a severe constraint for small shallow streams, it becomes seriously limiting for large river systems. Moreover, the riverbed morphology is only described on the transects so the approach tends to over-represent the lateral variability of the riverbed and under-represent it longitudinally (Ghanem *et al.*, 1994; Secretan *et al.*, 2001).

The flow field itself is empirically characterized at 5–6 different discharge values representing as much as possible the hydrological range relevant for the study. The PHABSIM approach, jointly with the measurements, allows one to employ simple semi-empirical flow models using 1D solution schemes (e.g. Chezy–Manning–Strickler formula, HEC-2, HEC-RAS) to represent the flow field in a more predictive manner. However, the limited number of physical processes included in such models (mainly the gravity/slope and bed resistance to flow) do not allow one to extrapolate accurately outside the flow range covered by the characterization protocol. Calibrating

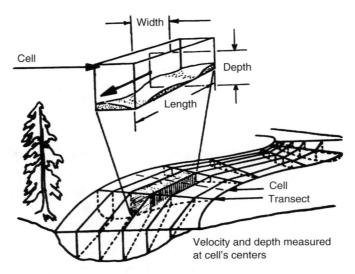

Figure 16.6 *Curvilinear 1D PHABSIM discretization scheme for characterizing abiotic factors.*

such models can be troublesome if one considers that the cross-section-based friction factor can evolve very significantly with the lateral extent of the flow domain (wetted perimeter) which depends on the discharge (or tide). Consequently, the calibration of such models needs to be revisited for different flow discharges or water levels. Another problem is raised by the requirement to gather homogeneous flow data sets over the entire flow domain with respect to discharge. This constraint is inherent to the model calibration process. For small catchments with highly variable hydrology, the discharge can change very significantly over a short period of time, and consequently jeopardize the measurement effort for the target discharge event.

Moreover, the low frequency of extreme events often makes it difficult to build flow models covering a wide flow range, especially for droughts. For floods, the problem is rather related to logistic and safety constraints (e.g. access to site, high velocities). These considerations together with the limits of validity of discharge-dependent friction factors seriously limit the range of extrapolation of such models. For very low droughts, this limitation can be troublesome when one aims to establish minimum conservation flow recommendations. Hopefully, habitat quality is often poor at extreme high or low flows, as defined by "bell" shaped HSIs. Therefore habitat predictions at these discharges are less dependent on predictions of depth and velocity.

Additional criticisms relate to complex flow structures such as return flows behind obstacles or transient behaviour associated with tides or flash floods that cannot be represented with such models. Moreover, potential modifications or alterations of the river morphology formed by enhancement works or natural river evolution cannot be modelled either. Considering that simple 1D approaches are highly dependent upon direct measurements of the current topography and flow conditions and provide low predictability, their utilization is restricted to flow domains that will not experience such morphological alterations.

One of the main reasons explaining the sustained utilization of this approach for more than 25 years, notably in the USA, is related to the legal role played by this model scheme and its necessary standardization. Most of the time, the results are used in the resolution of water allocation conflicts, in particular between ecological instream flow needs and other water resource utilizations, such as flow diversion for irrigation, drinking water supply or hydroelectricity purposes (Leclerc *et al.*, 1995). Unfortunately, the high standardization level of the method severely restricted its adaptation and retarded the utilization of more sophisticated and accurate hydrodynamic tools currently available (at least 2D). Another reason could be provided by disciplinary barriers: even though the developers of IFIM (Bovee and Milhous, 1978) intended to elaborate an ambitious and comprehensive modelling system including water quality, this pioneering effort in habitat modelling was mainly led by non-hydraulic researchers who focused as a priority on the development of the environmental and ecological aspects of the methodology with less consideration for the riverbed morphology and hydraulic representation. Finally, one must admit that, in the late seventies, very few 2D tools for modelling flow fields (considering drying–wetting processes) and visualizing results were then available and their use was seriously limited by the existent computer capabilities.

16.2.3.2 Two-dimensional modelling of abiotic factors

With the fast development of personal computing and CFD, full discretized 2D hydrodynamic simulation tools have become more popular over the last decade because of their enhanced description of complex hydraulic features, especially for high velocity areas, rapidly varied flow regimes and back eddies (Leclerc *et al.*, 1995; Ghanem *et al.*, 1996; Waddle *et al.*, 1997; Hardy, 1998; Heniche *et al.*, 1999; Secretan *et al.*, 2001; Hardy and Addley, 2003; Katopodis, 2003). Finite elements, differences or volumes form distinct but similar discretization schemes allowing one to represent numerically both the field data (digital terrain model) and the mathematical framework describing the flow equilibrium (conservation of mass and momentum). However, the non-structured aspect of the finite element or volume options is reputed to offer a better adaptability to local flow and riverbed features characterized by higher variable gradients, and consequently more accurate results (Figure 16.7).

As a 2D methodology necessitates more detailed field data, this requirement is sometimes perceived as a limitation for practical use because of the additional costs related to extensive field characterization strategies, in particular for topography and riverbed substrate. Nevertheless, the need for direct flow measurements (velocity, stage) is considerably reduced – compared to 1D approaches – considering that this data forms a limited sample of the flow characteristics, only necessary to calibrate and validate model predictions, and that it does not aim to represent the entire flow field as for 1D models. For the same reasons, the field campaign related to instream flow variables is reduced to typically two relatively distinct reference discharge events. This specification allows one to calibrate and validate the model with independent data sets and preserve to a certain extent the predictive power of the model outside the range of its establishment. In very particular contexts where additional forcing factors (e.g. seasonal variation of friction factors such as aquatic

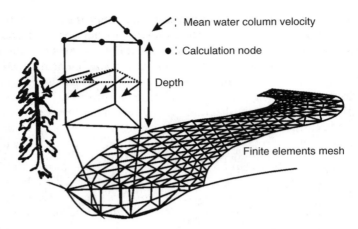

Figure 16.7 *Finite element discretization scheme in 2D habitat studies (after Leclerc, 2002).*

plants or ice cover) make the parameterization of the model more complex, more than two flow events representing a wider range of conditions can be necessary (Morin *et al.*, 2000).

Two-dimensional hydrodynamic models also allow one to simulate and, hence, to analyse the physical impact of important modifications in river morphology or riverbed substrate which are among the basic factors limiting the local habitat value. Transient behaviour of flow is naturally taken into account in the Saint Venant (shallow water equations) scheme. Such models also open the door to the application of sophisticated transport–diffusion simulation models which allow one to take into account a broader range of abiotic variables such as sediment transport, propagation and the fate of contaminants, thermal regime and so on. 2D simulation results also offer a much greater potential for computer graphics and meaningful interpretations, especially when coupled with Geographical Information Systems (GIS).

The implementation of these more sophisticated hydrodynamic models together with proper GIS tools for spatial analysis can be associated with more elaborate biological models that may, for instance, include key behavioural information, such as inter- or intra-specific competition or predator–prey avoidance relationships within the ecosystem (Hardy and Addley, 2003). Distance to cover, exclusion zones dictated by the presence of a predator species and hierarchical data can be incorporated in a study scheme when multi-dimensional hydraulic models are used. Even more sophisticated biological models may include the period of re-colonization by invertebrates of an area previously dried because of lower flows associated with dam operations. Other models may include bioenergetic components associated with feeding patterns or community characterization. This level of ecological sophistication is possible only if the habitat models are based on multi-dimensional (i.e. $\geq$2D) hydraulics. Implementing concepts of discontinuous spatial distribution of habitats such as their "patchiness" (Bovee, 1996) necessitates accurate distributed results allowing the calculation of gradients or an account of discontinuities.

In the last decade, ever-increasing desktop computing capabilities, new sophisticated data acquisition systems coupled with precise positioning such as the Global

Positioning System (GPS), micro-photogrammetry with the assistance of low flight aerial photos or laser scanning, remote sensing using satellite or airborne imagery, and better current measurements using Acoustic Doppler Current Profiler (ADCP) technology have also contributed to use of new generations of hydraulic models to their full potential (Hardy and Addley, 2003).

16.2.3.3 Time for 3D modelling?

In the future, 3D tools are likely to play a growing role in representing even more complex flow situations in the context of habitat studies (Leclerc, 2002). For example, helicoidal (secondary) flows in meanders, vertically stratified flows associated with thermal regime or salinity, and wind-driven flows in coastal areas can be represented accurately only by such models. However, transient behaviour associated with short period and local turbulence or complex flow patterns related to large substrate elements still remains beyond the current capabilities of most existing 3D models. Thus, it is very important to verify the underlying hypotheses posed to develop a particular model aimed at representing flows in a 3D fashion. These hypotheses can seriously limit their applicability, especially for habitat modelling purposes. For example, the incompressibility hypothesis limits the application to homogeneous water masses with respect to density. As with 2D models, hydrostatic pressure also restricts, to a certain extent, their application to rapidly varying flows. Finally, the extensive application of 3D codes to habitat modelling still needs corresponding empirical models (transfer functions) to describe species preferences in a 3D framework.

16.2.3.4 The need for drying–wetting capabilities

Wetting and drying is an important factor for assessing habitat (or ecological modelling) for many species, particularly plants. In this respect, there is great potential for CFD models to provide information on the timing and extent of inundation. Different factors can contribute to define the spatial extension of the wetted part of a given flow domain: tides, hydrological regime, and to a lesser extent, the magnitude and direction of winds. Thus, modelling flow variables with the shallow water equations, whatever the discretization scheme retained, requires non-linear *drying–wetting* capabilities to deal with transiently moving boundaries. This problem is addressed in Bates and Horritt (2004; Chapter 6 in this book). Heniche *et al.* (1999; 2002) and Leclerc *et al.* (1990a,b) also presented reviews of different approaches to solve this problem. Among several others, Hervouet (1992) and BOSS Int. (2003) propose drying–wetting simulators that can be readily used for habitat studies. The accuracy of the drying–wetting solution scheme may become an important issue as the method needs to solve the flow equations in partly wet cells (or elements) of the solution mesh (near shores). In specific situations where biological functions take place in such areas, accurate values of flow variables are necessary to represent such habitats properly. As the solution scheme may be weaker in these zones, due to the partial information taken into account by the model (dry nodes do not really contribute to the solution scheme), one should remain careful when interpreting such results.

16.2.3.5 Limited extent of modelling capabilities for secondary and turbulent flows

Whilst modern hydrodynamic simulators offer ever enhanced capabilities to represent complex flows, research/development is still needed to represent accurately the whole range of flow features present in natural rivers, especially with regard to turbulence and secondary currents related to spatial heterogeneity of river morphology and alluvial composition. If modelling of hydraulic characteristics around large substrate elements is theoretically possible for single grain elements (e.g. boulders), it remains obviously impracticable for entire flow reaches due to limited computing power and to the unavailability of appropriate boundary conditions. Upscaling is also an important aspect that limits this type of modelling.

One can represent the entire range of variability of river flows with the following equation where the different terms proposed are defined in Table 16.1. The development of mass and momentum equations with different space and time integration schemes leads to more or less reduced representations of the flow. However, most of these local flow patterns occur at scales of ecological importance and they play a critical role in the occurrence of microhabitats. Unfortunately, most hydrodynamic modelling schemes still fail to simulate such complexities. Does this mean that modelling microhabitats, even with the most sophisticated 3D tools, cannot offer any useful information for decision-making or for developing knowledge? Certainly not. But the need for more and more detailed results in the context of habitat studies

Table 16.1 *Spatio-temporal variability of flows of ecological relevance and hydrodynamic modelling schemes*

Processes	Variable notation	Description	Remarks
Mean velocity	$\overline{V}(x, y, z, t)$	Mean-term time-dependant	Obtained by time integration of turbulent components of flow
Turbulence	$V'(t)$	Short-term time-dependant	Accounted for in momentum equations by turbulent shear stresses (turbulence closures) but rarely represented
Vertical profile	$V''(z)$	Turbulent boundary layer; theoretically logarithmic	Accounted for by bottom resistance (1D, 2D) or explicitly represented by 3D models
Secondary flows	$V'''(x, y, z)$	Helicoidal meandering flow	Only explicitly represented in 3D (but allowed for in some 2D models)
Tertiary flows	$V''''(x, y, s)$	Local-scale dominant; substrate grain-size dependant	Hardly accounted for by transient turbulence closures (horizontal shear stresses), rarely simulated, even in 3D

will become an additional motivation for the future development and implementation of more elaborate simulation tools.

$$V(x,y,z,t) = \overline{V}(x,y,z,t) + V'(t) + V''(z) + V'''(x,y,z) + V''''(x,y) \quad (16.4)$$

16.2.4 Digital terrain modelling (DTM)

16.2.4.1 *Is measuring already a modelling process?*

As stated before, input data for habitat models is provided either by predictive hydraulic models, or by field measurements as represented by a DTM, but mostly by a mixed combination of both. The purpose of digital terrain modelling is similar and complementary to hydraulic modelling. Its main goal is to homogenize and represent the field data sets necessary to simulate hydraulic variables and, sometimes, other flow attributes such as water quality, temperature or sediment transport. The DTM can also provide input data directly to habitat models (e.g. riverbed substrate grain size and composition). Strictly speaking, terrain models are limited to measured field data. Important variables that relate to the DTM include topography, substrate attributes, and stage/discharge relationships at open boundaries. In addition to these variables, one can also take into account any given independent factor, natural or artificial (man-made), that influences the flow pattern such as the seasonal distribution of aquatic plants, debris accumulation, river engineering works and ice cover thickness and roughness.

The choice of a proper field characterization strategy is very important in order to allow accurate application of predictive models. In fact, terrain modelling starts immediately with the field work and our belief is that "Measuring is already modelling" (Leclerc, 2002). For example, characterizing the riverbed topography with lateral transects as is usual for 1D hydraulic models is a strategy that provides good lateral accuracy, but a very poor longitudinal description of the shapes and flow features in between.

A 2D random and/or adaptive distribution of measurement points based on local interpretation of the variability during the field campaign leads to a far better representation of river morphology and, consequently, more accurate results with respect to habitat variables. Nowadays, modern instrumentation for measuring topography and other physical variables coupled with accurate x, y, z positioning allows one to build a very accurate DTM for a fraction of the cost of older methods. This type of relatively coarse representation of the actual topography can be seen as a model.

The substrate grain-size distribution plays a very important role for either parameterising flow resistance from the river bed, or computing local habitat preference values. For flow resistance, local x, y properties of substrate roughness and macro-roughness in situations where sediment is mobilized and transported by the flow are necessary information to account for in a DTM underlying 2D or 3D hydraulic models. Several empirical formulae of the Manning–Chézy–Strickler type have been proposed in the literature to parameterize substrate characteristic size into roughness coefficients.

On the other hand, substrate information for habitat modelling may need to be represented in a particular format more adapted to model the instream flow needs of the fish or plant species under study. For example, some habitat models may require in addition to the grain size, the organic matter content of the sediment (e.g. aquatic plants). Some fish species avoid fine particles to spawn and seek clean and coarse substrates allowing oxygenation of the eggs during their incubation. Most salmonids require fairly large substrates (pebbles, cobbles, small boulders) as "home rocks" during their feeding period. Thus, a characterization protocol with respect to substrate should simultaneously take into account the needs for hydraulic modelling and those of the habitat models to be developed.

16.2.4.2 *The problem of heterogeneity of data sets*

Most physical factors can be measured using several instruments with variable precision and spatial distribution schemes. For topography, characterization with total stations, echosounders coupled with differential GPS (DGPS) positioning systems, or airborne remote sensing using photogrammetry or laser scanners can all provide good representations of limited parts of the riverbed shape (Goodwin and Hardy, 1999). However, most of the time, their intrinsic limitations (e.g. echosounding is only possible where a minimum depth exists) impose a need to mobilize more than one tool to achieve a complete characterization and, consequently, this generates more than one type of data. Moreover, these are sometimes heterogeneous with respect to their accuracy, their spatial distribution scheme or, at worst, their georeference.

Moreover, the density and spatial distribution of data points depends on the targeted variable with, as a consequence, heterogeneous data layers that cannot automatically be modelled in combination without some transfer procedures on a common compatible digital support, usually a mesh or a grid (Secretan and Leclerc, 1998; Secretan *et al.*, 2001), whatever the discretization scheme retained (finite difference, element or volume). This so-called receiving or client mesh is usually the one utilized for hydrodynamic simulations. Thus, geomatic tools adapted to deal with heterogeneity usually allow one to edit the data points and interpolate between them in order to homogenize data sets by a proper transfer onto the compatible mesh. Usually, a continuous representation of most field data sets is initially obtained by using the Delauney algorithm (Baker, 1987) which allows one to connect the data points together in a single unique way called a TIN (Triangular Irregular Network), which is in fact a linear finite element grid. These meshes are called "server meshes".

16.2.4.3 *Validating the DTM?*

Simulating flow fields with incorrect topographic data illustrates very well the expression "garbage in, garbage out". Error can arise either from inaccurate characterization strategies, or when modelling the field data (DTM). For this reason, it is preferable, if not essential, that the original data be preserved intact in the database and kept available for the subsequent modelling steps in order to allow one to get back to the field data, especially if adaptive mesh generation is used to improve the results or to take into account riverbed modifications or river enhancements (Secretan *et al.*, 2001).

Moreover, since any modelling activity should be submitted to proper validation procedures and fulfil minimum error criteria, it appears necessary to evaluate the DTM with respect to the quality of representation of the field data. If standardized procedures with respect to the DTM are not available, at least a simultaneous graphical representation and a visual comparison of original isolines and of those generated by the DTM can achieve this requirement (e.g. the MODELEUR software by Secretan et al., 1999). Even though this activity does not guarantee the accuracy of field data sets, at least one makes sure that the DTM reproduces the terrain as measured if not as is. Distribution of numerical error can also be analysed more accurately by making use of numerical methods adapted to error estimation.

16.2.5 Integrating habitat value spatially – The habitat availability

In the context of determining minimum conservation flows for fish, one way to consider the river attributes consists of looking at the wetted perimeter of a limited number of sections or transects and according to the shape of these sections, finding a discharge value for which there is no significant additional gain in the wetted perimeter (Reiser et al., 1989). In 2D, this concept can be extrapolated to an equivalent Wetted Area (WA(Q)) which corresponds analytically to the integration of the flow domain with respect to its wet or dry state (equation 16.5). In this approach, the riverbed surfaces only need to be wet to be considered as full habitats which, for obvious reasons related to the local habitat value (HSI(x, y)), may not be suitable, or effectively occupied by fish species. A minimum flow recommendation with respect to this method corresponds normally to point A on Figure 16.8:

$$\text{WA}(Q) = \int_D (\text{state} = \text{wet}) dA \cong \sum_N (\text{state} = \text{wet})_i \Delta A_i \quad (16.5)$$

where Q is the discharge and ΔA_i represents the area of a discretization element. This numerical formulation can vary according to the specific integration scheme utilized.

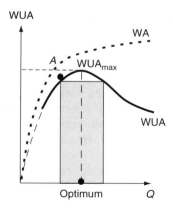

Figure 16.8 The concept of WA and habitat availability or weighted usable areas (WUA) as a function of discharge.

Similarly, habitat models aim to represent the distribution of habitats and their availability as a function of discharge in order to select the most appropriate minimum value required to maintain aquatic habitat availability to a suitable level, or to evaluate and compare future scenarios (e.g. enhancement works, climate change) with respect to historical or influenced conditions. With this approach, the wetted riverbed surfaces offer a given habitat value (HSI) that can be used as a weighting factor in the integration scheme. The habitat availability is being represented by the Weighted Usable Area (WUA) (equation 16.6). This is a function of flow discharge (WUA(Q)) which determines the spatial distribution of the main abiotic factors controlling the local habitat suitability index (i.e. HSI(x, y)). It is calculated by the numerical integration of HSI(x, y) over the flow domain, or a given sub-domain (e.g. a spawning ground comprised within the domain).

$$\text{WUA}(Q) = \int_D \text{HSI}(x,y)\,dA \cong \sum_N \text{HSI}_i \cdot \Delta A_i \qquad (16.6)$$

In this expression, the area and its HSI(x, y) value are taken into account linearly which means that a fairly large area of poor value could provide as much habitat as a limited area of excellent value. This algebraic behaviour has raised some criticisms in the literature arguing that, in terms of biological productivity, equal amounts of WUA, even though they seem equivalent in quantity, will not produce the same biomass as a result. One way of avoiding this drawback of the formulae is to integrate the HSI(x, y) variable for values corresponding to the best habitats only (e.g. $\geq \text{HSI}_{\min}(x, y)$) and neglect others (Capra et al., 1995).

Nevertheless, the concept of WUA is widely used to analyse the functional relationship between habitat availability and the flow discharge. Figure 16.8 illustrates this relationship. In such graphs, one can observe that typically the WUAs increase rapidly with discharge until a maximum habitat availability is achieved in the domain under study. With an additional increase of discharge, even though the wetted areas continue to spread out, the WUAs usually decline gradually, probably due to the fact that the currents become gradually too rapid, the depth too large, with corresponding increases in turbidity. However, in particular situations, this behaviour can become bi- or multi-modal, especially when more than one habitat feature occurs, each of them peaking at a different optimum flow (Leclerc et al., 1996).

Is the HSI value the only variable that can be used as the weighting factor to compute WUA? Some researchers (GENIVAR, 2003) proposed the use of the Probabilistic Habitat Index (PHI(x, y)) instead of HSIs to integrate the habitat availability over a flow domain. In fact, they integrate the areas where the PHI exceeds 0.5 in order to take into account a significant probability of presence which is greater than 50% in these zones. This means that only the best habitats are being considered in the integration scheme. This approach seems consistent with the probabilistic aspect of the PHI concept.

As to the fuzzy-based preference model proposed by Jorde et al. (2001), their suitability index is equivalent mathematically, if not conceptually, to the preference curve approaches (HSI) and, consequently, it can also be spatially integrated over the flow domain or sub-domains in the same manner as other indices to evaluate the habitat availability WUA(Q).

16.2.6 Towards habitat time series

As already stated, hydrological regime generates transient flow variables which in turn can improve or deteriorate the habitat features. This aspect is not often considered explicitly in habitat studies and is open for new ideas as to how to incorporate the time factor. At least, the combined considerations of hydrological seasons and sensitive phases of the fish's life cycle provide some indications for delineating proper periods of analysis.

One way to incorporate the time factor is simply to use the hydrographs, natural and/or influenced, for building habitat time series through the WUA(Q) functional relationship (Figures 16.8 and 16.9). Considering such a scheme, one can observe that the maximum amount of habitat should be achieved only during short periods of time when the optimum flow (Q_{opt}) is released from the drainage basin runoff. Any change in flow discharge from this specific point leads to a transient loss of habitat. As a corollary, any change of flow towards the optimum (increase or decrease) produces an improvement in the amount of WUA.

Again, some underlying hypotheses determine the interpretation of this simple scheme. First, one can observe that the functional relationship between the amount

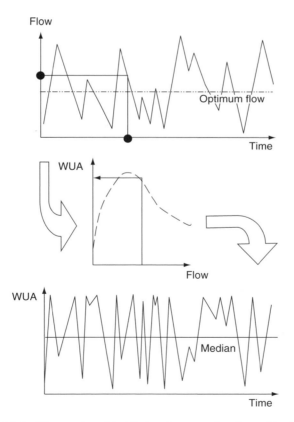

Figure 16.9 Time series of habitat regime as a function of hydrology.

of WUA and the discharge has two flow solutions for a single value of habitat availability. Are the two situations strictly equivalent? Probably not. For example, looking at the feeding ecology of salmonids, it is well known that the invertebrate drift load peaks during the rising limb of floods (Paraziewicz et al., 1996), due to the increasing velocities (shear stress) and to the escape behaviour of insects which take advantage of these flow sequences to move to other locations in order to colonize new habitats.

On the contrary, during the falling limb of the flood, most of the movement is complete as insects have already settled in their new location. As a consequence, the feeding behavior is not equivalent on both sides of the flood hydrograph. Moreover, one should mention that sediment transport processes are triggered by floods above a given flow threshold, and that the concentration of particles in the water column (suspension and saltation) increases with increasing flow. As a consequence, the visibility of prey diminishes and the predators can no longer distinguish between sand and drift particles. Thus, some weighting factor based on the rate of variation of flow $Q'(t)$ could theoretically be incorporated in the model in order to take into account these transient flow-dependant properties.

$$\text{WUA}(t) = f(Q(t), Q'(t)) \tag{16.7}$$

In a context of hydro-peaking, such transient considerations could be useful to set more efficient flow recommendations.

16.2.7 Spatial distribution and movements of habitats

Furthermore, the change in availability of habitats during a given flood hydrograph and/or peaking flow regime is accompanied by habitat "movements". In fact, the best habitats which were located in or close to the thalweg during low flow episodes will transfer to near the banks at higher flows while habitats in the thalweg will become less suitable because of velocities and depths that become too high compared to the preference ranges (Figure 16.10; Leclerc et al., 1994).

From an ecological point of view, a fundamental question arises: will the fish move with the shifts of the best habitats or will it stay on its "home rock" waiting for better days? In mathematical terms, this question is equivalent to considering whether the fish behaviour is Eulerian or Lagrangian. Obviously, the answer depends highly on the ecology of the fish species considered, and on their short-term behaviour. For example, the main strategy of young salmonids during their feeding period is to occupy the best habitat possible (home rock) and defend it against intraspecific or interspecific competitors. The strongest fish occupy sedentarily the best spots while the weakest individuals satisfy themselves by moving around trying to access opportunistically the best areas but with limited success.

The issue can also relate to the rate of change in flow conditions ($Q'(t)$). If the change takes place in a short period of time ($Q'(t) \gg$) and leads to less favourable conditions, the fish can possibly afford to shelter in the substrate interstices and wait until better conditions come back (Eulerian strategy). Otherwise, the fish could move

Ecohydraulics: A new interdisciplinary frontier for CFD 451

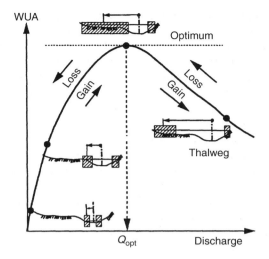

Figure 16.10 Habitat availability and movements with respect to flow discharge.

transiently to "better skies" (Lagrangian strategy) if the flow episodes last too long to allow sufficient feeding in the long term.

An attempt to quantify the movement of habitats was made by Leclerc *et al.* (1994). The method starts by partitioning the flow domain into sub-regions (patches) with continuous habitat distribution (Figure 16.11). Laterally, the thalweg separates the habitats on each side of the river section. Longitudinally, bed forms or flow facies play the same role. The mass centroid of each sub-region of the habitat is then located for each reference flow event, the weighting factor being represented by the local habitat value (HSI). The displacements of these centroids can be tracked in space with respect to the flow regime in order to quantify the rate of displacement (speed).

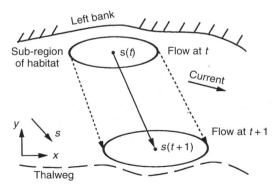

Figure 16.11 Displacement of sub-region of habitat with flow reduction – habitat feature moves to thalweg.

16.3 Validating habitat models in a context of uncertainties

Validation is usually recognized as an essential step in any modelling effort. As for any biological model, validation procedures constitute a great challenge and the lack of systematic and well-recognized methods still limits the credibility and confidence in the approach (Caissie and El Jabi, 2003). Model validation consists of ensuring that the choice of model parameters (mostly the biological parameters in habitat models) is adequate by comparing model results to observed values.

The most common strategy is to compare representative habitat availability values estimated by the model with the effective biomass produced within the river reach under study. This approach has produced mixed results in the past, most often non-significant. Few successes have been reported in the literature (Bovee, 1988; Jowett et al., 1998; see Guay et al., 2000 for additional discussion). The main reason for this limited success depends on the great deal of uncertainty characterizing the biomass variable itself which depends not only on the discharge, integrated over a long period of time (several years or the timescale of the ecosystem), but also on a wider range of secondary physical and/or physico-chemical factors (e.g. presence of deleterious contaminants). Considering that each of these factors behaves according to its own space/timescales, it becomes very difficult to pinpoint the specific value that plays the role of a limiting factor or of an explanatory variable with respect to the biomass variations. Moreover, it is often difficult to measure all those parameters in one single field survey.

In addition to abiotic factors, biomass is a very aggregative variable that is also a function of a number of independent biotic variables that belong to species population dynamics (recruitment, mortality) or interspecific relationships (competition, predation) among the aquatic communities. These processes can hardly be included in habitat models but some researchers are trying to build new frameworks dedicated to a higher synergy between the biotic and abiotic aspects of habitat modelling which also takes into account the dynamics of flow and corresponding habitat availability (Sabaton et al., 2003).

A more realistic approach to validate at least the habitat preference model without the need to characterize the fish biomass has been proposed by Boudreau et al. (1996) and is starting to be used extensively (e.g. Guay et al., 2000). Here, instead, measured densities of fish are compared to local habitat values (either HSI or PHI). In this case, the emphasis is put not on total biomass, but rather on attempting to reproduce the selective behaviour of individuals for a given habitat type. This method does not provide the level of accuracy expected from a conventional validation approach, but it reduces the uncertainty associated with model parameterization.

A great deal of uncertainty remains in defining habitat *preferenda* for a number of species and one can be interested in characterizing the typical error related to model parameterization schemes, especially when habitat modellers have to rely on subjective expert advice only. Duel et al. (2003) have proposed that one should perform sensitivity analysis of habitat models to *preferendum* parameters and input variables as defined by expert advice. Monte Carlo simulations can generate large samples and provide an estimate of model uncertainties as a function of various input variables or parameters. Various authors have mentioned that only 3–4 variables play a key role and associated relative uncertainties for preference values (HSI) are of the order of 0–0.2, with maximum values of the order of 0.4–0.6. Hopefully, minimal uncertainties are usually found for very high (e.g. ≈1) or very low habitat values which are the most interesting from an ecological point of view.

16.4 Upscaling habitat models with respect to river network or catchment

The issue of upscaling is very important given the requirements of the European WFD. CFD fits into a suite of available hydraulic modelling tools that range from high resolution, low coverage (3D CFD) to low resolution, broad coverage (1D hydraulic modelling). Most habitat model applications focus on the river reach scale which as a minimum should cover a limited number of morphological unit successions (pools, riffles, runs) representing to a certain extent the *river continuum* concept. Although this strategy provides meaningful information on the microhabitats, such a perspective does not allow one to analyse the ecological structure and dynamics at a larger scale belonging to the river network which corresponds to the catchment scale. The classical river reach approach provides typically a "sample" of information about river habitats with respect to the entire river system but does not offer any guarantee that such results represent accurately the river network ecology.

Several distinct river reaches characterized by different proportions of geomorphic features can be modelled to enlarge the representativeness of the study. One can also develop a model covering the entire river network, or at least the main watercourse (maximum river order), such as the St Lawrence River model (Morin *et al.*, 1997, 2000, 2003). However, the cost of such strategies often limits their applicability to large rivers with very important economic and ecological (e.g. fisheries management) issues that justify the related hydrographic characterization effort. Thus, one needs modelling strategies that allow one to upscale (extrapolate) the reach results to the river network scale.

Several authors have addressed this question of upscaling (e.g. Duel *et al.*, 2003; Paraziewicz, 2003). Some models consider as explanatory variables average abiotic factors at the reach level without considering their local distribution (0D; Lamouroux *et al.*, 1995, 2002; Lamouroux, 1998). These models are based on a prior aggregation of results (upscaling) obtained out of a wide sample of microhabitat models operating at a finer (1D, 2D) scale. By using average physical metrics (average reach Froude number and specific discharge, mean D_{50} diameter for substrate grain size) to describe the general abiotic conditions for a number of reaches, or even an entire hydrographic network, it is possible to obtain rough estimates of the habitat quality and availability at larger scales. These general metrics are then used to estimate habitat values for certain species or groups of similar species (guilds). It is noticeable that these parameters reflect a limnological classification of the river which considers their regime as either *lentic* which means slow currents, relatively deep, with finer sediments, or *lotic* which refers to facies with higher velocities, shallower depths and larger alluvial elements.

16.5 Habitat models in practical contexts

16.5.1 Developing new knowledge or managing water use?

Is it necessary to mention that a habitat modelling methodology, as most scientific tools, can be used either for research or for management purposes? Until now, habitat models have mostly been utilized for determining conservation flow values (see section 16.5.2) which correspond to management concerns. But a growing trend is observable which shows an increasing use of habitat modelling to support the development of new knowledge and integrate field information in a meaningful way, according to mass and momentum conservation laws.

In a context where major ecological study programmes are being conducted on rivers or other water bodies, distributed river flow models form the only approach allowing one to integrate several heterogeneous sets of field data, including morphology and hydrology. Empirical knowledge and field data can form very large databases on a given river, making it difficult to take into account the inherent complexity of this information. Geographical information systems are very useful tools to represent graphically, often in 3D, the various layers of terrain information, and to interpolate and combine them with simple mathematical functions. However, the algebraic capabilities of standard GIS tools remain limited and fall far short of those necessary to simulate river flows. With their predictive capabilities, steady state or transient river flow models can supply consistent and synoptic data for several abiotic variables that otherwise cannot be economically measured (e.g. transient temperature fields, sediment transport, wave generation) simultaneously everywhere in the flow domain.

Nevertheless, if such models provide unique integrated perspectives of the flow domain, they cannot supply information for parameters not included in the conceptual framework. In fact, they cannot take into account (reductionism) every aspect of physical reality (e.g. turbulence) either because some factors cannot yet be represented mathematically or because they do not contribute significantly to the analysis.

16.5.2 Determining conservation flows – Management plans

In many countries, conservation flow practices or regulation schemes aiming to maintain aquatic ecosystems or target fish species have been applied. In some countries, such practices have been applied throughout the 20th century, as in France where a minimum of 1/40th of the mean annual flow was enforced by public regulation all the year round in river reaches influenced by dams or diversion practices (Sabaton *et al.*, 2003).

Many approaches currently exist to establish water management schemes and priorities (Bérubé *et al.*, 2002; King *et al.*, 2002; Caissie and El Jabi, 2003). In Canada, for example, one of the guiding principles often used is the policy of no net loss of aquatic habitat, promulgated by the federal department of Fisheries and Oceans. In general, the guidelines aim to maintain sufficient flow in the river to allow for indigenous species to complete all their life stages and to maintain their population level. Ideally, these principles should also aim to maintain flow variability similar to that of the natural regime (the *physiomimesis* principle of Katopodis,

2003). Three broad categories of methods exist to achieve these objectives: *hydrological methods*, *hydraulic methods* and *habitat modelling* (Belzile et al., 1997). These three classes of approaches usually include (albeit to varying degrees) the notion of maintaining flow variability as determined by the natural regime and/or by anthropogenic impacts.

The simplest methods are mainly based on pure hydrological considerations (e.g. Tennant, 1976; USFW, 1981; Reiser et al., 1989; Russell, 1990; Bérubé et al., 2002; Caissie and El Jabi, 2003) and, even though most of them consider ecohydrological seasons, they do not take into explicit consideration the instream flow needs for ecosystems or fish species.

During the late 1970s, pure hydraulic methods based on the wetted perimeter were adopted in order to offer better insights to decision-makers. In addition to hydrological concerns, hydraulic methods can take into account (at least at the reach scale) the existing geomorphology and hydrodynamics in order to maintain a certain minimum wetted perimeter on given transects (Reiser et al., 1989; Caissie and El Jabi, 2003) or wetted area when 2D models are utilized. This approach hypothesizes that a riverbed parcel represents an available habitat as long as it is wet, which obviously overestimates the real amount of effective habitat.

Less reductionist habitat models aim to represent the distribution of suitable habitats and their total availability as a function of flow discharge. They are employed in order to select more accurately the most appropriate flow regime to maintain aquatic habitat for one (or many) aquatic species during some (or all) of their life stages. If the IFIM approach (Bovee, 1982) is the most commonly used method in North America and probably elsewhere in the world (Reiser et al., 1989), the new trend is oriented towards the use of more sophisticated 2D or even 3D discretized models of the river flow, as input data sources to habitat models.

These tools and metrics, especially the latest generation of habitat models, allow the most appropriate flow regimes for ecosystem health and conservation to be determined, in a context of multiple use of water resources. They also contribute to the rationalization of difficult decisions and to the resolution of water allocation conflicts between conservationists and water users.

16.5.3 Designing impact mitigation measures or habitat enhancements

Mitigation and compensation methods accompanying habitat restoration or impact mitigation programmes may include flow management, substrate maintenance and improvement, fishways, fish stocking and engineering structures such as weirs. The latter are often presented as a method to optimize the usage of a reduced flow because they can contribute to the maintenance of proper groundwater levels, and of human water uses such as navigation; they may improve the riverine landscape by maintaining at their full extension the wetted surfaces. They may also improve habitat availability for fish recruitment and overwintering (Britain, 2003).

Weirs can ensure a more constant wetted perimeter in upstream reaches and may help to concentrate flow, which can provide improved conditions under reduced discharge. However, Britain (2003) mentions that the conversion of a river system to a more lentic environment may lead to an increase in macrophyte production and may also favour modifications in the fish community structure (e.g. from a salmonid-dominated river to a cyprinid one). These structures may also be seen as obstacles to fish migration and increase fine sediment deposition. Some aquatic birds can benefit from such modifications, probably due to the increase in benthic productivity.

Weirs can be made of wood, concrete or rocks, according to the stated objectives, prevailing environmental and hydraulic conditions, as well as available material. A comprehensive follow-up program is advised upon implementation of such structures, in order to optimize flow management schemes or improve structure design when possible (Britain, 2003).

These intrusive methods do not traditionally rely on prior habitat modelling to justify their construction. The usual design criteria are often limited to the estimation of the wetted perimeter and the associated flow patterns. Weirs will often modify significantly the habitat structure from lotic to lentic, with lower velocities and higher sedimentation after construction. A coupling of the design process with habitat models is therefore a desirable objective because it may help to maximize the outcome of such structures on aquatic habitats. Traditional PHABSIM models may not be the ultimate tool for such coupling. Distributed 2D models offer the flexibility required to test the effectiveness of various structure designs.

The recent development of more sophisticated structures, such as adjustable sills, allows for an adaptive varying geometry and may permit a better modulation of the flow patterns. A periodic increase in wetted perimeter can be synchronized with spawning or migration requirements of certain fish species. These concepts appear feasible at reasonable costs with current technology and know-how.

References

Bates, P.D. and M.S. Horritt (2004). Modelling wetting and drying processes in hydraulic models. In Bates, P.D., Lane, S.N. and Ferguson, R.I. (eds), *Computational Fluid Dynamics: Applications in Environmental Hydraulics*. John Wiley & Sons, Chichester, UK.

Baker, T.J. (1987). Three-dimensional mesh generation by the triangulation of arbitrary point sets, AIAA Paper 87-1124-CP, 255–271.

Bechara, J., J. Morin and P. Boudreau (2003). Évolution récente de l'habitat du doré jaune, de la perchaude, du grand brochet et de l'achigan à petite bouche au lac Saint-François, fleuve Saint-Laurent. R640, INRS-Eau, Terre & Environnement; 74pp. Rapport remis à la ZIP du Haut Saint-Laurent. In French.

Belzile, L., P. Bérubé, V.D. Hoang and M. Leclerc (1997). Méthode écohydrologique de détermination des débits réservés pour la protection des habitats du poisson dans les rivières du Québec. Joint Report INRS-Eau – Groupe Conseil Génivar to Québec's Ministry of Environment and to Department of Fisheries and Oceans, Canada. Report INRS-Eau #494, 82pp. +8 appendices. January.

Bérubé, P., M. Leclerc and L. Belzile (2002). *Presentation of an Ecohydrological Method for Determining the Conservation Flow for Fish Habitats in Québec's Rivers (Canada)*. Proceedings 4th IAHR International Symposium on Ecohydraulics, Capetown, South Africa. March.

Boisclair, D. (2003). L'utilisation d'approches alternatives pour exprimer les préférences d'habitat. Oral communication. Workshop on the State-of-the-art of Habitat modelling and conservation flows, Québec City. March.

BOSS International (2003). URL: www.bossintl.co.uk/En/html/sms_overview.html.

Boudreau, P., G. Bourgeois, M. Leclerc, A. Boudreault and L. Belzile (1996). *Two-Dimensional Habitat Model Validation Based on Spatial Fish Distribution: Application To Juvenile Atlantic Salmon of the Moisie River (Québec, Canada)*. Proceedings of the 2nd International Conference on Habitats Hydraulics (Ecohydraulics 2000), Québec City, Vol. B: 365–380.

Bovee, K.D. (1978). *The Incremental Method of Assessing Habitat Potential for Cool Water Species, with Management Implications*. American Fisheries Society, Special Publication, **11**: 340–346.

Bovee, K.D. (1982). A guide to stream habitat analysis using the Instream Flow Incremental Method. Instream Flow Information Paper #12. FWS/OBS-82/86. USDS Fish and Wildlife Service, Office of Biological Services. Fort Collins, Co.

Bovee, K.D. (1988). *Use of IFIM to Evaluate Influences of Microhabitat Variability on a Trout Population in Four Colorado Streams*. 68th Annual Conference of the Western Association of Fish and Wildlife Agencies. Albuquerque, USA, 31pp.

Bovee, K.D. (1996). *Perspectives on Two-Dimensional River Habitat Models: The PHABSIM Experience*. Proceedings of the 2nd International Conference on Habitats Hydraulics (Ecohydraulics 2000), Québec City, Vol. B: 149–162.

Bovee, K.D. and R. Milhous (1978). Hydraulic simulation in instream flow studies: Theory and techniques. *US Fish and Wildlife Service FWS/OBS-78/33*.

Bovee, K.D., B.L. Lamb, J.M. Bartholow, C.B. Stalnaker, J. Taylor and J. Henriksen (1998). Stream habitat analysis using the Instream Flow Incremental Method. Information and Technology Report USGS/BRD/ITR-1998-0004, USGS.

Britain, John E. (2003). Weirs as a mitigation measure in regulated rivers – The Norwegian experience. *Canadian Water Resources Journal*, **28**(2): 217–230.

Caissie, D. and N. El Jabi (2003). Instream flow assessment: From holistic approaches to habitat modelling. *Canadian Water Resources Journal*, **28**(2): 173–184.

Capra, H., P. Breil and Y. Souchon (1995). A new tool to interpret magnitude and duration of fish habitat variations. *Regulated Rivers: Research and Management*, **10**: 281–289.

Duel, H., G.E.M. van der Lee, E.W. Penning and M.J. Baptist (2003). Habitat modelling of rivers and lakes in the Netherlands: An ecosystem approach. *Canadian Water Resources Journal*, **28**(2): 231–247.

GENIVAR, 2003. Centrale de l'Eastmain 1-A et dérivation Rupert. Détermination du régime de débits réservés écologiques dans la rivière Rupert. Rapport sectoriel. For Hydro-Québec. December. In French.

Ghanem, A., P. Steffler, F. Hicks and C. Katopodis (1994). *2D Finite Elements Flow Modelling of Fish Habitats*. In Vaskin et al. (eds) Proceedings of the 1st IAHR International Conference on habitat hydraulics. Norwegian Institute of Technology, Trondheim, Norway, pp. 84–89.

Ghanem, A., P. Steffler, F. Hicks and C. Katopodis (1996). Two-dimensional simulation of physical habitat conditions in flowing streams. *Regulated Rivers: Research and Management*, **12**: 185–200.

Ginot, V., Y. Souchon, H. Capra, P. Breil and S. Valentin (1998). Logiciel EVHA. EValuation de l'HAbitat physique des poissons en rivière. Version 2.0 Guide méthodologique. CEMAGREF BEA/1.HQ. Ministère de l'Environnement (France). Direction de l'Eau. 176pp.

Goodwin, P. and T.B. Hardy (1999). Integrated simulation of physical, chemical and ecological processes for river management. *Journal of Hydroinformatics*, **1**(1): 33–58.

Guay, J.C., D. Boisclair, D. Rioux, M. Leclerc, M. Lapointe and P. Legendre (2000). Development and application of numerical habitat models. *Canadian Journal of Fisheries and Aquatic Sciences*, **57**: 2057–2067.

Guay, J.C., D. Boisclair, M. Leclerc and M. Lapointe (2003). Assessment of the transferability of biological habitat models for juveniles of Atlantic salmon (*Salmo salar*). *Canadian Journal of Fisheries and Aquatic Sciences*, **60**: 1398–1408.

Hardy, T.B. (1998). The future of habitat modelling and instream flow assessment techniques. *Regulated Rivers: Research and Management*, **14**: 405–420.

Hardy, T.B. and R.C. Addley (2003). Instream flow assessment modelling: Combining physical and behavioral based approaches. Special issue of the *Canadian Water Resources Journal*. State-of-the-art in habitat modelling and conservation flows, **28**(2): 273–282.

Heniche, M., Y. Secretan, P. Boudreau and M. Leclerc (1999) A new finite element drying–wetting model for rivers and estuaries. *International Journal of Advances in Water Resources*, **38**(3): 163–172. January.

Heniche, M., Y. Secretan, P. Boudreau and M. Leclerc (2002). Dynamic tracking of flow boundaries in rivers with respect to discharge. *Journal of Hydraulic Research (IAHR)*, **40**(5): 589–602.

Hervouet, J.M. (1992). Solving shallow water equations with rapid flow and tidal flats. In Blain, W.R.R. and Cabrera, E. (eds), *Hydraulic Engineering Software. 4. Fluid Flow Modelling*, Elsevier Applied Science, London.

Jorde, K., M. Schneider, A. Peter and F. Zoellner (2001). *Fuzzy Based Models for the Evaluation of Fish Habitat Quality and Instream Flow Assessment*. Proceedings of the 2001 International Symposium on Environmental hydraulics. On the behalf of ISEH.

Jowett, I.G., J.W. Hayes, N. Deans and G.A. Eldon (1998). Comparison of fish communities and abundance in unmodified streams of Kahurangi National Park with other areas of New Zealand. *New Zealand Journal of Marine and Freshwater Research*, **32**: 307–322.

Katopodis, C. (2003). Case studies of instream flow modelling for fish habitat in Canadian Prairie Rivers. Special issue of the *Canadian Water Resources Journal*. State-of-the-art in habitat modelling and conservation flows, **28**(2): 199–216.

King, J.M., R.E. Tharme and M.S. de Villers (2002). Environmental flow assessment for rivers: Manual for the building block methodology. Freshwater Water Research Unit, University of Capetown, S.A. WTC Report No. TT 131/00, 340pp.

Lamouroux, N. (1998). Depth probability distributions in stream reaches. *Journal of Hydraulic Engineering*, **124**: 224–227.

Lamouroux, N. and Y. Souchon (2002). Simple prediction of instream habitat model outputs for fish habitat guilds in large streams. *Fresh Water Biology*, **47**: 1531–1542.

Lamouroux, N., Y. Souchon and E. Herouin (1995). Predicting velocity distributions in stream reaches. *Water Resources Research*, **31**: 2367–2375.

Leclerc, M. (2002). *Ecohydraulics, Last Frontier for Fluvial Hydraulics: Research Challenges and Multidisciplinary Perspectives*. Key-note. Proceedings of RiverFlow 2002 Conference. On the behalf of IAHR. Louvain-la-Neuve, Belgium. September.

Leclerc, M., J.F. Bellemare and S. Trussart (1990a). Simulation hydrodynamique de l'estuaire supérieur du fleuve Saint-Laurent (Canada) avec un modèle aux éléments finis couvrant-découvrant. *Revue canadienne de génie civil*, **17**(5): 739–751.

Leclerc, M., J.F. Bellemare, G. Dumas and G. Dhatt (1990b). A finite element model of estuarian and river flows with moving boundaries. *Advances in Water Resources*, **4**(13): 158–168.

Leclerc, M., P. Boudreau, J. Bechara, L. Belzile and D. Villeneuve (1994). Modélisation de la dynamique de l'habitat des ouananiches (*Salmo salar*) juvéniles de la rivière Ashuapmushuan (Québec, Canada). *Bulletin français des pêches et de la pisciculture*, **332**: 11–32.

Leclerc, M., A. Boudreault, J. Bechara and G. Corfa (1995). Two dimensional hydrodynamic modelling: a neglected tool in the Instream Flow Incremental Methodology. *Transactions of the American Fisheries Society*, **124**(5): 645–662.

Leclerc, M., J. Bechara, P. Boudreau and L. Belzile (1996). A numerical method for modelling the spawning habitat of landlocked salmon. *Regulated Rivers: Research and Management*, **12**: 273–285.

Leclerc, M., A. Saint-Hilaire and J. Bechara (2003). State-of-the-art and perspectives of habitat modelling for determining conservation flows. Special issue of the *Canadian Water Resources Journal*, **28**(2): 135–172.

Milhous, R. (1996). *Modelling of the Instream Flow Needs: The Link Between Sediment and Aquatic Habitat*. Proceedings of the 2nd International Conference on Habitats Hydraulics (Ecohydraulics 2000), Québec City, Vol. A: 319–330.

Morin, J., J. Lafleur, S. Côté, P. Boudreau and M. Leclerc (1997). Integrated hydrodynamic, waves and plants habitat modelling for restoration of shoreline on Clark Island, Lake St Francis. For Tecsult Environment Inc. INRS-Eau Report No. 503, 47pp.

Morin, J., P. Boudreau, Y. Secretan and M. Leclerc (2000). Pristine Lake Saint-Francois, St. Lawrence River: Hydrodynamic simulation and cumulative impact. *Journal of Great Lakes Research*, **26**(4): 384–401.

Morin, J., M. Mingelbier, J.A. Bechara, O. Champoux, Y. Secretan, M. Jean and J.J. Frenette (2003). Emergence of new explanatory variables in large rivers: The St. Lawrence Experience. Special issue of the *Canadian Water Resources Journal*. State-of-the-art in habitat modelling and conservation flows, **28**(2): 249–272.

Parasiewicz, P. (2003). Upscaling: Integrating habitat model into river management. Special issue of the *Canadian Water Resources Journal*. State-of-the-art in habitat modelling and conservation flows, **28**(2): 283–300.

Parasiewicz, P., S. Schmutz and O. Moog (1996). *The Effect of Managed Hydropower Peaking on the Physical Habitat, Benthos and Fish Fauna in the Bregenzerach, a Nival 6th Order River in Austria*. Proceedings of the 2nd International Conference on Habitats Hydraulics (Ecohydraulics 2000), Québec City, Vol. A: 684–697.

Reiser, D.W., T.A. Wescche and C. Estes (1989). Status of instream flow legislation and practices in North America. *Fisheries*, **14**(2): 22–29.

Russell, G.W. (1990). Determination of instream flow needs at hydroelectric project in the Northeast. In Bain, M.B. (ed.), *Ecology and Assessment of Warm Water Streams: Workshop Synopsis*. US Fish Wildlife Service, Biol. Rep. 90(5): 44pp. (Pages 36–37).

Sabaton, C., Y. Souchon, J.M. Lascaux, F. Vandewalle, P. Baran, H. Capra, V. Gouraud, F. Lauters, P. Lim, G. Merle and G. Paty (2003). *The "Guaranteed Flow Working Group": A French Feedback of IFIM Based on Habitat and Brown Trout Population Time Series Observations*. Proceedings of the IFIM User's Workshop. Fort Collins, CO. 1–5 June.

Schneider, M. and K. Jorde (2003). *Fuzzy-Rule Based Models for the Evaluation of Fish Habitat Quality and Instream Flow Assessment*. Proceedings of the IFIM User's Workshop. Fort Collins, CO. 1–5 June.

Scheiner, S.M. and J. Gurevitch (2001). *Design and Analysis of Ecological Experiments*. Second edition. Oxford University Press, 415pp.

Secretan, Y. and M. Leclerc (1998). *MODELEUR: A 2D Hydrodynamic GIS and Simulation Software*. Proceedings Hydroinformatics (on the behalf of IAHR). Copenhagen. August.

Secretan, Y., Y. Roy, Y. Granger and M. Leclerc (1999). MODELEUR/HYDROSIM – User's guide. MODELEUR Document 1.0a01. 235p. p.v., Janvier.

Secretan, Y., M. Leclerc, S. Duchesne and M. Heniche (2001). Une méthodologie de modélisation numérique de terrain pour la simulation hydrodynamique bidimensionnelle. *Revue des Sciences de l'eau*, **14**(2): 187–212.

Shirvell, C.S. (1986). Pitfalls of physical habitat simulation in the Instream Flow Incremental Methodology. Ca. *Technical Report of Fisheries and Aquatic Sciences 1460*.

Tennant, D.L. (1976). Instream flow regimens for fish, wildlife, recreation and related environmental resources. *Fisheries*, **1**(4): 6–10.

USFW (US Fish and Wildlife Service) (1981). Interim regional policy for New England stream flow recommendations. US Fish and Wildlife Service, Newton Corner, Massachusetts. 3pp.

Vannote, R.L., K.W. Minshall, J.R.S. Cummings and C.E. Cushing (1980). The river continuum concept. *Canadian Journal of Fisheries and Aquatic Sciences*, **37**: 130–136.

Waddle, T., K.D. Bovee and Z. Bowen (1997). *2D Habitat Modelling in the Yellowstone/Upper Missouri River System*. Proceedings of the NALMS. Instream and Environmental Flow Symposium, Houston Texas. December.

17
Towards risk-based prediction in real-world applications of complex hydraulic models

B.G. Hankin and K.J. Beven

17.1 Introduction

The application of computational flow models to complex environmental flows involves significant uncertainties. However, to date, there has been very little take-up of the challenge to carry out uncertainty assessments of such models in the research sphere, let alone in the 'real world'. A typical application within the water sector might ask for a sensitivity analysis to be carried out on complex modelling solutions. However, whilst sensitivity analyses can be useful, they do not generate predictive uncertainties that, if included, could indicate how wrong model predictions might be and therefore what the risk is to capital expenditure. A far better approach would include existing techniques, such as the Generalized Likelihood Uncertainty Estimation (GLUE) approach (Beven and Binley, 1992), at the design stage of a commissioned modelling package. The purpose of the chapter is to demonstrate these issues. The inadequacy of a real-life study, comprising a hugely complex system with very uncertain inputs in which there was only limited time to carry out a simple sensitivity analysis, is contrasted to an academic modelling study of pollutant transport for which a GLUE approach was undertaken, to illustrate the benefits of the uncertainty framework proposed. The complex real-world study comprised investigating the impact of aggregated spills from the four major sewer catchments in Preston on the designated Shellfish Waters in the Ribble Estuary using the Environment Agency's (EA) Ribble Estuary Model. The modelling was carried out during the Asset Management Programme (AMP3) process for upgrading water utility assets such as sewerage and Combined Sewer Overflows (CSOs) to assist in decision-making. The modelling work was not a formal part of the upgrade assessment process, but served as an extra tool available to water quality planners. Given the time scales involved, and since the EA's model was not commissiond to include a facility to carry out a predictive uncertainty analysis, only a brief sensitivity analysis was carried out, the limitations of which are all too apparent. The chapter then demonstrates GLUE by way of a 2D CFD application, and shows how predicted uncertainties and risk maps of model reliability can be used alongside model predictions.

17.2   Scenario 1: Wet weather impact of estimated sewer spills on the designated Shellfish Waters of the Ribble Estuary

This investigation formed an initial assessment of the impact of the aggregated spills from the CSOs from four major sewer catchments on the Shellfish Waters of the Ribble Estuary using the Ribble Estuary Model. The EA's Ribble Estuary Model was developed by external consultants (Kashefipour *et al.*, 2000a) and comprised five sections of river (three sections of the upper tidal Ribble plus the tidal Darwen and the tidal Douglas) that were all modelled using the 1D FASTER model (Flow and Solute Transport in Estuaries and Rivers – Kashefipour *et al.*, 2000b) and a larger portion of the estuary and estuary mouth that was modelled using DIVAST (Depth Integrated Velocities and Solute Transport – Falconer and Lin, 2000). DIVAST uses a third-order accurate QUICKEST numerical scheme (Lin and Falconer, 1997). It is not the intention here to give a detailed description of the numerical models involved, save to say that both models have been extensively tested in the laboratory and the field (Chapter 12).

The Ribble Estuary Model came pre-calibrated and partly verified in terms of hydrodynamics and water quality in terms of Faecal Coliform (FC) count. When it came to using it 'online' during the AMP3 process, there was little time to carry out large numbers of runs, and several scenarios were run instead, as is often the case. A risk map of predicted uncertainties would have been exceptionally useful alongside the predicted FC concentration map that came out of the study.

The combined spill volumes and spill durations for all the catchments on any particular date were estimated from sewer hydraulic model predictions produced by the water services provider based on 10 years of summer rainfall data. Only summer rainfall was used because the Bathing Water Standards (as opposed to Shellfish Waters which require all year round rainfall to be used) were initially considered as more important, and the English Bathing Season (BS) lasts between May and September. A number of scenarios were investigated using the predicted spills for the existing situation, and the future predicted spills once hypothetical improvement works have been undertaken to increase the amount of storage in the sewer network. These different storage scenarios are summarised in Table 17.1. The aim was to estimate how often the Shellfish Water Standard of 1500 FC/100 ml for 97% of the time anywhere in the designated water might be exceeded.

Table 17.1 Summary of scenarios investigated (the order of the event is simply the rank in terms of total spill volume over all combined events)

Scenario	Spill from catchment	Storage for estimated Spill per bathing season solution	Order of event from 10 years of summer rainfall data	Significant impact shellfish waters?	Significant impact on bathing waters?
1	All four catchments	Existing situation	31	Y	Southern-most
2	All four	3	31	Y	N
3	All four	3	21	Y	N
4	1	2	21	N	N
5	2	2	21	Y	N
6	2	2	31	N	N

Whilst many of the hydrodynamic and water quality parameters have been fixed in this study as the same as in the original wet weather scenarios from the Ribble Model Report (Kashefipour *et al.*, 2000a), further subjective decisions had to be made over which scenarios should be investigated. In what represents a typical approach to modelling complex systems when there is little time available, the scenarios attempted to strike a balance between conditions that could be perceived to yield pessimistic and optimistic predictions. For a GLUE-type analysis, it would be important to include these different physical situations, but by incorporating thousands of Monte Carlo simulations that build in uncertainty in parameters and boundary condition. This would require a huge number of simulations but perhaps this would be more commensurate with the enormous complexity of the system being investigated.

17.2.1 Pessimistic conditions (could result in over-estimation of impact)

A neap tide and a southerly wind were selected, since these are known from the Ribble Model Report (Kashefipour *et al.*, 2000a) to have significant impacts on the designated sites. The wet weather scenario on which the simulations were based only uses wet weather flow data for 24 hours from the contributing rivers (Ribble, Douglas and Darwen). For the scenarios that involve a storage solution in the sewer network, the average time of the spills contributing to each combined catchment event is used (discussed later). This could represent an over-estimation, as the storage solution could reduce the average spill duration for individual CSOs contributing to combined catchment events. Relative timing from sub-catchment, or effect of storage on relative timing is also an issue in reducing impacts, although there was insufficient time to make a completely linked model of the hydraulic sewer model and Ribble Estuary model.

17.2.2 Optimistic conditions (could result in under-estimation of impact)

Only water quality inputs from CSOs were investigated, with no inputs from Waste water Treatment Works (WwTWs), including storm tanks known to be major spillers in wet weather, or any consideration of the inputs from the tributaries of the estuary (including the Ribble, Douglas and Darwen). The total wet weather FC load from CSOs was estimated at 13.7% of the total load entering the estuary, before UV disinfection began at Preston sewage works (CREH, 1998). The percentage load that could be attributed to CSOs in wet weather, now that UV treatment has been implemented, is expected to be significantly larger than 14%. A CSO water quality indicator concentration of FC 3 000 000/100 ml was used throughout, whereas other studies have used values an order of magnitude larger than this (1×10^7).

Typically, the 21st or 31st worst combined catchment event (on one date) emanating from the sewer catchments is examined, rather than the worst event. This is because the hydraulic modelling uses 10 years of summer rainfall data, and the bathing water Environmental Quality Standard (EQS) is three spills per bathing season (hence allowing 30 spills, and designing sufficient storage to cope with the 31st spill). The EQS may be reduced to two spills, or one spill every five years depending on the nature of the impact and the soffit level of the outfall in relation to mean low water springs.

The rainfall data comes from a 10-year summer time series, and winter contributions are not considered. If a yearly rainfall time series was used to drive the sewer models, it is likely that the spill events would be more frequent, larger and for a longer duration.

17.2.3 Modelling assumptions

It is recognised that uncertainty in the Ribble Estuary Model predictions arises from multiple sources, including process uncertainties relating to approximations such as using a depth-averaged model, uncertainties in the original calibration and the resulting effective parameter values and uncertainties in the initial and boundary conditions. The calibration against direct field measurements that was undertaken during the construction of the model afforded a degree of confidence that the model represented a feasible simulator of the system.

It was assumed that the combined catchment spill events (sum of all CSO spills from a sewer catchment) and mean duration for each CSO spill over the whole catchment could be used to approximate the effects of different return period events. One of the wet weather scenarios described in the Ribble Model Report (Kashefipour *et al.*, 2000a) was taken as a basis for looking at these combined catchment spills. For this scenario, there are no sources of bacteria other than those from the CSO spills. The original Ribble Model Report states that the wet weather scenario used represented one of the worst-case wet weather scenarios in terms of the likely impact on the bathing waters. Other conditions for this scenario were a southerly wind, with release of spills for around 50 hours in a 134-hour simulation.

The time for 90% die-off of FC (T_{90}) was estimated from the original calibration to be 86 hours. Flows from rivers and WwTWs were set as for the series of baseline runs, but with zero-quality input. For wet weather baseline runs, this means using a 12-hour start up period with no quality input, 24 hours of Dry Weather Flow (DWF) data, 24 hours of wet weather data and the remainder of the simulation back to DWF. This study uses spills up to 24 hours in duration, whereas the previous maximum duration investigated was 7 hours.

For scenarios other than scenario 1 (which is for the estimated impact with no improvements to the CSOs), the following steps were carried out for each sewer catchment to assess the likely impact of the permitted spills following a three spill per BS per individual UID (Unsatisfactory Intermittent Discharge) solution:

Step 1: The spills were ranked in descending order of magnitude and the 31st spill volume was used as an estimate of the storage that would be required to meet 3 spills per BS for each CSO in isolation (an estimate was used where one spill = 24 hours which made data manipulation more simple when combining events, but represents a less-stringent definition of an event than the agreed definition of 12 hours).

Step 2: All of the spills other than the first 30 were removed, and the storage volume (or 31st spill volume) was also subtracted from the remaining spills. This could approximately represent the situation when all of the unsatisfactory CSOs (UCSOs) have undergone AMP3 improvements.

Step 3: All the dates from the 10 years of data on which these spills occurred were ranked in terms of the largest combined spill on any one date. This treats each catchment as one large UCSO with a storm discharge into the Ribble or major tributary.

Step 4: Following possible improvements, and allowing for 30 spills over 10 years from each UCSO, the 31st worst-amalgamated spill by date was determined for each catchment. This spill volume represents a true possibility given the scenario that all of the improved UCSOs are allowed to spill independently three times per year. The mean duration of all the spills contributing to the combined spill was also determined as another input to the Ribble model.

The total amount of hypothetical storage added to the different systems in step 2 is not given here, although the values used here were deliberately conservative owing to the event definition used (24 hours = 1 spill). The durations of spills used were the average of the events contributing to the combined spills. This is likely to be an overestimation following a spill solution. In what follows, only the results for scenarios 1 and 2 are shown to provide a flavour of how the solutions were presented to managers.

17.2.4 No storage solution

The 31st spill volumes for combined catchment events for 10 years of summer rainfall data were combined as inputs to the Ribble Model, assuming no storage solution imposed. Some of the predicted spills were in excess of 24 hours for one of the catchments, so in order to limit the total run time of the simulation the same volume of predicted spill was discharged over 24 hours. It should be noted that the volume spilled from one catchment was discharged over a very short time period and could skew the results. The exceedance of the 1500 counts/100 ml EQS is shown at grid reference SD 3673 2650 (the eastern boundary of the designated Shellfish Water) and the 11th milepost (inside the Shellfish Water).

The duration of the EQS exceedance on the eastern boundary of the Shellfish Water for this event was 36.02 hours. Figure 17.1a illustrates the spatial distributions of FC above 1500 counts/100 ml for 2 different times, with Figure 17.1b better representing the temporal behaviour. Considering Figure 17.1, it would appear that without improvement work, the Shellfish Waters are under threat, although there is no indication of the likely uncertainties in the predictions.

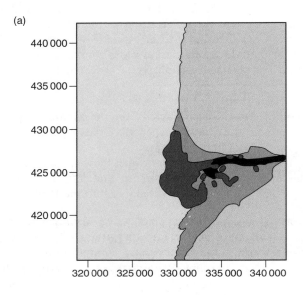

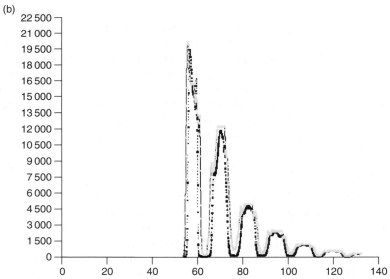

Figure 17.1 (a) Spatial distribution of FC concentrations for scenario 1: black implies FC counts > 1500 counts/100 ml, at time = 55 hours, and dark grey also implies FC counts > 1500/100 ml at time = 80 hours. The grey to the right is land, with lighter grey indicating dry cells, and light grey to the left of the figure indicating wet cells. (b) Exceedence of Shellfish Water Standard at SD 3673 2650 – solid line (light grey), and at the 11th milepost dashed line (black).

17.3 Scenario 2: Improvements to sewer storage assumed – Three spills per BS per UCSO storage assumed, for all catchments

Scenario 2 assumes that sufficient storage for the three spills per BS per UCSO solution has been put in place, and uses the 31st magnitude combined catchment event as an input to the Ribble Model. The modelling results show that the three spills storage solution could still result in exceedance of the Shellfish Water standard for FC (1500 cfu/100 ml) at the 11th milepost, although for a lesser duration than for scenario 1. This is without considering any other source of FC other than the WW load from the UCSOs, and assuming a conservative concentration of 3×10^6 cfu/100 ml FC.

The 31st worst catchment events were substituted for the spills in the wet weather scenario, and the resulting impact is demonstrated in Figure 17.2. Figure 17.2 indicates that the Shellfish Waters guideline value of 1500 FC counts/100 ml is exceeded during the second tidal cycle at the 11th milepost following input. Figure 17.2a gives the spatial distribution of FC counts in excess of the *earlier indicative standard* of 1100 counts/100 ml – Figure 17.2b should be referred to in order to compare against the 1500 counts/100 ml standard. The modelling suggested that three spills per season storage solutions would improve the quality by a substantial amount, but that this would not necessarily lead to a satisfactory solution in terms of the impact of the amalgamated spills on Shellfish Waters.

There are many possible scenarios that could be investigated using the Ribble Model, given the large number of variable parameters and assumptions in the modelling process. Clearly the significance of the findings of these Ribble Model

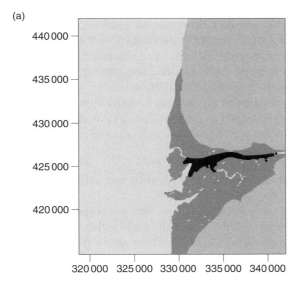

Figure 17.2 *(a) Snapshot of FC concentration distribution (where black indicates FC concentration > 1100 counts/100 ml) at T = 85 hours, dark grey indicates dry cells, and grey to the right is land, with light grey to left indicating open sea. (b) Time series output from Ribble model at 11th milepost for scenario 2. Faecal coliforms exceed the Shellfish Water Standard of 1500 counts/100 ml.*

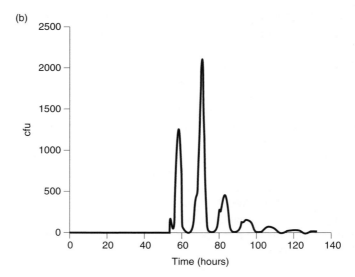

Figure 17.2 (Continued).

runs also depends on the CSO-generated percentage contribution to the total wet weather load contribution. The Kashefipour *et al.* (2000a) report estimated that wet weather loads to the system were around 14% of the total load, although this was before UV disinfection had commenced at Preston WwTWs. Therefore the percentage load from the CSOs would now be expected to be proportionately larger (although the total load would clearly be reduced).

Many assumptions were made in deriving the above FC concentration distributions, and so there are many possible parameter combinations that could have been equally justifiable but which might have yielded different behaviours that could have been acceptable on the basis of the limited calibration data. Yet all these behaviours remain unexplored, so studies such as these would benefit considerably from a full uncertainty analysis of the kind explained in later sections in this chapter. First, however, section 17.4 paints a qualitative picture of the importance of representing uncertainty.

17.4 The role of uncertainty in hydraulic modelling

Consider a flood wave propagating across a complex zone of inundation with uncertainty in the distributed topography and boundary roughness, and with buildings, trees, roads, railway lines affecting its progress. Consider the complex flow past a zone of poor initial mixing, resulting in a region of intense transverse shear that gives rise to coherent structures being shed chaotically and interacting with the underlying complex bed topography, which itself undergoes scour and deposition under higher flows. Consider flow past a strong region of shear during an over-bank flow, with strong secondary current history effects from upstream. Can these situations be modelled to give predictions of flow and contaminant transport which merit such confidence that only a single solution needs to be reported? Perhaps all those uncertainties in the bed topography, distributed effective roughness and initial conditions as defined using a simple flow meter upstream cancel out nicely, but this seems unlikely.

An alternative viewpoint is that a degree of process and parametric uncertainty needs to be accepted for these complex systems (Beven, 2002a,b). This can be evaluated in a relativistic framework through generating a set of weights based on performance measures for a set of model structures that exhibit a similar functionality to the system. These weights can then be used to generate prediction limits for different scenarios for that system, and can be updated in a Bayesian framework as more information becomes available. The Figure 17.3 illustrate how this uncertainty can be represented, although perhaps not in a way that is accessible to managers in the water sector.

Figure 17.3a is an attempt to show the uncertainty in a distributed flood model, modelling the peak of a flood wave. The crosses indicate the 10th and 90th percentile model predictions of water surface heights at different locations within a large river basin, with the data shown as circles. The figure indicates the range of more feasible model predictions for 1000 simulations of a finite element diffusive wave CFD application across a large Sicilian floodplain (Aronica *et al.*, 1998a). Feasible models are those models for which the key parameters are within the range of values that are accepted as being possible for the system. The sensitivity here is to the range of possible effective distributed roughness coefficients assigned in large blocks to the floodplain and river. Data constraints were limited to 17 measurements of the peak flood depth at distributed sites. The prediction limits are the 10th and 90th percentiles of all model predictions at each location. Figure 17.3b shows the results for the optimum parameterisation from Figure 17.3a, but incorporating assumed, spatially correlated topographic errors in order to represent a typical measurement error or (more likely) interpolation error arising from the construction of the finite element mesh (Aronica *et al.*, 1998b). Figure 17.4 shows predictions from a set of finite element models attempting to model the situation in the second question posed previously.

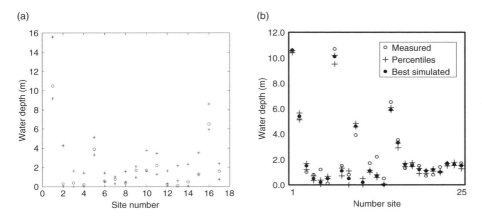

Figure 17.3 (a) Fuzzy prediction limits of depths in Imera Basin (after Aronica et al., 1998a, reproduced with kind permission of Elsevier Science). (b) Prediction limits for topographic uncertainty. (After Aronica et al., 1998b, reproduced with kind permission of A.A. Balkema Press.)

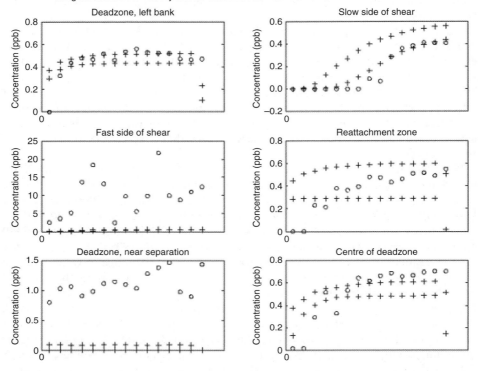

Figure 17.4 *Prediction limits of tracer concentration – after Hankin et al., 2001b. (Reproduced by permission of John Wiley & Sons Ltd.)*

The transport and mixing of a tracer in a river reach with a strong zone of lateral shear is examined in an uncertainty framework. Here, the crosses are predictions of tracer concentration time series in and around a zone of poor mixing, with the data as circles. The final scenario described earlier, for mixing in an overbank flow, is represented in Figure 17.5 using random particle tracking to model tracer dispersion in a study carried out in the Flood Channel Facility, Hydraulics Research, Wallingford. In all of these cases multiple realisations of each model have been run and compared with observational data. A numerical performance measure is calculated and used to weight the predictions of each model. Models that do not give good results (which are *non-behavioural*) are given a weight of zero and not considered further. The uncertainty bounds shown are prediction bounds of the weighted predictions over all *behavioural* models. This is the essence of the GLUE methodology, first reported by Beven and Binley (1992). All these figures show comparisons of the model predictions with point observational data but the methodology also allows all the distributed predictions of the model to be presented in this way. Thus, Figure 17.6, for example, shows a map of normalised weighted standard errors for the concentration field also represented temporally in Figure 17.4. The dispersion experiment used to derive these plots is described more fully in section 17.7. In order to produce this type of figure it is

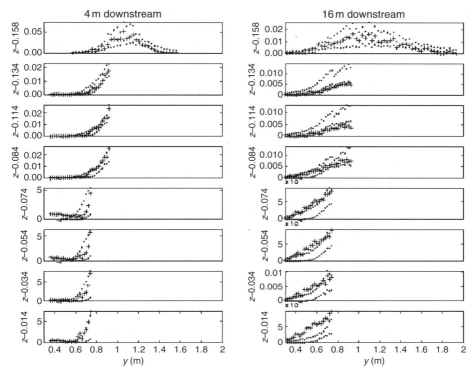

Figure 17.5 Observations (+) and 10 and 90 percentile predictions of tracer concentrations (.). Calibration of 4 different structure random walk models each using 1000 different combinations of vertical and transverse step sizes in each random particle tracking model at 4 m downstream. The models are then used predictively to estimate the dispersion at 16 m downstream using the set of weights derived from the model performances at 4 m. (Reproduced from Hankin and Beven, 1998b, with kind permission from Springer-Verlag GmbH & Co. KG.)

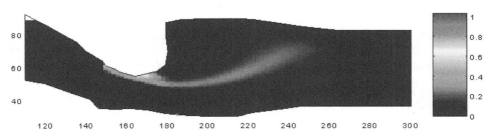

Figure 17.6 Map of prediction bounds. The normalised weighted deviation from the mean concentration field as an indicative measure of risk that model predictions are more uncertain. A measure of 0 (dark blue) represents locations where there is little deviation between all the different model predictions, and a measure of 1 means there is a large deviation, which gives rise to wider prediction limits. (See Plate 6.)

necessary to extend the weights derived for individual measurement sites to the entire concentration field. The next two sections describe (section 17.5) and illustrate (section 17.6) the GLUE approach upon which the above examples are based.

17.5 The GLUE procedure

The GLUE procedure (Beven and Binley, 1992; Beven and Freer, 2001; Beven et al., 2001) provides a fuzzy measure of the acceptability of different possible model structures and parameter sets. This is achieved by mapping the real systems into a model space, via multiple model simulations, using a performance measure which expresses our relative degree of belief that a set of parameter values within a chosen model structure provides a good simulator of the system. For complex surface water flow applications, the GLUE procedure has previously been used by Romanowicz et al. (1996) to produce flood risk maps on a section of the River Culm, by Aronica et al. (1998a), in the fuzzy calibration of a distributed FE flood inundation model, and by Hankin et al. (2001b) to calibrate a commercial CFD package in an uncertainty framework.

For this illustration of GLUE it is assumed that water quality concentrations have been measured at different spatial locations at different times. The first stage of the GLUE analysis involved calculating a set of measures indicating the goodness of fit of the various predictions to the observed data. One such measure, M_1, is given by equation (17.1), and is calculated for each spatio-temporal data set:

$$M_1^j = \frac{\sqrt{\sum_{t=1}^{N_t}(C_t - C_t^d)^2}}{\frac{\sum_{t=1}^{N_t} C_t^d}{N_t}} \qquad (17.1)$$

where C_t and C_t^d are the predicted and observed tracer concentrations at site j respectively, and N_t is the number of time steps at which values of water quality have been collected. The measures are combined from each site to give a total measure of goodness of fit over the whole flow domain using equation (17.2):

$$M_2 = \prod_{j=1}^{N_S} e^{-M_1^j} \qquad (17.2)$$

where N_S is the number of sites. At this point any models that do not reach a certain level of performance can be rejected as non-behavioural and given a performance measure of zero. The M_2 measures can then be rescaled over all behavioural simulations to give a set of weights that sum to 1 as:

$$\text{RPM}_k = \frac{M_2^k}{\sum_{j=1}^{N} M_2^k} \qquad (17.3)$$

These weights then represent a relative degree of belief in each simulation, or a relative (subjective) likelihood for each simulation. Since these measures do not provide direct estimates of the probability of predicting an observation given the model, as in classical likelihood theory, they can perhaps be better interpreted with the framework of fuzzy set theory as Relative Possibility Measures (RPM). The mapping of the real system into the model space, by means of the weights associated with the behavioural models, can also then be considered as a fuzzy (rather than probabilistic) mapping.

17.6 Demonstration of how GLUE could be used in CFD modelling with uncertainty

17.6.1 Transport model

This study investigated contaminant transport using a 2D CFD model in a much smaller flow domain than the Ribble Estuary; nonetheless, the techniques are easily transferable. The site comprised a large natural dead zone on a reach of the River Severn at Leighton near Shrewsbury (see Hankin *et al.*, 2001a,b, 2002 for more details). Intensive tracer measurements were made for a single flow of approximately $36\,m^3\,s^{-1}$. Tracer was injected continuously from the location indicated on Figure 17.7 for 220 minutes and then switched off. A continuous rate of tracer injection was achieved using a constant head bottle, and was switched on at 1.00 pm at a rate of injection of $1.817 \times 10^{-6}\,kg\,s^{-1}$. The tracer dispersion was monitored at site 6 at the centre of the dead zone using a pre-calibrated fluorometer operating in flow through mode using a sampling rate of 1 Hz. At a further five sites, distributed around the dead zone, small sample bottles were filled 10 minutes after tracer injection commenced and then every 15 minutes until 10 minutes after the tracer injection was terminated. Such distributed time series were available for FC counts/100 ml during

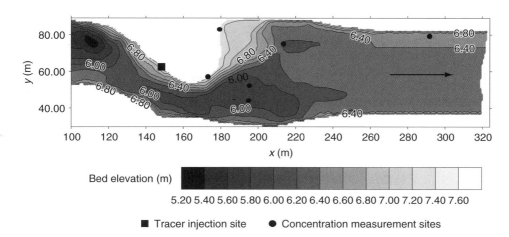

Figure 17.7 Bed elevation and details of experiment, after Hankin et al., 2001b. (Reproduced by permission of John Wiley & Sons Ltd.)

wet weather events on the Ribble. In addition to the concentration data, long period Acoustic Doppler Velocimeter (ADV) velocity time series were recorded at 4 locations in the shear and dead zones, also indicated in Figure 17.7.

The hydraulic scheme applied in this analysis is a 2D finite element model, TELEMAC-2D, as opposed to the finite difference-based DIVAST model used for the Ribble. TELEMAC-2D has also previously been extensively compared against field measurements for floodplain flows with similar strong regions of transverse shear – (Bates *et al.* (1995, 1997, 1998, 1999); Bates and Hervouet (1999); Horritt (2000)). TELEMAC-2D calculates the free surface hydraulics by solving the second-order partial differential equations for depth-averaged free surface flow derived from the full 3D Navier–Stokes equations. Details of the equations and the numerical techniques for solving them can be found in many of the above cited references, and also in Hankin *et al.* (2001b) but are omitted here as it is the calibration approach that is of interest.

The important feature of the CFD model set-up was that it was assumed that the local depth-averaged flow velocity ($<U>$), the local depth (H) and the local roughness (n) all exert a strong control on the eddy viscosity, ν_T, which governs the rate of turbulent momentum transfer in all directions for a zero-order closure model (e.g. Speziale, 1985). The relationship derived through physical reasoning in Hankin *et al.* (2001b) took the form given by equation (17.4):

$$\nu_T = K_{\nu n}<U>Hn \qquad (17.4)$$

where $K_{\nu n}$ is a calibration factor, having the inverse dimensions [$L^{1/3}T$] to Manning's n in order to make the equation dimensionally consistent. $K_{\nu n}$ was estimated from typical values of the other quantities in equation (17.4) to be in the range of 0.1–0.9. The coefficient was then treated as another calibration parameter, such that for any one simulation it would be constant throughout the flow domain, thus maintaining the same scaling of the eddy viscosity with the local Manning's coefficient. In this way, once the Monte Carlo simulations had been undertaken, any consistent scaling between these variables could be assessed for the model structures that attained a high performance measure. Furthermore, it was discovered that the scaling of the eddy viscosity in this way led to relatively low numerical diffusion (Hankin *et al.*, 2001b).

The boundary conditions used were an imposed flow rate at the upstream end of the reach, whilst at the outflow boundary a water surface elevation was imposed. On the other boundaries around the domain, a non-slip condition with zero flux was imposed. Combined with the grid topography, the known discharge can be used by TELEMAC-2D to generate a parabolic transverse velocity profile at the inlet cross section, given the assumption that the water surface slope is parallel to the bed slope at this point. This profile was allowed to evolve into a more natural profile over a lead-in section. The stage was fixed at the downstream outlet, allowing the model to specify both the velocities and distributed free surface elsewhere, but still giving a well-posed numerical problem.

Complete grid independence cannot be obtained for this type of complex boundary condition flow. The reason for this is that the topographic representation changes between grid resolutions, so the values of roughness parameters must also be changed in order to consistently predict distributed measurements (such as point measurements of velocities). This is because the roughness values, which are used in the depth-averaged model, are 'effective' parameters: they control the grid-scale rates of energy and momentum losses, but represent an integration of the effect that sub-grid processes are having on the flow as a whole (Hardy *et al.*, 1999); Chapter 10 in this book.

A single mesh having 3628 nodes was used throughout, and each simulation had to be run for 14 400 s with 1-s time step to cover the duration of the tracer experiment. The numerical tracer was injected at time $t = 600$, and switched off at time $t = 13\,800$ (duration 13 200 s, or 220 minutes). A one-second time step was used, in order to maintain a Courant number of approximately unity or less, and each model was run for just over the period of tracer injection. A single simulation took approximately 10 hours, and the 360 simulations were undertaken using a parallel LINUX PC facility.

17.6.2 Calibration

A large number of simulations were then undertaken, varying several important parameters between feasible ranges. The relevant parameters were the values of the Manning's roughness coefficient in the shear zone (3), dead zone (2) and main river (1), the coefficient K_{vn} in equation (17.4), and random error term added to the inflow discharge and outflow stage to represent uncertainty in these observations. The different parameter ranges and boundary condition data ranges and the order of generation are given in Table 17.2.

Table 17.2 Parameter generation and ranges

Order	Description	Range
1	Generate random Manning's coefficient for zone 3 (shear zone) in interval given in column three. Chow (1959) relates a value of Manning's n of $0.05\,\text{m}^{-1/3}\,\text{s}^{-1}$ to a stony stream. The Severn is a lowland river, where large values like this represent an over-estimate, so it is highly possible that if this is used as an upper threshold to the range, then most feasible situations are covered	$0.01 < n_3 < 0.05$
2	Generate random Manning's coefficient for zones 1 (main river) and 2 (dead zone) assuming that these are smaller than that of zone 3	$0.01 < n_2 < n_3$ $0.01 < n_1 < n_3$
3	Generate random value of K_{vn} in equation 17.4, and use this for all zones	$0.1 < K_{vn} < 0.9$
4	Generate random error term to add to upstream inflow. The error is set to be equal to the estimated measurement error	$Q = 36 \pm 0.5\,\text{m s}^{-1}$
5	Generate random error term for stage at downstream outflow. The error is set to be equal to the estimated measurement error	$H_0 = 7.3 \pm 0.01\,\text{m}$

These parameter and boundary condition ranges are crucial to the GLUE procedure. Parameter ranges must reflect values that are considered feasible given knowledge of the system. Often it is better to assume as wide a range of parameter values as observed in historical experiments, with no *a priori* assumption of probability distributions. Those parameter values which yield model predictions that give a relatively poor fit to observations will be rejected through some goodness of fit criterion which can be designed into an uncertainty management system. Of course, there are subjective judgements required in selecting both the initial range of parameters and the threshold for what constitutes a good level of fit, much like the decisions that are made over what level of statistical significance should be used in more conventional approaches. The process is fuzzy, and fits with the fuzziness of our understanding of the natural system, embodied as it is, for a few spatio-temporal measurements typically taken for a limited number of excitations of the system variables. As new data becomes available for a given model, so long as there is no redundancy in the new data, the behaviour of the modelled systems will be constrained further and more parameter combinations will be removed.

Not all of the parameter combinations yielded stable solutions. This could be due to amplification of approximations at certain locations in the numerical grid for certain parameter combinations. In fact only 118 of the 360 simulations undertaken were considered to be behavioural on visual inspection. Non-behavioural state encompasses the non-stable solutions and solutions which clearly do not reflect gross properties of the system – for instance some parameter combinations yielded no recirculation cells and were therefore rejected as recirculation was observed using tracers. The parameter combinations which gave rise to, for example, no recirculation were not looked at, although it is likely that certain groupings of parameters give rise to similar behaviour and this might well be a useful line of investigation for greater efficiency, should more simulations be carried out.

Figure 17.8 gives a representative sample of the 118 behavioural simulation velocity fields that were produced, with Figure 17.9 showing the corresponding tracer concentration fields. The dimensions and intensity of the recirculation were sensitive to variations in all these parameters, although there were also multiple model structures that yielded similar patterns, an indication of the problem of equifinality of models that are consistent with the calibration data available (Beven, 1993, 2002a,b; Beven and Freer, 2001). A similar equifinality phenomenon was found in a study of turbulent tracer mixing in the strong transverse shear layer of an overbank flow in the Flood Channel Experimental Facility, Wallingford (Hankin and Beven, 1998a,b) – as shown earlier in Figure 17.5. All of these simulations, corresponding to different parameter combinations, could be interpreted as relatively acceptable on the basis of the field data, with the proviso that we are not interested in the detailed flow structure within the shear zone (for which we have limited data). With improved resolution of data the range of parameter combinations that would retained within the GLUE procedure would be narrowed down, in fact it might be the case that higher-order models would need to be used – such as a second-order turbulence closure.

The TELEMAC-2D numerical scheme is second-order accurate, and approximate grid-independence for the mesh on this stretch of the Severn was demonstrated in an earlier paper (Hankin *et al.*, 2002), although complete mesh independence for such

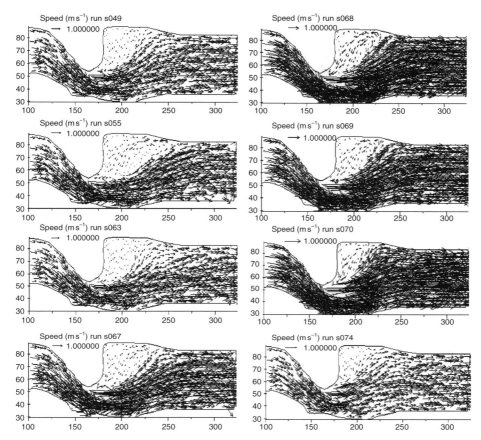

Figure 17.8 Depth-averaged flow vectors from some selected simulations, each having slightly different model structure (t = 14 400 s). (Reproduced from Hankin et al., 2001b, with kind permission of John Wiley & Sons Ltd.)

complex boundary flows is unlikely (Horritt, 2000). It was also found that the first-order estimate of numerical diffusion was smaller or the same order of magnitude as the eddy viscosity. It is therefore put forward that physical diffusion is not being swamped by the numerical diffusion. It could be added that the numerical approximation will be interacting on one level with parameter identification, but perhaps this is a further reason why such model structures ought not to be optimised.

17.6.3 Fuzzy calibration using GLUE

Having computed the Monte Carlo simulations, all that is left to do is estimate the weights or relative possibility measures that were described in section 17.4. It would have been preferable to have many more simulations, and for a model covering such a large domain as the Ribble Model, requiring a greater number of parameters, tens of thousands of simulations would be more suitable.

478   *Computational Fluid Dynamics*

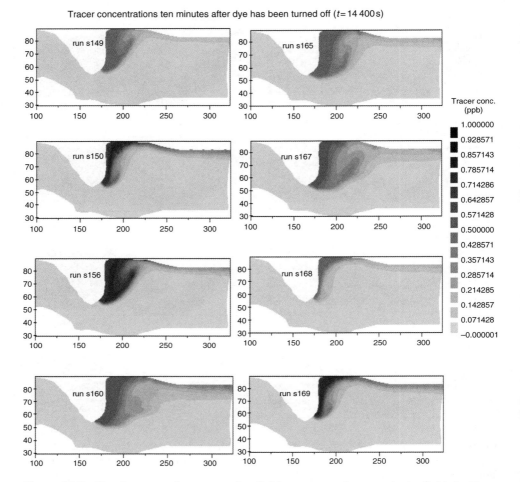

Figure 17.9 Depth-averaged concentration fields corresponding to velocity fields in Figure 17.8. (Reproduced from Hankin et al., 2001b, with kind permission of John Wiley & Sons Ltd.)

The next stage of the analysis could easily be repeated for the Ribble Model and involved interpolating the predicted tracer concentrations to the locations of the sampling points where the tracer time series were taken, using an inverse distance-weighted nearest neighbour interpolation. Figure 17.9 shows concentration fields at a time ten minutes after the tracer was turned off, as the un-contaminated water intruded back into the dead zone. Close inspection of Figure 17.9 indicates considerably different steady states, with run s150 having a smaller, but more concentrated retention of tracer and run s160 having perhaps the most extensive, but less concentrated area of tracer retention.

The RPMs were determined using equation (17.3) and were then used as weights to rank the set of model predictions. This is achieved by ordering the tracer concentration predictions at each site in order of magnitude, and forming cumulative RPM distributions for each site, as indicated by Figure 17.10. These cumulative

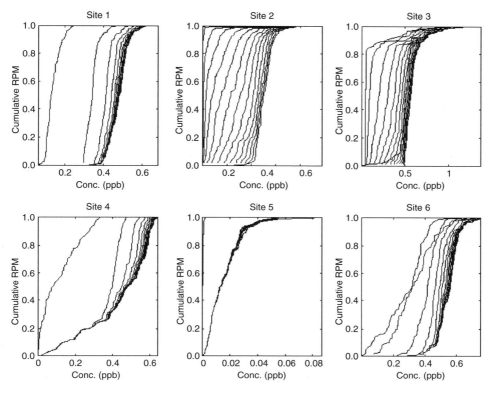

Figure 17.10 Cumulative relative possibility measures at each site with each curve representing the range of model fits at every time step (there were 15 time steps 15 minutes apart, the first one starting 10 minutes after dye injection). (Reproduced from Hankin et al., 2001b, with kind permission of John Wiley & Sons Ltd.)

distributions can then be used to derive percentiles of the more possible sets of parameter combinations, and in this instance the 10th and 90th percentiles are used to form prediction limits. These represent limits to the more feasible sets of predictions, given the constraints imposed upon behaviour by the tracer concentrations. These were given earlier in the chapter in Figure 17.4, where it is evident that the overall behaviour of the system is being captured quite well, although all of the numerical model simulations are failing to predict the tracer concentrations at sites 3 and 5. The subtleties of the tracer movement in the very shallow location at site 3 in close proximity to the flow separation from the bar simply could not be modelled with the grid scales used in the model. Site 5 was at the extreme right-hand side of the extent of the estimated region of strong shear (see Figure 17.7), and failure to predict at this location must be put down to some form of resolution error.

However, at the other four sites, including at the location where the fluorometer was used directly, the prediction limits represented the spatio-temporal behaviour of the observations well, considering the complexity of the system. The most difficult location to predict tracer concentration is in and around the shear zone, a problem which is reflected in the risk map given in Figure 17.6. The map indicates that the dispersion of the tracer is highly sensitive in this region of strong shear as might be expected. Given the under-prediction at site 5 and the over-prediction at site 4, it would appear that the predicted separation point and resulting shear zone are slightly askew. However, increasing the resolution of the finite element mesh is not necessarily a solution, as discussed in Lane and Richards (1998), because if the grid spacing for the finite element mesh was made overly fine relative to the topographic resolution then the flow solution will in part be a product of the nature of the topographic interpolation. If too coarse, the flow predictions become a product of the discretisation rather than the topography.

Long-period temporal and spatial unsteadiness was evident from the velocity time series that were collected on the day. This is inherent to the vortex shedding processes around a region of strong shear, and strong evidence for increased spatial variability in the eddy viscosity, if we are interested in modelling the interactions within the shear zone. ADV measurements were taken at site 6, where the flow was quiescent for long periods of time, of the order of minutes, followed by periods of rapid turbulent motion. This process clearly cannot be precisely represented through the use of a time-invariant eddy-viscosity closure. The behaviour was thought to be responsible for the rapid oscillations seen in the data at sites 3 and 5 and really calls for solvers with more sophisticated handling of temporal variations, such as Large Eddy Simulation (LES) (e.g. Bradbrook *et al.*, 2000). The oscillations were not reproduced by the zero-equation model, nor should they be expected to be.

17.7 GLUE approaches in a dynamic framework

Despite the limitations of the number of simulations that were computed for the Severn Study, the usefulness of the GLUE approach to generating predictive uncertainties has been demonstrated. The GLUE approach avoids optimising the structure of a complex distributed effective parameter model in an environment where boundary conditions and validation data will never be precisely defined for the entire flow field. This is all the more important for modelling studies such as the Ribble Estuary Model, where the spatio-temporal scales are very large, and the sensitivity to initial and boundary conditions can be amplified through the non-linear nature of the governing flow equations.

Once a set of core parameters have been identified such as in Table 17.3, and RPM weights have been generated for different combinations of these parameters, these can then be used in a predictive sense for modelling for example a new set of faecal coliform loadings perhaps for a new storm. This is achieved through re-running all the model structures (different parameter combinations) for the new storm and again ranking the predictions at each observation site by magnitude in terms of concentration. The RPM values that were estimated for each model structure using equations (17.1)–(17.3) for the calibration data are then again used to produce cumulative RPM plots for the newly ranked data. The 10th and 90th percentiles of these predictive concentration values are then used to produce another figure such as Figure 17.3 (e.g. Hankin and Beven, 1998b).

As more or improved data is obtained, the RPM can be updated in a Bayesian framework (e.g. Romanowicz *et al.*, 1996), and in the light of this new data any or all

Table 17.3 Principal process and parameter uncertainties in Ribble Estuary Model

Process uncertainty	Parameter uncertainty
CSO spill volume and duration	Pipe roughness, sediment, storage, weir height in CSO
Initial and boundary conditions for sewer model	Rainfall prediction model (typically Stormpac) parameters relating to quantity, spatial distribution and relief
Initial and boundary conditions for Ribble Estuary Model	Uncertainties of inflows, and tidal boundary conditions, effluent strength
Bed friction	Manning's *n* in 1D sections and 2D estuary section
Mixing rates	Distributed turbulent diffusion coefficients, and friction coefficients
Faecal coliform die-off rate and possible storage in bed sediments, with later re-suspension	T_{90}, sediment transport parameters
Wind-driven currents	Momentum transfer coefficients

of the model structures can be rejected. If all of the model structures are rejected, then it is likely that not all of the processes in the system can be represented by the model, and a more sophisticated model would need to be constructed.

The RPMs are plotted against the variable parameters and boundary conditions in Figure 17.11 in order to give an indication of the more feasible model structures, and whether there are any strong correlations. Of course, for such a parameter space, far more models need to be investigated in order for this part of the study to be more conclusive. The flat tops to these plots for five out of six of the first six plots indicate that the model structures across the parameter ranges investigated are equally acceptable on the basis of their fits to the data. This type of equifinality in quantitative performance measures has been found to be a common feature of distributed models of complex systems (Beven and Freer, 2001; Beven *et al.*, 2001). In this situation, an optimisation scheme that selected only the best-performing model structure would engender a degree of confidence in the model predictions that is not justifiable.

17.8 Implications of undertaking an uncertainty analysis for the Ribble Estuary Study

The previous sections showed how uncertainty bounds or ranges of feasible model behaviours can be generated when predicting distributed concentrations using a CFD model, so this can also be applied to the Ribble Estuary Model. The computational resources required to undertake such a study are no longer unaffordable, and as processor speed has increased rapidly, the large number of runs required for the uncertainty analysis would not consume a significant amount of time or resources. The time required to undertake the multiple runs carried out in the study described below was considerably reduced through using a set of 20 PCs linked via a hub. Such a system would cost as little as £20 000. Further to these requirements it would be

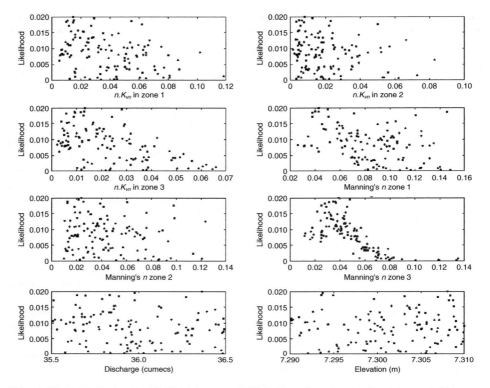

Figure 17.11 *Relative Possibility Measures (RPM) plotted against each of the variable parameters or variable boundary conditions. (Reproduced from Hankin et al., 2001b; with kind permission of John Wiley & Sons Ltd.)*

necessary to re-visit the calibration data that was used in setting up the original study to condition the model for the wet weather runs. Since the RPM weights depend upon the goodness of fit to the observational data, the quality of the uncertainty analysis also depends upon the quality and resolution of the original calibration data. Table 17.3 gives a list of the different uncertainties that would need to be considered in order to carry out the Shellfish Water study in a GLUE framework. In summary, to carry out the above study, there would be a need for more preparation of the model at the pre-application stage, preferably at the model build stage. The model would be commissioned to come pre-calibrated with sets of RPM weights, corresponding to different parameter combinations from Table 17.3, that could then be used for a study such as testing whether the UCSO spills during wet weather could affect the Shellfish Water Standards.

17.9 Conclusions

The GLUE technique would facilitate managers to understand more about model reliability and uncertainties in complex model predictions during the decision-making process. This comes at a greater computational and pre-application cost in terms of generation of Monte Carlo simulations and calculation of weights. The prediction limits generated using the GLUE approach may be very wide if there are insufficient data to constrain the model behaviour, and a manager may conclude that a new set of field measurements need to be taken, or that the model structure needs to be improved before a decision should be taken on the basis of the modelling.

The GLUE methodology is easily extended to multiple model structures, the inclusion of prior knowledge about effective values of model parameters and model evaluation over a range of different performance measures. Uncertainty estimates can be based only on prior estimates of sources of uncertainty in the case where no observational data is available for evaluating model performance measures. The evaluation of the potential models using observational data is, however, an important feature of the GLUE methodology, particularly in that it allows the rejection of non-behavioural models.

The resulting uncertainties in model predictions can enter directly into a risk-based decision-making process in which the costs of the potential outcomes also enter into the analysis (e.g. Bedford and Cooke, 2001). This would be an advance on current practice, but at the expense of requiring more computer time for the multiple model runs and more time for the analysis and decision-making processes. The former will increasingly become less of a problem, even for CFD problems, as cheap parallel computers become more widely used. In future, computer power is likely to continue to be extended more quickly than the observational data required to set up and evaluate CFD simulations of real systems. It will be possible to carry out Monte Carlo CFD experiments in a semi-automatic way. It should also be possible to develop decision support software that will take uncertainties in predictions into account automatically. The technical problems of this form of methodology will therefore become reduced over time.

However, there will still be important requirements of the user who will need to decide: What are the important sources of uncertainty associated with a particular problem (Table 17.3), what are the ranges or prior distributions associated with these sources of uncertainty, how should the observational data be compared against the model predictions and at what level of performance should a model be rejected as non-behavioural? These decisions will be problem-, model- and data-dependent and, as yet, little experience has been built up in treating environmental problems in this way. However, this should also improve in the future.

However, it may be felt by many decision-makers, that many of these problems will be eliminated as new model developments and data availability lead to improved model performance. Improved model resolution and geometry information (e.g. from LiDAR; Chapter 9), and better understanding of turbulent flow processes will then limit the expected uncertainties associated with CFD model simulations. This is, after all, what has happened in atmospheric modelling, ocean modelling and engineering design situations such as in aeronautics. It is not clear that this will actually be the case in hydraulic modelling. Some improvements in accuracy in solving different forms of the Navier–Stokes equations might indeed be expected but the difficulty of specifying the model boundary conditions and the limited data available for model evaluation will leave important degrees of freedom in representing real systems. Beven (2002a,b) has pointed out that this would be the case even if the 'perfect model' (in terms of both process representation and numerical solution) were known to be available. The perfect model is not a protection against equifinality or uncertainty for cases where limited local data are available for model implementation.

The extension of the GLUE approach to 3D CFD and the use of higher-order turbulent closure models would clearly incur additional processing requirements. In this case, the fluid dynamics become physically more realistic, so for example enhanced momentum transfer due to secondary currents (such as in helicoidal flow) no longer has to be lumped into the effective eddy viscosity as it is done with depth-averaged flow modelling. Depending on the closure model, this might make the eddy viscosity more identifiable, or in fact unnecessary to incorporate in the GLUE analysis. But it should be borne in mind that any parameters used in numerical modelling (be they parameters in second- or third-order closures) are still effective parameters as they are averaged over a numerical grid, and not modelled at the fluid dynamical 'continuum level'. Ultimately, the more physically realistic the flow model is, the more the predicted uncertainties will become dependent on boundary and initial conditions. However, remaining true to the rejection hypothesis approach of GLUE, although more complex models should lead to a greater understanding of a given system's behaviour, it is still likely that multiple complex models and complex model structures that give predictions will be found, which reproduce the observations equally well.

Complex computational models are informing billion-pound expenditure programs in the water industry, yet interpretative tools for assessing degree of belief in model predictions, or estimates of uncertainties of model predictions are not seen as integral to the process. In summary, the two key factors which might improve our decision-making are increased data collection for large systems modelling and the adoption of an uncertainty framework such as GLUE which provides a coherent framework for carrying out environmental modelling (Beven, 2002b). This is particularly pertinent at the moment with many multi-parameter river basin models being built to address the requirements of the Water Framework Directive.

Acknowledgements

Thanks to Clive Gaskell, Environment Agency, North West Region for permission to include the Ribble Estuary Model results. It should be noted that the CSO spill data used are representative only and were not the final figures used during the AMP3 programme.

References

Aronica, G., Hankin, B.G. and Beven, K.J., 1998a, Uncertainty and equifinality in calibrating distributed roughness coefficients in a flood propagation model with limited data. *Advances in Water Research*, 22(4), pp. 349–365.

Aronica, G., Hankin, B.G. and Beven, K.J., 1998b, *Topographic sensitivity and parameter uncertainty in the predictions of a complex distributed flood inundation model*, in V. Babovic and L.C. Larsen (eds), Proceedings of the 3rd International Conference on Hydroinformatics, A.A. Balkema Publishers, pp. 1083–1088.

Bates, P.D. and Hervouet, J.-M., 1999, *A new method in moving boundary hydrodynamic problems in shallow water*. Proceedings of the Royal Society of London, Series A. Mathematical Physical and Engineering Sciences, 455, pp. 3107–3128.

Bates, P.D., Anderson, M.G. and Hervouet, J.-M., 1995, *An initial comparison of two 2-dimensional finite element codes for river flood simulation*. Proceedings of the Institute of Civil Engineers, Water, Maritime and Energy, 112, pp. 238–248.

Bates, P.D., Horritt, M.S., Smith, C.N. and Mason, D., 1997, Integrating remote sensing observations of flood hydrology and hydraulic modelling. *Hydrological Processes*, 11, pp. 1777–1795.

Bates, P.D., Stewart, M.D., Siggers, G.B., Smith, C.N., Hervouet, J.M. and Sellin, H.J., 1998, *Internal and external validation of a two dimensional finite element code for river flood simulations*. Proceedings of the Institute of Civil Engineers, Water, Maritime and Energy, 130, pp. 279–298.

Bates, P.D., Wilson, C.A.M.E., Hervouet, J.-M. and Stewart, M.D., 1999, Two-dimensional finite element modelling of floodplain flow. *Houille Blanche-Revue International de L'eau*, 54(3–4), pp. 82–88.

Bedford, T. and Cooke, R., 2001, *Probabilistic Risk Analysis: Foundations and Methods*, Cambridge University Press, Cambridge.

Beven, K.J., 1993, Prophecy, reality and uncertainty in distributed hydrological modelling. *Advances in Water Research*, 16, pp. 41–51.

Beven, K.J., 2002a, Towards an alternative blueprint for a physically-based digitally simulated hydrologic response modelling system, *Hydrological Processes*, 16(2) pp. 189–206.

Beven, K.J., 2002b, *Towards a coherent philosophy for environmental modelling*, Proceedings of the Royal Society of London, A458, pp. 2464–2485.

Beven, K.J. and Binley, A., 1992, The future of distributed models: Model calibration and uncertainty prediction. *Hydrological Processes*, 6, pp. 279–298.

Beven, K.J. and Freer, J., 2001, Equifinality, data assimilation, and uncertainty estimation in mechanistic modelling of complex environmental systems, *Journal of Hydrology*, 249, pp. 11–29.

Beven, K.J., Freer, J., Hankin, B.G. and Schulz, K., 2001, *The use of generalised likelihood measures for uncertainty estimation in high order models of environmental systems*, in W.J. Fitzgerald, Richard L. Smith, A.T. Walden and Peter Young (eds), Proceedings of Newton Institute on Non-Linear and Non-stationary Signal Processing, Cambridge University Press, Cambridge, UK, pp. 144–183.

Bradbrook, K.F., Lane, S.N., Richards, K.S., Biron, P.M. and Roy, A.G., 2000. Large Eddy Simulation of periodic flow characteristics at river channel confluences. *Journal of Hydraulic Research*, 38, 207–216.

Centre for Research in Environmental Health (CREH), 1998, *Faecal Indicator Budgets Discharging to the Ribble Estuary*.

Chow, V.T., 1959, *Open-Channel Hydraulics*, New York, London: McGraw-Hill, xvii, p. 680.

Falconer, R.A. and Lin, B., 2000, *DIVAST User Reference Manual*, Cardiff Universtiy. See also http://www.cf.ac.uk/engin/research/water/2.1/2.1.1.6/erdf2.1.1.6.html.

Hankin, B.G. and Beven, K.J., 1998a, Modelling dispersion in complex open channel flows: (1) Equifinality of model structure, *Stochastic Hydrology and Hydraulics*, 12(6), pp. 377–396.

Hankin, B.G. and Beven, K.J., 1998b, Modelling dispersion in complex open channel flows: (2) Fuzzy calibration, *Stochastic Hydrology and Hydraulics*, 12(6), pp. 397–412.

Hankin, B.G. Kettle, H. and Beven, K.J., 2001a, Fuzzy mapping of momentum fluxes in complex shear flows with limited data, *Journal of Hydroinformatics*, 03.2, pp. 91–103.

Hankin, B.G., Hardy, R., Kettle, H. and Beven, K.J., 2001b, Using CFD in a GLUE framework to model the flow and dispersion characteristics of a natural fluvial dead zone, *Earth Surface Processes and Landforms*, 26, pp. 667–687.

Hankin, B.G., Holland, M.J., Beven, K.J. and Carling, P., 2002, Computational fluid dynamics modelling of flow and energy fluxes for a natural fluvial dead zone, *Journal of Hydraulic Research*, 40, pp. 389–401.

Hardy, R.J., Bates, P.D. and Anderson, M.G., 1999, The importance of spatial resolution in hydraulic models for floodplain environments, *Journal of Hydrology*, 216, pp. 124–136.

Horritt, M.S., 2000, Development of physically based meshes for two dimensional models of meandering channel flow, *International Journal of Numerical Methods in Engineering*, 47, pp. 2019–2037.

Kashefipour, S.M., Lin, B., Harris, E.L. and Falconer, R.A., 2000a, *Ribble Estuary Water Quality Modelling*, Environmental Water Management Research Centre, Cardiff University, Cardiff.

Kashefipour, S.M., Falconer, R.A. and Lin, B., 2000b, *FASTER Reference and User Manual*, Environmental Water Management Research Centre, Cardiff University, UK, 55pp.

Lane, S.N. and Richards, K.S., 1998, Two-dimensional modelling of flow processes in a multithread channel, *Hydrological Processes*, 12, pp. 1279–1298.

Lin, B. and Falconer, R.A., 1997, Tidal flow and transport modelling using the ULTIMATE QUICKEST scheme, *Journal of Hydraulic Engineering*, ASCE, 123(4), pp. 303–314.

Romanowicz, R., Beven, K.J. and Tawn, J., 1996, Bayesian calibration of flood inundation models, in D.E. Walling, P.D. Bates and M.G. Anderson (eds), *Floodplain Processes*, Wiley & Sons, pp. 333–360.

Speziale, C.G., 1985, Reynolds-stress closures in turbulence, *Journal of Fluid Mechanics*, pp. 107–153.

18

CFD for environmental design and management

G. Pender, H.P. Morvan, N.G. Wright and D.A. Ervine

In this chapter the off-the-shelf CFD software package CFX, mounted on standard design office computing facilities is used to simulate flows in natural river channels. Our purpose is not to develop the "best" computer model of a single river reach but, rather, to provide an insight into the quality of simulations that can be achieved by such tools in the hands of competent design engineers, with access to typical field data to define model boundary conditions and channel geometry. Those interested in a detailed validation of the code against laboratory data are referred to Morvan *et al.* (2001). To assess the quality of the simulations for the field application cases we have compared computer predictions for steady-flow simulations with field data, which, by its very nature, is unsteady. This is obviously an approximation.

Commercial CFD software is widely used in engineering design. It has the advantages of comprehensive validation against many hypothetical and laboratory test cases, a graphical user interface and a large number of different models for physical phenomena, such as the representation of turbulence. Many CFD codes on the market were developed for manufactured geometries and although there has been some use in civil engineering (Falconer and Ismail, 1997; Wright and Hargreaves, 2001) they require adaptation and interpretation if they are to be used effectively for

---

Computational Fluid Dynamics: Applications in Environmental Hydraulics
Edited by P.D. Bates, S.N. Lane and R.I. Ferguson © 2005 John Wiley & Sons, Ltd

natural river channels. However, this effort offers a number of advantages in river engineering where CFD can be used in the following ways:

- as a method of improving conveyance estimation to enhance one-dimensional and two-dimensional models;
- as a technique for assessing the performance of small-scale river engineering works and hydraulic structures such as river intakes/outtakes, weirs, fish passes and sluice gates;
- as a scientific tool for developing understanding of the fundamental fluid dynamics of flows in rivers.

The application of CFD to rivers raises a number of generic issues that will be introduced and demonstrated by the following test cases. These are:

- *Free surface representation*: Free surfaces have initially been represented as rigid lids (shear-free walls), which implies that the position of the free surface is known *a priori*. A first improvement (suggested by Morvan (2001)) has been to allow for the deformation of the lid which then becomes driven by local pressure or velocity variations. Ultimately it is possible to include a full calculation of the position of the free surface through a volume of fluid calculation (Hirt and Nichols, 1981), but this imposes significantly increased computing costs.

Additional issues arise in modelling the free surface stemming from the effect of the free surface on turbulence behaviour which leads to the maximum velocity being a distance below the free surface. This means that a standard symmetric or free-shear condition is not always appropriate. Suitable amendments are reported by Cokljat and Younis (1995).

- *Boundary conditions*: Most CFD codes were developed for situations in mechanical engineering where standard boundary conditions of fixed velocity or mass flow are sufficient. In-channel flows, conditions such as discharge with no determined depth or vice versa must be incorporated. This is not always straightforward.

- *Hydraulic roughness*: In a one-dimensional model, hydraulic roughness is included using an empirical coefficient applied globally, usually either the Chezy or the Manning coefficient. In three-dimensional modelling, a local definition of roughness is required. This represents a significant advantage, but great care must be taken to ensure that the fluid flow is correctly modelled, particularly in the near-wall region. So far in three-dimensional river modelling it has not been possible to construct a finite volume grid that resolves the flow field into the viscous layer adjacent to the wall. This is because the large number of grid points required at these scales to achieve this level of discretisation is greater than what can be achieved with standard computing facilities. This problem is common to other engineering applications and is overcome through the use of a "wall function" which is an algebraic formula applied in the near-wall region to bridge the gap between the near-wall laminar sub-layer and the fully turbulent part of the boundary layer. The approach uses the "wall function" to model the boundary layer flow instead of resolving the flow variables on a fine grid. The theory of wall functions was developed for smooth surfaces; as a result the functions require modification for the rough surfaces found in real rivers. The modification is normally based on an estimation of the local size of the roughness elements that allows for a shift of the velocity profile to account for non-represented roughness elements. Once this is done it is usually necessary to revise the position of the finite volume grid to ensure that the CFD computation starts where the wall function stops. The specification of the roughness in three-dimensional modelling, although more rigorous in principle, is clearly a very different exercise from that in one-dimensional and two-dimentional modelling, and constitutes a very difficult calibration exercise. More research and experience of application is required to ensure its rigour and to validate the technique.

- *Geometry*: There are two issues in developing an appropriate grid for a river channel: first, the correct representation of the boundary; and secondly, the appropriate level of grid refinement. The former is complex in the case of rivers where there is a high level of irregularity in the bathymetry and the amount of data is often restricted. This can lead to time-consuming work in building the geometry and, if the geometry is not adequately represented, to poor convergence of the numerical scheme. The move towards unstructured codes will significantly simplify this problem in future. The second condition requires an appropriate level of grid refinement to be established for a given simulation. In this process, the mesh is refined progressively until further refinements produce no noticeable change in the simulations. Only in this way can it be established that the solution is independent of the grid used. In some cases, computer resources limit the ability to achieve full grid independence. In such cases successive refinements must be examined to give an estimate of the level of grid independence that has been achieved. If this process of mesh refinement is accompanied by refinements in bathymetry then it can prove very difficult to fully satisfy the condition for mesh independence. In such cases the minimum requirement should be to examine the changes between successive grid refinements and use these to estimate likely errors in the final solution (Chapter 8). It should also be noted that in the absence of sufficient computer power, a calculation on a coarser grid should always be carried out and the evolution of the solution between two consecutive grids compared. This measure is essential to ensure a proper mathematical solution.

- *Turbulence*: Flows within a natural river will generally be turbulent. In one-dimensional and two-dimentional models of rivers, turbulent energy losses are usually accounted for through the use of a bulk parameter such as the Chezy or Manning coefficient. Three-dimensional models can obtain solutions with much more detail, and turbulence can be analysed on a local scale; however, the analysis of turbulence is not straightforward and a high level of expertise is required to use this feature effectively. In particular, it should be remembered that turbulence models remain "models" with specific applications, case-dependent constant values and inherent flaws. This needs to be borne in mind when interpreting the output of a simulation.

The purpose of the following examples is to assist in illustrating the above issues.

18.1 River Severn

The River Severn is the third longest river in Britain after the Thames and Wye. It is 206 km long, drains an area of 4330 km^2 and has a mean annual discharge of 63 m^3/s. It runs along the south-east border between England and Wales, and the Severn Estuary is a significant landmark on the map of Southern Britain.

The section modelled is located 20 km east of Shrewsbury. A single meander about 600 m long was instrumented throughout 1999, 2000 and 2001. At the instrumented reach the main channel is about 30 m wide, between 6 and 7 m deep with respect to the right flood plain but more than 9 m deep with respect to the left flood plain. The right flood plain is 180 m wide and 120 m long, and is bunded by an earth embankment to the south. The right flood plain has been lowered to extract material for the construction of the embankments and, as a result, is fairly flat and easily flooded. On the other hand the higher elevation of the left flood plain means that it is rarely inundated. The following describes the construction and verification of a CFD model of this reach using the CFX4 package produced by CFX International.

18.1.1 Field data

Topographic survey data was collected at the seven sections shown on Figure 18.1. Velocity data was also collected at these sections using a boat-mounted electromagnetic current meter. The figure also gives the location of a semi-permanent scaffolding tower on which a field Acoustic Doppler Velocimeter (ADV) was mounted. This provided point velocity data and turbulence measurements during out-of-bank flows.

During the period that the instruments were in place the reach experienced five flood events of around $100\,\text{m}^3/\text{s}$. These occurred in December 1999 ($102\,\text{m}^3/\text{s}$), March 2000 ($103\,\text{m}^3/\text{s}$), October 2000 ($95\,\text{m}^3/\text{s}$), November 2000 ($101\,\text{m}^3/\text{s}$) and February 2001 ($104\,\text{m}^3/\text{s}$). For the purpose of computer model verification the data collected during these events were consolidated into one data set. Data available for model verification therefore includes:

1. the water surface profile over most of the reach;
2. velocity profiles at cross sections 4, 5 and 7;
3. point velocity and turbulence measurements in the vicinity of the scaffolding tower.

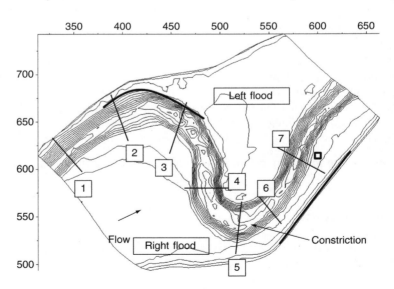

Figure 18.1 Test reach on the River Severn.

18.1.2 Geometry and mesh resolution

As mentioned previously there are two essential issues that need to be addressed when modelling natural channel geometries in three dimensions:

1. the numerical description of the geometry; and
2. the mesh resolution that can be achieved with the available hardware.

The numerical description of the geometry can only be as good as the topographical data permits. Since the practicalities of topographical data collection require the model geometry to be constructed using discrete points, it is often difficult to build a smooth surface using the sparse information available, especially with a structured grid, as required by CFX4. Consequently, inappropriate grid properties, such as poor element aspect ratio, large growth ratio and lack of orthogonality of the grid lines with the boundaries can lead to stability problems and high computing costs. The high computing costs arise due to the large number of iterations required for convergence with a poorly constructed mesh.

These problems were particularly acute for the site being modelled where the main channel shape varies significantly within the reach. At some locations its cross-sectional profile is "v"-shaped making it impractical to achieve a regular lateral mesh resolution throughout the depth. Indeed, whenever a mesh with a fine lateral resolution was generated, spurious elements were created at the intersection of the bed and the banks. This situation is complicated further by the plan curvature of the main channel, which required the use of the block structure to adequately define the main channel shape.

Once a reasonable representation of channel geometry has been obtained it is important to investigate the influence of grid resolution on the simulations. Normally, one would progressively refine the mesh to a point where successive refinement ceases to influence the solution; however, for the simulation of river channels available computer hardware is often a limiting factor in determining the number of elements that can be used. This was indeed the case here, where the strategy has been to generate a grid that enables simulations to be realised at a relatively low computational cost. As will be seen later, the authors believe that this results in reasonable simulations of the observed velocity field, although we accept that this judgement is subjective. The most refined grid consists of approximately 200 000 elements, which equates to about 20 000 elements for every 100 m of channel.

The first relatively coarse grid constructed for the River Severn consists of 97 732 elements. It has 10 elements positioned across the channel and 14 in the vertical, both with a growth ratio of 1.5. This gives a mean resolution in the solution domain that is 0.6 m vertically, 1.8 m laterally and 2.0 m longitudinally. The first element is typically located at a vertical and horizontal distance from the wall of 0.77 and 0.8 m, respectively.

Further refinement was undertaken to create a second grid consisting of 183 138 elements. This was achieved by vertical refinement to 15 nodes in the main channel. This represents an average resolution of 0.5 m vertically, 1.2 m laterally and 1.6 m longitudinally. The positioning of the first node relative to the bed and walls is now 0.4 m both horizontally and vertically. This represents a considerable improvement on the first grid but is still not sufficient to obtain a low y^+ value close to 300. Consequently the mesh is still too coarse to obtain a grid-independent solution;

however, given the purpose of the project as set out in the introduction, this mesh was used to generate the results presented in the table.

| | Model of the Severn | |
Location – direction	Mesh CFX S-1	Mesh CFX S-2
Main flood plain – lateral (on right side)	95 × 50 elements $r=1.0$; width × length	120 × 100 elements $r=1.0$; width × length
Flood plain – vertical (above bank level)	4 elements $r=1.0$	5 elements $r=1.0$
Main channel – lateral	10 elements $r=1.5*$	15 elements $r=3.0*$
Main channel – vertical (inside main channel only)	10 elements $r=1.5*$	12 elements $r=3.0*$
Main channel – longitudinal (with effects on all domains)	95 × 50 elements $r=1.0$ (along main channel)	425 elements $r=1.0$ (along main channel)
Total	97 732	183 138

*Indicates a bi-directional bias.

18.1.3 CFX grid characteristics used to model the Severn

The software package CFX provides a number of options for the discretisation of the advection terms in the Navier–Stokes equations. Simulations were conducted using the widely used second-order accurate Curvature Compensated Convection Transport (CCCT) scheme (Gaskell and Lau, 1998), and a first-order accurate hybrid representation of these terms. A comparison between the predictions at cross sections 3–5 for both these schemes indicated that the CCCT scheme predicts slightly higher velocities in the bend, but the differences are small and localised (about 5% of the maximum cross-sectional velocity). In a river engineering application this level of difference is minor.

In general, the use of higher-order schemes is recommended to minimise numerical diffusion (Easom, 2000); however, for the level of spatial resolution and accuracy required in this work, numerical diffusion was not a significant problem. This is important in practice, as the hybrid scheme is less computationally intensive and more robust than other methods. This is especially important with the multi-block approach used here as it simplifies the calculations conducted at the block interfaces by reducing the level of interdependence between the blocks.

18.1.3.1 Boundary conditions

The simple logarithmic velocity profile given in equation (18.1) was used to obtain the velocity profile at the inflow boundary.

$$U_{in} = C_{in} \ln(z/z_0) \quad \text{with} \quad z_0 = \frac{k_s}{30} \tag{18.1}$$

The parameters were adjusted to give a realistic velocity profile and yield the discharge obtained from the field data. C_{in} (equivalent to U_*/κ) was taken as 0.08 m/s. At the inlet to the main channel this gives velocities of up to 0.65 m/s, and a general profile which is close to the values measured in the field.

18.1.3.2 Turbulence

At the inlet turbulence kinetic energy, energy dissipation and the dissipation length scale were obtained using a shear velocity of 3.2 cm/s and applying equations (18.2), (18.3) and (18.4) (Morvan, 2001).

$$\overline{K_{in}} = \frac{U_*^2}{\sqrt{C_\mu}}\left(1 - \frac{2}{h}\right) \quad \Rightarrow \quad \overline{K_{in}} = 0.5\frac{U_*^2}{\sqrt{C_\mu}} \tag{18.2}$$

$$\varepsilon_{in} = \frac{K_{in}^{3/2}}{0.3L} \tag{18.3}$$

$$L = \frac{V_\varepsilon}{C_\mu\sqrt{2K}} \tag{18.4}$$

18.1.3.3 Outlet

At the outlet fully developed flow conditions are assumed and a mass flow boundary condition is implemented through the use of Neumann boundary conditions. It is essential to ensure that mass is properly conserved.

18.1.3.4 Representation of the boundary layer at the solid walls

At the walls the algebraic law given in equations (18.5) and (18.6) below was implemented.

$$\frac{U_\tau}{U_*} = y^+ \quad \text{if} \quad y^+ < y_0 = 11.63 \tag{18.5}$$

$$\frac{U_\tau}{U_*} = \frac{1}{\kappa}\ln(E(K_s^+)y^+) \quad \text{otherwise} \tag{18.6}$$

This law with a varying parameter E has been shown to be well suited to high turbulence flows (Noat, 1984; Morvan, 2001), such as those occurring in flooded rivers; however, because of the coarse grid resolution it is expected that the first node may be located too far from the wall. This could result in inaccuracies when merging the law of the wall with the fully turbulent flow condition. The wall roughness is implemented with separate roughness height values for the main channel and the flood plain.

18.1.3.5 Free surface

The free surface was modelled as a rigid lid, i.e. an impermeable shear-free wall. An initial estimate of the lid position was made using water surface elevation data collected from the site. During simulations the pressure on the lid was monitored and the lid position adjusted as a function of the local pressure, which is possible since steady state simulations are being undertaken.

18.1.3.6 Convergence

In the simulations discussed here the convergence rate was expected to be slow due to the large grid size and the large cell aspect ratio. It is not practical to implement the usual criteria of a reduction in mass residual to between 10^{-4} and 10^{-6} for such problems. Instead a reduction in the magnitude of the residuals by between three to four orders of magnitude (Sinha et al., 1996; Meselhe and Sotiropoulos, 2000) was adopted.

18.1.3.7 Turbulence model

Morvan et al. (2001) illustrated that when simulating overbank flows in rivers, channel geometry is more significant than turbulence in determining the velocity field. This is an important issue since the adoption of a relatively simple turbulence model permits computational effort to be directed to ensuring convergence of the calculation in the geometrically complex solution domain. In the results presented turbulence was therefore modelled using a standard k-ε model.

18.1.3.8 Simulations

The impact of the selected wall roughness values on results was investigated by comparing the predicted water surface profiles with the observed profiles for the December 1999 event. k_s values ranging from 0.005 to 0.2 m were applied throughout the model. Figure 18.2 provides a comparison of observed water surface profiles and CFX pressure outputs connected to water level. It can be seen that the predicted pressure on the rigid lid is relatively insensitive to variations in k_s. On the other hand, the simulated velocity field is poor for low k_s values and improves as k_s increases up to 0.1 m; beyond this the predicted velocity field changes little. The model converges best for k_s values around 0.1 m. In addition, the use of different k_s values in the main channel and on the flood plain has little impact on the model predictions. The following results were obtained using $k_s = 0.1$ m throughout the solution domain.

18.1.4 Comparison of model predictions with field observations of velocity and turbulence data

A comparison between the model Figure 18.3 and the field data collected at cross-section 5 in November 2000, Figure 18.4, indicates that the model predicts a velocity field similar to the observed field. A high velocity in the region of 0.45–0.80 m/s is visible in the upper left half of the channel, while velocities of about 0.75 m/s are seen to occur along the left bank. In the data from November 2000, these are slightly

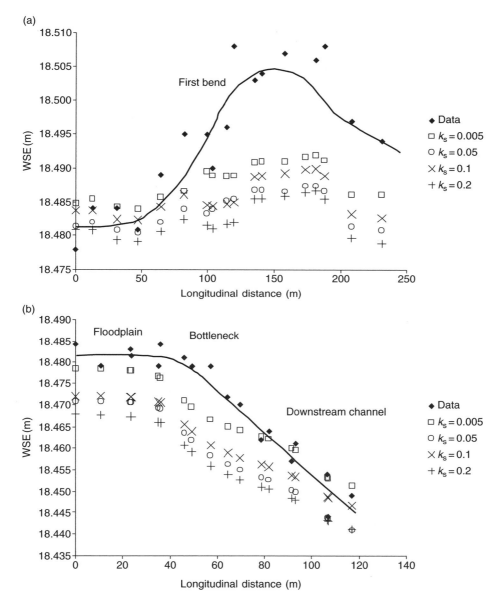

Figure 18.2 Comparison of water surface elevation between CFX models and field data (100 m³/s Event of December 1999): (a) along the upstream first bend; (b) along the right embankment.

inclined to the left, which is predicted by the CFX model. In the lower right part of the channel the measured velocities are 0.50 m/s, and are accurately predicted by the model. In the upper section they are reduced to 0.40 m/s in the field compared with 0.30 m/s from the model. For the event of December 2000, field measurements were

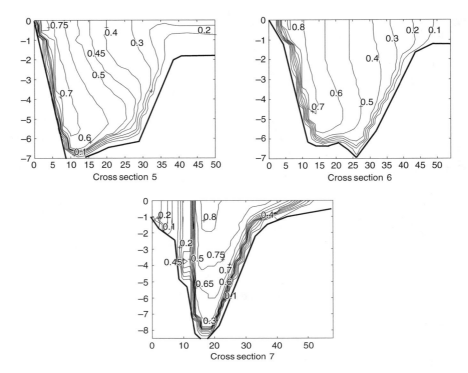

Figure 18.3 River Severn CFX velocity profile (m/s) at cross sections 5, 6 and 7 ($k_s = 0.100$ m, Grid CFX S-1, k-ε model).

made between cross sections 4 and 5 (Figure 18.5) and similar comments to those given earlier apply when comparing the predicted velocities at these sections (Figures 18.6 and 18.3). This suggests that the computer model reproduces the important mechanisms driving the velocity field.

The velocity field was also measured in detail at cross section 7 during the 100 m³/s flood flow event of December 1999. The comparison between the field data and predictions is not as good as those in the meandering section discussed above. This is probably due to a combination of the use of a uniform surface roughness value and the choice of a k-ε turbulence model. It is known that turbulent momentum exchange at the interface between the flood plain and the main channel is highly anisotropic and that this affects the predicted velocity profile in straight reaches. This is consistent with the calculated velocity maximum being lower than the measured value by about 15% in the main channel.

A further analysis compared the velocities measured over the water depth at precise locations at cross sections 4 and 5 in November and December 2000, and in February 2001 with the model's predictions (Figures 18.7 and 18.8). From these it is clear that, in general, the model under-predicts the velocity field and significant differences exist between field data and predictions at given locations. The order of the difference is about 20–30%. However, the shape of the velocity profiles is reasonably well reproduced at most locations and a significant part of the difference is at the walls.

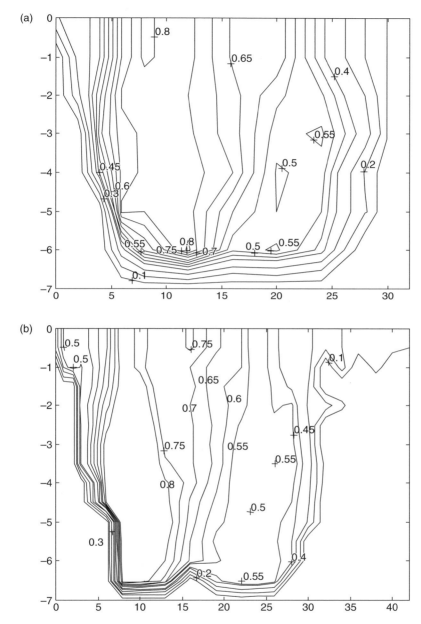

Figure 18.4 Measured velocity profile at cross section 5 in the Severn (m/s): (a) October 2000; and (b) November 2000.

Part of the difference between predictions and observations can be explained by the fact that one is attempting to compare a simulation for a steady flow of $100\,\text{m}^3/\text{s}$ with field measurements from different unsteady flows. Indeed, the differences could be arising from errors in the calculations of the discharges from the field data. An attempt was made

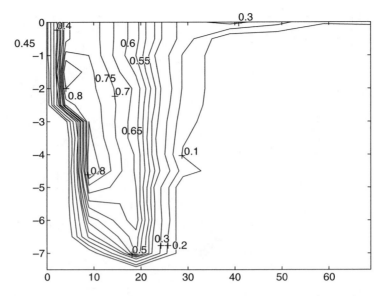

Figure 18.5 Measured velocity profile between cross sections 4 and 5 in the Severn (m/s; December 2000).

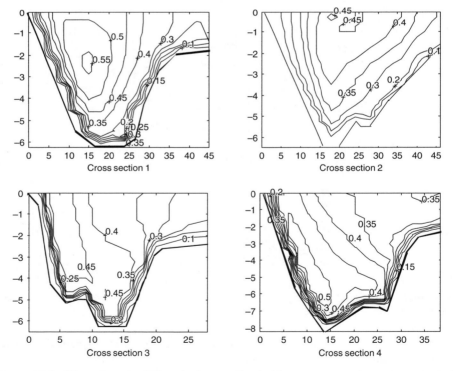

Figure 18.6 River Severn CFX velocity profile (m/s) at cross sections 1, 2, 3 and 4 ($k_s = 0.100$ m, Grid CFX S-1, k-ε model).

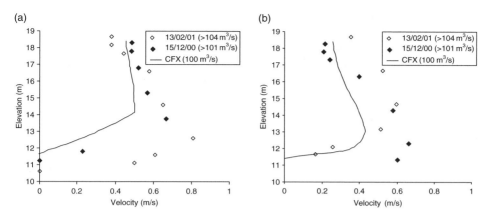

Figure 18.7 Comparison between field data and River Severn CFX model predictions at cross-section 4: (a) at 9.00 m from the left bank; and (b) at 19.00 m from the left bank.

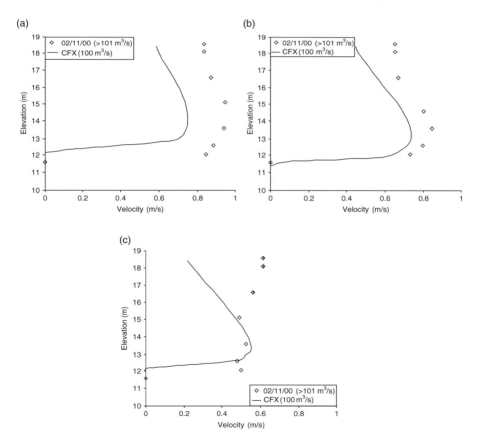

Figure 18.8 Comparison between field data and River Severn CFX model predictions at cross section 5: (a) at 8.00 m from the left bank; (b) at 12.00 m from the left bank; and (c) at 22.00 m from the left bank.

to resolve part of this uncertainty by running the model for a steady discharge of 120 m³/s. The results give closer comparisons between the velocity and the field data. This gives some indication of the sensitivity of model predictions to boundary conditions.

During the event of March 2000, detailed velocity and turbulence measurements were collected at the tower located on the right flood plain close to the downstream

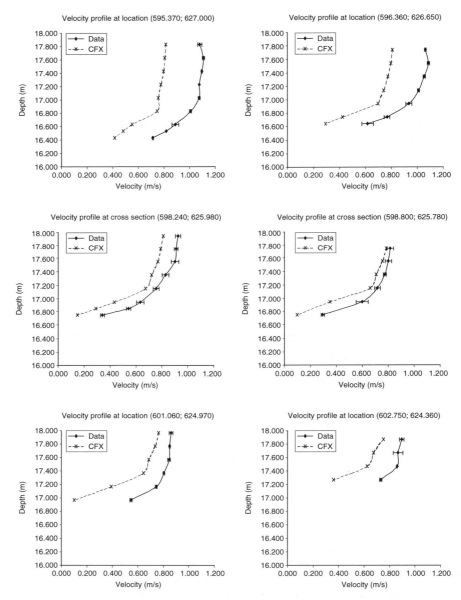

Figure 18.9 *Comparison between measured and predicted velocity profiles along the Severn main channel right bank, at the tower (m/s; March 2000).*

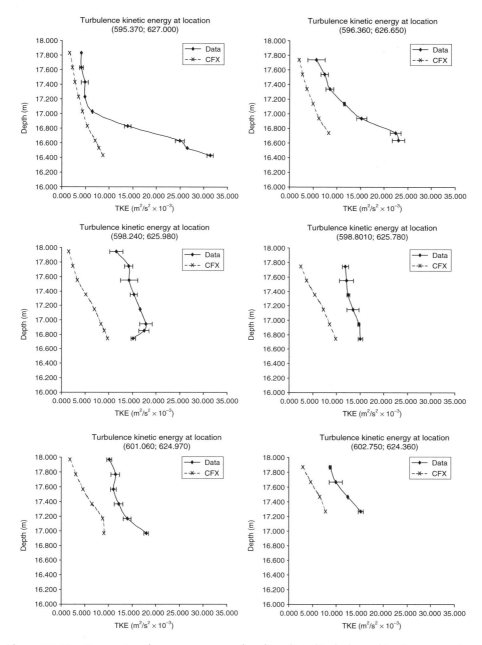

Figure 18.10 Comparison between measured and predicted turbulence kinetic energy along the Severn main channel right bank, at the tower (m^2/s^2; March 2000).

boundary (Figures 18.9 and 18.10). Unfortunately, these data were collected where the model produces the poorest predictions of velocity. What is clearly visible from the comparison of the model predictions with the data is that the velocity predictions are less accurate in the region of the channel–floodplain interface, with errors of up to 30%. Further away from the interface the predictions are in better agreement with the data (within 15%). In terms of Turbulence Kinetic Energy (TKE), the model presents a lower, somewhat uniform, TKE across the direction of the measurements, whereas the data indicates an increasing value of TKE from the main channel onto the flood plain. Errors of 100% are visible at the bed at most locations. This stems from an inadequate grid representation at the walls. Errors of a similar magnitude and pattern in the prediction of TKE have also been reported by Bradbook *et al.* (1998), although these comparisons are for a computer model of a laboratory experiment. The significance of this shortcoming will be reduced with improved computing power enabling the use of more nodes as shown in Morvan (2001).

18.2 CFD for habitat modelling in the River Idle

This application demonstrates the use of hydraulic modelling techniques in the study of river rehabilitation schemes and presents modelling work that has been carried out for the River Idle in north Nottinghamshire, UK.

The River Idle in north Nottinghamshire drains a catchment of 842 km^2 in the East Midlands of England. The river flows north from its origin at the confluence of the rivers Meden and Maun south of East Retford to its confluence with the River Trent at West Stockwith. The Idle would be tidal upstream of its confluence with the Trent as far as Mattersey, but a combination of embankments, sluices and a pumping station has eliminated this effect.

Several major floods led to the construction of a flood alleviation scheme in 1979. This consisted of the lowering of the bed by an average of 0.7 m, the construction of an enlarged trapezoidal cross section and the erection of flood embankments at varying distances from the channel.

The river was environmentally degraded as a result of a 1982 channelisation scheme. Plans for limited rehabilitation were drawn up in 1995 and, in the straighter reaches, the installation of in-stream deflectors was proposed as the most appropriate method for improving the diversity of bed habitat whilst still preserving an acceptable level of flood defence.

Three major factors in the rehabilitation were assessed with CFD, namely the effect on flooding potential, aquatic habitats and the stability of the modified channel. Numerical techniques have been used based on the application of one-dimensional and three-dimensional models. Results indicate benefits in the use of more complex models over the traditional approach, which relies wholly on a one-dimensional strategy. The methods described form an integrated approach to the assessment of rehabilitation designs.

In order to gain detailed insight into the local flow features created by the rehabilitation work, a three-dimensional model was implemented using the CFX software. A three-dimensional model gives better predictions through the use of the more detailed geometry and the more advanced discretisation and modelling of turbulence. It also provides predictions of the secondary currents generated around the deflectors.

Habitat requirements were simulated using the predicted depths and velocities together with the Instream Flow Incremental Methodology (IFIM). Here the habitat requirements of individual "target" species of fish are described by "preference curves", which relate habitat suitability to flow depth, velocity and the bed substrate (see Chapter 16, for a fuller discussion). A number of FORTRAN programs were written to take the predicted flow information from various hydraulic modelling packages, and compare the results with published preference curves for a number of species. The amount of habitat available for each of the species is based on a measure of Weighted Usable Area (WUA).

The pattern of erosion and deposition was predicted using the results from a CFX model. A further FORTRAN program was written to calculate approximate bed shear stresses. Predicted areas of erosion are plotted where the calculated bed shear stress exceeds a critical shear stress.

As with the River Severn simulations, sensitivity tests led to the choice of hybrid differencing, a roughness height of $k_s = 3.5D_{84}$ (Hodskinson, 1996), and a k-ε turbulence model with a mesh of 250 000 cells. Inlet conditions were developed from a uniform reach inserted upstream of the reach to be studied. Although three-dimensional CFD codes, including CFX, allow for the calculation of a free surface interface, it was decided not to implement this in order to reduce complexity and run times. The position of the free surface predicted by the one-dimensional model was used as a fixed boundary and the three-dimensional results checked for excessive variations in pressure which would represent the super-elevation of the water surface that would be expected on the outside of a bend.

Figures 18.11–18.13 show the velocities at 20% of the depth below the free surface for each of deflectors 3f, 3c and 6a. Examining these in comparison with field observations gives qualitative agreement. A more detailed analysis in comparison with measured data indicates that in most cases approximately 80% of the values predicted are within 50% of the values measured and 50% are within 20%. This was felt to be encouraging given the uncertainties of the measurements and the representation of roughness. It should be noted that agreement was much better where site velocities had been collected in two dimensions rather than one dimension.

From the plots it can be seen that the three-dimensional results show regions of recirculation downstream of the deflectors where secondary currents will be significant. Simulations were carried out without the turbulence model and these gave significantly worse results. This indicates that the inclusion of turbulence in a three-dimensional model can be significant and that adequate turbulence modelling is important. As turbulence is a three-dimensional phenomenon, this gives three-dimensional modelling an advantage over two dimensional or layered models.

A large number of methodologies are available for assessing the ecological suitability of a given hydraulic situation. The one chosen here was IFIM, which has been implemented in the software PHABSIM. The fundamental principle of IFIM is that the habitat requirements of individual species can be quantified in terms of a small number of habitat variables, and that these preferences for certain types of habitat can be expressed in terms of a suitability index.

The PHABSIM software contains several one-dimensional modelling procedures to provide the required velocities and depths for each flow cell. In this work

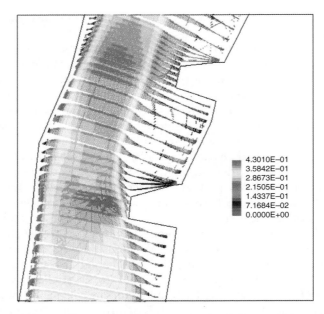

Figure 18.11 *Velocities 20% below surface at deflector 3f. (See Plate 7.)*

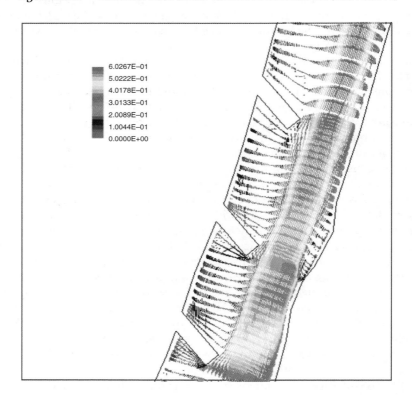

Figure 18.12 *Velocities 20% below surface at deflector 3c. (See Plate 8.)*

CFD for environmental design and management 505

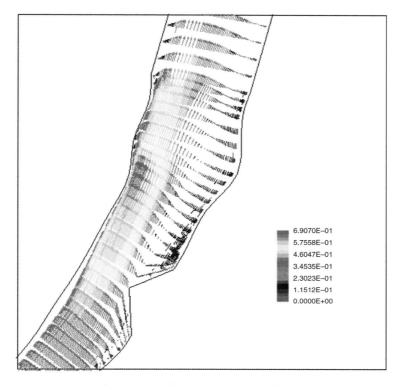

Figure 18.13 Velocities 20% below surface at deflector 6a. (See Plate 9.)

predictions from the local CFX model were composed with those from a one-dimensional model of the same reach. The depth, mean column velocity and a substrate code are combined with suitability indices. A suitability index gives a value between 0 and 1 for a particular species as a function of the variable in question. These are combined either by multiplication, averaging or taking a minimum value. The area of the cell under consideration then multiplies this combined value and all the values are summed for a given area to give the WUA. Areas are normally chosen to be indicative of the variety of habitats available in a given reach.

In order to produce WUAs from the one-dimensional and three-dimensional hydraulic models, two programs, COMBSIM1D and COMBSIM3D, were developed. For the one-dimensional model panel velocities are calculated to give some representation of the variance of velocity across the river. Figures 18.14 and 18.15 show the spatial plots of WUA for spawning chub for deflector 3c obtained for one-dimensional and three-dimensional models, respectively. They show noticeable differences between the one-dimensional and three-dimensional models for the post-installation cases with differing peak values. Some cases have shown differences in peak values of as much as an order of magnitude. This is to be expected and the three-dimensional model gives much greater insight into how the deflectors affect flow patterns and consequently habitat suitability in their vicinity. Figures 18.16 and 18.17 show total WUA for the reach surrounding deflector 3c for

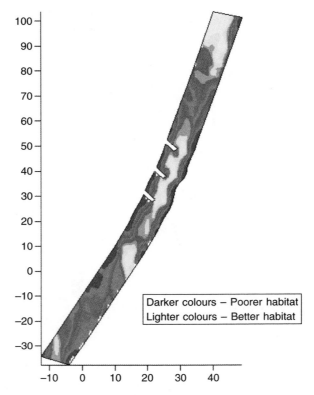

Figure 18.14 Spatial plots of WUA for spawning chub at deflector 3c for 1D model. (See Plate 10.)

one-dimensional and three-dimensional models, respectively, and do not reveal much variation between one dimension and three dimension, but it must be noted that these do represent global values with no account of local variations.

Some one-dimensional modelling of sediment transport was carried out and this indicated erosion at the deflectors. It did not give great detail in the areas around the deflectors, so bed shear stress calculated from the three-dimensional model was examined. The velocity profile was used to calculate the bed shear stress and then Shields criterion ascertained whether sediment movement would occur. Figure 18.18 shows the areas where erosion would occur for different grain sizes. The smaller areas correspond to the larger grain sizes. In these areas grains of a smaller size would also be transported. Qualitative examination of the erosion and deposition in the actual river two years after installation indicate that the numerical predictions are accurate.

This work demonstrated two main points:

1. The successful integration of a detailed hydraulic model into a habitat model.
2. The benefit of a three-dimensional model for local flow features, habitat prediction and potential erosion.

CFD for environmental design and management 507

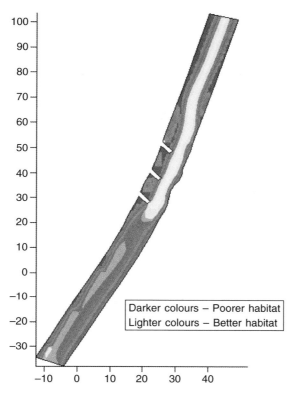

Figure 18.15 Spatial plot of WUA for spawning chub at deflector 3c for 3D model. (See Plate 11.)

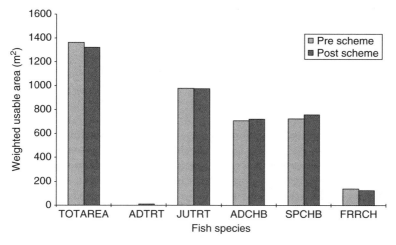

Figure 18.16 Total WUA for the reach surrounding deflector 3c (one-dimensional modelling results – initial flow).

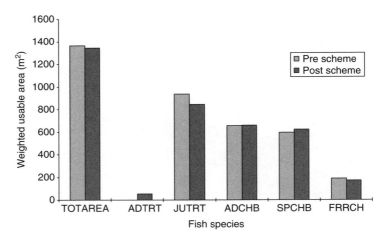

Figure 18.17 Total WUA for the reach surrounding deflector 3c (three-dimensional modelling results – initial flow).

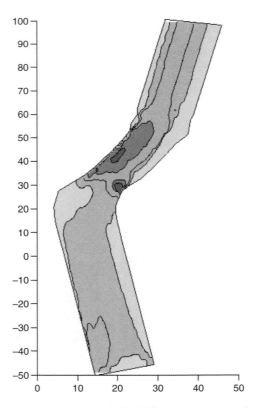

Figure 18.18 Erosion patterns for different grain sizes. (See Plate 12.)

18.3 Conclusions

Although standard design office computing equipment is sufficient to obtain what most practitioners would consider to be a reasonable resolution of the numerical grid, it was not possible to provide a rigorous demonstration of grid independence. This is a cause for some concern although we do not believe that it significantly influenced the predicted values in the cases presented here.

The velocity fields generated are considered to be adequate to inform design decisions. They were obtained using:

1. A hybrid treatment of convection;
2. A k-ε turbulence model; and
3. Adoption of convergence criteria of a reduction of residuals of at least three orders of magnitude.

References

Bradbook, K.F., Biron, P.M., Lane, S.N., Richards, K.S. and Roy, A.G. (1998). "Investigation of controls on secondary circulation in a simple confluence geometry using a three-dimensional numerical model". *Hydrological Processes*, 12, pp. 1371–1396.

Cokljat, D. and Younis, B.A. (1995). "Compound-channel flows – A parametric study using a Reynolds-stress transport closure". *Journal of Hydraulic Research*, 33, pp. 307–320.

Easom, G. (2000). "Improved turbulence models for computational wind engineering", PhD Thesis, The University of Nottingham, UK.

Falconer, R. and Ismail, A. (1997). "Numerical modelling of tracer transport in a contact tank". *Environmental International*, Elsevier Sciences, 23(6), pp. 763–773.

Gaskell, P.H. and Lau, A.K.C. (1988). "Curvature compensated convective transport: SMART". *International Journal for Numerical Methods in Fluids*, 8, pp. 617–641.

Hirt, C. and Nichols, B. (1981). "Volume of fluid methods for the dynamics of free boundaries". *Journal of Computational Physics*, 39, 201.

Hodskinson, A. (1996). "Computational fluid dynamics as a tool for investigating separated flow in river bends". *Earth Surface Processes and Landforms*, 21, pp. 993–1000.

Meselhe, E.A. and Sotiropoulos, F. (2000). "Three-dimensional numerical model for open-channels with free-surface variations". *Journal of Hydraulic Research*, 38(2), pp. 115–121.

Morvan, H.P. (2001). "Three-dimensional simulation of river flood flows", PhD Thesis, Department of Civil Engineering, University of Glasgow. http://www.nottingham.ac.uk/~evzhpm/.

Morvan, H., Pender, G., Wright, N.G. and Ervine, D.A. (2001). "Three-dimensional hydrodynamics of meandering compound channels". *Journal of Hydraulic Engineering*, ASCE, 128(7), 674–682.

Noat, D. (1984). "Response of channel flow to roughness heterogeneity". *Journal of Hydraulic Engineering*, ASCE, 110(11), pp. 1558–1587.

Sinha, S.K., Sotiropoulos, F. and Odgaard, A.J. (1996). "Numerical Model Studies for Fish Diversion at Wanapum/Priest Rapids Development – Part I: Three-Dimensional Numerical Model for Turbulent Flows through Wanapum Dam Tailrace Reach". *IIHR Limited Distribution Report No. 250*, Iowa Institute of Hydraulic Research, University of Iowa, Iowa City, Iowa, USA.

Wright, N. and Hargreaves, D. (2001). "The use of CFD in the evaluation of UV treatment systems". *Journal of Hydroinformatics*, 3(2), pp. 59–70.

Author index

Abba, A.   33
Ackers, P.   282
Addley, R.C.   433, 441, 443
Akanbi, A.A.   126
Akiyama, J.   122
Alcrudo, F.   132
Ali, S.   275
Alsdorf, D.E.   204
Anastasiou, K.   165
Andrews, E.D.   358, 366, 390
Anima, R.J.   389
Apsley, D.   230
Aradas, R.D.   166
Armanini, A.   221
Aronica, G.   203, 205, 206, 293, 294, 295, 469
Ashida, K.   81
Ashmore, P.   225
Ashworth, P.J.   220, 233, 250, 273
Atias, M.   159
Atkinson, P.   287

Baird, J.I.   273
Baker, T.J.   446
Baltsavias, E.P.   287
Balzano, A.   124, 126, 141
Bandyopadhyay, P.R.   329
Baptist, M.J.   84
Batchelor, G.K.   5, 20
Bates, P.D.   121, 122, 124, 127, 128, 129, 132, 135, 136, 137, 139, 165, 175, 186, 195, 197, 198, 199, 201, 202, 203, 260, 272, 275, 280, 282, 284, 285, 287, 290, 291, 293, 294, 295, 297, 406, 443, 474
Bathurst, J.C.   239
Bechara, J.A.   433
Bechteler, W.   139, 284
Beckett, G.   126
Bedford, T.   483
Beffa, C.   132
Behr, M.   126
Bekesi, G.   205
Belitz, K.   181
Belzile, L.   455
Benkhaldoun, F.   280, 297

Bennett, J.P.   364
Bennett, S.J.   86
Bergs, M.A.   240
Bernard, R.S.   24, 225
Bertram, H.-U.   397
Bérubé, P.   454, 455
Best, J.L.   225, 250, 251, 257, 258
Beus, S.S.   359
Beven, K.J.   4, 184, 186, 199, 205, 206, 292, 294, 295, 296, 461, 469, 470, 471, 472, 476, 481, 483, 484
Bijvelds, M.D.J.P.   4
Binley, A.   199, 205, 206, 294, 296, 461, 470, 472
Biron, P.M.   154, 225, 236, 250, 258
Bladé, E.   139, 284
Blom, A.   81, 83, 86
Blöschl, G.   291
Bocquillon, C.   194
Boisclair, D.   436
Bonnerot, R.   126
Booker, D.J.   34, 176, 185, 330, 332
Borthwick, A.G.L.   138
Bosmajian, G.   398
Bottasso, C.L.   126
Boudreau, P.   452
Bovee, K.D.   430, 433, 439, 441, 442, 452, 455
Boxall, J.B.   61, 62, 66
Boysan, F.   29, 31
Bradbrook, K.F.   4, 21, 33, 92, 139, 171, 176, 177, 182, 183, 194, 229, 230, 236, 250, 251, 253, 254, 255, 256, 257, 258, 329, 480, 502
Bradford, S.F.   130, 132
Bradshaw, P.   103, 329, 332
Brakenridge, G.R.   203
Bras, R.L.   249
Brath, A.   199
Bray, D.I.   37, 332
Bredehoeft, J.D.   179
Breuer, M.   415
Bridge, J.S.   86, 272
Briggs, M.J.   123
Britain, John E.   455, 456
Brocchini, M.   123, 132
Brookes, A.N.   131, 132

Brown, H.I.   188
Buffin-Bélanger, T.   330
Buffington, J.M.   75
Burke, R.W.   401
Burt, T.P.   275
Butler, J.B.   184, 195, 260, 289, 330, 332, 338

Caissie, D.   452, 454, 455
Caleffi, V.   230, 237
Campling, P.   205
Camps, A.J.   204
Cao, Z.X.   19, 241
Capra, H.   448
Carling, P.A.   219, 241
Carrier, G.F.   123
Carslaw, H.S.   55
Casey, M.   166
Castro, N.M.   275
Cebeci, T.   332
Cha, K.S.   126
Chakraborty, S.   195
Chan, C.T.   165
Chandler, J.H.   236
Chang, Y.C.   65
Chapra, S.C.   313
Charlton, M.E.   287
Chaudhury, M.H.   238
Chen, C.J.   112, 117
Chen, H.C.   98
Chen, J.M.   198
Chen, Y.   132, 133, 140, 307
Cheng, C.T.   201
Cheng, N.S.   75
Cheung, Y.   151
Chhun, V.H.   405
Chiew, Y.M.   75
Choi, S.-U.   402
Chow, V.T.   198, 199, 405, 475
Chrisohoides, A.   103
Christiaens, K.   205
Christian, C.D.   126
Christoph, G.H.   336
Chung, T.J.   148, 151, 162
Church, M.   85, 225, 332, 339, 340
Clausse, A.   284
Clifford, N.J.   240, 330, 332, 334
Cluer, B.L.   359, 390
Cobby, D.M.   197, 289, 291, 296
Coe, M.T.   296
Cokljat, D.   106, 111, 488
Colombini, M.   76, 103, 104
Connell, R.J.   132, 195
Constantinescu, G.S.   103
Cooke, R.   483
Cooper, V.A.   201

Cornelius, C.   227
Covelli, P.   140
Craft, T.J.   104, 105, 107
Crank, J.   55
Crapper, P.F.   201
Crosato, A.   76, 224
Crossley, A.J.   164
Crowder, D.W.   222, 259, 261
Crowe, J.C.   220
Cui, Y.   221
Culling, W.E.H.   195
Cundy, T.W.   130
Cunge, J.A.   164, 284, 291, 294
Czernuszenko, W.   180

Dainton, M.P.   207
Daly, B.J.   105
Dancey, C.L.   75
Daniell, T.M.   200–1
Darby, S.E.   249
Datko, M.   195
De Roo, A.P.J.   128, 129, 139, 203, 260, 275, 280, 284, 285, 290, 295
De Waal, J.P.   123
Defina, A.   121, 135, 136, 280, 297
Demuren, A.O.   65, 101, 103, 111, 242, 243
Dey, S.S.   195
Dhondia, J.F.   166, 284
di Silvio, G.   221
Dietrich, C.R.   294
Dietrich, W.E.   219, 224, 225, 233, 236, 240, 246, 248, 362
Diplas, P.   222, 259, 261
Dodd, N.   138
Dolan, R.   359, 365
Duan, J.G.   239, 249
Duan, Q.Y.   200
Dubi, A.M.   408
Duel, H.   452, 453
Duizendstra, H.D.   85
Dunn, C.   399, 401, 402, 407, 411
Durbin, P.A.   97, 98, 107, 117

Easom, G.   492
Egiazaroff, I.V.   81, 220
Einstein, H.A.   86
El-Hames, A.S.   181
El Jabi, N.   452, 454, 455
Elder, J.W.   25, 58, 310, 312
Ellis, J.   397
Elso, J.I.   222
Engelund, F.   76, 79, 336
Ervine, D.A.   273, 397
Estrela, T.   139, 284
Evans, R.A.   126

Falconer, R.A. 132, 133, 140, 165, 305, 306, 307, 312, 313, 462, 487
Fathi-Maghadam, M. 197, 399, 404, 418
Feldhaus, R. 280
Feng, Y.T. 126, 127, 278
Ferguson, R.I. 4, 37, 171, 176, 177, 220, 221, 222, 229, 230, 233, 243, 332
Ferziger, J.H. 227, 228, 230
Feyen, J. 205
Finson, M.L. 336
Fischer-Antze, T. 336, 338, 402, 415, 416, 417
Fischer, H.B. 2, 55, 57, 310, 312
Flavelle, P. 179, 181
Flood, M. 196
Florsheim, J.L. 219, 241
Fluent 31, 33
Forkel, C. 418
Franks, S.W. 201
Frasier, S.J. 204
Fredsøe, J. 79
Freeman, G.E. 400
Freer, J. 205, 472, 476, 481
Freeze, R.A. 2
French, J.R. 287
Fujita, K. 273
Fukuoka, S. 273

Gabel, S.L. 272
Gad-el-Hak, M. 329
Galappatti, R. 73
Garcia-Navarro 132
Garcia, M.H. 274, 336, 401, 415, 416
Garton, J.E. 405
Gaskell, P.H. 159, 492
Gatski, T.B. 31
Gee, D.M. 280
Ghanem, A. 439, 441
Ghanem, R. 195, 207
Ghisalberti, M. 274
Gibson, M.M. 106, 108, 109, 115, 116
Giller, P.S. 222
Gilvear, D.J. 223
Ginot, V. 439
Goldstein, S. 307, 308
Gomes-Pereira, L.M. 136, 196, 197, 287
Gomez, B. 332, 337, 338
Goodwin, P. 275, 446
Gopalakrishnan, T.C. 126
Graber, H.C. 204
Graf, J.B. 359, 365
Graf, W.H. 405
Graham, W.D. 207
Gray, W.G. 126, 127, 130, 138, 280, 297
Grayson, R.B. 201, 291, 292
Green, J.E.P. 405
Greenhill, R.K. 397

Greenspan, H.P. 123
Gresho, P.M. 169
Grimm, J.P. 61
Guay, J.C. 259, 435, 436, 452
Guo, Q.-C. 237
Gurevitch, J. 437
Gutelius, W. 196
Guymer, I. 62, 65

Hacking, I. 179
Hager, W. 194
Ham, D.G. 85
Hamed, M.E. 282
Hancu, S. 38
Hanjalic, K. 105
Hankin, B.G. 272, 294, 296, 471, 472, 473, 474, 476, 478
Hansen, E. 76
Hardy, R.J. 20, 128, 139, 140, 172, 173, 179, 185, 187, 227, 230, 235, 257, 260, 433, 441, 443, 446, 475
Hardy, T.B. 441
Hargreaves, D. 487
Harlan, R.L. 2
Harleman, D.R.F. 310
Harlow, F.H. 105
Hassan, M.A. 86
Havinga, H. 85
Hayes, J.W. 259
Hazel, J.E. 390
He, Q. 124
Henderson, F.M. 309
Heniche, M. 237, 441, 443
Hereford, R. 358, 359, 360, 365, 374, 375, 380, 384
Hervouet, J.-M. 121, 122, 129, 131, 133, 135, 137, 175, 198, 273, 277, 279, 280, 291, 297, 443, 474
Hey, R.D. 37, 225, 233, 289, 332
Higgitt, D.L. 195
Hildrew, A.G. 259
Hinze, J.O. 21, 27
Hirano, M. 81, 221
Hirt, C.W. 35, 126, 488
Hodgson, M.E. 287
Hodskinson, A. 4, 176, 177, 229, 230, 243, 329, 332, 333, 346, 503
Hoey, T.B. 188, 221, 222
Holly, F.M. 25, 132
Hong, H. 400
Hornberger, G.M. 275
Horritt, M.S. 128, 131, 132, 137, 139, 178, 186, 194, 196, 197, 199, 201, 203, 237, 274, 282, 285, 293, 294, 295, 398, 399, 403, 406, 443, 477
Horton, R. 403
Howard, A.D. 249, 359, 365
Howes, S. 186
HR Wallingford 305

Hsieh, T.Y.   198
Hu, W.   230
Hubbard, M.E.   138
Hughes, T.J.R.   131, 132
Huising, E.J.   196

Ikeda, S.   79, 333
Ingham, D.B.   28
Ip, J.T.C.   125, 141, 280, 297
Ismail, A.   487
Itakura, T.   224

Jackson, C.P.   181
Jaeger, J.C.   55
Jamet, P.   126
Janin, J.-M.   131, 133, 135
Jansen, P.Ph.   72, 73, 76
Jaw, S.Y.   112, 117
Jha, A.K.   132
Jin, Y.-C.   237
Johannesson, H.   24, 249
Johns, B.   126
Jolliffe, I.T.   201
Jones, D.   225
Jones, L.   205
Jones, W.P.   97
Jorde, K.   434, 437, 448
Jowett, I.G.   259, 452

Kaiser, W.   397
Kanazawa, M.   333
Kang, H.   402
Karanxha, A.   84
Karniadakis, G.E.   171, 175
Kashefipour, S.M.   313, 317, 462, 463, 464, 468
Kassem, A.A.   238
Katopodes, N.D.   126, 441, 454
Kawahara, M.   126, 280, 297
Kawahita, R.A.   128, 131, 132, 280, 297
Kearsley, L.H.   358
Keller, E.A.   219, 240, 241
Kennedy, G.J.A.   223
Kennedy, J.F.   79, 224
Kenworthy, S.T.   225, 236, 250, 255
Keulegan, G.H.   361, 362
Khalil, M.B.   62
King, I.P.   2, 121, 133, 165, 280
King, J.M.   454
Kirchner, J.W.   337, 338
Kirkbride, A.   330, 333
Kitamura, T.   407
Kjellesvig, H.M.   25, 34, 138, 344
Klaassen, G.J.   81
Kleinhans, M.G.   81, 86
Klemeš, V.   294

Knight, D.W.   272, 273, 274, 275, 282
Koblinsky, C.J.   204
Kohane, R.   275
Kolmogorov, A.N.   30, 102
Konikow, L.F.   179
Korman, J.   366
Kouwen, N.   197, 274, 289, 398, 399, 404, 405, 418
Kovacs, A.   79
Krabill, W.B.   287
Krone, R.B.   25, 314
Kutija, V.   400

Lallemand, B.   195
Lamb, H.   55
Lamb, R.   199, 205
Lamouroux, N.   433, 438, 439, 453
Lancaster, J.   259
Lancaster, S.T.   249
Lane, S.N.   4, 12, 20, 23, 24, 166, 171, 176, 177, 179, 180, 181, 182, 183, 184, 185, 186, 187, 194, 196, 198, 202, 224, 225, 227, 229, 230, 233, 234, 235, 236, 250, 251, 255, 257, 260, 287, 295, 330, 332, 333, 336, 337, 345, 346, 352, 480
Langtangen, H.P.   207
Lanzoni, S.   80
Lapointe, M.   221
Larsen, T.   407
Lau, A.K.C.   159, 492
Lauchlan, C.   85
Launder, B.E.   27, 28, 29, 62, 97, 98, 99, 105, 106, 107, 110, 111, 115, 116, 234, 329
Lawless, M.   330, 333
Leclerc, M.   133, 222, 258, 280, 297, 429, 430, 431, 434, 441, 443, 445, 446, 448, 450, 451
Lee, A.J.   233
Leonard, B.P.   159, 160, 169
Leopold, L.B.   358, 389
Leschin, M.F.   359
Leschziner, M.A.   242, 276
LeVeque, R.J.   164
Li, L.   195, 207
Li, R.M.   289
Li, S.P.   107
Lick, W.   314
Lilly, D.K.   33
Lima, J.T.   182
Lin, B.   132, 165, 305, 311, 462
Liu, P.L.F.   123, 339
Lohani, B.   196
Löhner, R.   121, 126, 138, 156, 158
Lopez, F.   274, 336, 401, 415, 416
Lotter, G.K.   397
Lucchitta, I.   358, 389
Luis, S.J.   179
Lynch, D.R.   126, 127, 130, 138, 280, 297

Ma, L.   20, 38, 124, 136, 241, 242, 278
MacArthur, R.C.   129
McConchie, J.   205
McCorquodale, J.A.   23
McCuen, R.H.   186
McLaughlin, D.   179
McLean, D.G.   85
McLean, S.R.   361, 362, 364
McLelland, S.J.   273, 330, 332
Maddock, I.P.   71, 222
Maheswaran, S.   224
Malalasekera, W.   148, 155
Manson, J.R.   227
Marelius, F.   180
Marks, K.J.   124, 287, 291
Marseguerra, M.   205
Mason, D.C.   196, 198, 290, 291, 296, 297
Mason, P.J.   114
Masterman, R.   397
Mathey, F.   415
Mead, R.   200
Mehta, A.J.   314
Meijer, D.   401
Melis, T.S.   365
Menduni, G.   75
Menenti, M.   198
Menter, F.R.   30, 31, 102
Mertens, W.   397
Mertes, L.A.K.   275
Meselhe, E.A.   4, 101, 180, 242, 274, 494
Meyer-Peter, E.   75, 363
Michiue, M.   81
Milhous, R.T.   222, 430, 433, 439, 441
Miller, A.J.   241
Mitchell, C.A.   122, 125, 128, 132, 274, 280
Mohammadi, B.   28, 31
Molinaro, P.   280
Molinas, A.   86
Moller, D.   204
Montgomery, D.R.   75
Monthe, L.   280, 297
Morin, J.   433, 442, 453
Morvan, H.P.   177, 227, 242, 487, 488, 493, 494, 502
Mosselman, E.   80
Moulin, C.   25, 121, 140
Moussa, R.   194
Muller, J.A.   313
Müller, R.   75, 363
Murr, A.   313
Musy, A.   205

Naden, P.S.   230
Nagata, N.   239, 240, 249
Nakagawa, H.   19, 23, 24, 101, 103, 117, 329, 401, 414

Namin, M.M.   165
Naot, D.   103, 106, 111, 339
Nassehi, V.   126
Nditiru, J.G.   200–1
Neary, V.S.   103
Nelder, J.A.   200
Nelson, J.M.   224, 238, 363, 364
Nepf, H.M.   197, 274, 339, 399, 401, 402, 410, 411, 418, 420, 424
Nezu, I.   19, 23, 24, 26, 101, 103, 117, 183, 225, 273, 329, 414
Nicholas, A.P.   4, 122, 125, 128, 132, 176, 184, 202, 221, 229, 235, 274, 280, 330, 332, 333, 334, 345, 346
Nichols, B.D.   35, 488
Nikora, V.L.   184, 330, 332, 334, 337, 338
Nikuradse, J.   100
Nino, Y.   86
Nishimura, T.   79
Noat, D.   493
Norton, W.R.   2, 165
Nowell, A.R.M.   332, 339, 340
Nuding, A.   397
Nujić, M.   132

Oberkampf, W.L.   194
Odgaard, A.J.   4, 240
Olsen, N.R.B.   25, 34, 62, 138, 165, 230, 239, 330, 336, 337, 344, 414
Onitsuka, K.   414
Oreskes, N.   178, 179, 181, 184
Orszag, S.A.   29, 331
Osnes, H.   207
Owen, M.   123

Paintal, A.S.   78
Palmer, G.N.   126
Paola, C.   221
Papanicolaou, A.N.   75, 86
Parasiewicz, P.   436, 453
Parchure, T.M.   314
Parker, G.   24, 79, 83, 86, 220, 221, 249
Parsons, D.R.   243
Parsons, J.G.   314
Partheniades, E.   25
Pasche, E.   397, 401, 402, 416, 417
Pastor, M.   131
Patankar, S.V.   62, 161, 360, 362, 415
Patel, V.C.   98, 99, 101, 103, 109, 110, 111, 329, 333
Pender, G.   273
Peregrine, D.H.   123
Perić, D.   126, 127, 278
Perić, M.   227, 228, 230
Petera, J.   126
Petryk, S.   398

Pettersson Reif, B.A.   97, 98, 117
Phillips, B.C.   73
Pilotti, M.   75
Pinder, G.E.   275
Piomelli, U.   92
Pironneau, O.   28, 31
Pitlick, J.   333
Plate, E.J.   396
Pletcher, R.H.   336
Popper, K.R.   179
Potochnik, A.   359, 360, 365
Powell, D.M.   80, 81
Powell, J.W.   359
Prandtl, L.   26, 232
Preston, R.W.   26, 312
Prinos, P.   4, 103, 105, 180, 329

Quecedo, M.   131
Quintas, L.   139, 284
Quraishi, A.A.   396

Rabus, B.   287
Rahuel, J.L.   221
Raithby, G.D.   331
Raleigh, R.F.   223
Rameshwaran, P.   230
Rastogi, A.K.   23, 24
Raupach, M.R.   401
Raw, M.J.   155, 162
Red-Horse, J.   207
Ree, W.O.   396, 398
Refsgaard, J.C.   200
Reiser, D.W.   447, 455
Reynolds, O.   92, 276
Rhoads, B.L.   225, 236, 250, 255
Ribberink, J.S.   81, 83
Richards, K.S.   4, 166, 179, 180, 181, 182, 184, 185, 186, 187, 219, 224, 232, 233, 236, 241, 250, 260, 332, 480
Ritchie, J.C.   196, 198
Roache, P.J.   166, 169, 172, 175, 176
Robert, A.   260, 330, 333
Robinson, R.A.J.   221
Rocca, F.   196
Rodi, W.   23, 24, 26, 31, 62, 65, 101, 103, 106, 108, 111, 112, 114, 115, 117, 224, 242, 276, 329, 402
Roe, P.L.   132
Roig, L.C.   121, 126, 133, 280
Rokni   31
Romanowicz, R.J.   139, 184, 205, 284, 294, 472, 480
Romeiser, R.   204
Rong, Y.   205
Rosso, R.   199
Rotta, J.C.   106, 107
Rouse, H.   363

Roy, A.G.   251, 257, 258
Roy, Y.   330
Rubin, D.M.   358, 362, 365, 389
Rufino, G.W.   196, 455
Rutherford, J.C.   25, 55, 58
Rutschmann, P.   194
Rylov, A.A.   180
Ryrie, S.C.   123

Saad, Y.   124, 126, 162
Sabaton, C.   452, 454
Samuels, P.G.   287
Sanders, B.F.   130, 132
Sanders, R.   196
Sanjiv, S.K.   180
Sargent, R.J.   398
Sauer, S.P.   275
Savic, I.J.   132
Scheiner, S.M.   437
Schlichting, H.   62, 339
Schmidt, J.C.   358, 359, 365, 390
Schneider, G.E.   155, 162
Schneider, M.   24, 225, 434, 437
Schultz, M.H.   162
Secretan, Y.   439, 441, 446, 447
Sekine, Malmaeus   86
Sellin, R.H.J.   273, 274, 397, 398
Seminara, G.   243
Sen, D.   280
Shao, X.J.   278
Sharma, B.I.   97, 98
Shaw, C.T.   171
Shaw, R.H.   401, 414, 415
Shields, A.   362
Shige-Eda, M.   122
Shima, N.   110, 111
Shimizu, Y.   224, 401
Shiono, K.   272, 273, 274, 275, 282, 311
Shirvell, C.S.   434
Shvidchenko, A.B.   75, 273
Simm, D.J.   122, 294
Sinha, S.K.   4, 100, 101, 330, 494
Slama, E. Ben   25, 121, 140
Slatton, K.C.   196
Sleigh, P.A.   165
Slingerland, R.L.   221
Smith, G.D.   149, 157, 158, 163, 164
Smith, G.H.S.   4, 176, 202, 229, 235, 330, 332, 333, 345, 346
Smith, J.D.   219, 224, 233, 236, 238, 240, 332, 333, 336, 337, 340, 361, 362, 363, 364
Smith, L.M.   29, 52, 53, 56, 59
Smith, N.D.   225
So, R.M.C.   110
Sofialidis, D.   4, 103, 105, 180, 329
Sorooshian, S.   201

Sotiropoulos, F.   99, 101, 103, 105, 109, 110, 111, 180, 242, 494
Souchon, Y.   433, 438, 439
Soulaimani, A.   124, 126
Southard, J.B.   81
Soutter, M.   205
Spalart, P.R.   92
Spalding, D.B.   27, 28, 29, 34, 62, 99, 161, 234
Speziale, C.G.   28, 31, 104, 107, 474
Squillace, P.J.   275
Srokosz, M.   297
Stanford, J.A.   275
Statzner, B.   75
Stelling, G.S.   166, 284
Stephenson, D.B.   201
Stevens, L.E.   384
Stewart, M.D.   275, 286
Stoesser, T.   124, 277, 278, 414, 415, 418
Stoksteth, S.   230, 330, 336, 337
Stolzenbach, K.D.   401
Strange, C.D.   223
Street, R.L.   92
Struiksma, N.   76, 83, 224
Sukhodolov, A.N.   250
Sun, T.   249
Sutherland, A.J.   73, 221
Sykes, R.I.   114
Synolakis, C.E.   123

Tabuchi, S.   196
Talbot, T.   221
Talmon, A.M.   79
Tamai, N.   251
Tani, J.   100
Taylor, C.   169
Taylor, G.I.   57, 59
Taylor, R.P.   336, 339
Tchamen, G.W.   128, 131, 132, 280, 297
Temmerman, L.   33
Temple, D.M.   289, 398, 399
Tennant, D.L.   455
Tezduyar, T.E.   126
Thacker, W.C.   123
Thackston, E.L.   313
Thannbichler, C.   84
Thomann, R.V.   313
Thomas, T.G.   21, 92, 278
Thompson, 1998, K.R.   204, 241, 358
Thorn, M.F.C.   314
Thorne, C.R.   225, 397
Tiffan, K.F.   258
Tingsanchali, T.   224
Titmarsh, G.W.   199
Topping, D.J.   358, 368, 389, 390
Toro-Escobar, C.M.   221
Toro, E.F.   132, 164

Trocano, T.G.   194
Tsujimoto, T.   336, 399, 401, 402, 407, 410, 415, 416, 424
Tubino, M.   80, 243
Tung, C.C.   126

Umetsu, T.   126, 280, 297
Uncles, R.J.   9
Usseglio-Polatera, J.M.   25
Usunoff, E.   185

Valentine, E.M.   60
van Bendegom, L.   79
Van Der Meer, J.W.   123
Van Haren, L.   129, 273, 277, 279, 280
Van Leer, B.   132
Van Niekerk, A.   221
van Rijn, L.C.   24, 25, 72, 73, 81, 86, 315, 364
Van Velzen, E.H.   401
Vandoormaal, J.P.   331
Vannote, R.L.   429
Vanoni, V.A.   237
Venere, M.   284
Ventikos, Y.   103, 105
Versteeg, H.   148, 155
Vieira, J.K.   311
Vismara, R.   223
Vivoni, E.R.   399, 401, 410, 411, 424
Vreugdenhil, C.B.   73

Waddle, T.   441
Wallbridge, S.   75
Walling, D.E.   124, 274, 280
Wallis, S.G.   60, 227
Wan, Q.   230
Wang, J.   333
Wang, S.S.Y.   165
Warburton, J.   195
Ward, J.V.   222, 275
Webb, R.H.   358, 365
Weerakoon, S.B.   251
Weltz, M.A.   197
Welz, R.   275
Westaway, R.M.   287
Whigham, P.A.   201
Whiting, P.J.   224, 225, 233, 246, 248
Wiberg, P.L.   233, 332, 333, 336, 337, 340, 362
Wicherson, R.J.   136, 197, 287
Wicks, J.   284
Wiele, S.M.   360, 364
Wiener, N.   207
Wilcock, P.R.   81, 220
Wilcox, D.C.   30, 31, 102, 117
Willets, B.B.   273, 274
Williams, J.J.R.   21, 92, 278

Wilson, C.A.M.E. 23, 61, 62, 63, 198, 230, 274, 280, 398, 399, 403
Wilson, N.R. 401, 414, 415
Wilson, R.T. 365
Wimmer, C. 196
Wintergerste, T. 166
Woessner, W.W. 275
Wood, I.R. 60
Wood, N. 219, 241
Wormleaton, P.R. 282, 397
Wright, N.G. 162, 487
Wu, B.S. 86
Wu, F.-C. 197, 398
Wu, J. 309
Wu, W. 101, 235, 236, 239
Wu, Y. 314
Wyer, M.D. 313

Yakhot, V. 29, 331
Yang, J.C. 198
Ye, J. 23
Yeatts, M. 359, 374
Yeh, K.-C. 224
Yoon, J.Y. 99, 103, 106, 111, 186, 313, 329, 488
Younis, B.A. 106, 111, 329, 488

Zedler, E.A. 92
Zhang, W. 130
Zhao, D.H. 132
Zheng, L.Y. 131
Zienkiewicz, O. 151
Zimmermann, C. 79
Zio, E. 205

Subject index

Abbott-Ionescu schemes 164
Abrasion 80
Abstraction management plans 430
Accuracy 164, 170, 174, 175, 179, 181, 188, 194, 198
Acoustic Doppler current profiler 260, 443
Acoustic Doppler Velocimeter 182, 202, 243, 335, 345, 403, 410, 424, 474, 480, 490
Active layer 81, 238
Active microwave sensors 203
Adaptation length 73
Adaptive curvilinear grid 240
Adaptive grid methods 2, 6, 128, 132, 138, 170, 172, 446
Adequacy 172, 181
Advection 42, 52, 56, 60, 285
Advection diffusion 58, 66, 311, 314, 418
Advection–diffusion equation 5, 25, 55, 73, 139, 140, 220, 310, 325, 360, 362
Aerial photography, *see* Airborne imagery
Aerial stereophotogrammetry, *see* Photogrammetry
Aggradation 83, 222
Airborne imagery 122, 443
Airborne laser altimetry, *see* Laser altimetry
Airborne SAR, *see* Synthetic Aperture Radar
Algebraic (zero-equation) models, *see* Zero-equation models
Algebraic stress model 31, 95, 108, 226, 278
Alternating Direction Implicit methods 165
Analytical solutions 5, 12, 135, 170, 194, 281
Angle of repose 363
Animation 187
Anisotropic Reynolds stress, *see* Reynolds shear stresses
Anisotropic turbulence 32, 65, 103, 106, 108, 219, 226, 402
Archaeological applications 358, 359, 365, 385
Arid environments 196
Armouring 221–2
Artificial viscosity 170–1
ASME criteria 187
Atlantic salmon 258
Atmospheric flows 198
Auxiliary relationships 5, 71, 85

Back-to-back meanders 255
Backward differencing 149, 163, 228
Backward-facing step 226, 329
Backwaters 372
Bacterial indicators 317
Bank erosion 217, 220, 237–9, 242, 249–50
Bank roughness 185, 235
Bank-storage effects 275
Bars 85, 233, 248, 359
Base level 359
Bathing water quality 313, 317, 464
Bathing Water Standards 462
Bathymetry, *see* Topography
Bayesian framework 205, 469, 480
Beach run-up 123
Beach topography 135
Bed exchange 221
Bed friction, *see* Roughness
Bed friction coefficient, *see* Roughness
Bed material 5, 193, 199, 231, 289
Bed roughness, *see* Roughness
Bed shear stress 23–4, 37, 85, 181, 186, 219, 224–5, 231, 233, 240, 314, 346, 362, 414, 418, 596
Bedforms 79, 85, 176–7, 364
Bedload 5, 72–3, 79, 83–4, 221–2, 333, 360, 363
Bedrock structure 365
Behavioural models 470
Benchmark solutions 170, 178
Benchmarking 7
Bend curvature 246
Bend migration 246
Bending stiffness 409
Bias 179, 181
Bifurcation 76, 236
Bioenergetics 432
Biomechanical properties 13, 408, 414
Blending 170–1
Block-structured grid 230
Blockage 177, 185, 336
Blocking pressure cells 419
Body-fitted coordinates 155, 229, 238, 242–3, 342
Boulder-bed streams 233
Boulders 259

Boundary condition   1, 2, 11, 20, 26, 30, 35–6, 43, 46, 99, 101–2, 126, 161–2, 170, 175–7, 179, 183, 186–8, 230, 235, 260, 276, 281, 285–6, 317–18, 319, 329, 463–4, 474–5, 481, 487–8
Boundary fitted coordinates, *see* Body-fitted coordinates
Boundary friction, *see* Roughness
Boundary roughness, *see* Roughness
Boundary shear stress, *see* Bed shear stress
Boundedness   159, 163
Boussinesq approximation   23, 27–8, 95, 97, 103–4, 112, 123, 232, 279, 307, 329
Braided rivers   9, 12, 261, 329, 333, 342, 345, 352
Bridge piers   217, 308
Brown trout   223
Buffer layer   98, 102
Buoyancy   8, 112, 115

Calibration   7, 13, 186, 188, 193–4, 199–201, 203–5, 219, 225, 291–2, 295–6, 317–19, 333, 396, 402, 422–4, 439, 464, 482, 489
Capacity-limited transport   73
CASI multispectral scanner   198
CASIMIR   438
Catchment hydrological models   199, 202, 205
CCHE2D   165
Cell aspect ratio   494
Central differencing   149, 158–9, 172, 228
CFX   31
Channel change   7, 220, 236, 260–1
Channel curvature, *see* Meanders
Channel diversion   217
Channel evolution   346
Channel fills   249
Channel–floodplain interface   502
Channel irregularities   389
Channel migration   217, 260–1
Channel routing   219
Channel shape   388
Characteristic-based methods   164
Chezy equation   24, 74–5, 78, 84, 231, 273, 289, 309, 395, 439, 488–9
Chinook salmon   258
Circulation zone   380
Coherence   179, 186–7, 281
Cohesive sediment   314, 322
Colebrook–White   309
Coliform decay rate   9, 313
Combined sewer overflows   51, 319, 462–5, 468, 481
COMBSIM1D   505
Competition   430
Competition predation   452
Complex geometries   171–2, 230, 260–1, 272, 295
Complexity   9

Compound channel flow   103, 124, 272–3, 275–6, 278–9, 296, 329, 406
Computational efficiency   366
Conceptual assessment   185, 187
Conceptual model   4, 10
Confluences   7, 24, 173, 176, 186, 226, 230, 240, 250–1, 253, 255, 259–60, 329, 365, 390, 502
Conservation of mass, *see* Mass conservation
Conservation of momentum, *see* Momentum conservation
Consistency   162
Contaminants   41, 46, 314, 442, 452, 468
Contamination of measurements   182
Continuity equation, *see* Mass conservation equation
Contour data   204, 287, 366
Control volume   154
Convection   94, 509
Convective terms   65, 170–1, 228
Convergence   63, 162–3, 173–4, 227, 489, 494
Convergence criteria   509
Convergence error   170
Convergence testing   6
Conveyance   272, 274, 281, 395–6, 488
Coriolis force   125, 276, 279, 307
Courant number   141, 174–5, 475
Crank–Nicholson scheme   164
Critical shear stress   75, 81, 220, 314, 323, 362
Curvature Compensated Convection Transport (CCCT)   492
Curvilinear spatial grid   220
Cyprinids   456

Dam break   121–3, 132–3, 279
Dam release   360, 364, 382
Darcy-Weisbach friction factor   219, 231, 289, 404
Data acquisition challenges   9, 297
Dead zone   5, 60, 272, 474–5, 478
Debris fan   365, 370, 390
Deflectors   502
Deformable mesh   124, 126–7, 278
Degradation   83, 222
Delauney method of triangulation   366, 446
Deposition   25, 80, 220–1, 227, 230, 239–40, 246, 249, 314, 360, 372, 375–6, 380–1, 388, 468
Deposits   359, 366, 372
Depth-averaged models   22, 125, 140, 185, 219, 224–7, 230–1, 250, 258, 279, 308, 360, 464, 474, 484
Differential GPS   446
Differential models   95
Diffusion   42, 52, 59, 108, 113, 170–1, 177, 228
Diffusion coefficient   54, 56, 61, 80
Diffusion wave model   8, 25, 260, 280, 284–5
Digital elevation data   122, 137, 196, 284
Digital photogrammetry, *see* Photogrammetry

Subject index

Digital terrain models 157, 441, 444
Dimensionality 261, 295
Direct Numerical Simulation 6, 20–1, 85, 91–2, 111, 226, 251
Dirichlet boundary condition 63, 66, 153, 156
Discordance 251, 253, 258
Discrepancies 138–9
Discrete element models 9, 336–7, 340, 418
Discretisation 7, 36, 129, 131, 148, 157, 163, 227, 277–8, 280, 289, 291, 336, 396, 430, 502
Dispersion 5, 26, 42, 52, 59–60, 140, 219, 480
Dispersion terms 23–5, 139, 185, 224–5, 308
Dissipation 101, 226, 419
Dissipation length scale 493
Dissipation rate 28, 99, 101, 104–8, 110–11, 113–14, 116–17, 308, 415
Dissipation rate tensor 107, 115
Dissipation tensors 115
DIVAST 165, 312, 313, 317, 462, 474
Divided channel method (DCM) 396
Domestic waste 313
Dominant discharge 221
Donor–acceptor scheme 35
Downstream conditions 36
Downstream fining 80–1, 221
Drag 9, 334, 339, 343, 395, 410, 413
Drag coefficient 2, 289, 337–9, 344–6, 352, 399, 404, 423
Drag force 289, 398–9, 402–4, 409–10, 414–16, 418–19
Dredging 314
Ducts 329
Dunes 79, 81, 310
Dynamic pressure 253

Earth's rotation 8, 306
Echosounding 446
Ecohydraulics 10
Ecological processes 7, 14, 51, 57, 314, 441
Eddies 441
Eddy diffusion 58, 227
Eddy viscosity 29, 95, 113, 232, 251, 288, 307–8, 310, 330, 361–2, 474, 480, 484
Eddy viscosity models, *see* Zero-equation turbulence models
Effective parameter values 294, 475, 484
Effective roughness 231, 233, 332, 336, 340
Effluent discharge 217
Effluent mixing 51
Egg deposition 223
Egg incubation 432
Einstein parameter 74
Electromagnetic current meter 182, 202
Element aspect ratio 491
Embankment failure 275
Empirical closure 72

Energy dissipation 414, 493
Energy losses 198, 274–5, 399, 413, 424
 see also Turbulence; Roughness
Entrainment 221, 240, 254, 315
Environmental quality standard 464–5
Equal mobility 80–1
Equifinality 82, 186, 205, 294, 476, 481
Equilibrium 99, 108, 234, 369
ERCOFTAC 166
Erosion 25, 72, 80, 220, 227, 230, 239, 246, 249, 314, 358, 359, 369, 373, 381, 390, 506
Error 137, 157, 164, 182, 188, 201, 292, 294, 446
Error propagation 202, 204
ERS 196
Estuaries 8, 305, 307, 462
Estuarine model 132
EU Water Framework Directive 11, 13, 430, 453, 484
Eulerian approach 156, 449
Evapotranspiration 275
EVHA 439
Exchange layer 81
Exclusion methods 131
Exner equation 71, 220, 238
Experimental approaches 10, 170, 178, 182, 184, 188, 296
Explicit Algebraic Stress Models, *see* Algebraic stress model
Exponential schemes 171
Exposure 80–1

Factor of safety 172
Factor perturbation 188
Faecal coliforms 9, 306, 313, 317, 319, 326, 462, 464, 466–8
False time step relaxation 174
Falsification 179
Fans 380
FASTER 313, 317, 462
Fathometer 366
Fickian diffusion 54
Fick's first law 53, 58, 310
Fine sediment 21, 25, 51
Finite difference methods 6, 140, 148, 151, 155, 158, 165, 221, 227–9, 276, 285
Finite element methods 6, 127–9, 133, 139–40, 151, 155, 162, 165, 186, 207, 227, 230, 276, 280, 291, 441, 446, 469, 474, 480
Finite volume methods 6, 127, 153–5, 161, 165, 170–1, 174, 227–30, 242, 276, 396, 414, 488
First-order schemes 170–2
Fish 10, 258–9, 455, 488
Fixed grid methods 128, 280
Fixed lid, *see* Rigid lid
Flexural rigidity 409, 411
Flood Channel Experimental Facility 476

Flood extent data 201, 294
Flood flow 198, 203
Flood hydrographs 199
Flood inundation 7–8, 122, 128, 201, 274, 290
Flood protection 305
Flood pulse 296
Flood risk 197
Flood risk analysis 219
Flood risk assessment 196
Flood risk maps 472
Flood routing 275
Flood routing models 281
Flood storage 395–6
Flood-warning 202
Flood wave routing 260, 272
Flooding 132–3, 364
Floodplain flow velocities 182, 424
Floodplain hydraulics 199
Floodplain infiltration 286
Floodplain inundation, *see* Flood inundation
Floodplain sediment deposition 124
Floodplains 83, 197, 200, 237, 260, 271–2, 310, 329, 336, 417, 420–1, 474
Flow abstraction 217
Flow curvature 219
Flow recirculation 251
Flow resistance, *see* Roughness
Flow reversal 101
Flow separation 219, 226, 234, 250, 257, 333
Flow structures 260–1, 277
Flow variability 454
Fluent 31
Flume experiments 197, 246
Forced empirical adequacy 184
Form drag 84, 233, 289, 336–7, 364, 401
Form roughness 41, 343, 352, 423
Forward differencing 149, 158, 229
Fractals 195, 260
Free surface 5, 11, 33–5, 101, 103, 106, 123, 133, 157, 176, 202–3, 219, 225–6, 230, 236, 239, 242, 250, 254, 271, 333, 349–50, 488, 494, 503
Fresh–saltwater interface 9
Friction, *see* Roughness
Friction coefficients, *see* Roughness coefficients
Friction factors 232, 440–1
Friction slope 218, 289
Frictional losses 124
Froude number 73, 286, 407, 433, 453
Fry 223
Fry emergence 223
Fully developed flow 36
Fully turbulent region 98, 109
Functional iteration 165
Functional regression 183
Fuzzy modelling 437, 448, 472–3, 476–7

Galerkin methods 131–2, 152
Generalized Likelihood Uncertainty Estimation 11, 205, 294, 461–3, 470, 472, 476–7, 480, 482, 484
Genetic algorithms 200
Geographical Information Systems 442, 454
Geometry, *see* Topography
Global Positioning System 3, 195–6, 288, 443
GMRES 162
Goodness-of-fit 188
Graded sediment 80–2
Gradually varied flow 222
Grain arrangements 176
Grain scales 177
Grain size 184, 221
Grain sorting 80
Gravel-bed river 9, 176, 185, 198, 222, 230, 233, 235, 249, 259–60, 332, 334, 336, 341, 345, 382
Gravitational effects on sediment transport 79
Grid Convergence Index 172–3
Grid generation 2, 127, 156, 291
Grid independence 6, 20, 128, 157, 162, 170, 172–4, 229, 251, 395, 474, 476, 489, 509
Grid quality 127, 138, 163, 230
Grid refinement 136, 156–7, 161, 228, 489
Grid resolution 20, 138, 173, 198, 228, 289, 332, 352, 366, 491
Groundwater 166, 196, 205, 275
Guild *preferenda* 438–9
Gullies 375–6, 384

Habitat enhancement 217
Habitat management 10
Habitat metrics 259
Habitat models 432, 442, 453–4, 506
Habitat prediction 506
Habitat preference curves 10, 445, 452
Habitat Probabilistic Index (HPI) 259
Habitat quality 437
Habitat suitability 222–3
Habitat Suitability Index 258–9, 433, 436, 440, 447, 448, 450, 452–3
Habitat
  Atlantic salmon 258
  bioenergetics 432
  brown trout 223
  CASIMIR 438
  Chinook salmon 258
  competition 430
  competition predation 452
  cyprinids 456
  egg deposition 223
  egg incubation 432
  fish 10, 258–9, 455, 488
  fry 223
  fry emergence 223

general 218, 220, 222, 258, 259, 261, 366, 454
Guild *preferenda* 438–9
habitat enhancement 217
habitat management 10
habitat metrics 259
habitat models 432, 442, 453–4, 506
habitat prediction 506
habitat preference curves 10, 445, 452
Habitat Probabilistic Index (HPI) 259
habitat quality 437
habitat suitability 222–3
Habitat Suitability Index 258–9, 433, 436, 440, 447, 448, 450, 452–3
humpback chub 358
Instream Flow Incremental Methodology 10, 430, 439, 441, 455, 503
instream flow needs 441, 446
macro-invertebrates 51, 259
membership functions 438
microhabitat 432–3
microtopographic variability 196, 234–5, 274, 287
numerical habitat value 438
over-wintering 432, 455
PHABSIM 10, 222, 259, 430, 439, 456, 503
physical habitat assessment 71
physical habitat modelling 431
predator avoidance 432, 442
preference curves 434–5, 448, 503
Probabilistic Habitat Index (PHI) 436, 448
refugia 259
salmonids 223, 432, 449, 456
spawning and habitat 223, 432, 446
stock recruitment 430
tree cover 433
utilization curve 435
Weighted Usable Area 222, 448–9, 503, 505
WUA, *see* Weighted Usable Area
Heavy materials 305, 314
HEC-2 439
HEC-RAS 164, 219, 281, 439
Helical flow 80, 246, 445
Heteroscedastic Maximum Likelihood Estimator (HMLE) 201
Hiding 80–1, 220
Higher-order discretisations 158–9, 228, 492
Highway runoff events 51
Hillslope hydrological models 184
HSI, *see* Habitat Suitability Index
Humpback chub 358
Hybrid schemes 171–2, 492
Hydraulic conductivity 196
Hydraulic radius 219
Hydraulic resistance 289–90, 396, 399
Hydraulically smooth 100
HYDRO3D 414–17, 420

Hydrology 255
Hydrostatic pressure 23, 123, 164, 224, 226–7, 253, 307, 443
Hyporheic zone 51, 275
Hysteresis 122

Ice cover 442
Image processing 196–7
Implicit methods 165, 175
In-channel sedimentation 217
Incipient motion 75
Incision 375
Inertial Navigation System 196, 288
Infiltration 126, 141
Inflow 5, 161, 170, 174, 176, 230, 243, 281, 287, 369, 481, 492
Initial condition 126, 170, 175, 281, 286, 370, 376, 464
Inlet, *see* Inflow
Instabilities 159, 177, 229–30
Instream Flow Incremental Methodology 10, 430, 438, 441, 455, 503
Instream flow needs 441, 446
Interferometric Synthetic Aperture Radar, *see* SAR interferometry
Interferometry, *see* SAR interferometry
Internal validation 188
Interpolated 127, 198, 228–9, 366, 478
Intertidal zone 196
Inundation, *see* Flood inundation
Inundation extent 203, 271, 295
Inundation modelling 293
ISIS 164, 219, 281, 284
Isotropic eddy-viscosity models, *see* Zero-equation turbulence closure
Isotropic turbulence 28, 96, 310
Iterative solution 173

Junction angle 225, 251, 253

K, *see* Turbulent kinetic energy
k-ε model 26–30, 62, 65, 97–8, 101, 104, 108–9, 114, 226, 243, 250–1, 257, 308, 402, 415–17, 424, 494, 496, 503, 509
Kinematic viscosity 56, 125, 307, 309
Kinetic transformation rate 310
Kolmogorov length scale 5, 91, 273
Kronecker delta 27, 95, 329
k-ω model 26–7, 29, 30–1, 102–3

Lag effects 73, 79
Lagrangian approaches 156, 239, 449
Laminar sublayer 98–9, 102, 488
Landcover 199
Large Eddy Simulation 6, 21, 32, 36, 92, 163, 226, 251, 257–8, 278, 416, 418, 424, 480

Laser altimetry 3, 8, 136–7, 196–8, 204, 287–91, 296, 425, 443, 483
Laser Doppler anemometer 62
Lateral shear 470
Law of the wall 99, 100–1, 109, 163, 232–3, 332, 339–40, 410, 424, 445, 492
Length scales 104
LiDAR, *see* Laser altimetry
Likelihood 473
Linear relaxation 174
LISFLOOD-FF 283–5
Local wall shear stress 98
Longitudinal dispersion 65, 311
Low Reynolds number model 98
Lumped parameters 199, 203

Macro-invertebrates 51, 259
Macrophyte production 455
Macrotidal estuary 321
Mangrove forests 305, 310
Manning equation, *see* Manning's *n*
Manning's *n* 23–4, 125, 139, 184, 188, 198–9, 224, 231–3, 273, 283–4, 289, 309, 318, 395, 398–400, 404–5, 416, 422, 425, 439, 445, 474–5, 481, 488–9
Mass conservation 6, 21, 25, 35, 53, 71, 121, 130, 133, 138, 141, 148, 160–1, 165, 185, 218, 239, 281, 360, 441
Mass conservation equation 124, 135, 221, 235–6, 250–1, 276, 279
Mass-conservative discretisation 140
Meander migration 222, 242
Meanders 7, 24, 61–2, 80, 101, 105, 173, 178, 181, 198, 219, 221, 225–7, 238–43, 246, 250–1, 259, 261, 273–4, 289, 307, 439, 496
Measure of fit 201
Measurements 183
MEI 399–400
Membership functions 438
Mesh generation, *see* Grid generation
Mesh independence, *see* Grid independence
Mesh quality, *see* Grid quality
Mesh refinement, *see* Grid refinement
Meshing
  adaptive curvilinear grid 240
  adaptive grid methods 2, 6, 128, 132, 138, 170, 172, 446
  block-structured grid 230
  body-fitted coordinates 155, 229, 238, 242–3, 342
  curvilinear spatial grid 220
  deformable mesh 124, 126–7, 278
  Delauney method of triangulation 366, 446
  Grid Convergence Index 172–3
  grid generation 2, 127, 156, 291
  grid independence 6, 20, 128, 157, 162, 170, 172–4, 229, 251, 395, 474, 476, 489, 509
  grid quality 127, 138, 163, 230
  grid refinement 136, 156–7, 161, 228, 489
  grid resolution 20, 138, 173, 198, 228, 289, 332, 352, 366, 491
  remeshing 127
  Triangular Irregular Network 366, 446
Meyer-Peter and Müller equation 75, 78, 80–1, 220
Microhabitat 432–3
Microtopographic variability 196, 234–5, 274, 287
Mid-channel bar 382
MIKE software 164–5, 219, 229, 281
Mineral oils 314
Mixing length hypothesis 27, 114, 308
Mixing lengths 27, 96, 114
Mixing processes 9, 51, 61, 250, 257, 305, 468, 470
Model assessment
  accuracy 164, 170, 174, 175, 179, 181, 188, 194, 198
  adequacy 172, 181
  ASME criteria 187
  benchmark solutions 170, 178
  benchmarking 7
  bias 179, 181
  coherence 179, 186–7, 281
  conceptual assessment 185, 187
  conceptual model 4, 10
  contamination of measurements 182
  convergence error 170
  convergence testing 6
  discrepancies 138–9
  error 137, 157, 164, 182, 188, 201, 292, 294, 446
  falsification 179
  functional regression 183
  goodness-of-fit 188
  internal validation 188
  measure of fit 201
  model evaluation 187
  Nash-Sutcliffe efficiency 201
  numerical accuracy 4, 148, 169, 170, 173
  numerical convergence 31
  ordinary least-squares regression 183
  orthogonality of the grid 491
  precision 179, 181, 183, 188
  reduced major axis regression 183
  reductionist tendencies 13
  regression analysis 181
  Richardson extrapolation 170, 172
  sensitivity 3, 184, 187, 346
  sensitivity analysis 7, 11, 185–8, 230, 461–2
  solution accuracy 6
  validation 2, 3, 6, 7, 12–13, 63, 148, 166, 171, 178–9, 182–3, 185–7, 194–5, 202–4, 207, 236, 243, 272, 280, 285, 291, 294–5, 297, 317, 322, 402, 418, 420, 425, 447, 452, 487, 496
  verification 6, 13, 148, 166, 171, 194, 227, 361

Model evaluation 187
Model nesting 166
Model resolution 483
Model stability 130
MODELEUR 447
Modulus of elasticity 399, 499
Molecular diffusion 53, 55, 58, 60, 176
Molecular viscosity 21, 26
Momentum conservation 4, 6, 21, 71, 121, 130, 133, 138, 141, 218, 281
Momentum correction factor 308
Momentum equations 125, 185, 231, 235–6, 279, 336–7, 340, 360, 418
Momentum exchange 273, 397, 407, 410, 412
Momentum sink 419
Monte Carlo methods 11, 204–7, 452, 463, 474, 477, 483
Morphodynamics 5, 218, 236
Moving boundary problems, *see* Wetting and drying
Mud flat 396
Multi-block approaches 138, 156, 230, 492
Multi-fraction models 222

Nash-Sutcliffe efficiency 201
Navier–Stokes equations 5, 10, 20–3, 26, 32, 38, 42, 91–4, 121, 124, 161, 193, 218, 223, 226, 241, 250, 275–6, 279, 306, 329, 402, 414–15, 424, 474, 483, 492
Near Channel Floodplain Storage model 285
Nearshore regions 138
Negative dynamic pressure 254, 257–8
Nested modelling 13
Neumann boundary conditions 493
Nikuradse equivalent sand grain roughness 233, 309, 318
No-slip condition 37, 98, 111, 474
Non-breaking waves 123
Non-cohesive sediments 315
Non-conservative formulations 9, 139
Non-equilibrium flows 108
Non-faecal bacteria 313
Non-linearity 76
Non-stable solutions 476
Non-stratified flows 112
Non-uniform flows 217, 289
Nonlinear eddy-viscosity models 104
Nonlinear two-equation models 108
Normal stresses 106
Normalised roughness height 99, 103, 234
Numerical accuracy 4, 148, 169, 170, 173
Numerical analysis 148
Numerical convergence 31
Numerical diffusion 8, 36, 159, 170–1, 177, 187, 229–30, 250, 260, 477, 492
Numerical dispersion 170

Numerical experiments 46
Numerical habitat value 438
Numerical instability 7–8, 31, 140
Numerical porosity 230
Numerical solution
  alternating direction implicit methods 165
  analytical solutions 5, 12, 135, 170, 194, 281
  artificial viscosity 170–1
  backward differencing 149, 163, 228
  blending 170–1
  boundary conditions 1, 2, 11, 20, 26, 30, 35–6, 43, 46, 99, 101–2, 126, 161–2, 170, 175–7, 179, 183, 186–8, 230, 235, 260, 276, 281, 285–6, 317–18, 319, 329, 463–4, 474–5, 481, 487–8
  boundedness 159, 163
  cell aspect ratio 494
  central differencing 149, 158–9, 172, 228
  characteristic-based methods 164
  computational efficiency 366
  consistency 162
  control volume 154
  convergence 63, 162–3, 173–4, 227, 489, 494
  convergence criteria 509
  Crank–Nicholson scheme 164
  Direct Numerical Simulation 6, 20–1, 85, 91–2, 111, 226, 251
  Dirichlet boundary condition 63, 66, 153, 156
  discrete element models 9, 336–7, 340, 418
  discretisation 7, 36, 129, 131, 148, 157, 163, 227, 277–8, 280, 289, 291, 336, 396, 430, 502
  donor–acceptor scheme 35
  downstream conditions 36
  Eulerian approach 156, 450
  exponential schemes 171
  false time step relaxation 174
  finite element methods 6, 127–9, 133, 139–40, 151, 155, 162, 165, 186, 207, 227, 230, 276, 280, 291, 441, 446, 469, 474, 480
  finite volume methods 6, 127, 153–5, 161, 165, 170–1, 174, 227–30, 242, 276, 396, 414, 488
  first-order schemes 170–2
  fixed grid methods 128, 280
  forward differencing 149, 158, 229
  functional iteration 165
  Galerkin methods 131–2, 152
  general 2, 6, 37, 127, 138, 148, 158, 169, 174
  higher-order discretisations 158–9, 228, 492
  hybrid schemes 171–2, 492
  implicit methods 165, 175
  initial condition 126, 170, 175, 281, 286, 370, 376, 464
  instabilities 159, 177, 229–30
  interpolated 127, 198, 228–9, 366, 478
  iterative solution 173
  mass-conservative discretisation 140

Numerical solution (*Continued*)
  model stability   130
  multi-block approaches   138, 156, 230, 492
  Neumann boundary conditions   493
  non-stable solutions   476
  numerical diffusion   8, 36, 159, 170–1, 177, 187, 229–30, 250, 260, 477, 492
  numerical instability   7–8, 31, 140
  numerical solvers   131
  numerical stability   124, 280
  numerical stiffness   100
  outflow   5, 161, 163, 170, 176, 230, 287, 493
  Peclet number   171–2, 228
  Petrov–Galerkin method   152
  phase error   170
  porosity methods   9, 34, 171, 198, 230, 236, 261, 336
  Preissmann schemes   164
  quadratic upwind solution   172
  QUICKEST   462
  Riemann solvers   132, 164–5
  rigid lid   33, 36, 37, 42, 101, 226, 236, 250–1, 494
  rounding errors   173
  Runge–Kutta methods   164
  second-order accurate methods   66, 171–2, 228
  SIMPLE   62, 161
  spatial discretisation   163, 172
  stopping criteria   170
  Streamwise Upwind/Petrov–Galerkin (SUPG)   132
  structured grids   138, 149, 154–6, 177, 229–30, 280
  Taylor expansion   228
  Taylor series   149–50, 158, 170–1
  temporal discretisation   163, 170
  truncation errors   158, 170–3, 207
  under-relaxation   174
  unstructured grids   128, 138, 154–5, 165, 171, 230, 280, 291
  upwind schemes   141, 159, 172, 228, 360
  volume of fluid   34–5, 38, 40, 124, 136, 225, 236, 278, 488
  von Neumann conditions   163
Numerical solvers   131
Numerical stability   124, 280
Numerical stiffness   100
Numerical techniques   281
Nutrient dynamics   430

Oldroyd derivative   104
Optimisation   188, 200–1
Ordinary least-squares regression   181
Orthogonality of the grid   491
Oscillations   131
Out-of-bank flows   203, 219, 275, 279, 281, 398, 468, 494

Outer-bank separation   246
Outfall sources   313
Outflow   5, 161, 163, 170, 176, 230, 287, 493
Overbank flows, *see* Out-of-bank flows
Overwintering   432, 455
Oxbows   249
Oxygenation   446

Parameter sensitivity   184
Parameter space   205
Parameter uncertainty   195
Parameter values   186, 464
Parameterization
  calibration   7, 13, 186, 188, 193–4, 199–201, 203–5, 219, 225, 291–2, 295–6, 317–19, 333, 396, 402, 422–4, 439, 464, 482, 489
  forced empirical adequacy   184
  general   2, 7, 177, 186, 194–5, 198, 201–2, 204, 207, 272, 285, 294, 329, 346, 396, 424, 452, 463, 469
  optimisation   188, 200–1
  parameter space   205
Partial differential equations   148, 151, 158, 161, 228, 430
Partially wet elements   131–2, 133, 135, 138
Particle image velocimetry   3
Peatlands   305
Pebble clusters   233
Peclet number   171–2, 228
Perceptual model   4
Perturbation   207
Petrov–Galerkin method   152
pH   222
PHABSIM   10, 222, 259, 430, 439, 456, 503
Phase error   170
Photogrammetry   3, 192, 257, 287, 289, 366, 446
Physical habitat assessment   71
Physical habitat modelling   431
Pipe flows   21, 56
Plant geometry   13
Plant stiffness   400
Polarisation   203
Polders   284
Pollutants   5, 21, 25, 41, 44, 51, 53, 305, 317, 461
Polynomial chaos   207
Pool riffle topography   221
Pools   71–80, 85, 223, 242–3, 246, 248, 439, 453
Porosity methods   9, 34, 171, 198, 230, 236, 261, 336
Post-positivist   188
Power-Law Schemes   62
Precision   179, 181, 183, 188
Predator avoidance   432, 442
Preference curves   434–5, 449, 503
Preissmann schemes   164
Pressure conditions   36, 105, 125, 161, 343–4

Pressure gradients 218, 250, 253, 306–7, 350
Probabilistic Habitat Index (PHI) 436, 448
Probabilistic sediment deposition rule 239
Protrusion 220
Pseudospectral techniques 20

Quadratic friction law 23, 231–2, 309
Quadratic upwind solution 172
Quasi-3D models 224
QUICKEST 462

Radar 204, 295, 297
Raster-based 186
Rating curves 286
Reattachment point 257
Recirculation zones 243, 246, 356, 364–6, 370, 374, 381, 385, 388, 390, 476
Redd destruction 223
Redistribution tensor 106
Reduced major axis regression 183
Reductionist tendencies 13
Refugia 259
Regression analysis 181
Relative roughness 235, 274, 333
Relative submergence 233
Reliability 179
Remeshing 127
Remote sensing
　active microwave sensors 203
　airborne imagery 122, 443
　CASI multispectral scanner 198
　contour data 204, 287, 366
　digital elevation data 122, 137, 196, 284
　ERS 196
　flood extent data 201, 294
　general 7, 11–13, 166, 177, 195–6, 198–9, 201, 203, 287, 290, 294, 296
　image processing 196–7
　Inertial Navigation System 196, 288
　laser altimetry 3, 8, 136–7, 196–8, 204, 287–91, 296, 425, 443, 483
　photogrammetry 3, 192, 257, 287, 289, 366, 446
　radar 204, 295, 297
　SAR interferometry 196, 204
　satellite imagery 122, 196, 201, 287, 295, 443
　satellite radar altimetry 204
　Shuttle Radar Topography Mission 287
　synthetic aperture radar 122
Renormalization group (RNG) turbulence closure 28–9, 38, 40, 42, 103, 226, 243, 257, 331
Reporting standards 6
Resistance laws 124, 126, 414
Restoration schemes 395
Resuspension 314
Reverse flow 36

Reynolds-averaging 5–6, 21–2, 32, 56–7, 92–3, 113, 123–4, 175, 223, 251, 276, 329
Reynolds number 28, 56, 85, 92, 110, 309, 339, 399
Reynolds shear stresses 22, 27–8, 32–3, 92, 94–5, 97, 104–5, 108–12, 116, 123, 225, 276, 306–8, 329, 401–2, 410, 413, 445
Reynolds-stress models 26, 31–2, 95, 104–5, 109, 226, 402
Reynolds stress tensor, *see* Reynolds shear stresses
Reynolds stresses, *see* Reynolds shear stresses
Richardson extrapolation 170, 172
Riemann solvers 132, 164–5
Riffle–pool systems 7, 40, 71, 80, 85, 219, 223, 233, 240–2, 243, 246, 261, 333, 439, 453
Rigid lid 33, 36, 37, 42, 101, 226, 236, 250–1, 494
Ripple factor 75–6
Ripples 79, 310
Risk-based approach 11
RMA2D 165
RNG models, *see* Renormalization group (RNG) turbulence closure
Rotation tensor 105
Rotational flows 178
Roughness
　bank roughness 185, 235
　blockage 177, 185, 336
　Chezy equation 24, 74–5, 78, 84, 231, 273, 289, 309, 395, 439, 488–9
　Colebrook–White 309
　Darcy-Weisbach friction factor 219, 231, 289, 404
　effective roughness 231, 233, 332, 336, 340
　form drag 84, 233, 289, 336–7, 364, 401
　form roughness 41, 343, 352, 423
　general 6–7, 9–11, 13, 37–8, 46, 61, 84, 99–100, 123, 124–5, 127, 133, 138, 183–5, 195, 198–9, 218–19, 230, 236, 239, 243, 250, 257, 276, 279, 285, 288, 309, 318, 329, 332, 334, 337–8, 342, 344–7, 351, 364, 397, 403, 413–14, 416, 417–18, 421, 423, 439, 445, 474, 481, 488–9
　Manning's n 23–4, 125, 139, 184, 188, 198–9, 224, 231–3, 273, 283–4, 289, 309, 318, 395, 398–400, 404–5, 416, 422, 425, 439, 445, 474–5, 481, 488–9
　Nikuradse equivalent sand grain roughness 233, 309, 318
　normalised roughness height 99, 103, 234
　quadratic friction law 23, 231–2, 309
　relative roughness 235, 274, 333
　resistance laws 124, 126, 414
　roughness coefficients 23, 61, 194, 199–200, 219, 224, 231, 281, 291, 295–6, 332, 361, 395, 398, 422, 424–5, 481
　roughness elements 183, 273, 337, 339, 399, 488

Roughness (*Continued*)
  roughness height   7, 24, 84, 99, 100, 133, 176–5, 184–5, 188, 232–5, 257, 332–3, 335, 338, 361, 491, 493, 503
  roughness parameter   10, 23–4, 74, 84, 182, 198, 231, 233, 236, 361, 420
  spatial correlations in roughness   198
  Strickler-type equation   233, 439
  wall roughness   20, 37, 102, 193, 493–4
Roughness coefficients   23, 61, 194, 199–200, 219, 224, 231, 281, 291, 295–6, 332, 361, 395, 398, 422, 424–5, 481
Roughness elements   183, 273, 337, 339, 399, 488
Roughness height   7, 24, 84, 99, 100, 133, 176–5, 184–5, 188, 232–5, 257, 332–3, 335, 338, 361, 491, 493, 503
Roughness parameter   10, 23–4, 74, 84, 182, 198, 231, 233, 236, 361, 420
Rounding errors   173
Rouse number   363
Runge–Kutta methods   164

St Venant equations   8, 135, 279–81, 284, 310
Salinity   9, 112, 317, 320, 322, 443
Salmonids   223, 432, 450, 456
Salt marsh   305, 396
Sand-bed rivers   9–10, 260
Sand grain roughness   102
Sand ridges   103–4
Sand supply   376
Sand transport processes   389
Sand troughs   103
SAR interferometry   196, 204
Satellite imagery   122, 196, 201, 287, 295, 443
Satellite radar altimetry   204
Saturated areas   201
Scalar quantities   175
Scale   3, 13
Schmidt number   25, 61, 113, 310
Scour   221, 250, 255–7, 468
Second-order accurate methods   66, 171–2, 228
Secondary circulation   24–5, 32, 41, 52, 63, 65, 80, 101, 103–4, 181, 183–4, 185, 198, 219, 225–7, 236, 239, 251, 253–5, 272–3, 285, 307–8, 396, 417, 443–5, 468, 484
Sediment continuity equation   71, 238, 249, 363
Sediment entrainment   9, 181, 188, 239
Sediment pulses   221
Sediment routing   220–1, 239
Sediment-sorting processes   237
Sediment transport
  abrasion   80
  active layer   81, 238
  armouring   221–2
  capacity-limited transport   73
  channel change   7, 220, 236, 260–1

  cohesive sediment   314, 322
  critical shear stress   75, 81, 220, 314, 323, 362
  degradation   83, 222
  deposition   25, 80, 220–1, 227, 230, 239–40, 246, 249, 314, 360, 372, 375–6, 380–1, 388, 468
  deposits   359, 366, 372
  downstream fining   80–1, 221
  Einstein parameter   74
  entrainment   221, 240, 254, 315
  equal mobility   80–1
  erosion   25, 71, 80, 220, 227, 230, 239, 246, 249, 314, 358, 359, 369, 372, 381, 390, 506
  exchange layer   81
  Exner equation   71, 220, 238
  fine sediment   21, 25, 51
  general   7, 9, 12, 14, 72, 139, 181, 217, 219, 236, 237, 238, 260, 261, 306, 314, 320, 346, 449, 506
  graded sediment   80–2
  gravitational effects on sediment transport   79
  hiding   80–1, 220
  in-channel sedimentation   217
  incipient motion   75
  incision   375
  Meyer-Peter and Müller equation   75, 78, 80–1, 220
  morphodynamics   5, 218, 236
  probabilistic sediment deposition rule   239
  sand transport processes   389
  scour   221, 250, 255–7, 468
  sediment continuity equation   71, 238, 249, 363
  sediment entrainment   9, 181, 188, 239
  sediment pulses   221
  sediment routing   220–1, 239
  sediment-sorting processes   237
  sediment transport equations   72–3, 76, 238
  Shields parameter   74–5, 78, 80, 82, 506
  siltation   314
  suspended load   5, 9, 73, 79, 219–20, 238, 305, 313, 324, 356, 360, 363, 389
  washload   72, 219–20
Sediment transport equations   72–3, 76, 238
Sensitivity   3, 184, 187, 346
Sensitivity analysis   7, 11, 185–8, 230, 461–2
Separation   29, 226, 253
Sewer   462–4
Sewerage   462
Shallow water equations   8, 123, 126, 128, 138–9, 141, 164–5, 274, 279–80, 296, 443
Shear   182, 184, 226, 410
Shear layers   246, 258, 272, 401, 406
Shear stress   24–5, 30, 231, 236, 276, 310, 347, 360–1, 362, 398, 400, 404, 449
Shear velocity   234
Shear zone   475, 479
Shellfish Waters   466–7

Shields parameter   74–5, 78, 80, 82, 506
Shoreline   136
Shoreline evolution   141
Shuttle Radar Topography Mission   287
Sidescan sonar   3
Siltation   314
SIMPLE   62, 161
Simulated annealing   201
Skin friction   84, 234, 289, 309, 364
SMART   159
SMS   165
SOBEK   164, 281
Soil hydraulic conductivity   199
Soil saturation   166
Solute transport   5, 60–1, 65, 205, 310, 326
Solution accuracy   6
Source terms   171
Spatial averaging   52
Spatial complexity   7
Spatial correlations in roughness   198
Spatial dimensionality   185
Spatial discretisation   163, 172
Spatial lumping   294
Spawning and habitat   223, 432, 446
Spurious oscillations   140
SSIIM   62, 414–17
Stable stratification   114
Stagnation point   253
Statistical turbulence modelling   92, 117
Stem wakes   401
Step-pool bed topography   233
Stereo-photogrammetry, see Photogrammetry
Stiffness   402, 407–8
Stochastic differential equations   207
Stock recruitment   430
Stopping criteria   170
Storage basins   284
Storage cell concept   284
Storm water discharges   317
Stranding   430
Stratification   113–15
Streamline convergence   101
Streamline curvature   101, 225–6, 250–1, 254–6
Streamline divergence   101
Streamwise Upwind/Petrov–Galerkin (SUPG)   132
Strickler-type equation   233, 439
Structured grids   138, 149, 154–6, 177, 229–30, 280
Structures   274, 308
Sub-critical flow   132, 281
Sub-grid process representation   33, 135–7, 141, 177, 185, 198–9, 229, 234–5, 260, 291, 332, 334, 336, 424
Subsurface flows   166, 286
Super-critical flow problems   132, 286
Super-elevation   253, 255

SUPG-like weighting function   141
Surface wind stress   8
Suspended load   5, 9, 73, 79, 219–20, 238, 305, 313, 324, 356, 360, 363, 389
Swamps   305
Symmetric   488
Synthetic Aperture Radar   122, 203, 287

Taylor expansion   228
Taylor series   149–50, 158, 170–1
TELEMAC-2D   165, 175, 474, 476
Temperature   112, 222
Temporal discretisation   163, 170
Thermal regime   442–3
Tidal currents   314
Tidal floodplains   123, 137, 305
Tidal flows   279, 321
Tidal range   319
Tides   279
Time error estimator   175
TKE, see Turbulent kinetic energy
TOPMODEL   201
Topographic complexity   225, 336
Topographic data   136, 198, 343, 352, 446, 490–1
Topographic discordance   225, 254
Topographic forcing   8, 240, 250, 256–7
Topographic parameterization   11, 177, 179
Topography   2, 3, 6, 9, 11–12, 20, 109, 122, 124–5, 127–8, 131–3, 141, 157, 165, 173, 176–7, 183–4, 187–8, 194–7, 199, 202, 219, 225, 227, 229–30, 236, 240, 274, 277–8, 280, 287, 294, 297, 321, 329, 337, 338, 366, 369, 445, 468, 483, 487, 489, 491, 494
Transient flow variables   449
Transport capacity   219–20
Transverse bed slope effects   78, 82
Transverse mixing   62
Tree cover   433
Trial functions   152
Triangular Irregular Network   366, 446
Tributary confluence, see Confluences
Tributary junctions, see Confluences
Truncation errors   158, 170–3, 207
Tsunamis   123, 279
Turbulence
  anisotropic turbulence   32, 65, 103, 106, 108, 219, 226, 402
  Boussinesq approximation   23, 27–8, 95, 97, 103–4, 112, 123, 232, 279, 307, 329
  dissipation   101, 226, 419
  dissipation length scale   493
  dissipation rate   28, 99, 101, 104–8, 110–11, 113–14, 116–17, 308, 415
  dissipation rate tensor   107, 115
  dissipation tensors   115
  eddy diffusion   58, 227

Turbulence (*Continued*)
  eddy viscosity   29, 95, 113, 232, 251, 288, 307–8, 310, 329, 361–2, 474, 480, 484
  eddy viscosity models, *see* Zero-equation turbulence models
  fully turbulent region   98, 109
  general   2, 5–6, 11, 13, 26, 31, 46, 52, 55–6, 63, 105, 123, 175, 193, 224, 226–7, 236, 242, 250, 258, 275, 310, 329, 333, 341, 350, 410, 424, 443–5, 487–9, 493–4, 502–3
  isotropic eddy-viscosity models, *see* Zero-equation turbulence closure
  isotropic turbulence   28, 96, 310
  k-ε model   26–30, 62, 65, 97–8, 101, 104, 108–9, 114, 226, 243, 250–1, 257, 308, 402, 415–17, 424, 494, 496, 503, 509
  kinematic viscosity   56, 125, 307, 309
  Kolmogorov length scale   5, 91, 273
  k-ω model   26–7, 29, 30–1, 102–3
  Large Eddy Simulation   6, 21, 32, 36, 92, 163, 226, 251, 257–8, 278, 416, 418, 424, 480
  law of the wall   99, 100–1, 109, 163, 232–3, 332, 339–40, 410, 424, 445, 492
  mixing length hypothesis   27, 114, 308
  mixing lengths   27, 96, 114
  nonlinear eddy-viscosity models   104
  nonlinear two-equation models   108
  renormalization group (RNG) turbulence closure   28–9, 38, 40, 42, 103, 226, 243, 257, 331
  Reynolds shear stresses   22, 27–8, 32–3, 92, 94–5, 97, 104–5, 108–12, 116, 123, 225, 276, 306–8, 329, 401–2, 410, 413, 445
  Reynolds Stress Models   26, 31–2, 95, 104–5, 109, 226, 402
  statistical turbulence modelling   92, 117
  turbulence closure   115, 276, 279, 296, 360
  turbulence-driven secondary flows   417
  turbulence generation   226, 414, 424
  turbulence length scale   96, 101, 106
  turbulence modelling   2, 20–2, 31, 85, 92, 95–6, 108, 166, 176, 184, 188, 226–7, 234, 258, 276–8, 288–9, 291, 308, 336, 340, 418, 424, 445, 474, 484
  turbulent diffusion   25, 57, 58, 68, 94, 105, 115, 311, 401
  turbulent diffusion coefficients   310–11, 481
  turbulent kinetic energy   28, 63, 95, 97, 99, 101–2, 104–10, 113–14, 116, 117, 163, 173, 234, 286, 308, 329, 331, 333–4, 336, 341, 345, 347–8, 415, 493, 502
  turbulent scalar fluxes   112–13
  two-equation models   26–31, 62, 65, 96–8, 101, 102–5, 108–9, 114, 226, 243, 250–1, 257, 308, 401–2, 415–17, 424, 494, 496, 503, 509
  viscous dissipation   93–4, 97–8, 105, 176
  vortex shedding   98, 175, 242
  wake effects   339
  wall function   9, 37, 62, 98–100, 106, 109, 177, 184, 229, 331–3, 334, 341, 342, 488
  zero-equation turbulence closure   31, 33, 95–6, 104, 108, 113, 226, 234, 238, 278, 288, 291, 308, 474
Turbulence closure   115, 276, 279, 296, 360
Turbulence-driven secondary flows   417
Turbulence generation   226, 414, 424
Turbulence length scale   96, 101, 106
Turbulence modelling   2, 20–2, 31, 85, 92, 95–6, 108, 166, 176, 184, 188, 226–7, 234, 258, 276–8, 288–9, 291, 308, 336, 340, 418, 424, 445, 474, 484
Turbulent diffusion   25, 57, 58, 68, 94, 105, 115, 311, 401
Turbulent diffusion coefficients   310–11, 481
Turbulent kinetic energy   28, 63, 95, 97, 99, 101–2, 104–10, 113–14, 116, 117, 163, 173, 234, 286, 308, 329, 331, 333–4, 336, 341, 345, 347–8, 415, 493, 502
Turbulent scalar fluxes   112–13
Turbulent shear stresses, *see* Reynolds shear stresses
Two-equation models   26–31, 62, 65, 96–8, 101, 102–5, 108–9, 114, 226, 243, 250–1, 257, 308, 401–2, 415–17, 424, 494, 496, 503, 509

Uncertainties   2, 3, 7, 9, 11, 13, 186, 193, 204–5, 207, 231, 233, 285, 296–7, 329, 424, 461, 463–4, 468, 481, 483
Uncertainty analysis
  Bayesian framework   205, 469, 480
  behavioural models   470
  equifinality   82, 186, 205, 294, 476, 481
  error propagation   202, 204
  factor perturbation   188
  general   195, 272, 291, 462, 468, 481
  Generalized Likelihood Uncertainty Estimation   11, 205, 294, 461–3, 470, 472, 476–7, 480, 482, 484
  Heteroscedastic Maximum Likelihood Estimator (HMLE)   201
  likelihood   473
  Monte Carlo methods   11, 204–7, 452, 463, 474, 477, 483
  parameter sensitivity   184
  parameter uncertainty   195
  uncertainties   2, 3, 7, 9, 11, 13, 186, 193, 204–5, 207, 231, 233, 285, 296–7, 329, 424, 461, 463–4, 468, 481, 483
Under-relaxation   174
Uniform flow   219, 231
Unstable stratification   114–15
Unsteady flow   218, 236, 237, 271, 449, 487, 497

Unstructured grids 128, 138, 154–5, 165, 171, 230, 280, 291
Upscaling 3, 182, 235, 420, 444, 453
Upstream junction corner 253
Upward displacement in numerical grids 235
Upward fining 83
Upwind schemes 141, 159, 172, 228, 360
Urban flood inundation 182
Utilization curve 435
UV disinfection 317

Validation 2, 3, 6, 7, 12–13, 63, 148, 166, 171, 178–9, 182–3, 185–7, 194–5, 202–4, 207, 236, 243, 272, 280, 285, 291, 294–5, 297, 317, 322, 402, 418, 420, 425, 447, 452, 487, 496
Variogram analysis 291
Vegetation
  bending stiffness 409
  biomechanical properties 13, 408, 414
  flexural rigidity 409, 411
  general 5, 8, 10, 13, 83–4, 141, 196–8, 204, 241, 274, 288–91, 296, 339, 358–9, 395–6, 398, 402, 404, 415–16, 418, 420, 424
  macrophyte production 455
  plant geometry 13
  plant stiffness 400
  redd destruction 223
  stem wakes 401
  stiffness 402, 407–8
  vegetative roughness 10, 84, 185, 235, 333, 336, 423
Vegetative roughness 10, 84, 185, 235, 333, 336, 423
Velocity dip 103
Velocity reversal hypothesis 240–2
Velocity shelters 259
Verification 6, 13, 148, 166, 171, 194, 227, 361
Viscous dissipation 93–4, 97–8, 105, 176
Visualisation 185, 187, 189
Volume of fluid 34–5, 38, 40, 124, 136, 225, 236, 278, 488
von Karman's constant 37, 232, 309, 332, 361
von Neumann conditions 163
Vortex shedding 98, 175, 242

Wake effects 339
Wall function 9, 37, 62, 98–100, 106, 109, 177, 184, 229, 331–3, 334, 341, 342, 488
Wall region 7
Wall roughness 20, 37, 102, 193, 493–4
Washload 72, 219–20
Waste material 8
Wastewater 317
Wastewater discharges 305
Wastewater treatment works (WwTW) 319
Water Framework Directive, *see* EU Water Framework Directive

Water quality
  bacterial indicators 317
  bathing water quality 313, 317, 464
  Bathing Water Standards 462
  coliform decay rate 9, 313
  combined sewer overflows 51, 319, 462–5, 468, 481
  contaminants 41, 46, 314, 442, 452, 468
  domestic waste 313
  environmental quality standard 464–5
  EU Water Framework Directive 11, 13, 430, 453, 484
  faecal coliforms 9, 306, 313, 317, 319, 326, 462, 464, 466–8
  general 3, 9, 165, 306, 310, 441, 462, 472
  heavy materials 305, 314
  highway runoff events 51
  non-faecal bacteria 313
  pH 222
  pollutants 5, 21, 25, 41, 44, 51, 53, 305, 317, 461
  sewer 462–4
  sewerage 462
  Shellfish Waters 466–7
  solute transport 5, 60–1, 65, 205, 310, 326
  storm water discharges 317
  UV disinfection 317
  waste material 8
  wastewater 317
  wastewater discharges 305
  wastewater treatment works (WwTW) 319
Water surface elevation effects, *see* Free surface
Water surface gradients, *see* Free surface
Water temperature 433
Wave action 314
Wave propagation 130, 132
Wave run-up 132
Weighted Usable Area 222, 448–9, 503, 505
Weirs 488
Wetlands 305
Wetted perimeter 222, 440, 447
Wetting and drying 6, 8, 121–3, 125, 127–9, 130–3, 139, 141, 165, 225, 237–8, 279–80, 284, 291, 295, 297
Whole catchment modelling 275
Wide swath sonar 3
Width-averaging 221
Wind-driven flows 443
Wind stress 125, 279
Woody debris 235
WUA, *see* Weighted Usable Area

Zero-equation models 96, 480
Zero-equation turbulence closure 31, 33, 95–6, 104, 108, 113, 226, 234, 238, 278, 288, 291, 308, 474
Zero-equation turbulence models 308